Lecture Notes in Statistics **150**

Edited by P. Bickel, P. Diggle, S. Fienberg, K. Krickeberg, I. Olkin, N. Wermuth, S. Zeger

T0214433

Springer Science+Business Media, LLC

Tadeusz Caliński
Sanpei Kageyama

Block Designs: A Randomization Approach

Volume I: Analysis

Springer

Tadeusz Caliński
Department of Mathematical and Statistical Methods
Agricultural University of Poznań
Wojska Polskeigo 28
PL-60-637 Poznań, Poland
calinski@owl.au.poznan.pl

Sanpei Kageyama
Department of Mathematics
Faculty of School Education
Hiroshima University
1-1-1 Kagamiyama
Higashi-Hiroshima 739, Japan
ksanpei@hiroshima-u.ac.jp

Library of Congress Cataloging-in-Publication Data
Calinski, T.
Block designs: a randomization approach / Tadeusz Calinski, Sanpei Kageyama.
p. cm.—(Lecture notes in statistics; 150)
Includes bibliographical references and indexes.
Contents: v. 1. Analysis
ISBN 978-0-387-98578-7 ISBN 978-1-4612-1192-1 (eBook)
DOI 10.1007/978-1-4612-1192-1
1. Block designs. I. Kageyama, Sanpei, 1945- II. Title. III. Lecture notes in statistics (Springer-Verlag); v. 150.

QA279.C35 2000
519.5—dc21 00-030762

Printed on acid-free paper.

Camera-ready copy provided by the authors.

9 8 7 6 5 4 3 2 1

ISBN 978-0-387-98578-7

To Maria Calińska and Masako Kageyama

Preface

In most of the literature on block designs, when considering the analysis of experimental results, it is assumed that the expected value of the response of an experimental unit is the sum of three separate components, a general mean parameter, a parameter measuring the effect of the treatment applied and a parameter measuring the effect of the block in which the experimental unit is located. In addition, it is usually assumed that the responses are uncorrelated, with the same variance. Adding to this the assumption of normal distribution of the responses, one obtains the so-called "normal-theory model" on which the usual analysis of variance is based. Referring to it, Scheffé (1959, p. 105) writes that "there is nothing in the 'normal-theory model' of the two-way layout ... that reflects the increased accuracy possible by good blocking." Moreover, according to him, such a model "is inappropriate to those randomized-blocks experiments where the 'errors' are caused mainly by differences among the experimental units rather than measurement errors." In view of this opinion, he has devoted one of the chapters of his book (Chapter 9) to randomization models, being convinced that "an understanding of the nature of the error distribution generated by the physical act of randomization should be part of our knowledge of the basic theory of the analysis of variance."

Similar conviction underlies the present book, the purpose of which is to expose the theory of block designs as developed from the Fisherian basic principles of experimentation. Although references to these principles can be found in many textbooks on experimental design, the randomization principle is seldom sufficiently discussed and explored in model building. As strongly emphasized by Kempthorne (1977, p. 6), "the matter of randomization is not covered at all, or is discussed with the suggestion that randomization is a good thing and that, having done it, one may pass immediately to linear models with independent Gaussian errors" (i.e., to the above normal-theory models).

Fortunately, some authors have been taking the randomization principle seriously into account in deriving linear models for the analysis of experimental data and, thanks to them, it has become possible to prepare this exposition. In fact, it has benefited substantially from some basic results obtained by two schools of research, to which references are made in Chapter 1 of the book. Particularly, the theories concerning randomization models given by O. Kempthorne, on the one side, and by J. A. Nelder, on the other, have influenced the present text very much. In preparing it, various ideas taken from other authors have also been helpful. In particular, the influence of S. C. Pearce has been very essential, as will be seen in many places.

The book is composed of two volumes. Volume I is devoted to the analysis of experiments in block designs. Volume II concerns the constructions of block designs. This order has been determined by the fact that one cannot understand the advantages of that or other design without knowing the model on which the analysis is based. The outline of the two volumes of the monograph is presented in the introduction chapter (Chapter 1). The reader is advised to read the whole chapter carefully, as it gives the main ideas on which the whole exposition relies. Chapter 2 comprises the main terminological notions and concepts used throughout the monograph. Readers more familiar with the subject can, perhaps, skip the first two sections of this chapter, but it is advisable to read the remaining two, to which references are frequently made in other chapters. The main part of Volume I is Chapter 3, presenting a randomization model for a general block design and its implications for the analysis. It offers ideas and methods essential for any further study of block designs within the randomization approach. Implications for considering properties of these designs and their classifications are discussed in Chapter 4. Those who are interested in extensions of the results presented in Chapter 3 to more advanced design structures are advised to read Chapter 5, devoted to nested block designs, to resolvable block designs in particular. It is hoped that readers finding the theory presented in Volume I interesting will also like to read Volume II, now under final preparation.

The book is aimed at an advanced audience, including students at the postgraduate level and research workers interested in designing and analyzing experiments with the full understanding of the principles.

This monograph does not pretend to give an exhaustive exposition of the theory of experimental design within the randomization approach, but it is hoped that at least it gives some perspectives for further research on the subject.

In the course of preparing the book, the help of several colleagues has been appreciated. In particular, thanks are due to Dr. Paweł Krajewski for his help in performing computations, and to Dr. Idzi Siatkowski and Dr. Takashi Seo for their assistance in the technical preparation of the manuscript. Last but not least, the authors are most grateful to Dr. John Kimmel, the editor, and to several reviewers for their many instructive comments and suggestions, and inspiring encouragement.

Poznań, Poland
Hiroshima, Japan
May 2000

Tadeusz Caliński
Sanpei Kageyama

Contents

Volume II: DESIGN

1
Introduction

1.1 Basic principles of experimental design

A statistical method suitable for analyzing data obtainable from an experiment can be derived mathematically when a theoretical model properly describing the construction of the experiment is available. More precisely, the method is to be derived from a model reflecting unambiguously the structure of the experimental material, the actual use of it in the experiment, and the design according to which treatments that are to be compared in the experiment are assigned to the units of the material (i.e., to experimental units or plots).

The modern approach to the design and analysis of experiments is usually credited to R. A. Fisher, who first introduced the analysis of variance (Fisher, 1925). It was invented as a method of dealing with data obtainable from agricultural experiments designed in randomized blocks, at first. Since then, great progress has been made both in the design of experiments and in their analysis, although the approach remains essentially the same. It is based on three principles introduced by Fisher (1925, Section 48) in his first formal exposition of experimental design. The basic principles are "replication," "randomization" and "local control" (see, e.g., Yates, 1965). Though originally invented in the context of agricultural field trials, the principles are applicable to experiments in many other areas of research as well.

The fundamental nature of those concepts can be explained as follows:

(a) Replication means using several experimental units for each of the treatments to provide a way of estimating the extent of errors affecting observations on the experimental material. Furthermore, by replicating the treatments, more precise estimates of the treatment effects or of comparisons among them are obtainable. In general, the standard error of the estimator of an effect or function of effects decreases mathematically if the number of replications increases. But on the other hand, the more units used, the less it is possible to ensure a homogeneity of the experimental material involved. This fact may be considered as disadvantageous, but sometimes the use of a heterogenous material is desirable, if the heterogeneity represents the natural dispersion of the population to which inferences are to be drawn from the experiment. In this sense, replications contribute to the representativity of the experimental material.

(b) Randomization is usually thought of as a device for eliminating bias of measurements caused by systematic variation of the experimental units. This unbiasedness is really the most appreciable purpose of randomizing the units before the assignment of treatments to them. On that way, the experimenter is trying to ensure that treatments are not continually favored or handicapped by extraneous sources of variation, either in the material or in the environment, over which one has no control or chooses not to have any. However, another not-less-important aim of randomization is to induce randomness in the effects of the existing variation among experimental units, effects not attributable to the treatments and usually called "residuals." With the randomization, a properly derived statistical analysis of the experimental results becomes valid, assuming that in the derivation the executed randomization is fully taken into account.

(c) Local control of the material used in the experiment can be achieved by recognizing the main "gradient" of the existing variation and stratifying the material accordingly, i.e., by grouping experimental units into "blocks," such that the units are as uniform within the blocks as possible. The stratification of the material may be performed in one direction, which gives rise to the formation of ordinary blocks, or in two, leading to the so-called row-and-column designs (the Latin squares and Youden squares being their most common examples), or it may be of a higher way layout. Blocks of experimental units so obtainable can be used by applying a suitable design, a "block design," aimed at making the experiment more efficient. Any increase in efficiency or sensitivity of the experiment will depend on a successful stratification of the material, on which the grouping of experimental units into blocks is based.

In designing an experiment, all three principles are to be taken into account and considered carefully from the point of view of both the purpose of the experiment and the conditions under which the experiment is to be conducted. Depending on the circumstances, one of the principles may play a more important role than the others, but none of them is to be neglected. When implementing the principles in practice, one has to be aware of certain contradictions existing among them, which depending on the experimental context become more or less apparent. For example, to obtain a good level of local control, some constraints on randomization may be needed and a freedom in using replications may become limited. The art of experimental design consists in finding an appropriate balance between possibly contradicting postulates following from the principles, always with the aim and the conditions of the experiment in mind.

In formulating models for the analysis of experimental results, a common agreement prevails on how the replication of treatments and the local control by grouping of experimental units affect the model. But, surprisingly, there are different views on how randomization is to be taken into account. In most of the literature, the models discussed are based on the assumption that the residuals are distributed around zero with equal variance, and that they are uncorrelated, i.e., independent under normality assumption, this being secured by the appropriate randomization (see, e.g., the discussion in Pearce, 1983, Chapter 2). Such a model, so-called "the assumed linear model," does not strongly take account of the randomization implemented in designing and carrying out the experiment (as noticed, e.g., by Zyskind, 1975, p. 647). Several authors, however, have been more critical in appreciating the effects of randomization, and they have tried to build models fully recognizing the randomization techniques employed. Among these more realistic approaches, of particular interest here are two main lines, different in attitude, though not completely disjoint in results, along which the randomization is incorporated into the model building. One, as indicated by Scheffé (1959, Chapter 9), originates from Neyman (1923, 1935) and was later extended by Kempthorne (1952, 1955) and his followers, and the other was initiated by Nelder (1954, 1965a), under some influence of Anscombe (1948). Most of the references concerning the first development, attributed to what is called by Speed and Bailey (1987) the Iowa school, can be found in the papers by Kempthorne (1975, 1977), Zyskind (1975) and White (1975) [see also the discussion by Speed (1990) and Rubin (1990) on the occasion of publishing the translation (in English) of a section of Neyman's (1923) original paper (in Polish)], whereas the most essential references concerning the second approach have been given by Bailey (1981, 1991). Notice should also be taken of other, more skeptical views on the role of randomization in model-based inference, such as that exposed in the paper by Thornett (1982).

The randomization approach adopted in this book will become evident from the next three sections. Because they concern the most simple designs, the completely randomized design and the randomized blocks, it will be easy to see and appreciate the important role played by randomization in building a model for a satisfactory statistical analysis of experimental data. To master the approach and the subject of the book, the reader is advised to read these sections carefully.

1.2 A randomization model for the completely randomized design

Suppose that v treatments are to be compared in an experiment in which n out of N ($n \leq N$) available experimental units (plots) can be used. It is assumed that no reason exists for imposing any stratification of the units, so that the principle of local control does not play its role here. The experimenter, however, has to decide on the numbers of units to be used for individual treatments, i.e., on treatment replications. These numbers, denoted by $r_1, r_2, ..., r_v$, may be chosen equal for all v treatments, or they may be differentiated, e.g., in relation to some scheme of relative importance of the treatments (Pearce, 1983, Section 2.1). The only constraint here is that $r_1 + r_2 + \cdots + r_v = n$.

The randomization principle is implemented in this experimental context in a simple way, which may be described as follows. The N available units are arbitrarily numbered with $\xi = 1, 2, ..., N$, and then these labels are randomized; i.e., a permutation of them is chosen at random. It is assumed that any permutation can be chosen with equal probability. Once a permutation is selected, the units are renumbered, with $j = 1, 2, ..., N$, according to the positions of their original labels in the chosen permutation.

To deal with units subjected to such random renumbering, it is useful to employ the following N^2 indicator random variables:

$$f_{j\xi} = \begin{cases} 1 & \text{if the unit originally labeled } \xi \text{ receives} \\ & \text{by the randomization the label } j, \\ 0 & \text{otherwise,} \end{cases} \tag{1.2.1}$$

for $j, \xi = 1, 2, ..., N$. (Such variables were originally introduced by Kempthorne, 1952, Section 8.2.) These variables are constrained by the conditions

$$\sum_{\xi=1}^{N} f_{j\xi} = 1 \text{ for any } j \quad \text{and} \quad \sum_{j=1}^{N} f_{j\xi} = 1 \text{ for any } \xi$$

and have, under the equal probability assumption, the expectation

$$\mathrm{E}(f_{j\xi}) = \Pr\{f_{j\xi} = 1\} = N^{-1} \quad \text{for any } j \text{ and } \xi.$$

Furthermore, from the obvious relation

$$\begin{aligned} \mathrm{E}(f_{j\xi} f_{j'\xi'}) &= \Pr\{f_{j\xi} f_{j'\xi'} = 1\} = \Pr\{f_{j\xi} = 1, f_{j'\xi'} = 1\} \\ &= \Pr\{f_{j'\xi'} = 1 | f_{j\xi} = 1\} \Pr\{f_{j\xi} = 1\}, \end{aligned}$$

it follows that

$$\mathrm{E}(f_{j\xi} f_{j'\xi'}) = \begin{cases} N^{-1}\delta_{\xi\xi'} & \text{if } j = j', \\ (N-1)^{-1}N^{-1}(1 - \delta_{\xi\xi'}) & \text{if } j \neq j', \end{cases}$$

where $\delta_{\xi\xi'}$ is the Kronecker delta, taking the value 1 if $\xi = \xi'$ and 0 otherwise (see Scheffé, 1959, Section 9.1).

The random variables $\{f_{j\xi}\}$ will now be used to derive a suitable model for analyzing data from an experiment conducted under the described randomization. Following the strategy introduced by Nelder (1965a), it will be useful to consider first a model appropriate for analyzing experimental data under the assumption that all N units receive the same treatment (which of them is immaterial at this stage). The concept of such a "null" experiment, used at the preliminary stage, makes the derivation of the final model more straightforward. It can, however, be adopted further only under the assumption that the v treatments under consideration are similar in the sense that the variation of the responses among the available N experimental units does not depend on the treatment received. This assumption is usually referred to as the unit-treatment additivity assumption (Nelder, 1965b, p. 168; see also White, 1975, p. 560), or the assumption of "additivity in the strict sense" (Hinkelmann and Kempthorne, 1994, p. 148).

For this null experiment, let the "true" (expected) response on the unit originally labeled ξ be denoted by μ_ξ, and let it be denoted by m_j if after the randomization the unit receives the label j. Then, the latter can be expressed as a function of the former by the formula

$$m_j = \sum_{\xi=1}^{N} f_{j\xi}\mu_\xi$$

or, because of the identity $\mu_\xi = \mu_. + (\mu_\xi - \mu_.)$, where $\mu_. = N^{-1}\sum_{\xi=1}^{N}\mu_\xi$, by the formula

$$\begin{aligned} m_j &= \sum_{\xi=1}^{N} f_{j\xi}[\mu_. + (\mu_\xi - \mu_.)] \\ &= \mu_. + \sum_{\xi=1}^{N} f_{j\xi}(\mu_\xi - \mu_.) = \mu + \eta_j, \end{aligned}$$

the last equality following from the notation $\mu = \mu_.$ and $\eta_j = \sum_{\xi=1}^{N} f_{j\xi}$ $(\mu_\xi$ $-\mu_.)$. Note that while μ is a constant parameter, η_j is a random variable. It will be called the "unit error" (see also Scheffé, 1959, p. 293; White, 1975, p. 563; Hinkelmann and Kempthorne, 1994, p. 150). The first and second moments of the random variables $\{\eta_j\}$ are determined by those of $\{f_{j\xi}\}$. It follows that

$$\mathrm{E}(\eta_j) = \sum_{\xi=1}^{N}(\mu_\xi - \mu_.)\mathrm{E}(f_{j\xi}) = 0$$

and that

$$\mathrm{E}(\eta_j\eta_{j'}) = \sum_{\xi=1}^{N}\sum_{\xi'=1}^{N}(\mu_\xi - \mu_.)(\mu_{\xi'} - \mu_.)\mathrm{E}(f_{j\xi}f_{j'\xi'}),$$

from which

$$\mathrm{Var}(\eta_j) = N^{-1}\sum_{\xi=1}^{N}(\mu_\xi - \mu_.)^2 \quad \text{for any } \; j$$

and

$$\mathrm{Cov}(\eta_j, \eta_{j'}) = -N^{-1}(N-1)^{-1}\sum_{\xi=1}^{N}(\mu_\xi - \mu_.)^2 \quad \text{for any } \; j \neq j',$$

as

$$\sum_{\xi=1}^{N}\sum_{\xi'=1}^{N}(\mu_\xi - \mu_.)(\mu_{\xi'} - \mu_.)(1 - \delta_{\xi\xi'}) = -\sum_{\xi=1}^{N}(\mu_\xi - \mu_.)^2.$$

But, introducing the variance component

$$\sigma_U^2 = (N-1)^{-1}\sum_{\xi=1}^{N}(\mu_\xi - \mu_.)^2,$$

the second moments can also be written as

$$\text{Var}(\eta_j) = N^{-1}(N-1)\sigma_U^2 = (1-N^{-1})\sigma_U^2 \quad \text{for any } j$$

and

$$\text{Cov}(\eta_j, \eta_{j'}) = -N^{-1}\sigma_U^2 \quad \text{for any } j \neq j',$$

exactly as in the "derived linear model" of Hinkelmann and Kempthorne (1994, Sections 6.3.1 and 6.3.2), being called so to distnguish from the usual assumed linear model (mentioned in Section 1.1). Thus, the responses $\{m_j\}$ have for any j the model

$$m_j = \mu + \eta_j, \tag{1.2.2}$$

with $\text{E}(m_j) = \mu$ and $\text{Cov}(m_j, m_{j'}) = (\delta_{jj'} - N^{-1})\sigma_U^2$.

It should, however, be noticed that when observing the responses of the units in reality, any observation may be affected by a "technical error" (Neyman, 1935, pp. 110-114, 145), an error caused by variations in experimental technique and other extraneous factors, also perhaps caused by the measuring instrument or by the observer, but completely independent of the randomization of the unit labels (see also Kempthorne, 1952, pp. 132 and 151; Scheffé, 1959, p. 293; Ogawa, 1961, 1963; Hinkelmann and Kempthorne, 1994, p. 152). This error is also called the measurement error by some authors, e.g., White (1975, p. 569). Denoting by e_ξ the technical error by which the variable observed on the unit originally labeled ξ differs from its true response, μ_ξ, and denoting the difference by e_j if in the course of the randomization the unit receives the label j, the model (1.2.2) can be extended to

$$y_j = \mu + \eta_j + e_j \quad (j = 1, 2, ..., N), \tag{1.2.3}$$

where y_j is the variable observed on the jth (randomized) unit in the null experiment, μ and η_j are defined as in (1.2.2), and $e_j = \sum_{\xi=1}^{N} f_{j\xi}e_\xi$. It may usually be assumed (see, e.g., Scheffé, 1959, pp. 292-296) that the technical errors $\{e_\xi\}$ are uncorrelated random variables, each with expectation zero and a finite variance, and that they are independent of the $\{f_{j\xi}\}$. Then, it is easy to show that $\text{E}(e_j) = 0$, $\text{Var}(e_j) = N^{-1}\sum_{\xi=1}^{N}\text{Var}(e_\xi) = \sigma_e^2$ (say), $\text{Cov}(e_j, e_{j'}) = 0$ if $j \neq j'$ and $\text{Cov}(\eta_j, e_{j'}) = 0$ for any j and j'. On account of these properties and those established for (1.2.2), the first and second moments of the variables (1.2.3) are

$$\text{E}(y_j) = \mu \quad \text{and} \quad \text{Cov}(y_j, y_{j'}) = (\delta_{jj'} - N^{-1})\sigma_U^2 + \delta_{jj'}\sigma_e^2$$

for any j and any pair j, j', respectively.

It should be emphasized here that these moments are the same for all j and for all pairs j, j', which means that they do not depend on the chosen permutation of the labels $\{1, 2, ..., N\}$. Thus, the population of the N experimental units can be regarded as "homogeneous," in the sense that the observed responses of the units may be considered as observations on exchangable random variables $\{y_1, y_2, ..., y_N\}$, i.e., as such that their moments are invariant for any choice of the permutation of labels. (Moreover, as shown by Thornett, 1982, Theorem 4, the invariance applies to the joint distribution of the variables, not just to their moments; see also, Bailey, 1981, Section 2.2, 1991, Section 2).

So far, only the conceptual null experiment has been considered, which means that the model (1.2.3) is valid for all N experimental units under the assumption that each of them has been exposed to the same treatment.

Now, returning to the situation described at the beginning of this section, the question of interest is how the units are to be chosen for the different treatments and how the application of them changes the model given in (1.2.3). If the treatments are to be assigned to experimental units at random, according to the randomization principle underlying the present completely randomized design, and the treatment replications are chosen as r_i for the ith treatment ($i = 1, 2, ..., v$), then it is sufficient to take from the population of N randomized units the first r_1, i.e., those labeled $1, 2, ..., r_1$, for the application of the 1st treatment, the next r_2, i.e., those labeled $r_1 + 1, r_1 + 2, ..., r_1 + r_2$, for the application of the 2nd treatment, and so on, until finally the r_v units labeled $n - r_v + 1, n - r_v + 2, ..., n$ for the application of the vth treatment. [In fact, because the randomized population of units is, as stated above, homogeneous in the sense that the observations on the unit responses are represented by exchangeable random variables, any *a priori* chosen order of assigning the randomized units (more precisely their randomized labels) to the treatments is acceptable, provided it is not depending in any way on the potential responses of the units. However, this independence is virtually guaranteed by the randomization. See also Thornett (1982, pp. 140 and 143).]

As to the model of the observed responses for a real experiment, the preliminary model (1.2.3) is to be adjusted by incorporating the index i, i.e., by writing

$$y_j(i) = \mu(i) + \eta_j(i) + e_j(i) \tag{1.2.4}$$

for all units receiving the treatment i ($i = 1, 2, ..., v$). This model, however, would create substantial problems for the analysis, if not simplified further. The matter concerns the moments of the random variables $\{\eta_j(i)\}$ and $\{e_j(i)\}$. To derive a manageable analysis of variance from the model (1.2.4), it is necessary to adopt the usual "additivity" assumption, "additivity in the broad sense" according to Hinkelmann and Kempthorne (1994, p. 152), which in the present context means to assume that the variances and covariances of $\{\eta_j(i)\}$ as well as of $\{e_j(i)\}$ do not depend on i, i.e., on the treatment applied. The assumption on the moments of $\{\eta_j(i)\}$ is exactly the same as the unit-treatment additivity assumption of Nelder (1965b, Section 3) and of White (1975, Section 5). The latter author seems also to accept the technical error-treatment additivity (White, 1975, Section 7), whereas the former author does not take the errors $\{e_j(i)\}$ into account at all.

Under the adopted additivity assumption, the index i can be omitted from $\eta_j(i)$ and $e_j(i)$ in (1.2.4), giving

$$y_j(i) = \mu(i) + \eta_j + e_j \quad (i = 1, 2, ..., v;\ j = 1, 2, ..., n), \tag{1.2.5}$$

with

$$\mathrm{E}[y_j(i)] = \mu(i) = N^{-1} \sum_{\xi=1}^{N} \mu_\xi(i) \tag{1.2.6}$$

and

$$\mathrm{Cov}[y_j(i), y_{j'}(i')] = (\delta_{jj'} - N^{-1})\sigma_U^2 + \delta_{jj'}\sigma_e^2, \tag{1.2.7}$$

independently of i and i', where the parameter $\mu_\xi(i)$ in (1.2.6) denotes the "true" response of the ξth unit to the ith treatment. Now, writing the observed variables in form of a vector

$$\boldsymbol{y} = [y_1(1), ..., y_{r_1}(1), y_{r_1+1}(2), ..., y_{r_1+r_2}(2), ..., y_{n-r_v+1}(v), ..., y_n(v)]',$$

the model (1.2.5) can be expressed as

$$\boldsymbol{y} = \boldsymbol{\Delta}'\boldsymbol{\tau} + \boldsymbol{\eta} + \boldsymbol{e}, \tag{1.2.8}$$

where $\boldsymbol{\Delta}'$, of the form $\boldsymbol{\Delta}' = \mathrm{diag}[\mathbf{1}_{r_1} : \mathbf{1}_{r_2} : \cdots : \mathbf{1}_{r_v}]$, with $\mathbf{1}_{r_i}$ representing a column vector of r_i unit elements, is the "design matrix for treatments," a matrix relating the vector $\boldsymbol{y}$ to the vector $\boldsymbol{\tau} = [\mu(1), \mu(2), ..., \mu(v)]'$ of the so-called "treatment parameters" (also called "treatment effects"), and where $\boldsymbol{\eta} = [\eta_1, \eta_2, ..., \eta_n]'$ and $\boldsymbol{e} = [e_1, e_2, ..., e_n]'$ are the vectors of experimental unit errors and technical errors, respectively. The accompanying properties can be written as

$$\mathrm{E}(\boldsymbol{y}) = \boldsymbol{\Delta}'\boldsymbol{\tau}, \quad \mathrm{Cov}(\boldsymbol{y}) = (\boldsymbol{I}_n - N^{-1}\mathbf{1}_n\mathbf{1}_n')\sigma_U^2 + \boldsymbol{I}_n\sigma_e^2, \tag{1.2.9}$$

the expectation, $\mathrm{E}(\boldsymbol{y})$, following from (1.2.6), and the covariance (dispersion) matrix, $\mathrm{Cov}(\boldsymbol{y})$, following from (1.2.7). Note that because of the additivity assumption, the latter does not depend on the treatments. Otherwise, i.e., without this assumption, it would be impossible to define the treatment effects as in (1.2.5), which has already been pointed out by several authors, e.g., Bailey (1991, p. 30).

Comparing the model with the relevant assumed linear model used, e.g., by Pearce (1983, Section 2.1), it is to be noticed that (1.2.8) differs from Pearce's (2.1.1) by the inclusion of the technical errors, $\boldsymbol{e}$, and by a different form of the covariance matrix; that used by Pearce is of the simple form $\boldsymbol{I}_n\sigma^2$, where σ^2 is the variance of random "residuals" resulting from various sources of errors. So, the assumed linear model is a simplification of the present linear model derived from the adopted randomization theory. That the simple model is, nevertheless, somehow justified for making comparisons between treatments will be seen below.

Once the model is settled, it is interesting to ask for which functions of the expectation vector $\boldsymbol{\Delta}'\boldsymbol{\tau}$ the best linear unbiased estimates can be obtained. To answer this question, use can be made of the following theorem from Zyskind (1967, Theorem 3). (It will be helpful here to refer to Appendix A.)

Theorem 1.2.1. *Under the assumptions* $\mathrm{E}(\boldsymbol{y}) = \boldsymbol{X}\boldsymbol{\beta}$ *and* $\mathrm{Cov}(\boldsymbol{y}) = \boldsymbol{V}$, *a known linear function* $\boldsymbol{w}'\boldsymbol{y}$ *is the best linear unbiased estimator (BLUE) of its expectation,* $\boldsymbol{w}'\boldsymbol{X}\boldsymbol{\beta}$, *if and only if the condition* $(\boldsymbol{I}_n - \boldsymbol{P}_X)\boldsymbol{V}\boldsymbol{w} = \boldsymbol{0}$ *holds, where* $\boldsymbol{P}_X$ *denotes the orthogonal projector on* $\mathcal{C}(\boldsymbol{X})$, *the column space of* $\boldsymbol{X}$, *i.e., if and only if the vector* $\boldsymbol{V}\boldsymbol{w}$ *belongs to* $\mathcal{C}(\boldsymbol{X})$.

Proof. Let $\boldsymbol{u}'\boldsymbol{y}$ be any known linear function different from $\boldsymbol{w}'\boldsymbol{y}$, such that $\mathrm{E}(\boldsymbol{u}'\boldsymbol{y}) = \boldsymbol{w}'\boldsymbol{X}\boldsymbol{\beta}$ for all $\boldsymbol{\beta}$. Then, it can be written as $\boldsymbol{u}'\boldsymbol{y} = \boldsymbol{w}'\boldsymbol{y} + (\boldsymbol{u}'\boldsymbol{y} - \boldsymbol{w}'\boldsymbol{y}) = \boldsymbol{w}'\boldsymbol{y} + \boldsymbol{d}'\boldsymbol{y} = (\boldsymbol{w}+\boldsymbol{d})'\boldsymbol{y}$, where $\boldsymbol{d} = \boldsymbol{u} - \boldsymbol{w}$ is such that $\boldsymbol{d}'\boldsymbol{X} = \boldsymbol{0}'$, i.e., that $\boldsymbol{d} = (\boldsymbol{I}_n - \boldsymbol{P}_X)\boldsymbol{a}$ for some nonzero vector $\boldsymbol{a}$ [which means that $\boldsymbol{d} \in \mathcal{C}^{\perp}(\boldsymbol{X})$, the orthogonal complement of $\mathcal{C}(\boldsymbol{X})$]. Hence, the variance of $\boldsymbol{u}'\boldsymbol{y}$ can be written as $\mathrm{Var}(\boldsymbol{u}'\boldsymbol{y}) = (\boldsymbol{w}+\boldsymbol{d})'\boldsymbol{V}(\boldsymbol{w}+\boldsymbol{d}) = \boldsymbol{w}'\boldsymbol{V}\boldsymbol{w} + 2\boldsymbol{d}'\boldsymbol{V}\boldsymbol{w} + \boldsymbol{d}'\boldsymbol{V}\boldsymbol{d}$. It follows that $\mathrm{Var}(\boldsymbol{u}'\boldsymbol{y}) < \boldsymbol{w}'\boldsymbol{V}\boldsymbol{w}$ if and only if $2\boldsymbol{d}'\boldsymbol{V}\boldsymbol{w} + \boldsymbol{d}'\boldsymbol{V}\boldsymbol{d} < 0$, and this inequality holds for any vector $\boldsymbol{d}$ such that either $\boldsymbol{d}'\boldsymbol{V}\boldsymbol{d} = 0$ and $\boldsymbol{d}'\boldsymbol{V}\boldsymbol{w} < 0$ or $\boldsymbol{d}'\boldsymbol{V}\boldsymbol{w} > 0$ and $\boldsymbol{d}'\boldsymbol{V}\boldsymbol{w}/\boldsymbol{d}'\boldsymbol{V}\boldsymbol{d} < -1/2$, unless $\boldsymbol{d}'\boldsymbol{V}\boldsymbol{w} = 0$. But the latter equality holds for any $\boldsymbol{d} \in \mathcal{C}^{\perp}(\boldsymbol{X})$ if and only if $(\boldsymbol{I}_n - \boldsymbol{P}_X)\boldsymbol{V}\boldsymbol{w} = \boldsymbol{0}$. Thus, the necessity of the condition is proved. To see its sufficiency, note that if it holds, then $\mathrm{Var}(\boldsymbol{u}'\boldsymbol{y}) \geq \mathrm{Var}(\boldsymbol{w}'\boldsymbol{y})$ for any vector $\boldsymbol{d} = \boldsymbol{u} - \boldsymbol{v} \in \mathcal{C}^{\perp}(\boldsymbol{X})$, the equality holding if and only if, in addition, $\boldsymbol{d}'\boldsymbol{V}\boldsymbol{d} = 0$. However, the latter holds (together with $\boldsymbol{d}'\boldsymbol{V}\boldsymbol{w} = 0$) if and only if either $\boldsymbol{u} = \boldsymbol{w}$ or $\boldsymbol{u}'\boldsymbol{V}\boldsymbol{w}/\boldsymbol{w}'\boldsymbol{V}\boldsymbol{w} = \boldsymbol{u}'\boldsymbol{V}\boldsymbol{w}/\boldsymbol{u}'\boldsymbol{V}\boldsymbol{u} = 1$, i.e., the correlation between $\boldsymbol{u}'\boldsymbol{y}$ and $\boldsymbol{w}'\boldsymbol{y}$ is equal to 1, which means that these estimators are "almost surely" identical. (See also Rao, 1973, Section 5a.2.) □

This proof is a modified version of that given originally by Zyskind (1967). Theorem 1.2.1 can also be seen as following from Theorem 5a.2(i) of Rao (1973).

Applying Theorem 1.2.1 to the model (1.2.8), with properties (1.2.9), it can be seen that as $\boldsymbol{P}_{\Delta'} = \boldsymbol{\Delta}'(\boldsymbol{\Delta}\boldsymbol{\Delta}')^{-1}\boldsymbol{\Delta} = \boldsymbol{\Delta}'\boldsymbol{r}^{-\delta}\boldsymbol{\Delta}$, where $\boldsymbol{r}^{\delta} = \mathrm{diag}[r_1, r_2, ..., r_v]$ and $\boldsymbol{r}^{-\delta} = (\boldsymbol{r}^{\delta})^{-1}$, the condition of the theorem is here

$$(\boldsymbol{I}_n - \boldsymbol{\Delta}'\boldsymbol{r}^{-\delta}\boldsymbol{\Delta})[(\boldsymbol{I}_n - N^{-1}\mathbf{1}_n\mathbf{1}_n')\sigma_U^2 + \boldsymbol{I}_n\sigma_e^2]\boldsymbol{w} = \boldsymbol{0}.$$

It is satisfied if and only if the vector $\boldsymbol{w}$ belongs to the column space of $\boldsymbol{\Delta}'$. In fact, because $(\boldsymbol{I}_n - \boldsymbol{\Delta}'\boldsymbol{r}^{-\delta}\boldsymbol{\Delta})\mathbf{1}_n = \mathbf{1}_n - \boldsymbol{\Delta}'\boldsymbol{r}^{-\delta}\boldsymbol{\Delta}\mathbf{1}_n = \mathbf{1}_n - \boldsymbol{\Delta}'\mathbf{1}_v = \mathbf{1}_n - \mathbf{1}_n = \boldsymbol{0}$ (as $\boldsymbol{\Delta}\mathbf{1}_n = \boldsymbol{r}$, the vector of treatment replications), the equality reduces to $(\boldsymbol{I}_n - \boldsymbol{\Delta}'\boldsymbol{r}^{-\delta}\boldsymbol{\Delta})\boldsymbol{w} = \boldsymbol{0}$, which holds if and only if $\boldsymbol{w}$ belongs to $\mathcal{C}(\boldsymbol{\Delta}')$.

Thus, for any $v \times 1$ vector $\boldsymbol{s}$, the function $\boldsymbol{s}'\boldsymbol{\Delta}\boldsymbol{y}$ is the BLUE of the parametric function $\boldsymbol{s}'\boldsymbol{\Delta}\boldsymbol{\Delta}'\boldsymbol{\tau} = \boldsymbol{s}'\boldsymbol{r}^{\delta}\boldsymbol{\tau}$. But, because for any $v \times 1$ vector $\boldsymbol{c}$, a vector $\boldsymbol{s}$ exists such that $\boldsymbol{c}' = \boldsymbol{s}'\boldsymbol{r}^{\delta}$ (as $\boldsymbol{r}^{\delta}$ is nonsingular), for any parametric function $\boldsymbol{c}'\boldsymbol{\tau}$, there exists the BLUE $\boldsymbol{c}'\boldsymbol{r}^{-\delta}\boldsymbol{\Delta}\boldsymbol{y}$. It may easily be noticed that $\boldsymbol{\Delta}\boldsymbol{y}$ ($= \boldsymbol{T}$ in the usual notation) is the column vector of treatment totals and $\boldsymbol{r}^{-\delta}\boldsymbol{\Delta}\boldsymbol{y}$ is that of treatment means. Hence, as $\boldsymbol{c}$ may be any vector, each treatment parameter has as its BLUE the corresponding treatment mean. The meaning of this outcome is that $\mu(i)$, the true response to the ith treatment averaged over the N units (plots) under consideration, is best (in the sense of minimum variance) linearly and unbiasedly estimated by the observed mean response averaged over the r_i units to which the ith treatment is applied. This result can shortly be written as

$$\hat{\boldsymbol{\tau}} = \boldsymbol{r}^{-\delta}\boldsymbol{\Delta y} = \boldsymbol{r}^{-\delta}\boldsymbol{T}. \tag{1.2.10}$$

The covariance matrix of this estimator is

$$\begin{aligned} \text{Cov}(\hat{\boldsymbol{\tau}}) &= \boldsymbol{r}^{-\delta}\boldsymbol{\Delta}\text{Cov}(\boldsymbol{y})\boldsymbol{\Delta}'\boldsymbol{r}^{-\delta} \\ &= \boldsymbol{r}^{-\delta}(\sigma_U^2 + \sigma_e^2) - \mathbf{1}_v\mathbf{1}_v'(\sigma_U^2/N). \end{aligned} \tag{1.2.11}$$

Hence, the BLUE of $\boldsymbol{c}'\boldsymbol{\tau}$ can be written as $\widehat{\boldsymbol{c}'\boldsymbol{\tau}} = \boldsymbol{c}'\hat{\boldsymbol{\tau}}$ and its variance as

$$\text{Var}(\widehat{\boldsymbol{c}'\boldsymbol{\tau}}) = \boldsymbol{c}'\boldsymbol{r}^{-\delta}\boldsymbol{c}(\sigma_U^2 + \sigma_e^2) - \boldsymbol{c}'\mathbf{1}_v\mathbf{1}_v'\boldsymbol{c}(\sigma_U^2/N). \tag{1.2.12}$$

The variance (1.2.12) reduces to $\boldsymbol{c}'\boldsymbol{r}^{-\delta}\boldsymbol{c}(\sigma_U^2 + \sigma_e^2)$ if $\boldsymbol{c}'\boldsymbol{\tau}$ is a contrast, i.e., if $\boldsymbol{c}$ is such that $\boldsymbol{c}'\mathbf{1}_v = 0$.

Now, it may be interesting to compare these results with those obtainable under the assumed linear model often appearing in the literature, i.e., for which $\text{Cov}(\boldsymbol{y}) = \boldsymbol{I}_n\sigma^2$. It follows from Theorem 1.2.1 that, under such a model, $\boldsymbol{w}'\boldsymbol{y}$ is the BLUE of its expectation if and only if $\boldsymbol{w}$ belongs to $\mathcal{C}(\boldsymbol{X})$. An estimator $\boldsymbol{w}'\boldsymbol{y}$ satisfying this condition is called a simple least-squares estimator (SLSE). Whether an SLSE can be at the same time the BLUE of its expectation under a more general model is a question considered in the next theorem from Zyskind (1967, Theorem 4).

Theorem 1.2.2. *Under the assumptions* $\text{E}(\boldsymbol{y}) = \boldsymbol{X\beta}$ *and* $\text{Cov}(\boldsymbol{y}) = \boldsymbol{V}$, *a linear function* $\boldsymbol{w}'\boldsymbol{y}$ *is both the SLSE and the BLUE of its expectation,* $\boldsymbol{w}'\boldsymbol{X\beta}$, *if and only if the conditions* $(\boldsymbol{I}_n - \boldsymbol{P}_X)\boldsymbol{w} = \mathbf{0}$ *and* $(\boldsymbol{I}_n - \boldsymbol{P}_X)\boldsymbol{Vw} = \mathbf{0}$ *hold simultaneously, i.e., if and only if both vectors* $\boldsymbol{w}$ *and* $\boldsymbol{Vw}$ *belong to* $\mathcal{C}(\boldsymbol{X})$.

Proof. This result follows immediately from Theorem 1.2.1 and the definition of an SLSE given above. □

From Theorem 1.2.2 another result of Zyskind (1967, Corollary 4.2) follows.

Corollary 1.2.1. *Under the assumptions of Theorem* 1.2.2, *if* $\boldsymbol{w}'\boldsymbol{y}$ *is the SLSE of* $\text{E}(\boldsymbol{w}'\boldsymbol{y}) = \boldsymbol{w}'\boldsymbol{X\beta}$, *then it is also the BLUE of* $\boldsymbol{w}'\boldsymbol{X\beta}$ *if and only if* $(\boldsymbol{I}_n - \boldsymbol{P}_X)\boldsymbol{Vw} = \mathbf{0}$, *i.e., if and only if* $\boldsymbol{Vw}$ *belongs to* $\mathcal{C}(\boldsymbol{X})$.

This result can also be expressed as in Rao and Mitra (1971, Section 8.2, Corollary 3).

Corollary 1.2.2. *Under the assumptions of Theorem* 1.2.2, *any SLSE of its expectation is also the BLUE of that if and only if, for any vector* $\boldsymbol{s}$, $\boldsymbol{VXs}$ *belongs to* $\mathcal{C}(\boldsymbol{X})$, *i.e., if and only if* $\mathcal{C}(\boldsymbol{VX}) \subset \mathcal{C}(\boldsymbol{X})$.

Thus, applying Corollary 1.2.2 to the model (1.2.8), with properties (1.2.9), note that

$$\text{Cov}(\boldsymbol{y})\boldsymbol{\Delta}' = (\boldsymbol{I}_n - N^{-1}\boldsymbol{1}_n\boldsymbol{1}_n')\boldsymbol{\Delta}'\sigma_U^2 + \boldsymbol{\Delta}'\sigma_e^2 = \boldsymbol{\Delta}'[(\boldsymbol{I}_v - N^{-1}\boldsymbol{1}_v\boldsymbol{r}')\sigma_U^2 + \boldsymbol{I}_v\sigma_e^2],$$

which shows that, for any vector $\boldsymbol{s}$, $\boldsymbol{s}'\boldsymbol{\Delta y}$ is both the SLSE and the BLUE of $\boldsymbol{s}'\boldsymbol{r}^{\delta}\boldsymbol{\tau}$. Hence, also (1.2.10) is both the SLSE and the BLUE of $\boldsymbol{\tau}$, in the sense that each element of $\hat{\boldsymbol{\tau}}$ is such an estimator of the corresponding element of $\boldsymbol{\tau}$. However, whereas under (1.2.9) the covariance matrix of (1.2.10) is as in (1.2.11), the assumed linear model, with $\text{Cov}(\boldsymbol{y}) = \boldsymbol{I}_n\sigma^2$, provides $\text{Cov}(\hat{\boldsymbol{\tau}}) = \boldsymbol{r}^{-\delta}\sigma^2$ (see, e.g., Pearce, 1983, p. 33), which, when taking $\sigma^2 = \sigma_U^2 + \sigma_e^2$, coincides with (1.2.11) if and only if N goes to infinity, i.e., when the experimental units are selected from a practically infinite population. But, if the estimation is restricted to contrasts of treatment parameters only, then there is no difference between using model (1.2.8), based on the randomization theory, or making use of the assumed linear model adopted, e.g., by Pearce (1983, Section 2.1).

1.3 A randomization model for randomized blocks

A typical design in which all three basic principles discussed in Section 1.1 play their roles is the classic "randomized-blocks" design. To control, in accordance with principle (c), the local variation over the experimental material or area, the experimental units are grouped into blocks, such that the units (plots) are less variable within a block than within the experimental material or area as a whole (see, e.g., Pearce, 1983, p. 41). If each block contains as many units as treatments to be compared in the experiment, then the blocks can be used in the classic way as the so-called randomized blocks. The advantage of these blocks is that in each of them there is room for comparing all the v (say) treatments, assigning each of them to one unit of the block. According to principle (b), the assignment is to be made at random by randomizing the units within each block. But, as discussed later, it is also advisable to randomize the blocks themselves. Evidently, the number of blocks used, b, say, coincides here with the number of replications of each of the treatments; therefore the number is to be considered from the point of view of principle (a).

In accordance with the approach adopted in this text, the main interest is now in deriving a model for this classic design, considering it from the randomization point of view. Following Nelder (1954), it will be assumed that the experimenter randomizes the available blocks, of number N_B, say, and then randomizes the experimental units (plots) within the blocks. To explain this procedure mathematically, suppose that there are N_B blocks originally labeled $\xi = 1, 2, ..., N_B$ and that each of them contains v units, which are originally labeled $\pi = 1, 2, ..., v$. The index may also be written as $\pi(\xi)$, if it is desirable to refer it to the ξth block. The randomization of blocks can then be understood as a random selection of a permutation of numbers $\{1, 2, ..., N_B\}$, and a renumeration of the blocks with $j = 1, 2, ..., N_B$, according to the positions of their original labels in the permutation selected. Similarly, the randomization of units within the ξth block can be seen as selecting at random a permutation of the numbers $\{1, 2, ..., v\}$ and renumbering the units of the block with $\ell = 1, 2, ..., v$, according to the positions of their original labels in the permutation randomly chosen. As in the case of the completely randomized design, it will be assumed that any permutation of block labels can be chosen with equal probability, and that any permutation of unit labels within a block can be selected with equal probability. Furthermore, it will be assumed that the randomizations of units within the blocks are independent among the blocks and independent of the randomization of blocks.

To describe the essential results of these randomization procedures, it is helpful to introduce the following random variables:

$$f_{j\xi} = \begin{cases} 1 & \text{if the block originally labelled } \xi \text{ receives} \\ & \text{by the randomization the label } j, \\ 0 & \text{otherwise,} \end{cases} \tag{1.3.1}$$

for $j, \xi = 1, 2, ..., N_B$, and

$$g_{\ell\pi(\xi)} = \begin{cases} 1 & \text{if the unit originally labelled } \pi(\xi) \text{ receives} \\ & \text{by the randomization the label } \ell, \\ 0 & \text{otherwise,} \end{cases} \tag{1.3.2}$$

for $\ell, \pi(\xi) = 1, 2, ..., v$. It may be noticed that the first set of N_B^2 variables associated with the randomization of blocks has been defined in the same way as the N^2 variables (1.2.1) defined in association with the randomization of units in the completely randomized design. Hence, the properties of (1.3.1) can be derived similarly as those of (1.2.1). Thus, the conditions are

$$\sum_{\xi=1}^{N_B} f_{j\xi} = 1 \text{ for any } j \text{ and } \sum_{j=1}^{N_B} f_{j\xi} = 1 \text{ for any } \xi,$$

and the first and second moments are

$$\mathrm{E}(f_{j\xi}) = N_B^{-1} \quad \text{for any } j \text{ and } \xi$$

and

$$\mathrm{E}(f_{j\xi} f_{j'\xi'}) = \begin{cases} N_B^{-1}\delta_{\xi\xi'} & \text{if } j = j', \\ (N_B - 1)^{-1} N_B^{-1}(1 - \delta_{\xi\xi'}) & \text{if } j \neq j'. \end{cases}$$

As to the $v^2 N_B$ random variables defined in (1.3.2), their properties can be derived by similar arguments. It appears that

$$\sum_{\pi(\xi)=1}^{v} g_{\ell\pi(\xi)} = 1 \text{ for any } \ell \text{ and } \sum_{\ell=1}^{v} g_{\ell\pi(\xi)} = 1 \text{ for any } \pi(\xi)$$

and that $\mathrm{E}(g_{\ell\pi(\xi)}) = v^{-1}$ for any ℓ and $\pi(\xi)$,

$$\mathrm{E}(g_{\ell\pi(\xi)}g_{\ell'\pi'(\xi)}) = \begin{cases} v^{-1}\delta_{\pi(\xi)\pi'(\xi)} & \text{if } \ell = \ell', \\ (v-1)^{-1}v^{-1}(1-\delta_{\pi(\xi)\pi'(\xi)}) & \text{if } \ell \neq \ell', \end{cases}$$

and

$$\mathrm{E}(g_{\ell\pi(\xi)}g_{\ell'\pi'(\xi')}) = v^{-2} \quad \text{if } \xi \neq \xi',$$

the last property following from the independence of randomizations in different blocks. Also, because the randomization within any block is independent of the randomization of blocks, it follows that

$$\mathrm{E}(f_{j\xi}g_{\ell\pi(\xi')}) = v^{-1}N_B^{-1},$$

whether $\xi = \xi'$ or $\xi \neq \xi'$.

As in Section 1.2, the random variables (1.3.1) and (1.3.2) and their properties will now be used in building an appropriate model for the analysis of data from an experiment conducted in blocks randomized according to the described procedures. Again, first a null experiment model will be derived, and then it will be extended to provide the final model of the really observed variables.

So, at the beginning, it is assumed that all vN_B units receive the same treatment. Let the true response of the unit labeled $\pi(\xi)$ be denoted by $\mu_{\pi(\xi)}$, and let it be denoted by $m_{\ell(j)}$ if by the randomization the block originally labeled ξ receives the label j and the unit in this block originally labeled π receives the label ℓ. The relation between $\mu_{\pi(\xi)}$ and $m_{\ell(j)}$ can be expressed by the formula

$$m_{\ell(j)} = \sum_{\xi=1}^{N_B} \sum_{\pi(\xi)=1}^{v} f_{j\xi}g_{\ell\pi(\xi)}\mu_{\pi(\xi)}.$$

But, introducing the identity

$$\mu_{\pi(\xi)} = \mu_{\cdot(\cdot)} + (\mu_{\cdot(\xi)} - \mu_{\cdot(\cdot)}) + (\mu_{\pi(\xi)} - \mu_{\cdot(\xi)}),$$

where

$$\mu_{\cdot(\xi)} = v^{-1} \sum_{\pi(\xi)=1}^{v} \mu_{\pi(\xi)} \quad \text{and} \quad \mu_{\cdot(\cdot)} = N_B^{-1}\sum_{\xi=1}^{N_B}\mu_{\cdot(\xi)},$$

the relation can be written as

$$m_{\ell(j)} = \mu + \beta_j + \eta_{\ell(j)} \quad \text{for any } \ell \text{ and } j, \tag{1.3.3}$$

where $\mu = \mu_{\cdot(\cdot)}$ is a constant parameter, whereas

$$\beta_j = \sum_{\xi=1}^{N_B} f_{j\xi}(\mu_{\cdot(\xi)} - \mu_{\cdot(\cdot)}) \tag{1.3.4}$$

and

$$\eta_{\ell(j)} = \sum_{\xi=1}^{N_B} \sum_{\pi(\xi)=1}^{v} f_{j\xi} g_{\ell\pi(\xi)} (\mu_{\pi(\xi)} - \mu_{\cdot(\xi)}) \tag{1.3.5}$$

are random variables, (1.3.4) representing the block random effect and (1.3.5) representing the unit error (called the soil error by Neyman, 1935, p. 112).

From the properties of (1.3.1) and (1.3.2), given above, the following moments are obtainable:

$$\mathrm{E}(\beta_j) = 0 \quad \text{and} \quad \mathrm{E}(\eta_{\ell(j)}) = 0,$$

and

$$\mathrm{Cov}(\beta_j, \beta_{j'}) = \begin{cases} N_B^{-1} \sum_{\xi=1}^{N_B} (\mu_{\cdot(\xi)} - \mu_{\cdot(\cdot)})^2 & \text{if } j = j', \\ -N_B^{-1}(N_B - 1)^{-1} \sum_{\xi=1}^{N_B} (\mu_{\cdot(\xi)} - \mu_{\cdot(\cdot)})^2 & \text{if } j \neq j', \end{cases}$$

$$\mathrm{Cov}(\eta_{\ell(j)}, \eta_{\ell'(j')})$$
$$= \begin{cases} N_B^{-1} v^{-1} \sum_{\xi=1}^{N_B} \sum_{\pi(\xi)=1}^{v} (\mu_{\pi(\xi)} - \mu_{\cdot(\xi)})^2 & \text{if } j = j' \text{ and } \ell = \ell', \\ -N_B^{-1} v^{-1}(v-1)^{-1} \sum_{\xi=1}^{N_B} \sum_{\pi(\xi)=1}^{v} (\mu_{\pi(\xi)} - \mu_{\cdot(\xi)})^2 & \text{if } j = j' \text{ and } \ell \neq \ell', \\ 0 & \text{if } j \neq j', \end{cases}$$

and

$$\mathrm{Cov}(\beta_j, \eta_{\ell(j')}) = 0 \quad \text{whether} \quad j = j' \text{ or } j \neq j'.$$

Moreover, introducing the variance components (Nelder, 1977, p. 52)

$$\sigma_B^2 = (N_B - 1)^{-1} \sum_{\xi=1}^{N_B} (\mu_{\cdot(\xi)} - \mu_{\cdot(\cdot)})^2, \tag{1.3.6}$$

related to blocks, and

$$\sigma_U^2 = N_B^{-1}(v-1)^{-1} \sum_{\xi=1}^{N_B} \sum_{\pi(\xi)=1}^{v} (\mu_{\pi(\xi)} - \mu_{\cdot(\xi)})^2, \tag{1.3.7}$$

related to units within blocks, the variances and covariances of the random variables (1.3.4) and (1.3.5) can also be written in the more convenient forms

$$\mathrm{Cov}(\beta_j, \beta_{j'}) = \begin{cases} N_B^{-1}(N_B - 1)\sigma_B^2 & \text{if } j = j', \\ -N_B^{-1}\sigma_B^2 & \text{if } j \neq j', \end{cases}$$

and

$$\mathrm{Cov}(\eta_{\ell(j)}, \eta_{\ell'(j')}) = \begin{cases} v^{-1}(v-1)\sigma_U^2 & \text{if } j = j' \text{ and } \ell = \ell', \\ -v^{-1}\sigma_U^2 & \text{if } j = j' \text{ and } \ell \neq \ell', \\ 0 & \text{if } j \neq j'. \end{cases}$$

Thus, it has been shown that the responses $\{m_{\ell(j)}\}$ have the model (1.3.3) with

$$\mathrm{E}(m_{\ell(j)}) = \mu \tag{1.3.8}$$

and

$$\mathrm{Cov}(m_{\ell(j)}, m_{\ell'(j')}) = (\delta_{jj'} - N_B^{-1})\sigma_B^2 + \delta_{jj'}(\delta_{\ell\ell'} - v^{-1})\sigma_U^2. \tag{1.3.9}$$

The results (1.3.8) and (1.3.9) coincide exactly with the first and second moments given by Nelder (1954) for his y_{ij} variables, which are comparable with the present $m_{\ell(j)}$ variables when replacing i by j and j by ℓ. His constants σ^2, ρ_1 and ρ_2 can be defined with the variance components (1.3.6) and (1.3.7) as

$$\sigma^2 = \frac{N_B - 1}{N_B}\sigma_B^2 + \frac{v-1}{v}\sigma_U^2,$$

$$\rho_1 = \left(\frac{N_B - 1}{N_B}\sigma_B^2 - \frac{1}{v}\sigma_U^2\right) \Big/ \left(\frac{N_B - 1}{N_B}\sigma_B^2 + \frac{v-1}{v}\sigma_U^2\right)$$

and

$$\rho_2 = -\frac{1}{N_B}\sigma_B^2 \Big/ \left(\frac{N_B - 1}{N_B}\sigma_B^2 + \frac{v-1}{v}\sigma_U^2\right).$$

With them, the covariance formula (1.3.9) obtains the form

$$\mathrm{Cov}(m_{\ell(j)}, m_{\ell'(j')}) = \begin{cases} \sigma^2 & \text{if } j = j' \text{ and } \ell = \ell', \\ \rho_1\sigma^2 & \text{if } j = j' \text{ and } \ell \neq \ell', \\ \rho_2\sigma^2 & \text{if } j \neq j'. \end{cases}$$

See also Bailey (1991, Example 2).

However, as in building the completely randomized design model, it seems justified to incorporate in the present model the technical error term, $e_{\pi(\xi)}$, by which the variable observed on the unit originally labeled $\pi(\xi)$ differs from its true response, $\mu_{\pi(\xi)}$. Denoting this difference by $e_{\ell(j)}$, if by the randomization the unit receives the label $\ell(j)$, the extended model can be written in the form

$$\begin{aligned} y_{\ell(j)} &= m_{\ell(j)} + e_{\ell(j)} \\ &= \mu + \beta_j + \eta_{\ell(j)} + e_{\ell(j)} \quad \text{for any } j \text{ and } \ell, \end{aligned} \tag{1.3.10}$$

where $e_{\ell(j)} = \sum_{\xi=1}^{N_B}\sum_{\pi(\xi)=1}^{v} f_{j\xi} g_{\ell\pi(\xi)} e_{\pi(\xi)}$. Assuming, as in Section 1.2, that the technical errors $\{e_{\pi(\xi)}\}$ are uncorrelated random variables, each with expectation zero and a finite variance, and that they are independent of the random variables $\{f_{j\xi}\}$ and $\{g_{\ell\pi(\xi)}\}$, it can easily be shown that the $\{e_{\ell(j)}\}$ are random variables with the properties

$$\mathrm{E}(e_{\ell(j)}) = 0 \text{ and } \mathrm{Var}(e_{\ell(j)}) = (N_B v)^{-1} \sum_{\xi=1}^{N_B} \sum_{\pi(\xi)=1}^{v} \mathrm{Var}(e_{\pi(\xi)}) = \sigma_e^2 \text{ (say)}$$

constant over all $\ell(j)$, and $\mathrm{E}(e_{\ell(j)}e_{\ell'(j')}) = 0$ for all $\ell(j) \neq \ell'(j')$, as well as with

$$\mathrm{E}(e_{\ell(j)}\beta_{j'}) = 0 \quad \text{and} \quad \mathrm{E}(e_{\ell(j)}\eta_{\ell'(j')}) = 0$$

for all $\ell(j)$ and j' and for all $\ell(j)$ and $\ell'(j')$, respectively. Thus, the first and second moments of the random variables $\{y_{\ell(j)}\}$ defined in (1.3.10) have the forms

$$\mathrm{E}(y_{\ell(j)}) = \mu \quad \text{for all } \ell \text{ and } j \tag{1.3.11}$$

and

$$\begin{aligned} \mathrm{Cov}(y_{\ell(j)}, y_{\ell'(j')}) &= (\delta_{jj'} - N_B^{-1})\sigma_B^2 \\ &\quad + \delta_{jj'}(\delta_{\ell\ell'} - v^{-1})\sigma_U^2 + \delta_{jj'}\delta_{\ell\ell'}\sigma_e^2. \end{aligned} \tag{1.3.12}$$

Now, returning to the principles underlying the randomized block design, two questions are to be considered. First, how the randomized units within the randomized blocks are to be used to assign the treatments to them and, second, how the model (1.3.10) is to be adjusted to the fact that more than one treatment is applied in the experiment, each of the v treatments to different units.

As to the first question, it should be noticed that the moments expressed in (1.3.11) and (1.3.12) for the null experiment do not depend on the labels $j(= 1, 2, ..., N_B)$ and $\ell(= 1, 2, ..., v)$ specifying the units after the randomization. This independence means that the population of the N_B randomized blocks can be regarded as "homogeneous" and the set of v units randomized within a block can be regarded as such, in the sense that the observed responses of the units may, under the same treatment, be considered as observations on random variables $\{y_{\ell(j)}\}$, exchangeable within a block and jointly among the blocks (see Thornett, 1982, Example 2). Therefore, to implement the principle of randomization, it is sufficient to assign the ℓth treatment ($\ell = 1, 2, ..., v$) to the ℓth randomized unit in the jth randomized block, taking the first b of them (i.e., blocks labelled $1, 2, ..., b$ in result of the randomization), if b is the number of replications chosen by the researcher. (In fact, because of the indicated homogeneity of the randomized blocks and of the units randomized within blocks, any *a priori* chosen order in which the randomized blocks as well as the randomized units enter the experiment is acceptable, provided it does not depend in any way on the potential responses of the units. This independence, however, is virtually secured by the randomizations underlying the model.)

The second question can be answered as in Section 1.2. Adopting the additivity assumption that the variances and covariances of the random variables $\{\beta_j\}$, $\{\eta_{\ell(j)}\}$ and $\{e_{\ell(j)}\}$ do not depend on the treatment applied, i.e., that the unit-treatment and the technical error-treatment additivities hold, the adjustment of the model (1.3.10) to the real situation of comparing several treatments in the experiment can be made by changing the constant term only. Thus, the final model gets the form

$$y_{\ell(j)}(i) = \mu(i) + \beta_j + \eta_{\ell(j)} + e_{\ell(j)} \tag{1.3.13}$$

$(i, \ell = 1, 2, ..., v;\ j = 1, 2, ..., b)$, with

$$\mathrm{E}[y_{\ell(j)}(i)] = \mu(i) = (N_B v)^{-1} \sum_{\xi=1}^{N_B} \sum_{\pi(\xi)=1}^{v} \mu_{\pi(\xi)}(i), \tag{1.3.14}$$

where $\mu_{\pi(\xi)}(i)$ is the "true" response of the πth unit in the ξth block to the ith treatment, and with

$$\begin{aligned}\mathrm{Cov}[y_{\ell(j)}(i), y_{\ell'(j')}(i')] &= (\delta_{jj'} - N_B^{-1})\sigma_B^2 \\ &+ \delta_{jj'}(\delta_{\ell\ell'} - v^{-1})\sigma_U^2 + \delta_{jj'}\delta_{\ell\ell'}\sigma_e^2,\end{aligned} \tag{1.3.15}$$

independently of i and i'.

It should, however, be noticed that if the random assignment of treatments to units within blocks is implemented exactly as described above, i.e., if the ℓth treatment is applied to the ℓth randomized unit in the jth randomized block, for $j = 1, 2, ..., b$, then the model (1.3.13) can be written in a more familiar way as

$$y_{ij} = \tau_i + \beta_j + \eta_{ij} + e_{ij} \quad (i = 1, 2, ..., v;\ j = 1, 2, ..., b), \tag{1.3.16}$$

where $y_{ij} = y_{\ell(j)}(i), \eta_{ij} = \eta_{\ell(j)}$ and $e_{ij} = e_{\ell(j)}$ for $i = \ell$, and where

$$\tau_i = \mu(i). \tag{1.3.17}$$

Now, writing the observed variables, defined in (1.3.16), in the form of a vector

$$\boldsymbol{y} = [y_{11}, ..., y_{v1}, y_{12}, ..., y_{v2}, ..., y_{1b}, ..., y_{vb}]', \tag{1.3.18}$$

the model can be expressed as

$$\boldsymbol{y} = \boldsymbol{\Delta}'\boldsymbol{\tau} + \boldsymbol{D}'\boldsymbol{\beta} + \boldsymbol{\eta} + \boldsymbol{e}, \tag{1.3.19}$$

where $\boldsymbol{\Delta}' = \mathbf{1}_b \otimes \boldsymbol{I}_v$ is the design matrix for treatments and $\boldsymbol{D}' = \boldsymbol{I}_b \otimes \mathbf{1}_v$ is that for blocks, $\boldsymbol{\tau} = [\tau_1, \tau_2, ..., \tau_v]'$ is the vector of treatment parameters, $\boldsymbol{\beta} = [\beta_1, \beta_2, ..., \beta_b]'$ is the vector of block random effects, $\boldsymbol{\eta} = [\eta_{11}, ..., \eta_{v1}, \eta_{12}, ..., \eta_{v2}, ..., \eta_{1b}, ..., \eta_{vb}]'$ is the vector of unit errors, $\boldsymbol{e} = [e_{11}, ..., e_{v1}, e_{12}, ..., e_{v2}, ..., e_{1b}, ..., e_{vb}]'$ is the vector of technical errors, and the symbol $\otimes$ is used to denote the Kronecker product of the matrices. The accompanying properties of the model can now be written as

$$\mathrm{E}(\boldsymbol{y}) = \boldsymbol{\Delta}'\boldsymbol{\tau} \tag{1.3.20}$$

and

$$\begin{aligned}\mathrm{Cov}(\boldsymbol{y}) &= (\boldsymbol{I}_b - N_B^{-1}\mathbf{1}_b\mathbf{1}_b') \otimes \mathbf{1}_v\mathbf{1}_v'\sigma_B^2 \\ &+ \boldsymbol{I}_b \otimes (\boldsymbol{I}_v - v^{-1}\mathbf{1}_v\mathbf{1}_v')\sigma_U^2 + \boldsymbol{I}_{bv}\sigma_e^2,\end{aligned} \tag{1.3.21}$$

(1.3.20) following from (1.3.14) and (1.3.17), and the form of (1.3.21) from (1.3.15).

So, the model is completed and ready for statistical exploration. But it may be noticed first that (1.3.19) differs from the usually assumed linear model considered, e.g., by Pearce (1983, Chapter 3) in three aspects. In the assumed linear model, (a) $\boldsymbol{\beta}$ is a vector of constant block parameters, (b) $\boldsymbol{\eta}$ has the covariance matrix of the form $\boldsymbol{I}_{bv}\sigma^2$ [σ^2 should not be confused with the variance used by Nelder (1954) and considered above], and (c) the technical errors are not included. The consequences of the differences between the models will be discussed below.

Now, it is interesting to see for which functions of the expectation vector $\boldsymbol{\Delta}'\boldsymbol{\tau}$ the BLUEs exist. Again, Theorem 1.2.1 will be helpful in finding the answer. Applying it, one finds that the necessary and sufficient condition for $\boldsymbol{w}'\boldsymbol{y}$ ($\boldsymbol{w} \neq \boldsymbol{0}$) to be the BLUE of $\mathrm{E}(\boldsymbol{w}'\boldsymbol{y}) = \boldsymbol{w}'\boldsymbol{\Delta}'\boldsymbol{\tau} = \boldsymbol{w}'(\mathbf{1}_b \otimes \boldsymbol{I}_v)\boldsymbol{\tau}$ is the equality

$$[(\boldsymbol{I}_b - b^{-1}\mathbf{1}_b\mathbf{1}_b') \otimes \boldsymbol{I}_v][(\boldsymbol{I}_b - N_B^{-1}\mathbf{1}_b\mathbf{1}_b') \otimes \mathbf{1}_v\mathbf{1}_v'\sigma_B^2 \\ + \boldsymbol{I}_b \otimes (\boldsymbol{I}_v - v^{-1}\mathbf{1}_v\mathbf{1}_v')\sigma_U^2 + \boldsymbol{I}_{bv}\sigma_e^2]\boldsymbol{w} = \boldsymbol{0}.$$

It holds for any values of σ_B^2, σ_U^2 and σ_e^2 if and only if

$$[(\boldsymbol{I}_b - b^{-1}\mathbf{1}_b\mathbf{1}_b') \otimes \mathbf{1}_v\mathbf{1}_v']\boldsymbol{w} = \boldsymbol{0} \quad \text{and} \quad [(\boldsymbol{I}_b - b^{-1}\mathbf{1}_b\mathbf{1}_b') \otimes \boldsymbol{I}_v]\boldsymbol{w} = \boldsymbol{0}$$

hold simultaneously. It can easily be shown that the condition is satisfied if and only if the vector $\boldsymbol{w}$ belongs to the column space of the matrix $[\mathbf{1}_b \otimes \boldsymbol{I}_v]$.

Thus, for any $v \times 1$ vector $\boldsymbol{c}$ and any scalar s, the function $(s\mathbf{1}_b' \otimes \boldsymbol{c}')\boldsymbol{y}$ is the BLUE of the parametric function $(s\mathbf{1}_b' \otimes \boldsymbol{c}')(\mathbf{1}_b \otimes \boldsymbol{I}_v)\boldsymbol{\tau} = (sb)\boldsymbol{c}'\boldsymbol{\tau}$. In particular, taking $s = b^{-1}$, the function $(b^{-1}\mathbf{1}_b' \otimes \boldsymbol{c}')\boldsymbol{y}$ is for any $v \times 1$ vector $\boldsymbol{c}$ the BLUE of $\boldsymbol{c}'\boldsymbol{\tau}$. From the form of the vector $\boldsymbol{y}$, given in (1.3.18), it is evident that the estimator can be written as

$$\widehat{\boldsymbol{c}'\boldsymbol{\tau}} = b^{-1}\sum_{j=1}^{b}\boldsymbol{c}'\boldsymbol{y}_j = \boldsymbol{c}'\left(b^{-1}\sum_{j=1}^{b}\boldsymbol{y}_j\right), \tag{1.3.22}$$

where $\boldsymbol{y}_j = [y_{1j}, y_{2j}, ..., y_{vj}]'$. Because the result holds for any $\boldsymbol{c}$, it implies that the BLUEs exist also for the parameters $\{\tau_i\}$, and that they have the form

$$\hat{\boldsymbol{\tau}} = b^{-1}\sum_{j=1}^{b}\boldsymbol{y}_j = b^{-1}\boldsymbol{T}, \tag{1.3.23}$$

where $\boldsymbol{T} = \boldsymbol{\Delta}\boldsymbol{y}$ is the vector of treatment totals. It confirms, what is commonly known, that $\tau_i = \mu(i)$, the average true response to the ith treatment, defined in (1.3.14), is in a randomized-blocks experiment best linearly and unbiasedly estimated by the observed mean response averaged over the b units to which the ith treatment has been applied, one unit in each of the b blocks. The covariance matrix of the estimator (1.3.23) is

$$\begin{aligned}\mathrm{Cov}(\hat{\boldsymbol{\tau}}) &= b^{-2}\boldsymbol{\Delta}\mathrm{Cov}(\boldsymbol{y})\boldsymbol{\Delta}' \\ &= b^{-1}[(1 - b/N_B)\mathbf{1}_v\mathbf{1}_v'\sigma_B^2 + (\boldsymbol{I}_v - v^{-1}\mathbf{1}_v\mathbf{1}_v')\sigma_U^2 + \boldsymbol{I}_v\sigma^2].\end{aligned} \tag{1.3.24}$$

It follows from (1.3.22), (1.3.23) and (1.3.24) that for any $v \times 1$ vector $\boldsymbol{c}$ the BLUE of $\boldsymbol{c}'\boldsymbol{\tau}$ can be written $\widehat{\boldsymbol{c}'\boldsymbol{\tau}} = \boldsymbol{c}'\hat{\boldsymbol{\tau}}$, and that its variance is

$$\begin{aligned}\operatorname{Var}(\widehat{\boldsymbol{c}'\boldsymbol{\tau}}) &= b^{-1}[(1 - b/N_B)\boldsymbol{c}'\mathbf{1}_v\mathbf{1}_v'\boldsymbol{c}\sigma_B^2 \\ &\quad + \boldsymbol{c}'(\boldsymbol{I}_v - v^{-1}\mathbf{1}_v\mathbf{1}_v')\boldsymbol{c}\sigma_U^2 + \boldsymbol{c}'\boldsymbol{c}\sigma_e^2]. \qquad (1.3.25)\end{aligned}$$

It reduces to

$$\operatorname{Var}(\widehat{\boldsymbol{c}'\boldsymbol{\tau}}) = b^{-1}\boldsymbol{c}'\boldsymbol{c}(\sigma_U^2 + \sigma_e^2),$$

if the parametric function $\boldsymbol{c}'\boldsymbol{\tau}$ is a contrast.

Now, to compare these results with those obtainable under the assumed linear model, it should be noticed that if in (1.3.19) $\boldsymbol{\beta}$ is considered as a vector of (constant) parameters, then $\mathrm{E}(\boldsymbol{y}) = \boldsymbol{\Delta}'\boldsymbol{\tau} + \boldsymbol{D}'\boldsymbol{\beta}$ and, with $\operatorname{Cov}(\boldsymbol{y}) = \boldsymbol{I}_{bv}\sigma^2$, as in the assumed linear model, the condition of Theorem 1.2.1 is

$$[\boldsymbol{I}_{bv} - \boldsymbol{P}_{[\Delta':D']}]\boldsymbol{w} = \mathbf{0}. \qquad (1.3.26)$$

It is not difficult to find that, with $\boldsymbol{\Delta}' = \mathbf{1}_b \otimes \boldsymbol{I}_v$ and $\boldsymbol{D}' = \boldsymbol{I}_b \otimes \mathbf{1}_v$, the matrix in (1.3.26) can be written as

$$\boldsymbol{I}_{bv} - \boldsymbol{P}_{[\Delta':D']} = (\boldsymbol{I}_b - b^{-1}\mathbf{1}_b\mathbf{1}_b') \otimes (\boldsymbol{I}_v - v^{-1}\mathbf{1}_v\mathbf{1}_v'). \qquad (1.3.27)$$

Hence, the condition is satisfied by any vector $\boldsymbol{w} = \boldsymbol{w}_1 \otimes \boldsymbol{w}_2$ such that $\boldsymbol{w}_1$ is proportional to $\mathbf{1}_b$, $\boldsymbol{w}_2$ is proportional to $\mathbf{1}_v$, or both. In particular, it is satisfied by a vector $\boldsymbol{w} = b^{-1}\mathbf{1}_b \otimes \boldsymbol{c}$ for any $v \times 1$ vector $\boldsymbol{c}$. This result implies that $b^{-1}(\mathbf{1}_b' \otimes \boldsymbol{c}')\boldsymbol{y} = b^{-1}\boldsymbol{c}'\boldsymbol{T}$ is the BLUE of its expectation

$$(b^{-1}\mathbf{1}_b' \otimes \boldsymbol{c}')(\boldsymbol{\Delta}'\boldsymbol{\tau} + \boldsymbol{D}'\boldsymbol{\beta}) = \boldsymbol{c}'\boldsymbol{\tau} + b^{-1}\boldsymbol{c}'\mathbf{1}_v\mathbf{1}_b'\boldsymbol{\beta}. \qquad (1.3.28)$$

The estimator $b^{-1}\boldsymbol{c}'\boldsymbol{T}$ is, certainly, the same as that in (1.3.22), but it cannot, in general, be denoted by $\widehat{\boldsymbol{c}'\boldsymbol{\tau}}$, because it estimates the function given in (1.3.28), not just the function $\boldsymbol{c}'\boldsymbol{\tau}$. Also, the variance of this estimator is simply

$$\operatorname{Var}(b^{-1}\boldsymbol{c}'\boldsymbol{T}) = b^{-1}\boldsymbol{c}'\boldsymbol{c}\sigma^2, \qquad (1.3.29)$$

which differs from (1.3.25). So, the same estimator $b^{-1}\boldsymbol{c}'\boldsymbol{T}$ has, in general, the first two moments different in the two compared models. But, if $\boldsymbol{c}'\boldsymbol{\tau}$ is a contrast, then the moments are under both models exactly the same, with $\sigma^2 = \sigma_U^2 + \sigma_e^2$. In fact, e.g., Pearce (1983, Section 3.3) considers the estimation of contrasts only. Thus, it is seen that the assumed linear model and that discussed here give the same results for estimation of contrasts in the context of a design in randomized blocks.

This equivalence follows also directly from Theorem 1.2.2, by which, for any vector $\boldsymbol{c}$, the function $b^{-1}(\mathbf{1}_b' \otimes \boldsymbol{c}')\boldsymbol{y} = b^{-1}\boldsymbol{c}'\boldsymbol{T}$ is both the SLSE and the BLUE of $\boldsymbol{c}'\boldsymbol{\tau}$ under the model (1.3.19) with properties (1.3.20) and (1.3.21), and by which the same is true under the assumed linear model, if $\boldsymbol{c}'\boldsymbol{\tau}$ is a contrast. Whether this result is true for block designs in general cannot be answered now. The question will be discussed in Chapter 3.

1.4 More on the randomization of blocks

When describing the design in randomized blocks, authors of the classic experimental design books point out the necessity of a random assignment of treatments to units within blocks (see, e.g., Fisher, 1935, Section 22; Kempthorne, 1952, Section 9.1; Cochran and Cox, 1957, Section 4.2; Cox, 1958, Chapter 5; Scheffé, 1959, Section 9.1; Finney, 1960, Section 3.1; Montgomery, 1984, Section 5.1). For example, Finney (1960, p. 23) writes: "The procedure for any block is to select one plot at random for the first treatment, another at random for the second treatment, and so on, and to use a new random order in each block." This rule of intra-block randomization can be illustrated as shown in Figure 1.1.

Before randomization

Block 1				Block 2				Block 3				Block 4				Block 5			
1	2	3	4	1	2	3	4	1	2	3	4	1	2	3	4	1	2	3	4

After randomization

Block 1				Block 2				Block 3				Block 4				Block 5			
3	*4*	*2*	*1*	*4*	*3*	*1*	*2*	*2*	*1*	*4*	*3*	*1*	*2*	*4*	*3*	*1*	*3*	*2*	*4*

Fig. 1.1 (Caliński, 1996b, Fig. 1). Randomization for an experiment with four treatments and five blocks. Here the unit labels are randomly permuted within blocks taken in the order given by their original labels.

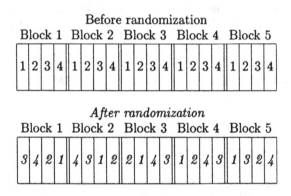

Notably, however, one can hardly find in the classic books any advice to randomize the blocks themselves, i.e., to assign the block labels at random. It seems that in the classic approach to the design in randomized blocks this kind of randomization was not taken seriously into account, at least as long as complete blocks, i.e., of the number of units equal to the number of treatments, were considered. The reason might be that in the randomized block design the same set of treatments is allocated in each of the blocks, and so the blocks do not differ with regard to the treatments contained. It should however be noticed that blocks of units are formed in the experiment not only with the purpose of partial elimination of the influence of the inherent variation among the experimental units on the comparisons of treatments, but also to reduce possible interactions of that variation with any lack of uniformity among the units in the technical conduct of the experiment. In particular, the blocks help to cope with the effects of time trends in experimental techniques or recordings, if they extend over a number of work periods, e.g., days (as noticed by Cochran and Cox, 1957, Section 2.42; and discussed by Pearce, 1983, Section 2.5). From this point of view, it is reasonable to consider the order in which the blocks enter the experiment, or are subjected to some management in the technical or observational sense. Then, it appears that the randomization of blocks, or, more precisely, the randomization of block labels, makes sense.

One of the first researchers to adopt this point of view was Anscombe (1948). It was later formalized by Nelder (1954), who also generalized this approach (Nelder, 1965a). An instruction given by Nelder is "choose a block at random and reorder its members at random ...; repeat the procedure with one of the remaining blocks chosen at random ..., and so on" (Nelder, 1954, Section 2). This advice means to assign labels to units within a block at random, but also to assign labels to blocks at random. The randomization procedures described and used in Section 1.3 have been based on this instruction, which can be illustrated as shown in Figure 1.2. Because it may not appeal to everybody, it might be useful to see the consequences of not randomizing the blocks among themselves.

Before randomization

Block 1				Block 2				Block 3				Block 4				Block 5			
1	2	3	4	1	2	3	4	1	2	3	4	1	2	3	4	1	2	3	4

After randomization

Block *3*				Block *1*				Block *5*				Block *4*				Block *2*			
2	*1*	*4*	*3*	*3*	*4*	*2*	*1*	*1*	*3*	*2*	*4*	*1*	*2*	*4*	*3*	*4*	*3*	*1*	*2*

Fig. 1.2 (Caliński, 1996b, Fig. 2). Randomization for an experiment with four treatments and five blocks. Here, the unit labels are randomly permuted within blocks taken in the order given by their randomly permuted labels *3*, *1*, *5*, *4*, *2*. (For the sake of comparison, it is assumed that the random permutations within blocks have been obtained exactly as in Fig. 1.1.)

Because the objection to randomizing block labels is most likely to appear when the number of available blocks is equal to the number of replications the researcher wants to use, it will be fair enough to consider the case of $b = N_B$.

Suppose, then, that, according to the classic approach, only the randomization of units within blocks is performed (as in Fig. 1.1). Then, the random variables (1.3.1) are not taken into account and the model is built solely by using the random variables (1.3.2). Thus, the unit labeled $\pi(\xi)$ receives, after the randomization, the label $\ell(\xi)$, for $\xi = 1, 2, ..., b$ and $\ell = 1, 2, ..., v$. The true response of the unit may be denoted by $\mu_{\pi(\xi)}$ before randomization and by $m_{\ell(\xi)}$ after. The relation between these two quantities is of the form

$$m_{\ell(\xi)} = \sum_{\pi(\xi)=1}^{v} g_{\ell\pi(\xi)}\mu_{\pi(\xi)},$$

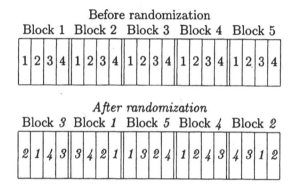

which, after incorporating the identity used in connection with the model (1.3.3), converts to

$$m_{\ell(\xi)} = \mu + \beta_\xi + \eta_{\ell(\xi)} \quad \text{for any } \ell \text{ and } \xi, \tag{1.4.1}$$

where $\mu = \mu_{\cdot(\cdot)}$ and $\beta_\xi = \mu_{\cdot(\xi)} - \mu_{\cdot(\cdot)}$ are constant parameters, the former being defined as in (1.3.3) and the latter representing the block effects, whereas

$$\eta_{\ell(\xi)} = \sum_{\pi(\xi)=1}^{v} g_{\ell\pi(\xi)}(\mu_{\pi(\xi)} - \mu_{\cdot(\xi)}) \tag{1.4.2}$$

is a random variable, representing the unit error. For (1.4.2), it can be found, from the properties of $\{g_{\ell\pi(\xi)}\}$, that

$$\mathrm{E}(\eta_{\ell(\xi)}) = 0$$

and that

$$\mathrm{Cov}(\eta_{\ell(\xi)}, \eta_{\ell'(\xi')}) = \begin{cases} v^{-1}\sum_{\pi(\xi)=1}^{v}(\mu_{\pi(\xi)} - \mu_{\cdot(\xi)})^2 & \text{if } \xi = \xi' \text{ and } \ell = \ell', \\ -v^{-1}(v-1)^{-1}\sum_{\pi(\xi)=1}^{v}(\mu_{\pi(\xi)} - \mu_{\cdot(\xi)})^2 & \text{if } \xi = \xi' \text{ and } \ell \neq \ell', \\ 0 & \text{if } \xi \neq \xi'. \end{cases}$$

For further considerations, it will be helpful to introduce the variance components

$$\sigma^2_{U,\xi} = (v-1)^{-1} \sum_{\pi(\xi)=1}^{v} (\mu_{\pi(\xi)} - \mu_{\cdot(\xi)})^2, \quad \xi = 1, 2, ..., b \tag{1.4.3}$$

(Scheffé, 1959, p. 300). Then, the variances and covariances of the random variables (1.4.2) may be written as

$$\mathrm{Cov}(\eta_{\ell(\xi)}, \eta_{\ell'(\xi')}) = \begin{cases} v^{-1}(v-1)\sigma^2_{U,\xi} & \text{if } \xi = \xi' \text{ and } \ell = \ell', \\ -v^{-1}\sigma^2_{U,\xi} & \text{if } \xi = \xi' \text{ and } \ell \neq \ell', \\ 0 & \text{if } \xi \neq \xi'. \end{cases}$$

From these results, it follows that the model (1.4.1) is accompanied with the properties

$$\mathrm{E}(m_{\ell(\xi)}) = \mu + \beta_\xi \quad \text{and} \quad \mathrm{Cov}(m_{\ell(\xi)}, m_{\ell'(\xi')}) = \delta_{\xi\xi'}(\delta_{\ell\ell'} - v^{-1})\sigma^2_{U,\xi}.$$

But, by incorporating, as in Sections 1.2 and 1.3, the technical error term, $e_{\pi(\xi)}$, the model (1.4.1) for the variable observed in the null experiment extends to

$$\begin{aligned} y_{\ell(\xi)} &= m_{\ell(\xi)} + e_{\ell(\xi)} \\ &= \mu + \beta_\xi + \eta_{\ell(\xi)} + e_{\ell(\xi)} \quad \text{for any } \xi \text{ and } \ell, \end{aligned} \tag{1.4.4}$$

where $e_{\ell(\xi)} = \sum_{\pi(\xi)=1}^{v} g_{\ell\pi(\xi)} e_{\pi(\xi)}$. Under the same relevant assumptions concerning the random variables $\{e_{\pi(\xi)}\}$ as those adopted in Section 1.3, the first and the second moments of the variables $\{y_{\ell(\xi)}\}$ defined in (1.4.4) obtain the forms

$$\mathrm{E}(y_{\ell(\xi)}) = \mu + \beta_\xi \quad \text{for any} \quad \ell(\xi)$$

and

$$\mathrm{Cov}(y_{\ell(\xi)}, y_{\ell'(\xi')}) = \delta_{\xi\xi'}(\delta_{\ell\ell'} - v^{-1})\sigma_{U,\xi}^2 + \delta_{\xi\xi'}\delta_{\ell\ell'}\sigma_{e,\xi}^2, \tag{1.4.5}$$

for all pairs $\ell(\xi)$ and $\ell'(\xi')$, where $\sigma_{e,\xi}^2 = v^{-1}\sum_{\pi(\xi)=1}^{v} \mathrm{Var}(e_{\pi(\xi)})$.

Now, adjusting the model (1.4.4) to the real situation of an experiment in which v treatments are compared, the model is to be written in the form

$$y_{\ell(\xi)}(i) = \mu(i) + \beta_\xi(i) + \eta_{\ell(\xi)}(i) + e_{\ell(\xi)}(i). \tag{1.4.6}$$

But this form can be simplified under the usual unit-treatment and technical error-treatment additivity assumption, implying that the variances and covariances of the random components of the model do not depend on the treatments applied, which in the present context means that the unit-treatment additivities within the blocks and the usual technical error-treatment additivity hold. Under this assumption, (1.4.6) can be written as

$$\begin{aligned} y_{\ell(\xi)}(i) &= \mu(i) + \beta_\xi(i) + \eta_{\ell(\xi)} + e_{\ell(\xi)} \\ &= \mu(i) + \beta_\xi + \gamma_\xi(i) + \eta_{\ell(\xi)} + e_{\ell(\xi)}, \end{aligned} \tag{1.4.7}$$

where

$$\mu(i) = \mu_{\cdot(\cdot)}(i) = (bv)^{-1}\sum_{\xi=1}^{b}\sum_{\pi(\xi)=1}^{v} \mu_{\pi(\xi)}(i),$$

as in (1.3.14), is the true overall response of the units when exposed to the application of the ith treatment,

$$\beta_\xi = \beta_\xi(\cdot) = \mu_{\cdot(\xi)}(\cdot) - \mu_{\cdot(\cdot)}(\cdot)$$

is the block main effect, and

$$\gamma_\xi(i) = \beta_\xi(i) - \beta_\xi(\cdot) = \mu_{\cdot(\xi)}(i) - \mu_{\cdot(\cdot)}(i) - \mu_{\cdot(\xi)}(\cdot) + \mu_{\cdot(\cdot)}(\cdot)$$

is the treatment–block interaction. The accompanying properties of the model (1.4.7) are now

$$\mathrm{E}[y_{\ell(\xi)}(i)] = \mu(i) + \beta_\xi + \gamma_\xi(i), \tag{1.4.8}$$

with the constraints (following from the definitions of parameters)

$$\sum_{\xi=1}^{b}\beta_\xi = 0, \quad \sum_{\xi=1}^{b}\gamma_\xi(i) = 0 \;\text{ for any }\; i, \quad \sum_{i=1}^{v}\gamma_\xi(i) = 0 \;\text{ for any }\; \xi \tag{1.4.9}$$

and

$$\text{Cov}[y_{\ell(\xi)}(i), y_{\ell'(\xi')}(i')] = \text{Cov}[y_{\ell(\xi)}, y_{\ell'(\xi')}] \quad \text{for all } i \text{ and } i',$$

as given in (1.4.5).

However, with the random assignment of treatments to units within blocks implemented in the usual way, that is by assigning the ℓth treatment to the ℓth randomized unit in the ξth nonrandomized block (i.e., by taking $i = \ell$), the model (1.4.7) may be written in a more familiar form as

$$y_{i\xi} = \tau_i + \beta_\xi + \gamma_{i\xi} + \eta_{i\xi} + e_{i\xi}, \tag{1.4.10}$$

where $y_{i\xi} = y_{\ell(\xi)}(i), \tau_i = \mu(i), \gamma_{i\xi} = \gamma_\xi(i), \eta_{i\xi} = \eta_{\ell(\xi)}$ and $e_{i\xi} = e_{\ell(\xi)}$, for $i = \ell$.

Also, writing the observed variables in a vector form, as in (1.3.18), and the corresponding components of the model (1.4.10) in form of vectors $\boldsymbol{\tau}$, $\boldsymbol{\beta}$, $\boldsymbol{\gamma}$, $\boldsymbol{\eta}$ and $\boldsymbol{e}$, the model can be expressed in the matrix notation as

$$\boldsymbol{y} = \boldsymbol{\Delta}'\boldsymbol{\tau} + \boldsymbol{D}'\boldsymbol{\beta} + \boldsymbol{\gamma} + \boldsymbol{\eta} + \boldsymbol{e}, \tag{1.4.11}$$

where $\boldsymbol{\Delta}'$ and $\boldsymbol{D}'$ are defined as in (1.3.19), and its properties written as

$$\text{E}(\boldsymbol{y}) = \boldsymbol{\Delta}'\boldsymbol{\tau} + \boldsymbol{D}'\boldsymbol{\beta} + \boldsymbol{\gamma}, \tag{1.4.12}$$

with $\mathbf{1}_b'\boldsymbol{\beta} = 0$, $(\mathbf{1}_b' \otimes \boldsymbol{I}_v)\boldsymbol{\gamma} = \mathbf{0}$, $(\boldsymbol{I}_b \otimes \mathbf{1}_v')\boldsymbol{\gamma} = \mathbf{0}$ and

$$\begin{aligned}\text{Cov}(\boldsymbol{y}) &= \text{diag}[(\boldsymbol{I}_v - v^{-1}\mathbf{1}_v\mathbf{1}_v')\sigma_{U,1}^2 : (\boldsymbol{I}_v - v^{-1}\mathbf{1}_v\mathbf{1}_v')\sigma_{U,2}^2 : \\ &\qquad \cdots : (\boldsymbol{I}_v - v^{-1}\mathbf{1}_v\mathbf{1}_v')\sigma_{U,b}^2] \\ &\quad + \text{diag}[\boldsymbol{I}_v\sigma_{e,1}^2 : \boldsymbol{I}_v\sigma_{e,2}^2 : \cdots : \boldsymbol{I}_v\sigma_{e,b}^2]\end{aligned} \tag{1.4.13}$$

following from (1.4.8), (1.4.9) and (1.4.5).

Note, however, that because the matrix $[\boldsymbol{\Delta}' : \boldsymbol{D}' : \boldsymbol{I}_{bv}]$ in (1.4.12) is of rank bv, equal to the number of experimental units used in the experiment, nothing is left for the estimation of the variance components appearing in (1.4.13). In other words (Scheffé, 1959, p. 130), "there are no degrees of freedom left for an error sum of squares." Therefore, the model (1.4.11) is to be simplified by assuming that $\boldsymbol{\gamma} = \mathbf{0}$. This assumption, together with the previously adopted additivity assumption, imposes a kind of "additivity in the broad sense" (similar to that in Hinkelmann and Kempthorne, 1994, Section 9.2.6). In terms of the model (1.4.11), the additivity in the broad sense means that the treatment–block interactions $\{\gamma_{i\xi}\}$ are all zero and that the variances and covariances of the unit errors $\{\eta_{i\xi}\}$ and of the technical errors $\{e_{i\xi}\}$ do not depend in any way on the treatments, which is equivalent to saying that the unit-treatment additivity holds for the complete set of available units (not just within the blocks) and that the usual technical error-treatment additivity holds.

Adopting this broad additivity assumption, (1.4.11) can be written as

$$\boldsymbol{y} = \boldsymbol{\Delta}'\boldsymbol{\tau} + \boldsymbol{D}'\boldsymbol{\beta} + \boldsymbol{\eta} + \boldsymbol{e}, \tag{1.4.14}$$

with

$$\mathrm{E}(\boldsymbol{y}) = \boldsymbol{\Delta}'\boldsymbol{\tau} + \boldsymbol{D}'\boldsymbol{\beta},$$

subject to the constraint

$$\mathbf{1}_b'\boldsymbol{\beta} = 0, \tag{1.4.15}$$

and with the covariance matrix (1.4.13) remaining unchanged.

Now, it will be interesting to see whether the model (1.4.14) gives rise to the same BLUEs as those obtained under the model (1.3.19), in which $\boldsymbol{\beta}$ is a random vector, not of constant (fixed) parameters as in (1.4.14).

A function of interest is $b^{-1}(\mathbf{1}_b'\otimes\boldsymbol{c}')\boldsymbol{y}$ for a $v\times 1$ vector $\boldsymbol{c}$. It has the expectation as in (1.3.27), but on account of (1.4.15), it reduces to $\mathrm{E}[b^{-1}(\mathbf{1}_b'\otimes\boldsymbol{c}')\boldsymbol{y}] = \boldsymbol{c}'\boldsymbol{\tau}$. So, the function $b^{-1}(\mathbf{1}_b'\otimes\boldsymbol{c}')\boldsymbol{y}$ is unbiased for $\boldsymbol{c}'\boldsymbol{\tau}$. But is it the BLUE? To answer this question, Theorem 1.2.1 can be applied again. Its necessary and sufficient condition is now in the form

$$[\boldsymbol{I}_{bv} - \boldsymbol{P}_{[\boldsymbol{\Delta}':\mathbf{D}']}]\mathrm{Cov}(\boldsymbol{y})(b^{-1}\mathbf{1}_b\otimes\boldsymbol{c}) = \mathbf{0},$$

with $\mathrm{Cov}(\boldsymbol{y})$ given in (1.4.13). On account of the form of $\boldsymbol{I}_{bv} - \boldsymbol{P}_{[\boldsymbol{\Delta}':\mathbf{D}']}$, shown in (1.3.27), and because of the formula (1.4.13), the condition can be written as

$$\begin{bmatrix} (\boldsymbol{I}_v - v^{-1}\mathbf{1}_v\mathbf{1}_v')\boldsymbol{c}(\sigma^2_{E,1} - b^{-1}\sum_{\xi=1}^b \sigma^2_{E,\xi}) \\ (\boldsymbol{I}_v - v^{-1}\mathbf{1}_v\mathbf{1}_v')\boldsymbol{c}(\sigma^2_{E,2} - b^{-1}\sum_{\xi=1}^b \sigma^2_{E,\xi}) \\ \vdots \\ (\boldsymbol{I}_v - v^{-1}\mathbf{1}_v\mathbf{1}_v')\boldsymbol{c}(\sigma^2_{E,b} - b^{-1}\sum_{\xi=1}^b \sigma^2_{E,\xi}) \end{bmatrix} = \mathbf{0},$$

where

$$\sigma^2_{E,\xi} = \sigma^2_{U,\xi} + \sigma^2_{e,\xi} \quad (\xi = 1, 2, ..., b). \tag{1.4.16}$$

It is satisfied if and only if either $\boldsymbol{c}$ is proportional to the vector $\mathbf{1}_v$ or the equality

$$\sigma^2_{E,1} = \sigma^2_{E,2} = \cdots = \sigma^2_{E,b} \tag{1.4.17}$$

holds.

Thus, it has been shown that

(i) the parametric function $v^{-1}\mathbf{1}_v'\boldsymbol{\tau}$, i.e., the overall true response averaged over the v treatments, has the BLUE in the form of the general mean

$$v^{-1}\widehat{\mathbf{1}_v'\boldsymbol{\tau}} = (bv)^{-1}(\mathbf{1}_b'\otimes\mathbf{1}_v')\boldsymbol{y};$$

(ii) a parametric function $c'\tau$, where $c'1_v = 0$, i.e., a contrast of treatment parameters, has the BLUE in the form

$$\widehat{c'\tau} = b^{-1}(1_b' \otimes c')y = b^{-1}c'T \tag{1.4.18}$$

if and only if the equality (1.4.17) holds, i.e., the joint error variances (1.4.16) are constant for all b blocks.

If the condition (1.4.17) holds, the covariance matrix of y obtains a more simple form,

$$\mathrm{Cov}(y) = I_b \otimes (I_v - v^{-1}1_v1_v')\sigma_U^2 + I_b \otimes I_v\sigma_e^2, \tag{1.4.19}$$

where σ_U^2 is the common variance component (for all ξ) of the form $\sigma_U^2 = b^{-1}\sum_{\xi=1}^{b}\sigma_{U,\xi}^2$ and $\sigma_e^2 = b^{-1}\sum_{\xi=1}^{b}\sigma_{e,\xi}^2$.

It may be noticed that (1.3.21) approaches the form given in (1.4.19) if the block variance σ_B^2, defined in (1.3.6), tends to zero. From (1.4.19), the variance of the BLUE (1.4.18) of a contrast $c'\tau$ is $\mathrm{Var}(\widehat{c'\tau}) = b^{-1}c'c(\sigma_U^2 + \sigma_e^2)$, which coincides with (1.3.29) when taking $\sigma^2 = \sigma_U^2 + \sigma_e^2$.

From these results, it can be concluded that if the blocks themselves are not randomized, i.e., if the randomization is confined to units within blocks only, then a valid model, which provides BLUEs for contrasts $c'\tau$, is obtainable under the assumption that

(a) the complete additivity holds, which includes the assumption that the treatment–block interactions are all zero;

and, in addition, under the assumption that

(b) the joint error variances (1.4.16) are constant not only for all treatments, but also for all blocks.

The assumptions (a) and (b) were already considered as crucial a long time ago by Kempthorne (1952, p. 166), who wrote (p. 150) that "we can justify the usual process of estimating the errors of treatment comparisons if we can assume that $\sigma_{E,\xi}^2$ is constant (for all blocks and treatments) and that there are no block–treatment interactions."

Finally, it may also be concluded that the assumed linear model used, e.g., by Pearce (1983, Chapter 3) is valid for estimating contrasts $c'\tau$ in a randomized-blocks experiment if the assumptions (a) and (b) above are satisfied.

For more discussion on the problem of the existence of BLUEs under different randomization models, see Caliński (1996b), where a controversial paper by Kaiser (1989) is reviewed.

1.5 Outline of this monograph

As could be noticed from the preceding sections, the main feature of the introduction to the present text is the emphasis put on the role of randomization in the model building and on the consequences of the randomization procedures for the estimation of parametric functions of interest. This approach will continue to be a distinguishing characteristic of the whole book, but especially of its first five chapters, through which the general theory is developed.

Chapter 2, supplementing the first, gives an introduction to the terminology related to block designs, together with some auxiliary results that will be used in this book.

The most important for presenting the general theory of the analysis of a block design is Chapter 3. In that chapter, the methods of deriving a model and of examining its properties, which have been used in Sections 1.2–1.4, are used in the context of a general block design. Results obtained there are important for the analysis of any block design, for investigating its properties and for providing a basis for the classification of block designs. The general theory obtained in that chapter can be considered as fundamental for any further study of the existing block designs and for constructing desirable new ones. It is hoped that the way of presenting the theory will be helpful in clarifying certain confusions existing in the literature of the subject.

Some of the confusions or misunderstandings concern various notions related to basic properties of block designs, such as the balance and the efficiency of a block design. These notions are reexamined in Chapter 4. Conclusions drawn there are decisive for the organization of the remaining content of the book.

One additional chapter supplements the presentation of the general theory. The theoretically and practically important notion of nested block designs and the concept of resolvability of a block design are investigated and discussed in Chapter 5, where relevant extensions of the theory are obtained. Resolvable block designs have certain desirable properties, making the designs particularly suitable in some domains of application. By considering these designs, the generality of the present theory becomes apparent, which may itself be of interest for possible further extensions.

Whereas the first five chapters, composing Volume I of the monograph, concern mainly a general theory of the analysis of block designs, the remaining five chapters, i.e., those of Volume II, are devoted to the description of various constructional approaches and methods (Chapter 6) and to the presentation of different classes and subclasses of block designs (Chapters 7-10). The organization of the material of this volume of the book into chapters and sections reflects the classification adopted in the present text. It is based on the efficiency property of a block design. As shown in Chapter 3, the most natural balance of a block design is that connected with its efficiency for estimating contrasts of treatment parameters. The degree of balance of a design is determined by the number of distinct efficiency factors. In general, the smaller the number the more balanced the design. Thus, a design with one efficiency factor has the highest degree of balance. From this point of view, the block designs are classified according to the number of distinct efficiency factors they have. However, in this classification, proposed in Chapter 4, of particular interest are designs having one of the distinct efficiency factors of unit value. Such designs allow at least some of the contrasts of treatment parameters to be estimated with full efficiency. Chapter 7 contains designs of this type, whereas Chapter 8 contains those with no unit efficiency factor. In each of the two chapters (7 and 8), distinction is made between designs with one, two, three and more different efficiency factors. Within these classes, four types of designs are considered: those which are equireplicate and proper, those which are not equireplicate but are proper, those which are equireplicate but are not proper and, finally, those which are neither equireplicate nor proper ("equireplicate" meaning that all treatments are equally replicated, whereas "proper" referring to designs with all blocks of equal size). In most applications, equireplicate and proper are the favorable designs and, therefore, they are considered in broader extent than the other designs. In both of these chapters, carefully chosen examples illustrate the applicability of the discussed classes of designs.

It has appeared desirable to treat certain classes of designs separately. As already mentioned, a special class is that of resolvable designs. They are treated in details in Chapter 9, again in subdivision into sections for designs with one, two and three efficiency factors. Because resolvable designs from their nature are equireplicate, only subsections for proper and not proper designs are included. Among the resolvable designs of particular interest are the so-called α-designs. They are, therefore, considered distinctly in the last section of Chapter 9.

Taking also into account the interest stirred by some other distinctive designs, a separate chapter, Chapter 10, is included to cover those special designs. The so-called variance-balanced designs if they are not equireplicate need a particular consideration; also, some resistant designs are worth separate attention, and designs that are not binary or are not connected need to be discussed separately.

Finally, it is necessary to say something about the background that the reader is expected to possess when wishing to follow all arguments and fully appreciate the results given in the two volumes of the book. First of all, it is supposed that the reader is familiar with the basic literature on experimental design, such as the books by Pearce (1983) and John (1987). So, it is assumed that the terminology and the mathematics used there are known to the reader. Therefore, no appendix on linear algebra is included in the present book. The readers wishing to learn the basic vector and matrix algebra needed here are referred to Pearce (1983) or John (1987) [or John and Williams (1995)], particularly to the appendices there (also to Raghavarao, 1971, Appendices A.3 and A.4). But, because at several places in derivations of analytical methods use is made of projections on subspaces, which are not covered by those appendices, it has become necessary to give here, at the end of Volume I, an appendix on subspaces and projections (Appendix A). Also, because in describing constructional methods some more advanced mathematical tools are used, it may be helpful for some readers to read the other three appendices given at the end of Volume II, that on finite fields (Appendix B) and those on finite geometry (Appendix C) and orthogonal Latin squares (Appendix D). Certainly, readers well acquainted with the book by Raghavarao (1971), to which references are also often made throughout the present text, will not need to master the content of the last three apppendices before reading Volume II of the book starting from Chapter 6.

2
Basic Terminology and Preliminaries

2.1 General terminological remarks

Originally, as introduced by Fisher (1925, 1926), block designs were used in agricultural experiments with the aim of allowing all treatments to be compared within similar conditions. For this aim, the experimental units (plots) were arranged in compact sets, each comprising possibly uniform units in a number equal to the number of treatments. Such sets of units were called blocks, and because the treatments were assigned to units within blocks at random, the design of the experiment was called "randomized blocks" (see, e.g., Cochran and Cox, 1957, Section 4.2). The design has also been called "the randomized-blocks design" (see, e.g., Scheffé, 1959, Section 4.2), "the randomized block design" (see, e.g., Finney, 1960, Section 3.1) or "the randomized complete block design" (see, e.g., Federer, 1955, Chapter 5; Margolin, 1982, p. 289; Hinkelmann and Kempthorne, 1994, p. 251). Here, the abbreviation RBD will be used throughout for this design. Remember, however, that now the RBD is no more a unique design using blocks of experimental units.

In the history of the development of experimental designs, the system of blocking has been extended in various ways. Two developments are of particular interest. One of them concerns the introduction of a two-way blocking system that has led to the so-called row-and-column designs, the most notable example being "the Latin-square design" (see, e.g., Scheffé, 1959, Section 5.1; Finney, 1960, Section 3.3). The other development was connected with the difficulty of comprising all experimental treatments under study in each of the blocks formed according to the principle of within-block homogeneity. This difficulty occurred originally in the context of crop variety trials, in which usually a large number of treatments, i.e., varieties, are to be compared. A solution was found by

Yates (1936a, 1936b), who suggested the use of "balanced incomplete blocks," which have fewer units (plots) per block than the total number of treatments or varieties. This idea gave rise to an enormous development of the theory and practice of block designs. With this development, many new ideas and terms have emerged. They have been introduced to describe the various properties of block designs, which have become much more flexible in their construction once the main condition of the randomized blocks, that all treatments are to be compared in each block, was abandoned. Many of the new terms are connected merely with the constructional aspects, but several terms concern certain important statistical properties of the designs. All of these terms, and those of the latter type in particular, deserve unambiguous definitions and explanations. In the subsequent sections an attempt will be made (following Caliński, 1993a) to give clear meaning to some of the most basic terms and concepts that will be used in the rest of the book. Some other terms will be explained and discussed later, after the presentation of a general theory of the statistical analysis for experiments in block designs. It is, however, assumed from the beginning that the reader is well acquainted with such elementary terms as experimental unit (plot), treatment, replication, etc., that is, of terms already used in Chapter 1. These terms are well explained in basic texts on experimental design, one of the most elucidating being the book by Cox (1958).

At this point, one should also be aware of the distinction between the term "randomized block design" and the term "block design." The former is reserved solely for the classic design composed of "complete" blocks, which are of the size coinciding with the number of treatments, allowing the treatments to be allocated in the blocks according to the rules described in Section 1.3. The term block design is of a broader sense. It covers all designs using experimental units grouped into blocks. Although in these designs randomization is applied according to the same principle as that discussed in Section 1.3, the adjective "randomized" will not be used for naming designs other than the classic RBD (to keep with the well-established traditional terminology, as suggested by Pearce, 1983, p. 96; see also Street, 1996, p. 760).

It is essential to distinguish between the description of a block design indicating the allocation of treatments in blocks, and the actual assignment of treatments to units within blocks, obtained after completing the randomizations involved. The former may be given either in the form of a plan, as frequently shown in classic books on experimental designs, such as, e.g., that by Cochran and Cox (1957), or in form of the "incidence matrix," as applied by Raghavarao (1971) and by Pearce (1983). As to the latter, its presentation is important for building a linear model on which the statistical analysis is to be based. In this process, the appropriate design matrices, for blocks and for treatments, are used to indicate relations between the observed responses of units in different blocks to the treatments applied, on the one side, and the corresponding parameters or random variables for which inferences are to be drawn, on the other (see Section 1.3 for the RBD). These matrices will be defined precisely in the next section, together with main characteristics and properties of block designs.

2.2 Basic descriptive tools and terms

To introduce the concept of a general block design, suppose that v treatments are to be compared in an experiment on n experimental units (plots) arranged in b groups called blocks. A design determining the allocation of the treatments to units grouped in such blocks is called a block design.

As already mentioned in Section 2.1, any block design can be described by its $v \times b$ incidence matrix $\boldsymbol{N} = [n_{ij}]$, with a row for each treatment and a column for each block, where n_{ij} is the number of units in the jth block receiving the ith treatment ($i = 1, 2, ..., v$; $j = 1, 2, ..., b$). Therefore, a block design may be called by its incidence matrix, and thus, the expression "a design $\boldsymbol{N}$" will be used sometimes for convenience. To relate the design to variables observable on the experimental units, two other matrices are needed. Their role will become evident, particularly when modeling the analysis. Let $\boldsymbol{D}$ be a $b \times n$ matrix, with a row for each block and a column for each unit, and with elements 0 and 1, an element being 1 if the unit to which the column corresponds is in the block corresponding to the row, and 0 otherwise. The transposed matrix $\boldsymbol{D}'$ is called the "design matrix for blocks." Similarly, let $\boldsymbol{\Delta}$ be a $v \times n$ matrix, with a row for each treatment and a column for each unit, and with elements 0 and 1, an element being 1 if the unit to which the column corresponds receives the treatment corresponding to the row, and 0 otherwise. The transposed matrix $\boldsymbol{\Delta}'$ is then called the "design matrix for treatments." [Such design matrices have already been used in the model (1.3.19).] Furthermore, let $\boldsymbol{k} = [k_1, k_2, ..., k_b]'$ be the vector of block sizes and $\boldsymbol{r} = [r_1, r_2, ..., r_v]'$, the vector of treatment replications. Then, it can easily be checked that the following relations concerning the matrices $\boldsymbol{N}, \boldsymbol{D}'$ and $\boldsymbol{\Delta}'$ hold:

$$\boldsymbol{\Delta D}' = \boldsymbol{N}$$

and

$$\begin{array}{lll} \boldsymbol{D}'\mathbf{1}_b = \mathbf{1}_n, & \boldsymbol{D}\mathbf{1}_n = \boldsymbol{k} = \boldsymbol{N}'\mathbf{1}_v, & \boldsymbol{DD}' = \boldsymbol{k}^\delta = \text{diag}[k_1, k_2, ..., k_b], \\ \boldsymbol{\Delta}'\mathbf{1}_v = \mathbf{1}_n, & \boldsymbol{\Delta}\mathbf{1}_n = \boldsymbol{r} = \boldsymbol{N}\mathbf{1}_b, & \boldsymbol{\Delta\Delta}' = \boldsymbol{r}^\delta = \text{diag}[r_1, r_2, ..., r_v], \end{array} \tag{2.2.1}$$

where

$$n = \mathbf{1}_b'\boldsymbol{k} = \mathbf{1}_v'\boldsymbol{r} = \mathbf{1}_v'\boldsymbol{N}\mathbf{1}_b.$$

In the analysis of data obtained from an experiment in a block design, several matrices related to those given above are helpful. In particular, use is often made of the $n \times n$ projection matrix

$$\boldsymbol{\phi} = \boldsymbol{I}_n - \boldsymbol{D}'\boldsymbol{k}^{-\delta}\boldsymbol{D} \tag{2.2.2}$$

(the symbol $\boldsymbol{\phi}$ being adopted from Pearce, 1983, p. 59). Note that

$$\boldsymbol{\phi}' = \boldsymbol{\phi}, \ \boldsymbol{\phi\phi} = \boldsymbol{\phi}, \ \boldsymbol{\phi}\mathbf{1}_n = \mathbf{0}, \ \boldsymbol{\phi D}' = \mathbf{O} \tag{2.2.3}$$

and that $\boldsymbol{\phi}$ is of rank $n - b$. The matrix $\boldsymbol{\phi}$ appears often in conjunction with the matrix $\boldsymbol{\Delta}$, in particular, in the product

$$\boldsymbol{\Delta\phi} = \boldsymbol{\Delta} - \boldsymbol{\Delta D'k^{-\delta}D} = \boldsymbol{\Delta} - \boldsymbol{Nk^{-\delta}D}. \tag{2.2.4}$$

Its transpose, $\boldsymbol{\phi\Delta'}$, is an $n \times v$ matrix that may be called the "adjusted design matrix for treatments," adjusted in the sense of eliminating block contributions.

An analogous matrix,

$$\boldsymbol{\phi}_* = \boldsymbol{I}_n - \boldsymbol{\Delta' r^{-\delta}\Delta}, \tag{2.2.5}$$

of rank $n - v$ and with the properties

$$\boldsymbol{\phi}'_* = \boldsymbol{\phi}_*,\ \boldsymbol{\phi}_*\boldsymbol{\phi}_* = \boldsymbol{\phi}_*,\ \boldsymbol{\phi}_*\boldsymbol{1}_n = \boldsymbol{0},\ \boldsymbol{\phi}_*\boldsymbol{\Delta}' = \mathbf{O}, \tag{2.2.6}$$

gives

$$\boldsymbol{D\phi}_* = \boldsymbol{D} - \boldsymbol{D\Delta' r^{-\delta}\Delta} = \boldsymbol{D} - \boldsymbol{N' r^{-\delta}\Delta}. \tag{2.2.7}$$

Its transpose, $\boldsymbol{\phi}_*\boldsymbol{D}'$, is an $n \times b$ matrix that may be called the "adjusted design matrix for blocks," in the sense of eliminating treatment contributions.

Another important matrix, both in the analysis of data and in studying design properties, is the $v \times v$ matrix

$$\boldsymbol{\Delta\phi\Delta'} = \boldsymbol{r}^{\delta} - \boldsymbol{Nk^{-\delta}N'} = \boldsymbol{C} \ \text{(say)}, \tag{2.2.8}$$

called the "coefficient matrix" (Pearce, 1983, p. 59) or, simply, the "C-matrix" (Raghavarao, 1971, p. 49) or the "(reduced) intra-block matrix," because of its relation to the so-called "(reduced) intra-block equations," i.e., normal equations leading to the least-squares intra-block estimates of the treatment parameters (Kempthorne, 1952, Section 6.3; Chakrabarti, 1962; John, 1980, Section 2.2). It is also called the "information matrix" (John, 1987, p. 8; Nigam, Puri and Gupta, 1988, p. 11; Shah and Sinha, 1989, p. 2). In the present text, the term C-matrix and the symbol $\boldsymbol{C}$, exchangeably with $\boldsymbol{\Delta\phi\Delta'}$, will be used, though later it will become convenient to use more specified symbols, $\boldsymbol{C}_1 = \boldsymbol{\Delta\phi}_1\boldsymbol{\Delta}'$, where $\boldsymbol{\phi}_1 \equiv \boldsymbol{\phi}$, $\boldsymbol{C}_2 = \boldsymbol{\Delta\phi}_2\boldsymbol{\Delta}'$, where $\boldsymbol{\phi}_2 = \boldsymbol{D'k^{-\delta}D} - n^{-1}\boldsymbol{1}_n\boldsymbol{1}'_n$, and $\boldsymbol{C}_3 = \boldsymbol{\Delta\phi}_3\boldsymbol{\Delta}'$, where $\boldsymbol{\phi}_3 = n^{-1}\boldsymbol{1}_n\boldsymbol{1}'_n$.

The particular role of the C-matrix, i.e., of $\boldsymbol{C} \equiv \boldsymbol{C}_1$, and of $\boldsymbol{C}_2$ and $\boldsymbol{C}_3$, will become apparent in Chapters 3 and 4, and it will be explored in many places of the book.

Now, several definitions will be given that concern some special cases of $\boldsymbol{N}, \boldsymbol{r}$ and $\boldsymbol{k}$, common in applications (as recalled in Caliński, 1993a, Section 2).

Definition 2.2.1. A block design is said to be binary if the elements of its incidence matrix, $\boldsymbol{N} = [n_{ij}]$, admit the values 0 and 1 only. Otherwise, it is called a nonbinary design.

Definition 2.2.2. A block design is said to be proper (or equiblock-sized) if its block sizes are all equal, i.e., $k_1 = k_2 = \cdots = k_b$. Otherwise, it is called a nonproper (or unequal block-sized) design.

Definition 2.2.3. A block design is said to be equireplicate (or equireplicated) if its treatment replications are all equal, i.e., $r_1 = r_2 = \cdots = r_v$. Otherwise, it is called a nonequireplicate (or nonequireplicated) design.

Definition 2.2.4. A block design is said to be disconnected if its set of treatments can be divided into two or more subsets, such that for any pair of treatments belonging to different subsets, treatment i and treatment i', say, $n_{ij} > 0$ implies $n_{i'j} = 0$ for each j. Otherwise, it is called a connected design.

It may be mentioned here that the majority of the literature on block designs concerns binary designs, as can be seen from the book by Raghavarao (1971), and among these designs, the connected proper equireplicate designs are the most common (see also, e.g., John, 1980). However, situations occur when certain contrasts (one or more) have deliberately to be totally confounded with blocks, in order to permit the gaining of the advantages of small blocks for the benefit of the remaining contrasts of interest. This idea of confounding, originally introduced by Yates (1935) for factorial schemes of treatments, gives rise to disconnected block designs. (Their practical sense can be seen by referring, e.g., to Finney, 1960, p. 67; also to Pearce, 1983, Sections 3.7 and 8.3.)

As to Definition 2.2.4, it should be noticed that there are several equivalent definitions of connectedness (see, e.g., Raghavarao, 1971, p. 49). A definition that can easily be seen as equivalent to Definition 2.2.4 is the following, given originally by Bose (1950), and repeated by Raghavarao (1971, Definition 4.2.3) and Ogawa (1974), also recalled as Definition 1.1 in Caliński and Kageyama (1996b).

Definition 2.2.5. A block design is said to be connected if, given any two treatments i and i', it is possible to construct a chain of treatments $i = i_0, i_1, ..., i_m = i'$ such that every consecutive two treatments in the chain occur together in a block. Otherwise, it is called a disconnected design.

Another equivalent and convenient way of defining disconnectedness (used, e.g., by Baksalary and Tabis, 1985) can be written in two exchangeable forms, as follows.

Definition 2.2.6a. A block design is said to be disconnected of degree $g-1$ if, after an appropriate ordering of rows and columns, its $v \times b$ incidence matrix $\boldsymbol{N}$ can be written as $\boldsymbol{N} = \text{diag}[\boldsymbol{N}_1 : \boldsymbol{N}_2 : \cdots : \boldsymbol{N}_g]$, with $g > 1$, where all $v_\ell \times b_\ell$ matrices $\boldsymbol{N}_\ell$ are incidence matrices of connected subdesigns, $\ell = 1, 2, ..., g$, in accordance with the corresponding partitions $v = v_1 + v_2 + \cdots + v_g$ and $b = b_1 + b_2 + \cdots + b_g$. Otherwise, it is called a connected design (i.e., when $g = 1$).

Definition 2.2.6b. A block design is said to be disconnected of degree $g-1$ if the joint design matrix $[\boldsymbol{\Delta}' : \boldsymbol{D}']$ is of rank $v + b - g$, with $g > 1$. Otherwise, it is called a connected design (i.e., when $g = 1$).

Definition 2.2.6a seems to be more appealing from a practical point of view.

Remark 2.2.1. Note that for a disconnected design of degree $g-1$, the equality $\boldsymbol{k}^{\delta} = \text{diag}[k_1\boldsymbol{I}_{b_1} : k_2\boldsymbol{I}_{b_2} : \cdots : k_g\boldsymbol{I}_{b_g}]$ means that the block sizes within each connected subdesign are constant (equal to k_ℓ for the ℓth subdesign). Then

$$\boldsymbol{N}\boldsymbol{k}^{-\delta} = \tilde{\boldsymbol{k}}^{-\delta}\boldsymbol{N}, \tag{2.2.9}$$

where

$$\tilde{\boldsymbol{k}}^{\delta} = \text{diag}[k_1\boldsymbol{I}_{v_1} : k_2\boldsymbol{I}_{v_2} : \cdots : k_g\boldsymbol{I}_{v_g}]. \tag{2.2.10}$$

Some application of Remark 2.2.1 will be seen in Chapter 3.

A property of block designs that may be considered as particularly desirable, though restrictive, is the orthogonality of a design. This term can be defined in various ways (see Preece, 1977). At the present stage, the following definition, using the matrices (2.2.4) and (2.2.7), will be convenient.

Definition 2.2.7. A block design is said to be orthogonal if the adjusted design matrices for blocks and for treatments, $\boldsymbol{\phi}\boldsymbol{\Delta}'$ and $\boldsymbol{\phi}_*\boldsymbol{D}'$, are mutually orthogonal, i.e., if the condition

$$\boldsymbol{\Delta}\boldsymbol{\phi}\boldsymbol{\phi}_*\boldsymbol{D}' = \mathbf{O} \tag{2.2.11}$$

holds. Otherwise, it is called nonorthogonal.

As will become more apparent later, the sense of the condition (2.2.11) is that it allows to estimate the pure effects of the treatments independently of the pure effects of the blocks. This condition is in particular met by the classic RBD, as for this design [after an appropriate ordering of the observed variables, as in (1.3.18)]

$$\boldsymbol{D}' = \boldsymbol{I}_b \otimes \mathbf{1}_v, \quad \boldsymbol{\Delta}' = \mathbf{1}_b \otimes \boldsymbol{I}_v, \quad \boldsymbol{\phi} = \boldsymbol{I}_b \otimes (\boldsymbol{I}_v - v^{-1}\mathbf{1}_v\mathbf{1}_v'),$$
$$\boldsymbol{\phi}_* = (\boldsymbol{I}_b - b^{-1}\mathbf{1}_b\mathbf{1}_b') \otimes \boldsymbol{I}_v,$$

giving the adjusted design matrices

$$\boldsymbol{\phi}\boldsymbol{\Delta}' = \mathbf{1}_b \otimes (\boldsymbol{I}_v - v^{-1}\mathbf{1}_v\mathbf{1}_v') \quad \text{and} \quad \boldsymbol{\phi}_*\boldsymbol{D}' = (\boldsymbol{I}_b - b^{-1}\mathbf{1}_b\mathbf{1}_b') \otimes \mathbf{1}_v,$$

which evidently satisfy the condition (2.2.11).

Sometimes, it may be more convenient to use the following equivalent definition.

Definition 2.2.8. A block design is said to be orthogonal if the condition

$$\boldsymbol{\phi}\boldsymbol{\Delta}'\boldsymbol{r}^{-\delta}\boldsymbol{N} = \mathbf{O} \tag{2.2.12}$$

or, equivalently, the condition

$$\boldsymbol{\phi}_*\boldsymbol{D}'\boldsymbol{k}^{-\delta}\boldsymbol{N}' = \mathbf{O} \tag{2.2.13}$$

holds. Otherwise, it is called nonorthogonal.

It can easily be proved (which is left to the reader) that Definitions 2.2.7 and 2.2.8 are equivalent.

Also, the condition (2.2.11) can readily be rewritten as

$$\boldsymbol{N}\boldsymbol{k}^{-\delta}\boldsymbol{N}'\boldsymbol{r}^{-\delta}\boldsymbol{N} = \boldsymbol{N}, \tag{2.2.14}$$

which is the orthogonality condition originally given by Chakrabarti (1962, p. 19). On the other hand, because $\mathcal{C}(\boldsymbol{\phi}\boldsymbol{\Delta}')$, the column space of $\boldsymbol{\phi}\boldsymbol{\Delta}'$, is the orthogonal complement of $\mathcal{C}(\boldsymbol{D}')$ in $\mathcal{C}[\boldsymbol{\Delta}' : \boldsymbol{D}']$, and $\mathcal{C}(\boldsymbol{\phi}_*\boldsymbol{D}')$ is the orthogonal complement of $\mathcal{C}(\boldsymbol{\Delta}')$ in $\mathcal{C}[\boldsymbol{\Delta}' : \boldsymbol{D}']$ (see Appendix A, particularly Theorem A.2.5), the condition (2.2.11) can equivalently be written as

$$\mathcal{C}^{\perp}(\boldsymbol{D}') \cap \mathcal{C}[\boldsymbol{\Delta}' : \boldsymbol{D}'] \perp \mathcal{C}^{\perp}(\boldsymbol{\Delta}') \cap \mathcal{C}[\boldsymbol{\Delta}' : \boldsymbol{D}'],$$

which is the orthogonality condition from Darroch and Silvey (1963). Thus, the two well-established definitions of orthogonality are equivalent for block designs.

In addition, it may be noticed that the conditions (2.2.12) and (2.2.13) can equivalently be written as $\boldsymbol{\phi}\boldsymbol{\phi}_*\boldsymbol{D}' = \mathbf{O}$ and as $\boldsymbol{\phi}_*\boldsymbol{\phi}\boldsymbol{\Delta}' = \mathbf{O}$, respectively, the meaning of the former being that $\mathcal{C}(\boldsymbol{\phi}_*\boldsymbol{D}') \subset \mathcal{C}(\boldsymbol{D}')$, and of the latter that $\mathcal{C}(\boldsymbol{\phi}\boldsymbol{\Delta}') \subset \mathcal{C}(\boldsymbol{\Delta}')$. Thus, Definition 2.2.8 is equivalent to the definition of orthogonality given by Seber (1980, p. 42).

Finally, as it will be proved later (Corollary 2.3.4), if a block design is connected, then its incidence matrix $\boldsymbol{N}$ satisfies any of the orthogonality conditions [the equality (2.2.14) in particular] if and only if it is of the form $\boldsymbol{N} = n^{-1}\boldsymbol{r}\boldsymbol{k}'$, given by Pearce (1970, 1983, p. 95).

2.3 Some auxiliary results

In this section, several simple results that will be often used subsequently are given. They mainly concern the matrices defined in Section 2.2, the matrices (2.2.4) and (2.2.8) in particular.

Lemma 2.3.1. *The matrix $\boldsymbol{C} = \boldsymbol{\Delta}\boldsymbol{\phi}\boldsymbol{\Delta}'$ is positive semidefinite and of rank equal to that of $\boldsymbol{\phi}\boldsymbol{\Delta}'$, (say) $h \leq v-1$.*

Proof. The result follows from (2.2.3), which allows the matrix $\boldsymbol{C}$ to be written as $\boldsymbol{C} = \boldsymbol{\Delta}\boldsymbol{\phi}(\boldsymbol{\Delta}\boldsymbol{\phi})'$, where $(\boldsymbol{\Delta}\boldsymbol{\phi})'\mathbf{1}_v = \boldsymbol{\phi}\boldsymbol{\Delta}'\mathbf{1}_v = \mathbf{0}$ on account of (2.2.1) and (2.2.3), and from the well-known fact that for any (real) matrix $\boldsymbol{A}$, the matrix $\boldsymbol{A}\boldsymbol{A}'$ is nonnegative definite and of rank equal to that of $\boldsymbol{A}$ (see, e.g., Scheffé, 1959, Appendix II, Theorem 7). □

Lemma 2.3.2. *The rank of the matrix $\boldsymbol{C}$, denoted by h, is equal to $v-1$ if the block design is connected, and is equal to $v-g$ if the design is disconnected of degree $g-1$.*

Proof. From Lemma 2.3.1, $\text{rank}(\boldsymbol{C}) = \text{rank}(\boldsymbol{\phi}\boldsymbol{\Delta}')$. From Appendix A.2, it follows that $\mathcal{C}(\boldsymbol{\phi}\boldsymbol{\Delta}')$ is the orthogonal complement of $\mathcal{C}(\boldsymbol{D}')$ in $\mathcal{C}[\boldsymbol{\Delta}' : \boldsymbol{D}']$ and that the orthogonal projector on this complement is $\boldsymbol{P}_{\phi\Delta'} = \boldsymbol{P}_{[\Delta' : D']} - \boldsymbol{P}_{D'}$.

Hence, $\text{rank}(\boldsymbol{\phi\Delta'}) = \text{rank}(\boldsymbol{P}_{\phi\Delta'}) = \text{tr}(\boldsymbol{P}_{\phi\Delta'}) = \text{rank}[\boldsymbol{\Delta}' : \boldsymbol{D}'] - \text{rank}(\boldsymbol{D}')$. Evidently, $\text{rank}(\boldsymbol{D}') = b$, and, on account of Definition 2.2.5, $\text{rank}[\boldsymbol{\Delta}' : \boldsymbol{D}'] = v+b-1$ if the design is connected, giving $h = v-1$ then. Otherwise, on account of Definition 2.2.6,

$$[\boldsymbol{\Delta}' : \boldsymbol{D}'] = \begin{bmatrix} \boldsymbol{\Delta}_1' & \mathbf{O} & \cdots & \mathbf{O} & \boldsymbol{D}_1' & \mathbf{O} & \cdots & \mathbf{O} \\ \mathbf{O} & \boldsymbol{\Delta}_2' & \cdots & \mathbf{O} & \mathbf{O} & \boldsymbol{D}_2' & \cdots & \mathbf{O} \\ \cdot & \cdot & \cdots & \cdot & \cdot & \cdot & \cdots & \cdot \\ \mathbf{O} & \mathbf{O} & \cdots & \boldsymbol{\Delta}_g' & \mathbf{O} & \mathbf{O} & \cdots & \boldsymbol{\Delta}_g' \end{bmatrix},$$

which shows that $\text{rank}[\boldsymbol{\Delta}' : \boldsymbol{D}'] = \sum_{\ell=1}^{g}(v_\ell + b_\ell - 1) = v+b-g$, giving $h = v-g$, if the disconnectedness is of degree $g-1$. □

The characterization of disconnectedness given in Lemma 2.3.2 is equivalent to Definition 1 of Baksalary, Dobek and Kala (1980).

Lemma 2.3.3. *If the incidence matrix $\boldsymbol{N}$ of a block design is of rank $v - \rho$ and its coefficient matrix $\boldsymbol{C}$ is of rank h, then*

$$\boldsymbol{Nk}^{-\delta}\boldsymbol{N}' = \boldsymbol{r}^{\delta}\left(\sum_{i=\rho+1}^{h} \mu_i \boldsymbol{s}_i\boldsymbol{s}_i' + \sum_{i=h+1}^{v} \boldsymbol{s}_i\boldsymbol{s}_i'\right)\boldsymbol{r}^{\delta} \tag{2.3.1}$$

and

$$\boldsymbol{C} = \boldsymbol{r}^{\delta} - \boldsymbol{Nk}^{-\delta}\boldsymbol{N}' = \boldsymbol{r}^{\delta}\left(\sum_{i=1}^{\rho} \boldsymbol{s}_i\boldsymbol{s}_i' + \sum_{i=\rho+1}^{h} \varepsilon_i \boldsymbol{s}_i\boldsymbol{s}_i'\right)\boldsymbol{r}^{\delta}, \tag{2.3.2}$$

where $0 = \mu_1 = \cdots = \mu_\rho < \mu_{\rho+1} \le \cdots \le \mu_h < \mu_{h+1} = \cdots = \mu_v = 1$ are the eigenvalues of $\boldsymbol{Nk}^{-\delta}\boldsymbol{N}'$ with respect to $\boldsymbol{r}^{\delta}$, i.e., are the roots of the determinantal equation $|\boldsymbol{Nk}^{-\delta}\boldsymbol{N}' - \mu\boldsymbol{r}^{\delta}| = 0$, and the vectors $\boldsymbol{s}_1, ..., \boldsymbol{s}_\rho, \boldsymbol{s}_{\rho+1}, ..., \boldsymbol{s}_h, \boldsymbol{s}_{h+1}, ..., \boldsymbol{s}_v$ are the corresponding $\boldsymbol{r}^{\delta}$-orthonormal eigenvectors of $\boldsymbol{Nk}^{-\delta}\boldsymbol{N}'$ with respect to $\boldsymbol{r}^{\delta}$, i.e., are such that $\boldsymbol{Nk}^{-\delta}\boldsymbol{N}'\boldsymbol{s}_i = \mu_i\boldsymbol{r}^{\delta}\boldsymbol{s}_i$ and $\boldsymbol{s}_i'\boldsymbol{r}^{\delta}\boldsymbol{s}_{i'} = \delta_{ii'}$ (the Kronecker delta) for $i, i' = 1, 2, ..., v$, with $\boldsymbol{s}_v = n^{-1/2}\mathbf{1}_v$, and where $0 < \varepsilon_i = 1 - \mu_i$ for $i = \rho+1, \rho+2, ..., h$.

Proof. The results follow from the singular value decomposition of $\boldsymbol{N}$ with respect to $\boldsymbol{r}^{-\delta}$ and $\boldsymbol{k}^{-\delta}$ (see Rao and Mitra, 1971, p. 7 and Section 6.3), i.e., from the expression

$$\boldsymbol{r}^{-\delta}\boldsymbol{Nk}^{-\delta} = \sum_{i=\rho+1}^{v} \mu_i^{1/2}\boldsymbol{s}_i\boldsymbol{t}_i', \tag{2.3.3}$$

where $\mu_{\rho+1}, \mu_{\rho+2}, ..., \mu_v$ are the nonzero eigenvalues of $\boldsymbol{Nk}^{-\delta}\boldsymbol{N}'$ with respect to $\boldsymbol{r}^{\delta}$ and, simultaneously, of $\boldsymbol{N}'\boldsymbol{r}^{-\delta}\boldsymbol{N}$ with respect to $\boldsymbol{k}^{\delta}$, and the vectors $\boldsymbol{s}_{\rho+1}, \boldsymbol{s}_{\rho+2}, ..., \boldsymbol{s}_v$ are the corresponding $\boldsymbol{r}^{\delta}$-orthonormal eigenvectors of $\boldsymbol{Nk}^{-\delta}\boldsymbol{N}'$

with respect to $\boldsymbol{r}^\delta$, whereas the vectors $\boldsymbol{t}_{\rho+1}, \boldsymbol{t}_{\rho+2}, ..., \boldsymbol{t}_v$ are the corresponding $\boldsymbol{k}^\delta$-orthonormal eigenvectors of $\boldsymbol{N}'\boldsymbol{r}^{-\delta}\boldsymbol{N}$ with respect to $\boldsymbol{k}^\delta$, the $\boldsymbol{r}^\delta$-orthogonality implying that

$$\boldsymbol{r}^{-\delta} = \sum_{i=1}^{v} \boldsymbol{s}_i \boldsymbol{s}_i',$$

and from the fact that the matrices $\boldsymbol{N}\boldsymbol{k}^{-\delta}\boldsymbol{N}'$ and $\boldsymbol{C}$ are both nonnegative definite (see Lemma 2.3.1), which implies that $0 < \mu_i \leq 1$ and $0 \leq \varepsilon_i < 1$ for $i = \rho+1, \rho+2, ..., v$. That at least one of the eigenvalues $\mu_{\rho+1}, \mu_{\rho+2}, ..., \mu_v$ is equal to 1 follows from the equality

$$\boldsymbol{N}\boldsymbol{k}^{-\delta}\boldsymbol{N}'\boldsymbol{1}_v = \boldsymbol{r} \ (= \boldsymbol{r}^\delta \boldsymbol{1}_v), \quad \text{or} \quad \boldsymbol{N}'\boldsymbol{r}^{-\delta}\boldsymbol{N}\boldsymbol{1}_b = \boldsymbol{k} \ (= \boldsymbol{k}^\delta \boldsymbol{1}_b),$$

which hold on account of (2.2.1). The unit eigenvalue so obtained is denoted by μ_v, the corresponding eigenvectors being then $\boldsymbol{s}_v = n^{-1/2}\boldsymbol{1}_v$, as that of $\boldsymbol{N}\boldsymbol{k}^{-\delta}\boldsymbol{N}'$ with respect to $\boldsymbol{r}^\delta$, and $\boldsymbol{t}_v = n^{-1/2}\boldsymbol{1}_b$, as that of $\boldsymbol{N}'\boldsymbol{r}^{-\delta}\boldsymbol{N}$ with respect to $\boldsymbol{k}^\delta$. □

Remark 2.3.1. On account of (2.3.3), expressions analogous to (2.3.1) and (2.3.2) can be given for $\boldsymbol{N}'\boldsymbol{r}^{-\delta}\boldsymbol{N}$ and $\boldsymbol{k}^\delta - \boldsymbol{N}'\boldsymbol{r}^{-\delta}\boldsymbol{N}$, respectively, the nonzero eigenvalues of $\boldsymbol{N}'\boldsymbol{r}^{-\delta}\boldsymbol{N}$ with respect to $\boldsymbol{k}^\delta$ being exactly the same as those of $\boldsymbol{N}\boldsymbol{k}^{-\delta}\boldsymbol{N}'$ with respect to $\boldsymbol{r}^\delta$, i.e., the nonzero roots of $|\boldsymbol{N}\boldsymbol{k}^{-\delta}\boldsymbol{N}' - \mu\boldsymbol{r}^\delta| = 0$ being simultaneously the nonzero roots of $|\boldsymbol{N}'\boldsymbol{r}^{-\delta}\boldsymbol{N} - \mu\boldsymbol{k}^\delta| = 0$. In fact, the analogous expressions can be written as the spectral decompositions

$$\boldsymbol{N}'\boldsymbol{r}^{-\delta}\boldsymbol{N} = \boldsymbol{k}^\delta\Bigg(\sum_{j=\rho_*+1}^{h_*} \mu_j^* \boldsymbol{t}_j^* \boldsymbol{t}_j^{*\prime} + \sum_{j=h_*+1}^{b} \boldsymbol{t}_j^* \boldsymbol{t}_j^{*\prime}\Bigg)\boldsymbol{k}^\delta$$

and

$$\boldsymbol{k}^\delta - \boldsymbol{N}'\boldsymbol{r}^{-\delta}\boldsymbol{N} = \boldsymbol{k}^\delta\Bigg(\sum_{j=1}^{\rho_*} \boldsymbol{t}_j^* \boldsymbol{t}_j^{*\prime} + \sum_{j=\rho_*+1}^{h_*} \varepsilon_j^* \boldsymbol{t}_j^* \boldsymbol{t}_j^{*\prime}\Bigg)\boldsymbol{k}^\delta,$$

with the eigenvalues $\{\mu_j^*\}$ in the first decomposition and $\{\varepsilon_j^*\}$ in the second related to those in (2.3.1) and (2.3.2), respectively, by the equalities $\mu_j^* = \mu_{j-b+v} = \mu_i$ and $\varepsilon_j^* = \varepsilon_{j-b+v} = \varepsilon_i$ for $j = \rho_*+1, \rho_*+2, ..., h_*$, where $\rho_* = b - v + \rho$ and $h_* = b - v + h$, and with the eigenvectors $\boldsymbol{t}_1^*, ..., \boldsymbol{t}_{\rho_*}^*, \boldsymbol{t}_{\rho_*+1}^*, ..., \boldsymbol{t}_{h_*}^*, \boldsymbol{t}_{h_*+1}^*, ..., \boldsymbol{t}_b^*$ being such that $\boldsymbol{N}'\boldsymbol{r}^{-\delta}\boldsymbol{N}\boldsymbol{t}_j^* = \mu_j^*\boldsymbol{k}^\delta\boldsymbol{t}_j^*$ and $\boldsymbol{t}_j^{*\prime}\boldsymbol{k}^\delta\boldsymbol{t}_{j'}^* = \delta_{jj'}$ for $j, j' = 1, 2, ..., b$, with $\boldsymbol{t}_b^* = n^{-1/2}\boldsymbol{1}_b$, and related to $\{\boldsymbol{t}_i\}$ in (2.3.3) by the equalities $\boldsymbol{t}_j^* = \boldsymbol{t}_{j-b+v} = \boldsymbol{t}_i$ for $j = \rho_*+1, \rho_*+2, ..., b$.

Note that the decomposition of $\boldsymbol{k}^\delta - \boldsymbol{N}'\boldsymbol{r}^{-\delta}\boldsymbol{N}$ in Remark 2.3.1 is obtained by using the equality

$$\boldsymbol{k}^{-\delta} = \sum_{j=1}^{b} \boldsymbol{t}_j^* \boldsymbol{t}_j^{*\prime}.$$

Also note that from (2.3.3) the relation between $\{s_i\}$ and $\{t_i\}$ for $i = \rho+1, \rho+2, ..., v$ [equal to $\{t_j^*\}$ for $j = \rho_* + 1, \rho_* + 2, ..., b$] is given by the formulae

$$s_i = \mu_i^{-1/2} r^{-\delta} N t_i \quad \text{and} \quad t_i = \mu_i^{-1/2} k^{-\delta} N' s_i.$$

Lemma 2.3.3 and Remark 2.3.1 show that the eigenvalues $\{\mu_i\}$ and $\{\varepsilon_i\}$, and the corresponding eigenvectors $\{s_i\}$, which will be used at many places in this book, can be obtained in various ways. The usual procedure would be as follows.

(1) Take the incidence matrix N of the design to be considered.

(2) Calculate the matrix $Nk^{-\delta}N'$.

(3) Find eigenvalues $\{\mu_i\}$ of $Nk^{-\delta}N'$ with respect to r^δ as the roots of the equation

$$|Nk^{-\delta}N' - \mu r^\delta| = 0$$

and order them as in Lemma 2.3.3, remembering that $\mu_1 = \mu_2 = \cdots = \mu_\rho = 0$ and $\mu_{h+1} = \mu_{h+2} = \cdots = \mu_v = 1$, and noting that multiplicities for $\rho < i < h$ may also occur. If required, obtain $\varepsilon_i = 1 - \mu_i$ for $i = \rho+1, \rho+2, ..., h$.

(4) Except for $\mu_v = 1$, find for each μ_i the corresponding eigenvector s_i by solving the equations

$$Nk^{-\delta}N's_i = \mu_i r^\delta s_i, \quad i = 1, 2, ..., v-1,$$

remembering that $s_v = n^{-1/2}\mathbf{1}_v$. Make them all to be r^δ-orthonormal. Certainly, for multiplicities in $\{\mu_i\}$ the solutions will not be unique.

(5) If there are multiplicities in $\{\mu_i\}$, divide the set of corresponding eigenvectors $\{s_i\}$ into subsets, so that each of them corresponds to a different distinct eigenvalue. A useful notation for such division will be introduced in Section 3.4.

However, cases may occur in which it would be more convenient to solve, instead of the determinantal equation given under (3), the equation

$$|N'r^{-\delta}N - \mu k^\delta| = 0.$$

Because only its nonzero roots are of interest, and these are the same as those of the former equation, any of these two can be used to obtain the nonzero μ_i's. Also, it may happen that, instead of calculating the eigenvectors $\{s_i\}$, it would be easier to obtain its duals, the eigenvectors $\{t_i\}$. Then, the results following Remark 2.3.1 may become helpful.

Another change in the above usual procedure may result from the fact that often one can avoid going through step (3), and find both the eigenvalues and the eigenvectors simultaneously, as solutions of the equations considered in step (4). This is particularly true when some evident pattern exists in the matrix $Nk^{-\delta}N'$, allowing its eigenvectors with respect to r^δ to be found almost immediately.

Example 2.3.1. Consider a block design (taken from Caliński, 1971, Section 5) with the incidence matrix

$$N = \begin{bmatrix} 2 & 0 & 2 & 1 & 1 \\ 1 & 1 & 1 & 1 & 0 \\ 1 & 1 & 1 & 0 & 1 \\ 2 & 1 & 2 & 1 & 1 \end{bmatrix}.$$

From it, $r = [6, 4, 4, 7]', k = [6, 3, 6, 3, 3]'$ and

$$Nk^{-\delta}N' = \frac{1}{3}\begin{bmatrix} 6 & 3 & 3 & 6 \\ 3 & 3 & 2 & 4 \\ 3 & 2 & 3 & 4 \\ 6 & 4 & 4 & 7 \end{bmatrix}.$$

From the pattern of this matrix, the following choice of r^{δ}-orthonormal eigenvectors of it with respect to $r^{\delta} = \text{diag}[6, 4, 4, 7]$ can easily be made:

$$\begin{array}{ll} s_1 = [1, 1, 1, -2]'/\sqrt{42}, & s_2 = [4, -3, -3, 0]'/\sqrt{168}, \\ s_3 = [0, 1, -1, 0]'/\sqrt{8}, & s_4 = [1, 1, 1, 1]'/\sqrt{21}. \end{array}$$

Check that they satisfy the condition $s_i' r^{\delta} s_{i'} = \delta_{ii'}$ for $i, i' = 1, 2, 3, 4$, and that

$$\begin{array}{ll} Nk^{-\delta}N's_1 = \mathbf{0}, & Nk^{-\delta}N's_2 = (1/12)r^{\delta}s_2, \\ Nk^{-\delta}N's_3 = (1/12)r^{\delta}s_3, & Nk^{-\delta}N's_4 = r^{\delta}s_4. \end{array}$$

These equalities show that the corresponding eigenvalues of $Nk^{-\delta}N'$ with respect to r^{δ} are $\mu_1 = 0$, $\mu_2 = \mu_3 = 1/12$ and $\mu_4 = 1$. It seems that this way of obtaining the required eigenvectors and eigenvalues is here easier than when following exactly the usual procedure described above.

From Lemma 2.3.3, several useful corollaries also follow.

Corollary 2.3.1. *A block design satisfies Fisher's inequality $v \leq b$ if any of the following two equivalent conditions hold:*

(a) *the matrix $Nk^{-\delta}N'$ has no zero eigenvalue;*

(b) *the matrix C has no unit eigenvalue with respect to r^{δ}.*

Proof. Because $\text{rank}(N) = v - \rho \leq b$, the conditions follow immediately from Lemma 2.3.3. □

Because both (a) and (b) in Corollary 2.3.1 are equivalent to the condition that the rows of the incidence matrix N are linearly independent, the inequality $v \leq b$, originally established by Fisher (1940) for a certain class of incomplete block designs, may be replaced by the row independence condition and called as follows (see Baksalary et al., 1980; Baksalary and Puri, 1988).

Definition 2.3.1. A block design is said to satisfy Fisher's condition if the rows of its incidence matrix are linearly independent.

Corollary 2.3.2. *In relation to the definitions of connectedness (given in Section 2.2) the following conditions are useful:*

(a) *A block design is connected if and only if its matrix* $\boldsymbol{N}\boldsymbol{k}^{-\delta}\boldsymbol{N}'$ *has only one unit eigenvalue with respect to* $\boldsymbol{r}^{\delta}$.

(b) *A block design is disconnected of degree* $g-1$ *if and only if its matrix* $\boldsymbol{N}\boldsymbol{k}^{-\delta}\boldsymbol{N}'$ *has* $g\ (=v-h)$ *unit eigenvalues with respect to* $\boldsymbol{r}^{\delta}$. *All corresponding eigenvectors,* $\boldsymbol{s}_{h+1}, \boldsymbol{s}_{h+2}, ..., \boldsymbol{s}_v$ *in* (2.3.1)[*and in* (2.3.3)], *belong then to the subspace* $\mathcal{C}\{\text{diag}[\mathbf{1}_{v_1} : \mathbf{1}_{v_2} : \cdots : \mathbf{1}_{v_g}]\}$ (*using the notation of Definition* 2.2.6a).

(c) *A block design is disconnected of degree* $g-1$ *if and only if its matrix* $\boldsymbol{N}'\boldsymbol{r}^{-\delta}\boldsymbol{N}$ *has* $g\ (=v-h)$ *unit eigenvalues with respect to* $\boldsymbol{k}^{\delta}$. *All corresponding eigenvectors,* $\boldsymbol{t}_{h+1}, \boldsymbol{t}_{h+2}, ..., \boldsymbol{t}_v$ *in* (2.3.3), *belong then to the subspace* $\mathcal{C}\{\text{diag}[\mathbf{1}_{b_1} : \mathbf{1}_{b_2} : \cdots : \mathbf{1}_{b_g}]\}$.

Proof. These results follow immediately from Lemma 2.3.3 on account of Lemma 2.3.2 and Definition 2.2.6. Details of the proofs of (b) and (c) are left to the reader. On account of Remark 2.3.1, (c) can be obtained either similarly as (b), or from (b) by the use of (2.3.3). □

Corollary 2.3.3. *A block design is orthogonal if and only if the multiplicity of the unit eigenvalue of its matrix* $\boldsymbol{C}$ *with respect to* $\boldsymbol{r}^{\delta}$ *is equal to the rank of the matrix, i.e.,* $\rho = h$ *in* (2.3.2).

Proof. The orthogonality condition (2.2.12) or (2.2.14) can readily be rewritten as $\boldsymbol{C}\boldsymbol{r}^{-\delta}\boldsymbol{N} = \mathbf{O}$, which, on account of (2.3.2) and (2.3.3), can also be expressed as

$$\boldsymbol{r}^{\delta}\Big(\sum_{i=\rho+1}^{h} \varepsilon_i \mu_i^{1/2} \boldsymbol{s}_i \boldsymbol{t}_i'\Big)\boldsymbol{k}^{\delta} = \mathbf{O}.$$

This condition, however, on account of the notation in Lemma 2.3.3, implies that $\rho = h$. The sufficiency of this is obvious. □

In view of Lemmas 2.3.2 and 2.3.3, Corollary 2.3.3 is equivalent to Theorem 2 of Baksalary et al. (1980).

Corollary 2.3.4. *A connected block design is orthogonal if and only if its incidence matrix is of the form*

$$\boldsymbol{N} = n^{-1}\boldsymbol{r}\boldsymbol{k}' \tag{2.3.4}$$

(Pearce, 1983, p. 95).

Proof. From Lemma 2.3.2 and Corollary 2.3.3, a necessary and sufficient condition for the orthogonality is $\rho = v-1$. But on account of (2.3.3), this means that $\boldsymbol{r}^{-\delta}\boldsymbol{N}\boldsymbol{k}^{-\delta} = n^{-1}\mathbf{1}_v\mathbf{1}_b'$, which is equivalent to (2.3.4). □

Remark 2.3.2. If (2.3.4) holds, the matrix $\boldsymbol{C}$ is of the form (Pearce, 1983, p. 94)

$$\boldsymbol{C} = \boldsymbol{r}^{\delta} - n^{-1}\boldsymbol{r}\boldsymbol{r}', \tag{2.3.5}$$

which becomes $\boldsymbol{C} = r(\boldsymbol{I}_v - v^{-1}\mathbf{1}_v\mathbf{1}_v')$ if and only if, in addition, the design is equireplicate.

It follows from this remark that for all equireplicate connected orthogonal block designs, i.e., designs with $\boldsymbol{N} = v^{-1}\mathbf{1}_v\boldsymbol{k}'$, the C-matrix is the same for given r and v, whether it is the classical RBD or a design in which each block contains the complete set of v treatments once or many times, not necessarily the same for all blocks. It covers also a design consisting of just one block, with $\boldsymbol{N} = r\mathbf{1}_v$. Moreover, any connected orthogonal block design with the same vector $\boldsymbol{r}$ has the same C-matrix (2.3.5), independently of the number of blocks and their sizes.

The following results are related to a matrix $\boldsymbol{C}^-$, a generalized inverse (a g-inverse) of the matrix $\boldsymbol{C}$, i.e., to a matrix satisfying the condition

$$\boldsymbol{C}\boldsymbol{C}^-\boldsymbol{C} = \boldsymbol{C} \tag{2.3.6}$$

(see Rao and Mitra, 1971, Section 2.2).

Lemma 2.3.4. *Regardless of the choice of* $\boldsymbol{C}^-$,

$$\boldsymbol{C}\boldsymbol{C}^-\boldsymbol{\Delta}\boldsymbol{\phi} = \boldsymbol{\Delta}\boldsymbol{\phi}. \tag{2.3.7}$$

Proof. The equality (2.3.7) follows directly from Lemma 2.2.6(b) of Rao and Mitra (1971) when taking $\boldsymbol{A} = \boldsymbol{\phi}\boldsymbol{\Delta}'$. (See also Pearce, 1983, Lemma 3.2.D.) □

Lemma 2.3.5. *The matrix* $\boldsymbol{\phi}\boldsymbol{\Delta}'\boldsymbol{C}^-\boldsymbol{\Delta}\boldsymbol{\phi}$ *is invariant for any choice of* $\boldsymbol{C}^-$, *and it is idempotent, symmetric and of* rank h = rank($\boldsymbol{C}$).

Proof. This result follows from Lemma 2.2.6(d) of Rao and Mitra (1971), on account of (2.2.3), and from Lemmas 2.3.4 and 2.3.1. (See also Pearce, 1983, Lemma 3.2.E.) □

From Lemma 2.3.3, the following useful result is evidently obtainable, by referring to (2.3.6).

Corollary 2.3.5. *One possible choice of* $\boldsymbol{C}^-$ *is the matrix*

$$\sum_{i=1}^{\rho} \boldsymbol{s}_i\boldsymbol{s}_i' + \sum_{i=\rho+1}^{h} \varepsilon_i^{-1}\boldsymbol{s}_i\boldsymbol{s}_i',$$

where the notation is as in Lemma 2.3.3.

A matrix that will often be applied is

$$\boldsymbol{\psi} = \boldsymbol{\phi} - \boldsymbol{\phi}\boldsymbol{\Delta}'\boldsymbol{C}^-\boldsymbol{\Delta}\boldsymbol{\phi} = \boldsymbol{\phi}(\boldsymbol{I}_n - \boldsymbol{\Delta}'\boldsymbol{C}^-\boldsymbol{\Delta})\boldsymbol{\phi}. \tag{2.3.8}$$

It can easily be shown, on account of (2.2.3), that

$$\boldsymbol{\psi}\boldsymbol{\phi} = \boldsymbol{\psi},\ \boldsymbol{\psi}\mathbf{1}_n = \mathbf{0},\ \boldsymbol{\psi}\boldsymbol{D}' = \mathbf{O}$$

and, applying Lemmas 2.3.4 and 2.3.5, that

$$\boldsymbol{\psi}' = \boldsymbol{\psi},\ \boldsymbol{\psi}\boldsymbol{\psi} = \boldsymbol{\psi} \quad \text{and} \quad \boldsymbol{\psi}\boldsymbol{\Delta}' = \mathbf{O}. \tag{2.3.9}$$

A further result concerning the matrix (2.3.8) is the following lemma.

Lemma 2.3.6. *The matrix $\boldsymbol{\psi}$ is invariant for any choice of $\boldsymbol{C}^-$, and it is of rank (and trace) equal to $n - b - h$.*

Proof. This lemma follows immediately from Lemma 2.3.5, from (2.3.8) and from the readily seen result that $\mathrm{tr}(\boldsymbol{\phi}) = n - b$, on account of (2.2.2) and (2.2.3). □

Now, from (2.2.1), Lemma 2.3.5, the formulae (2.2.3), (2.2.6) and (2.3.9), and on account of Appendix A.2, the following corollary can be given.

Corollary 2.3.6. *The matrices $\boldsymbol{D}'\boldsymbol{k}^{-\delta}\boldsymbol{D}$, $\boldsymbol{\phi}\boldsymbol{\Delta}'\boldsymbol{C}^-\boldsymbol{\Delta}\boldsymbol{\phi}$, $\boldsymbol{\phi}$, $\boldsymbol{\phi}_*$ and $\boldsymbol{\psi}$ are all orthogonal projectors.*

Thus, the following notation can be used:

$$\boldsymbol{D}'\boldsymbol{k}^{-\delta}\boldsymbol{D} = \boldsymbol{P}_{D'}, \tag{2.3.10}$$

$$\boldsymbol{\phi}\boldsymbol{\Delta}'\boldsymbol{C}^-\boldsymbol{\Delta}\boldsymbol{\phi} = \boldsymbol{P}_{\phi\Delta'}, \tag{2.3.11}$$

$$\boldsymbol{\phi} = \boldsymbol{I}_n - \boldsymbol{P}_{D'} = \boldsymbol{P}_{(D')^\perp}, \tag{2.3.12}$$

$$\boldsymbol{\phi}_* = \boldsymbol{I}_n - \boldsymbol{P}_{\Delta'} = \boldsymbol{P}_{(\Delta')^\perp} \tag{2.3.13}$$

and

$$\boldsymbol{\psi} = \boldsymbol{P}_{(D')^\perp} - \boldsymbol{P}_{\phi\Delta'}. \tag{2.3.14}$$

Formula (2.3.14) shows, by referring to Theorem A.2.4, that $\boldsymbol{\psi}$ is the orthogonal projector on the subspace $\mathcal{C}^\perp(\boldsymbol{\phi}\boldsymbol{\Delta}') \cap \mathcal{C}^\perp(\boldsymbol{D}')$, as it follows from $\boldsymbol{P}_{(D')^\perp}\boldsymbol{P}_{\phi\Delta'} = \boldsymbol{P}_{\phi\Delta'}\boldsymbol{P}_{(D')^\perp} = \boldsymbol{P}_{\phi\Delta'}$ (see Rao and Mitra, 1971, Theorem 5.1.3). On the other hand, because $\mathcal{C}^\perp(\boldsymbol{A}') = \mathcal{N}(\boldsymbol{A})$, the null space of $\boldsymbol{A}$ (see Definition A.1.14), the subspace on which $\boldsymbol{\psi}$ projects can, equivalently, be written as

$$\mathcal{N}(\boldsymbol{\Delta}\boldsymbol{\phi}) \cap \mathcal{N}(\boldsymbol{D}) = \mathcal{N}\begin{bmatrix} \boldsymbol{\Delta}\boldsymbol{\phi} \\ \boldsymbol{D} \end{bmatrix}. \tag{2.3.15}$$

But, as $\boldsymbol{\Delta}\boldsymbol{\phi} = \boldsymbol{\Delta} - \boldsymbol{\Delta}\boldsymbol{D}'\boldsymbol{k}^{-\delta}\boldsymbol{D}$, the subspace (2.3.15) is in fact equal to the null space of $[\boldsymbol{\Delta}' : \boldsymbol{D}']'$, and thus is the orthogonal complement of $\mathcal{C}[\boldsymbol{\Delta}' : \boldsymbol{D}']$, i.e., $\mathcal{C}^\perp[\boldsymbol{\Delta}' : \boldsymbol{D}']$. By a similar argument, it can be shown that $\boldsymbol{\psi}_* = \boldsymbol{P}_{(\Delta')^\perp} - \boldsymbol{P}_{\phi_* D'}$ is the orthogonal projector on the subspace $\mathcal{C}^\perp(\boldsymbol{\Delta}') \cap \mathcal{C}^\perp(\boldsymbol{\phi}_*\boldsymbol{D}')$, which in turn is equal to $\mathcal{C}^\perp[\boldsymbol{D}' : \boldsymbol{\Delta}']$, and thus $\boldsymbol{\psi}_* = \boldsymbol{\psi}$. This important result will be used later on.

2.4 Concurrences and two notions of balance

As already mentioned in Section 2.2, and as will be explained more thoroughly in the next two chapters, the C-matrix of a block design to a great extent determines the statistical properties of the design. In connection with this, it can be expected that the pattern of that matrix must be taken into account in any considerations concerning a concept of balance. Two such concepts will be considered in this section, following the terminology and notation adopted in Caliński (1993a, Section 3) and used also in Caliński and Kageyama (1996b, Sections 2.1 and 2.2).

2.4.1 Balance based on equality of weighted concurrences

If a design is proper and equireplicate, then its C-matrix is clearly of the form

$$\boldsymbol{C} = r\boldsymbol{I}_v - k^{-1}\boldsymbol{N}\boldsymbol{N}', \tag{2.4.1}$$

where r is the common treatment replication number and k is the common block size. Any further simplification of the matrix (2.4.1) depends on the pattern of the $v \times v$ matrix $\boldsymbol{N}\boldsymbol{N}' = [\sum_j n_{ij}n_{i'j}]$, called the "concordance matrix" (by John, 1980, p. 30) or "concurrence matrix" (by Pearce, 1963, p. 353). This matrix has certain informative properties, particularly interesting for binary designs. It can be seen that for any binary design, the ith diagonal element of $\boldsymbol{N}\boldsymbol{N}'$ is equal to r_i, the number of blocks in which the ith treatment "occurs," and the off-diagonal element in the ith row, and the i'th column is equal to the number of blocks in which the ith and the i'th treatments "concur," i.e., are both present (Pearce, 1983, p. 4). Hence, the off-diagonal elements are usually called "concurrences" (Pearce, 1976, p. 106), and are denoted sometimes by $\lambda_{ii'}$ (see, e.g., John, 1980, p. 13; John, 1987, p. 17). Thus, the pattern of the C-matrix of a binary proper design depends on $\{r_i\}$ and $\{\lambda_{ii'}\}$, and that of a binary proper and equireplicate design depends on the latter only. Obviously, a disconnected design cannot have all $\lambda_{ii'}$ positive, whereas the classic RBD has all r_i and all $\lambda_{ii'}$ equal to b, the number of blocks.

These considerations lead to the following concept of balance.

Definition 2.4.1. A block design is said to be pairwise balanced if the off-diagonal elements of its concurrence matrix, $\boldsymbol{N}\boldsymbol{N}'$, are all equal.

Originally, this definition was introduced by Bose and Shrikhande (1960) for binary block designs. They called any such design with a common concurrence λ a "pairwise balanced design of index λ" (see Rahgavarao, 1971, p. 36). But Definition 2.4.1 does not need to be restricted to binary designs only. In the present general form, it is from Hedayat and Federer (1974). These authors, however, have shown that the pairwise balance is neither necessary nor sufficient for a block design to be balanced in a statistical sense. Therefore, the pairwise balance is sometimes called combinatorial balance (see, e.g., Puri and Nigam, 1977b, p. 1174). In fact, the pattern of the C-matrix depends directly on the pattern of the matrix $\boldsymbol{NN'}$ in case of a proper design only. (This pattern may give some justification for using the adjective "proper".) Moreover, if the design is proper and binary, then equal concurrences also imply equal occurrences, i.e., equal treatment replications, and, hence, equal diagonal elements of the matrix $\boldsymbol{C}$. This result can be seen from the equality

$$k\boldsymbol{r} = \boldsymbol{r} + (v-1)\lambda\mathbf{1}_v, \tag{2.4.2}$$

obtained when taking $\boldsymbol{NN'}\mathbf{1}_v$ for any proper block design of block size k (left-hand side) and for any such design that in addition is binary and pairwise balanced (right-hand side).

Designs of this type, i.e., proper binary equireplicate pairwise balanced designs, were introduced by Yates (1936a), though their combinatorial idea was known much earlier (see Caliński and Kageyama, 1996b, Section 0 and the relevant references there). These designs are traditionally defined as follows (see, e.g., Raghavarao, 1971, Definition 4.3.2; Rasch and Herrendörfer, 1986, Definition 3.14a).

Definition 2.4.2. An arrangement of v treatments in b blocks, each containing $k(< v)$ treatments, is said to be a balanced incomplete block (BIB) design if it meets the following conditions:

1. Every treatment occurs at most once in a block.
2. Every treatment occurs in exactly r blocks.
3. Every two treatments concur in exactly $\lambda(> 0)$ blocks.

The quantities v, b, r, k and λ appearing in Definition 2.4.2 are usually called the parameters of a BIB design. Evident relations among them are $vr = bk(= n)$ and, from (2.4.2), $\lambda(v-1) = r(k-1)$. Note that from (2.4.2), the condition 2 is redundant. The BIB designs will be considered in detail in Chapters 6 and 8.

To extend the concept of balance based on the equality of concurrences from proper designs to designs of unequal block sizes, one may wish to consider the equality of the off-diagonal elements in $\boldsymbol{Nk}^{-\delta}\boldsymbol{N'}$ instead of that in $\boldsymbol{NN'}$, because if the off-diagonal elements of $\boldsymbol{Nk}^{-\delta}\boldsymbol{N'}$ are equal, so too are the off-diagonal elements of the C-matrix, this being in general not true for the off-diagonal elements of $\boldsymbol{NN'}$. Pointing to the difference between $\boldsymbol{Nk}^{-\delta}\boldsymbol{N'}$ and $\boldsymbol{NN'}$, Pearce (1964, 1976) introduced the term "weighted concurrences" for the off-diagonal elements of the former, just as "concurrences" had been used for those elements of the latter. Using a matrix $\boldsymbol{W}$ with those weighted concurrences as its off-diagonal elements and with all its diagonal elements equal to zero, he could write $\boldsymbol{Nk}^{-\delta}\boldsymbol{N'} = \boldsymbol{p}^{\delta} + \boldsymbol{W}$ and $\boldsymbol{C} = \boldsymbol{q}^{\delta} - \boldsymbol{W}$, denoting by $\boldsymbol{p}^{\delta}$ a diagonal matrix of the same diagonal elements as those of the matrix $\boldsymbol{Nk}^{-\delta}\boldsymbol{N'}$, and using $\boldsymbol{q}^{\delta} = \boldsymbol{r}^{\delta} - \boldsymbol{p}^{\delta}$. The diagonal elements of $\boldsymbol{q}^{\delta}$ were called by Pearce (1971) the "quasi-replications" (see also Pearce, 1983, p. 66). Because $\boldsymbol{q} = \boldsymbol{q}^{\delta}\mathbf{1}_v = \boldsymbol{W}\mathbf{1}_v$, just as $\boldsymbol{r} = \boldsymbol{Nk}^{-\delta}\boldsymbol{N'}\mathbf{1}_v$, the equality of the weighted concurrences can easily be seen to imply the equality of the quasi-replications and, hence, of the diagonal elements of $\boldsymbol{C}$.

Thus, the property of pairwise balance in binary proper designs can be extended to all connected block designs.

Definition 2.4.3. A connected block design is said to be totally balanced (or of Type T_0) if the off-diagonal elements of its matrix $\boldsymbol{N}\boldsymbol{k}^{-\delta}\boldsymbol{N}'$ (i.e., the weighted concurrences) are all equal.

The term total balance (Type T_0) used here is taken from Pearce (1976, Section 4.A, and 1983, Section 5.2). It should not be confused with the notion of total balance (Type T) used by him earlier (Pearce, 1963, Section 3.2), and after him by some other authors, e.g., Graf-Jaccottet (1977) and Nigam et al. (1988, Section 2.6.2). Also one must note that Definition 2.4.3 is equivalent to Definition 4.3.1 of Raghavarao (1971), on account of Theorem 1 of Puri and Nigam (1977b). This definition of balance, originally used by Tocher (1952) and Rao (1958), is also equivalent to that of "balanced block (BB) designs" considered by Kageyama (1974), but it does not coincide with that used by Kiefer (1958) for his BB designs, except when the design is binary and proper.

The following result related to Definition 2.4.3 is from Rao (1958).

Theorem 2.4.1. *A connected block design is totally balanced (in the sense of Definition* 2.4.3) *if and only if the* $v-1$ *nonzero eigenvalues of its* C*-matrix are all equal.*

Proof. Because the off-diagonal elements of the matrix $\boldsymbol{N}\boldsymbol{k}^{-\delta}\boldsymbol{N}'$ and those of the matrix $\boldsymbol{W}$, defined above, are the same, the condition of Definition 2.4.3 can be written as $\boldsymbol{W} = w(\mathbf{1}_v\mathbf{1}_v' - \boldsymbol{I}_v)$, w being some positive scalar, which holds if and only if

$$\boldsymbol{N}\boldsymbol{k}^{-\delta}\boldsymbol{N}' = \boldsymbol{r}^{\delta} - \theta(\boldsymbol{I}_v - v^{-1}\mathbf{1}_v\mathbf{1}_v') \tag{2.4.3}$$

or, equivalently, the C-matrix is of the form

$$\boldsymbol{C} = \theta(\boldsymbol{I}_v - v^{-1}\mathbf{1}_v\mathbf{1}_v'), \tag{2.4.4}$$

where $\theta = vw$. Evidently, as a connected block design has rank$(\boldsymbol{C}) = v-1$ (Lemma 2.3.2), the C-matrix has the form (2.4.4) if and only if it has a unique nonzero eigenvalue, θ, of multiplicity $v-1$. □

Remark 2.4.1. In general, $\theta = [n - \text{tr}(\boldsymbol{N}\boldsymbol{k}^{-\delta}\boldsymbol{N}')]/(v-1) \leq n/v$, whereas $\theta = (n-b)/(v-1)$ if the design is binary, whereas $\theta = r$ for an equireplicate connected orthogonal block design (see Remark 2.3.2). Therefore, θ has been called by Pearce (1964, 1976, 1983, pp.74-75) the "effective replication."

That the property of pairwise balance and its generalization (Definition 2.4.3) apply to connected block designs only, can be seen directly from Definition 2.2.4. Now, the question is, whether the concept of balance expressed in Definition 2.4.3 can be generalized further, to cover disconnected block designs as well.

It follows from Definition 2.2.6a that the C-matrix of a disconnected design of degree $g-1$ can be written as

$$\boldsymbol{C} = \text{diag}[\boldsymbol{C}_1 : \boldsymbol{C}_2 : \cdots : \boldsymbol{C}_g], \tag{2.4.5}$$

with

$$\boldsymbol{C}_\ell = \boldsymbol{r}_\ell^\delta - \boldsymbol{N}_\ell \boldsymbol{k}_\ell^{-\delta} \boldsymbol{N}_\ell' \quad (\ell = 1, 2, ..., g), \tag{2.4.6}$$

as the C-matrix of the ℓth connected subdesign with the incidence matrix $\boldsymbol{N}_\ell$. This presentation leads to the following extension of Definition 2.4.3.

Definition 2.4.4. A block design with disconnectedness of degree $g-1$ (connected, when $g = 1$) is said to be balanced (totally, when $g = 1$) if the off-diagonal elements of its matrices $\boldsymbol{N}_\ell \boldsymbol{k}_\ell^{-\delta} \boldsymbol{N}_\ell'$, $\ell = 1, 2, ..., g$, are all equal.

The concept of balance expressed in Definition 2.4.4 coincides with that originally used by Vartak (1963), who also proved the following theorem.

Theorem 2.4.2. *A block design is balanced (in the sense of Definition 2.4.4) if and only if the nonzero eigenvalues of its C-matrix are all equal. There are $v-g$ of them, if the degree of disconnectedness is $g-1$.*

Proof. In a manner similar to that for a connected block design, the matrix (2.4.6) can be written as $\boldsymbol{C}_\ell = \boldsymbol{q}_\ell^\delta - \boldsymbol{W}_\ell$, where $\boldsymbol{W}_\ell$ has all its diagonal elements equal to zero, the off-diagonal elements equal to those of $\boldsymbol{N}_\ell \boldsymbol{k}_\ell^{-\delta} \boldsymbol{N}_\ell'$, and $\boldsymbol{q}_\ell = \boldsymbol{W}_\ell \boldsymbol{1}_{v_\ell}$. Thus, the condition of Definition 2.4.4 holds if and only if

$$\boldsymbol{C}_\ell = \theta(\boldsymbol{I}_{v_\ell} - v_\ell^{-1} \boldsymbol{1}_{v_\ell} \boldsymbol{1}_{v_\ell}'),$$

for all ℓ, and this holds if and only if each $\boldsymbol{C}_\ell$ has the same unique nonzero eigenvalue, θ, of multiplicity $v_\ell - 1$ or, equivalently, if and only if the C-matrix (2.4.5) has a unique nonzero eigenvalue, θ, of multiplicity $v-g$, equal to rank$(\boldsymbol{C})$ by Lemma 2.3.2. □

2.4.2 Balance based on proportionality of weighted concurrences

The concept of balance that has led to Definition 2.4.4 originates from the concept of pairwise balance (Definition 2.4.1) based on the equality of all concurrences, $\{\lambda_{ii'}\}$. An alternative concept of balance originates from the idea of the concurrences being not equal but proportional to the products of the relevant treatment replications, i.e., it is based on the condition $\lambda_{ii'} = \zeta r_i r_{i'}$, where ζ is some positive scalar, constant for all i and i'. For a proper binary design, this condition can be written (with $\boldsymbol{r}^{2\delta} = \boldsymbol{r}^\delta \boldsymbol{r}^\delta$) as

$$\boldsymbol{N}\boldsymbol{N}' = \boldsymbol{r}^\delta + \zeta(\boldsymbol{r}\boldsymbol{r}' - \boldsymbol{r}^{2\delta}),$$

which, however, implies that the design is equireplicate. Thus, for proper binary block designs, the two concepts of balance, that based on equal concurrences and that based on proportional concurrences, both lead to BIB designs. But in general, these two concepts differ.

In a manner similar to that previously used in connection with the concept of balance based on the equality of concurrences, the alternative concept of balance, based on the proportionality of concurrences, can suitably be generalized by considering the weighted concurrences as those which are to be proportional to the products of the relevant treatment replications. For general connected block designs, this concept originates from Jones (1959) and can be expressed as follows.

Definition 2.4.5. A connected block design is totally balanced in the sense of Jones if the off-diagonal elements of its matrix $\boldsymbol{N}\boldsymbol{k}^{-\delta}\boldsymbol{N}'$ are proportional to the products of the relevant treatment replications, i.e., $\sum_{j=1}^{b} n_{ij}n_{i'j}/k_j = zr_ir_{i'}$, where z is some positive scalar, constant for any pair i, i' $(= 1, 2, ..., v)$.

This concept of balance was implicitly used by Nair and Rao (1948) and called "total balance" by Caliński (1971). Graf-Jaccottet (1977) introduced the term "balanced in the Jones sense" or "J-balanced."

The condition of Definition 2.4.5 can also be written as

$$\boldsymbol{N}\boldsymbol{k}^{-\delta}\boldsymbol{N}' = (1 - \varepsilon)\boldsymbol{r}^{\delta} + (\varepsilon/n)\boldsymbol{r}\boldsymbol{r}', \quad \text{where } \varepsilon = zn. \tag{2.4.7}$$

Comparison of this equation with the condition (2.4.3) of Definition 2.4.3 shows that the two definitions coincide if and only if the design is equireplicate, as originally noticed by Williams (1975).

It follows from (2.4.7) that the C-matrix of a connected block design totally balanced in the sense of Jones (Definition 2.4.5) is of the form

$$\boldsymbol{C} = \varepsilon(\boldsymbol{r}^{\delta} - n^{-1}\boldsymbol{r}\boldsymbol{r}'), \tag{2.4.8}$$

where $\varepsilon = [n - \text{tr}(\boldsymbol{N}\boldsymbol{k}^{-\delta}\boldsymbol{N}')]/(n - n^{-1}\boldsymbol{r}'\boldsymbol{r}) \leq 1$. This condition is in fact equivalent to that of Definition 2.4.5 and may be traced back to Jones (1959) (see also Caliński, 1971, p. 281). Now, the following result is easily obtainable (see Puri and Nigam, 1975a, 1975b; Caliński, 1977, Corollary 7).

Theorem 2.4.3. *A connected block design is totally balanced in the sense of Jones (Definition 2.4.5) if and only if the $v - 1$ nonzero eigenvalues of its C-matrix with respect to $\boldsymbol{r}^{\delta}$ are all equal.*

Proof. Because (2.4.7) holds if and only if (2.4.8) holds, the latter must be shown to hold if and only if ε is the unique nonzero eigenvalue of the C-matrix with respect to $\boldsymbol{r}^{\delta}$, with multiplicity $v - 1$. The necessity is obvious, whereas the sufficiency follows from the representation (2.3.2). □

To generalize this concept of balance to disconnected block designs, the following definition can be adopted.

Definition 2.4.6. A block design with disconnectedness of degree $g - 1$ (connected, when $g = 1$) is balanced in the sense of Jones (totally, when $g = 1$)

if the off-diagonal elements of its matrices $\boldsymbol{N}_\ell \boldsymbol{k}_\ell^{-\delta} \boldsymbol{N}_\ell'$, $\ell = 1, 2, ..., g$, are proportional to the products of the relevant treatment replications, with the same proportionality factor.

Now, Theorem 2.4.3 can be extended as follows.

Theorem 2.4.4. *A block design is balanced in the sense of Jones (Definition 2.4.6) if and only if the nonzero eigenvalues of its C-matrix with repect to $\boldsymbol{r}^\delta$ are all equal. There are $v - g$ of them, if the degree of disconnectedness is $g - 1$.*

Proof. This result can be proved similarly to Theorem 2.4.2, by applying Theorem 2.4.3 to each connected part (subdesign) of the design and remembering that $\text{rank}(\boldsymbol{C}) = v - g$. □

Remark 2.4.2. It follows from Theorem 2.4.4 and Corollary 2.3.3 that an orthogonal block design is balanced in the sense of Jones (Definition 2.4.6), independently of its disconnectedness. The unique nonzero eigenvalue of its C-matrix with respect to $\boldsymbol{r}^\delta$ is 1, with multiplicity $v - g$ if the disconnectedness is of degree $g - 1$. Furthermore, it follows that an equireplicate orthogonal block design is simultaneously balanced in the sense of Definition 2.4.4, with the unique nonzero eigenvalue of $\boldsymbol{C}$ equal to r, of multiplicity, again, $v - g$.

The various definitions considered here have been formulated by reference to some constructional features of the design, without mentioning any statistical property. Certainly, these properties are affected by the design features. The effects depend, however, on the assumed model, and the statistical meaning of any concept of balance will, therefore, be discussed later (in Chapter 4) when the model is specified (in Chapter 3).

Acknowledgement. Much of this chapter has been reprinted, with some changes and additions, from Sections 2 and 3 of the article by Caliński (1993a), under kind permission from Elsevier Science – NL, Sara Burgerhartstraat 25, 1055 KV Amsterdam, The Netherlands.

3
General Block Designs and Their Statistical Properties

3.1 A randomization model for a general block design

According to one of the basic principles of experimental design, the randomization principle (see Section 1.1), the experimental units (plots) are to be randomized before they enter the experiment. Suppose that to apply a general block design, in the sense of Section 2.2, randomization is performed as described by Nelder (1954), i.e., by randomly permuting blocks within a total area of them and by randomly permuting units within the blocks. Then, assuming the usual unit-treatment additivity (in the sense of Nelder, 1965b, p. 168; see also White, 1975, p. 560; Bailey, 1981, p. 215, 1991, p. 30; Kala, 1991, p. 7; Hinkelmann and Kempthorne, 1994, p. 251), and, as usual, that the technical errors are uncorrelated, each with zero expectation and a finite variance, and that they are independent of the unit responses to treatments (see Neyman, 1935, pp. 110-114 and 145; Kempthorne, 1952, p. 132 and Section 8.4; Ogawa, 1961, 1963; Hinkelmann and Kempthorne, 1994, Section 9.2.6), the model of the variables observed on the n units actually used in the experiment can be written in matrix notation as in (1.3.19), i.e., as

$$\boldsymbol{y} = \boldsymbol{\Delta}'\boldsymbol{\tau} + \boldsymbol{D}'\boldsymbol{\beta} + \boldsymbol{\eta} + \boldsymbol{e}, \tag{3.1.1}$$

where $\boldsymbol{y}$ is an $n \times 1$ vector of observed variables, $\boldsymbol{\tau}$ is a $v \times 1$ vector of treatment parameters, $\boldsymbol{\beta}$ is a $b \times 1$ vector of block random effects, $\boldsymbol{\eta}$ is an $n \times 1$ vector of unit errors and $\boldsymbol{e}$ is an $n \times 1$ vector of technical errors, the matrices $\boldsymbol{\Delta}'$ and $\boldsymbol{D}'$ being defined as in Section 2.2. Properties of the model (3.1.1) can be established by following its derivation from the randomizations involved, as it has been performed for randomized blocks, the classic RBD, in Section 1.3.

3.1.1 Derivation of the model

Suppose that N_B blocks are available, originally labeled $\xi = 1, 2, ..., N_B$, and that block ξ contains K_ξ units (plots), originally labeled $\pi = 1, 2, ..., K_\xi$. The label may also be written as $\pi(\xi)$, if it is desirable to refer it to block ξ. The randomization of blocks can then be understood as choosing at random a permutation of numbers $\{1, 2, ..., N_B\}$ and then renumbering the blocks, with $j = 1, 2, ..., N_B$, according to the positions of their original labels taken in the permutation randomly chosen. Similarly, the randomization of units within block ξ can be seen as selecting at random a permutation of numbers $\{1, 2, ..., K_\xi\}$ and renumbering the units of the block, with $\ell = 1, 2, ..., K_\xi$, according to the positions of their original labels taken in the permutation so selected. It will be assumed here that any permutation of block labels can be selected with equal probability, as well as that any permutation of unit labels within a block can be selected with equal probability. Furthermore, it will be assumed that the randomizations of units within the blocks are among the blocks independent, and that they are also independent of the randomization of blocks.

The above randomization procedure reflects the practical instruction given by Nelder (1954), recalled in Section 1.4 of the present book. It can be accomplished whether the available blocks are of equal or unequal number of their units. The purpose of this randomization is not only to "homogenize" the within-block variability among the experimental units, but also to "average out" these possibly heterogeneous variances from different blocks to a common value (as noticed, e.g., by White, 1975, Sections 2 and 7).

To describe the essential results induced by these randomizations, it is helpful to introduce the following random (indicator) variables:

$$f_{j\xi} = \begin{cases} 1 & \text{if the block originally labeled } \xi \text{ receives} \\ & \text{label } j \text{ after the randomization of blocks,} \\ 0 & \text{otherwise,} \end{cases} \tag{3.1.2}$$

for $j, \xi = 1, 2, ..., N_B$, and

$$g_{\ell\pi(\xi)} = \begin{cases} 1 & \text{if the unit in block } \xi \text{ originally labeled } \pi(\xi) \\ & \text{receives label } \ell \text{ after the randomization} \\ & \text{of units in that block,} \\ 0 & \text{otherwise,} \end{cases} \tag{3.1.3}$$

for $\ell, \pi(\xi) = 1, 2, ..., K_\xi$. It may be noticed that the N_B^2 variables (3.1.2) associated with the randomization of blocks are constrained by the conditions

$$\sum_{\xi=1}^{N_B} f_{j\xi} = 1 \text{ for any } j, \quad \text{and} \quad \sum_{j=1}^{N_B} f_{j\xi} = 1 \text{ for any } \xi,$$

and that, under the equal probability assumption, their first and second moments are

$$\mathrm{E}(f_{j\xi}) = N_B^{-1} \quad \text{for any } j \text{ and } \xi,$$

and

$$\mathrm{E}(f_{j\xi} f_{j'\xi'}) = \begin{cases} N_B^{-1}\delta_{\xi\xi'} & \text{if } j = j', \\ (N_B - 1)^{-1} N_B^{-1}(1 - \delta_{\xi\xi'}) & \text{if } j \neq j' \end{cases}$$

(exactly as in Section 1.3). As to the $\sum_{\xi=1}^{N_B} K_\xi^2$ random variables defined in (3.1.3), it follows that

$$\sum_{\pi(\xi)=1}^{K_\xi} g_{\ell\pi(\xi)} = 1 \text{ for any } \ell, \quad \text{and} \quad \sum_{\ell=1}^{K_\xi} g_{\ell\pi(\xi)} = 1 \text{ for any } \pi(\xi),$$

and that

$$\mathrm{E}(g_{\ell\pi(\xi)}) = K_\xi^{-1} \quad \text{for any } \ell \text{ and } \pi(\xi),$$

$$\mathrm{E}(g_{\ell\pi(\xi)} g_{\ell'\pi'(\xi)}) = \begin{cases} K_\xi^{-1}\delta_{\pi(\xi)\pi'(\xi)} & \text{if } \ell = \ell', \\ (K_\xi - 1)^{-1} K_\xi^{-1}(1 - \delta_{\pi(\xi)\pi'(\xi)}) & \text{if } \ell \neq \ell', \end{cases}$$

and

$$\mathrm{E}(g_{\ell\pi(\xi)} g_{\ell'\pi'(\xi')}) = (K_\xi K_{\xi'})^{-1} \quad \text{if } \xi \neq \xi',$$

the last property following from the independence of the randomizations in different blocks. Also, because the randomization within any block is independent of the randomization of blocks, it follows that

$$\mathrm{E}(f_{j\xi} g_{\ell\pi(\xi')}) = K_{\xi'}^{-1} N_B^{-1},$$

whether $\xi = \xi'$ or $\xi \neq \xi'$.

The random variables (3.1.2) and (3.1.3) will now be used to derive a suitable model for analyzing data from an experiment conducted under the described randomizations (as originally shown in Caliński and Kageyama, 1988, 1991). Following the strategy introduced by Nelder (1965a), and already used in Sections 1.2 and 1.3, it will be useful first to consider a model appropriate for analyzing experimental data under the assumption that all $\sum_{\xi=1}^{N_B} K_\xi$ units receive the same treatment. The concept of such a null experiment will be adopted in the preliminary stage, and then it will be extended to produce the final model of the variables observed in a real experiment. This extension will need, however, the assumption that the treatments under consideration are similar (or additive) in the sense that the variation of the responses among the available experimental units does not depend on the treatment received (see Section 1.2, where the concept of a null experiment is adopted for the first time in the present text).

For this null experiment, let the response of the unit labeled $\pi(\xi)$ be denoted by $\mu_{\pi(\xi)}$, and let it be denoted by $m_{\ell(j)}$ if by the randomizations the block originally labeled ξ receives the label j and the unit originally labeled π in this block receives the label ℓ. The relation between $\mu_{\pi(\xi)}$ and $m_{\ell(j)}$ can then be expressed by the formula

$$m_{\ell(j)} = \sum_{\xi=1}^{N_B} \sum_{\pi(\xi)=1}^{K_\xi} f_{j\xi} g_{\ell\pi(\xi)} \mu_{\pi(\xi)}.$$

But, introducing the identity

$$\mu_{\pi(\xi)} = \mu_{\cdot(\cdot)} + (\mu_{\cdot(\xi)} - \mu_{\cdot(\cdot)}) + (\mu_{\pi(\xi)} - \mu_{\cdot(\xi)}),$$

where (according to the usual dot notation for arithmetic averages)

$$\mu_{\cdot(\xi)} = K_\xi^{-1} \sum_{\pi(\xi)=1}^{K_\xi} \mu_{\pi(\xi)} \quad \text{and} \quad \mu_{\cdot(\cdot)} = N_B^{-1} \sum_{\xi=1}^{N_B} \mu_{\cdot(\xi)},$$

one can write the relation as

$$m_{\ell(j)} = \mu + \beta_j + \eta_{\ell(j)} \tag{3.1.4}$$

for any ℓ and j, where $\mu = \mu_{\cdot(\cdot)}$ is a constant parameter, whereas

$$\beta_j = \sum_{\xi=1}^{K_\xi} f_{j\xi}(\mu_{\cdot(\xi)} - \mu_{\cdot(\cdot)}) \tag{3.1.5}$$

and

$$\eta_{\ell(j)} = \sum_{\xi=1}^{N_B} \sum_{\pi(\xi)=1}^{K_\xi} f_{j\xi} g_{\ell\pi(\xi)}(\mu_{\pi(\xi)} - \mu_{\cdot(\xi)}) \tag{3.1.6}$$

are random variables, (3.1.5) representing the block random effect and (3.1.6) the unit error.

From the properties of the variables (3.1.2) and (3.1.3), derived above, the following moments of (3.1.5) and (3.1.6) are obtainable:

$$\mathrm{E}(\beta_j) = 0, \qquad \mathrm{E}(\eta_{\ell(j)}) = 0,$$

$$\mathrm{Cov}(\beta_j, \beta_{j'}) = \begin{cases} N_B^{-1} \sum_{\xi=1}^{N_B} (\mu_{\cdot(\xi)} - \mu_{\cdot(\cdot)})^2 & \text{if } j = j', \\ -N_B^{-1}(N_B - 1)^{-1} \sum_{\xi=1}^{N_B} (\mu_{\cdot(\xi)} - \mu_{\cdot(\cdot)})^2 & \text{if } j \neq j', \end{cases}$$

$$\mathrm{Cov}(\eta_{\ell(j)}, \eta_{\ell'(j')})$$

$$= \begin{cases} N_B^{-1} \sum_{\xi=1}^{N_B} K_\xi^{-1} \sum_{\pi(\xi)=1}^{K_\xi} (\mu_{\pi(\xi)} - \mu_{\cdot(\xi)})^2 & \text{if } j = j' \text{ and } \ell = \ell', \\ -N_B^{-1} \sum_{\xi=1}^{N_B} K_\xi^{-1}(K_\xi - 1)^{-1} \sum_{\pi(\xi)=1}^{K_\xi} (\mu_{\pi(\xi)} - \mu_{\cdot(\xi)})^2 & \text{if } j = j' \text{ and } \ell \neq \ell', \\ 0 & \text{if } j \neq j', \end{cases}$$

and

$$\mathrm{Cov}(\beta_j, \eta_{\ell(j')}) = 0, \quad \text{whether} \quad j = j' \ \text{or} \ j \neq j'.$$

Now, introducing the variance components

$$\sigma_B^2 = (N_B - 1)^{-1} \sum_{\xi=1}^{N_B} (\mu_{\cdot(\xi)} - \mu_{\cdot(\cdot)})^2 \quad \text{and} \quad \sigma_U^2 = N_B^{-1} \sum_{\xi=1}^{N_B} \sigma_{U,\xi}^2 ,$$

where

$$\sigma_{U,\xi}^2 = (K_\xi - 1)^{-1} \sum_{\pi(\xi)=1}^{K_\xi} (\mu_{\pi(\xi)} - \mu_{\cdot(\xi)})^2,$$

and the weighted harmonic average K_H defined by the equality

$$K_H^{-1} = N_B^{-1} \sum_{\xi=1}^{N_B} K_\xi^{-1} \sigma_{U,\xi}^2 / \sigma_U^2,$$

the variances and covariances of the random variables $\{\beta_j\}$, defined in (3.1.5), and $\{\eta_{\ell(j)}\}$, defined in (3.1.6), can also be written in the more convenient forms

$$\text{Cov}(\beta_j, \beta_{j'}) = \begin{cases} N_B^{-1}(N_B - 1)\sigma_B^2 & \text{if } j = j', \\ -N_B^{-1}\sigma_B^2 & \text{if } j \neq j', \end{cases}$$

and

$$\text{Cov}(\eta_{\ell(j)}, \eta_{\ell'(j')}) = \begin{cases} K_H^{-1}(K_H - 1)\sigma_U^2 & \text{if } j = j' \text{ and } \ell = \ell', \\ -K_H^{-1}\sigma_U^2 & \text{if } j = j' \text{ and } \ell \neq \ell', \\ 0 & \text{if } j \neq j'. \end{cases}$$

Thus, it has been shown that the responses $\{m_{\ell(j)}\}$ have the model (3.1.4) with

$$\text{E}(m_{\ell(j)}) = \mu, \tag{3.1.7}$$

$$\text{Cov}(m_{\ell(j)}, m_{\ell'(j')}) = (\delta_{jj'} - N_B^{-1})\sigma_B^2 + \delta_{jj'}(\delta_{\ell\ell'} - K_H^{-1})\sigma_U^2. \tag{3.1.8}$$

The results (3.1.7) and (3.1.8) are comparable with the first and second moments given by Nelder (1954) (as shown in Section 1.3).

To continue the derivation, it should be noticed (as in Sections 1.2 and 1.3) that when observing the responses of the units in reality, any observation may be affected by a technical error. Let $e_{\pi(\xi)}$ denote the technical error by which the variable observed on the unit originally labeled $\pi(\xi)$ differs from its true response, $\mu_{\pi(\xi)}$. Denoting then this difference by $e_{\ell(j)}$ if because of the randomization the unit receives the label $\ell(j)$, the model of the variable observed on that unit in the null experiment can be written, as in (1.3.10), in the form

$$y_{\ell(j)} = m_{\ell(j)} + e_{\ell(j)} = \mu + \beta_j + \eta_{\ell(j)} + e_{\ell(j)} \tag{3.1.9}$$

for any j and ℓ, where

$$e_{\ell(j)} = \sum_{\xi=1}^{N_B} \sum_{\pi(\xi)=1}^{K_\xi} f_{j\xi} g_{\ell\pi(\xi)} e_{\pi(\xi)}.$$

As usual, it will be assumed that the technical errors $\{e_{\pi(\xi)}\}$ are uncorrelated random variables, each with expectation zero and a finite variance, and that they are independent of the random variables $\{f_{j\xi}\}$ and $\{g_{\ell\pi(\xi)}\}$. Then, as can easily be shown,

$$\mathrm{E}(e_{\ell(j)}) = 0, \quad \mathrm{Var}(e_{\ell(j)}) = N_B^{-1} \sum_{\xi=1}^{N_B} K_\xi^{-1} \sum_{\pi(\xi)=1}^{K_\xi} \mathrm{Var}(e_{\pi(\xi)}) = \sigma_e^2$$

and

$$\mathrm{Cov}(e_{\ell(j)}, e_{\ell'(j')}) = 0$$

for all $\ell(j) \neq \ell'(j')$, as well as

$$\mathrm{Cov}(e_{\ell(j)}, \beta_{j'}) = 0 \quad \text{and} \quad \mathrm{Cov}(e_{\ell(j)}, \eta_{\ell'(j')}) = 0$$

for all $\ell(j)$ and j', and for all $\ell(j)$ and $\ell'(j')$, respectively.

On account of these properties and those established for (3.1.4), the first and second moments of the random variables $\{y_{\ell(j)}\}$ defined in (3.1.9) have the forms

$$\mathrm{E}(y_{\ell(j)}) = \mu, \tag{3.1.10}$$

$$\mathrm{Cov}(y_{\ell(j)}, y_{\ell'(j')}) = (\delta_{jj'} - N_B^{-1})\sigma_B^2 + \delta_{jj'}(\delta_{\ell\ell'} - K_H^{-1})\sigma_U^2 + \delta_{jj'}\delta_{\ell\ell'}\sigma_e^2 \tag{3.1.11}$$

for all $\ell(j)$ and $\ell'(j')$. The formulae (3.1.10) and (3.1.11) are comparable with (1.3.11) and (1.3.12), respectively, the only difference being that now K_H in (3.1.11) stands in place of v appearing in (1.3.12).

Now, as in Section 1.3, two questions are to be considered. First, how the randomized units within the randomized blocks are to be used to assign treatments to them and, second, how the model (3.1.9) is to be adjusted accordingly.

As to the first question, it should be noticed that the moments expressed in (3.1.7) and (3.1.8) for the null experiment do not depend on the labels received by the blocks and their units in the course of the randomizations. This independence means that the N_B randomized blocks can be regarded as "homogeneous" and that the set of units randomized within a block can be regarded as such, in the sense that the observed responses of the units may, under the same treatment, be considered as observations on random variables $\{y_{\ell(j)}\}$ exchangeable within a block and also jointly in sets among the blocks, provided that any such set is of a size not exceeding the smallest K_ξ. In fact, to make this concept more feasible, one would prefer to have all K_ξ equal (as demanded by Bailey, 1981, Section 2.2), but this is not necessary for the derived model to be at least notionally acceptable.

Because of the homogeneity of blocks and the homogeneity of units within blocks, in the sense given above, the randomization principle can be applied to a block experiment designed according to a chosen incidence matrix $\boldsymbol{N}$ (see Section 2.2) by adopting the following rule. The b columns of $\boldsymbol{N}$ are assigned to b out of the N_B available blocks of experimental units by assigning the jth column of $\boldsymbol{N}$ to the block labeled j after the randomization. Then, the treatments indicated by the nonzero elements of the jth column of $\boldsymbol{N}$ are assigned to the experimental units of the block labeled j, in numbers defined by the corresponding elements of $\boldsymbol{N}$ (i.e., the ith treatment to n_{ij} units) and in order determined by the labels the units of the block have received after the randomization. This rule implies not only that $b \leq N_B$, but also that the units in the available blocks are in sufficient numbers with regard to the vector $\boldsymbol{k} = \boldsymbol{N}'\mathbf{1}_v$, which means that either the choice of $\boldsymbol{N}$ is to be conditioned by the constraint that none of its k_j's exceeds the smallest K_ξ or an adjustment of $\boldsymbol{N}$ is to be made after the randomization of blocks (as suggested by White, 1975, Section 4).

These conditions impose some constraints on designing the experiment. The difficulty they may cause depends on how much the researcher is restricted in choosing appropriate experimental material. If a given material to be used in the experiment is already structured somehow into blocks, then the researcher may be constrained to choose a design with blocks sufficiently small to fit in with any among those given. On the other hand, if there is much freedom in preselecting and forming the experimental material, then the researcher may prepare it in accordance with the design deliberately chosen for the experiment. This kind of freedom depends essentially on the nature of the experimental material.

If the blocks are naturally formed in advance, then before performing the randomization, one would need to discard blocks having not enough units to match the chosen design or, on the contrary, the design "would have to be modified by deletion of a number of the design units" (i.e., elements of the incidence matrix), as suggested by White (1975, p. 558). In some situations, none of these adjustments might be practical. In such a case, the model considered here would not be relevant.

A completely different situation occurs when the blocks can freely (or almost freely) be formed by the researcher. For example, in a laboratory experiment it may not be feasible to initiate the experiment with all of its experimental units on one day, but it might be reasonable to extend it over several days. Then, the researcher can form distinct blocks of experimental units for different days according to the chosen design. Also, in an agricultural field experiment, after choosing a piece of land appropriate for the purpose of the experiment, the researcher is to some extent free in dividing the land into suitable blocks of plots (see Pearce, 1983, Section 2.5). Note that in both of these examples, the randomization of blocks, their labels, is evidently not depending on whether the blocks are to be formed of equal or unequal number of units. The randomly chosen permutation of block labels decides only the order in which the design blocks enter the experiment, either in time (the first example) or in space (the second example). In any of such cases, the blocks physically used in the experiment can easily be adjusted to the design requirements.

Although in most cases equal block sizes are favorable, instances occur when blocks of unequal sizes are unavoidable or desirable, as noticed by Pearce (1964). Also, White (1975, p. 569) pointed out that "available experimental units often come in unequal-sized blocks as, e.g., in clinical trials." According to Kageyama (1976b), block designs with unequal block sizes may be particularly useful in large experiments both in industry and agriculture. In fact, designs with two block sizes, k_1 and k_2, where $k_2 = k_1 + 1$, are often used in cereal variety trials, as indicated by Patterson and Hunter (1983). From the preceding discussion, it should become clear under which circumstances the present randomization model is relevant to experiments with unequal block sizes.

Example 3.1.1. Consider an experiment in which three treatments, one of them regarded as standard, are to be compared by application to certain animals. Suppose that the animals (experimental units) are available in sufficient number, but they are by nature not uniform with regard to potential responses to the treatments. It can, however, be assumed that some of the animals are more alike than others, so that there is a possibility of joining them into groups (blocks) on the basis of the existing similarities. Suppose then that the researcher has succeeded in forming five such blocks, each comprising at least four units. Performing the randomization of blocks and units within blocks, in the way described earlier, the blocks and units have been relabeled as shown in Figure 3.1.

Before randomization

Block 1					Block 2					Block 3				Block 4				Block 5			
1	2	3	4	5	1	2	3	4	5	1	2	3	4	1	2	3	4	1	2	3	4

After randomization

Block *2*					Block *4*					Block *1*				Block *5*				Block *3*			
3	*1*	*4*	*2*	*5*	*3*	*5*	*2*	*1*	*4*	*3*	*2*	*4*	*1*	*1*	*4*	*2*	*3*	*4*	*1*	*3*	*2*

Fig. 3.1. Randomization for an experiment in which at most five blocks with no more than four units are to be used, to meet the conditions $b \leq N_B = 5$ and $\max_j k_j \leq \min_\xi K_\xi = 4$.

The experimental material so prepared is now well suited to an experiment designed according to the incidence matrix

$$\boldsymbol{N} = \begin{bmatrix} 2 & 2 & 2 & 2 & 2 \\ 1 & 1 & 1 & 1 & 1 \\ 1 & 1 & 1 & 1 & 1 \end{bmatrix}.$$

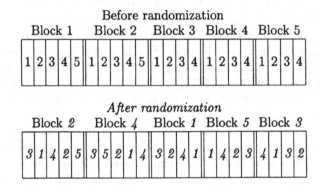

Let the treatments, to which the rows of the matrix correspond, be denoted by A, B and C, the first treatment, A, being regarded as the standard. Applying the assignment rule described above, treatment A is to be assigned to all units that by the randomization have received labels 1 and 2. Treatment B is to be assigned to all units that have received label 3, whereas treatment C should be assigned to all units that have received label 4. Note that in this layout, 2 out of 22 available units are not taken to the experiment. This shoud not cause much concern in view of the advantages of the above orthogonal design. Sometimes more units are to be discarded, for that or other reasons. For example, one can think of a situation when because of some constraint, e.g., economical, the experimenter decides to use in the experiment only 16 of the above units. Then, the block by the randomization labeled 5 would not be used, though still rightly considered as part of the population to which inferences are to be drawn from the experiment.

Example 3.1.2. Remaining in the same experimental context of Example 3.1.1, suppose now that the available experimental material is smaller, and that the researcher has been able to form only four blocks, two of them composed of three and the other two of only two units, sufficiently uniform for the purpose of the experiment. The results of the randomization of the labels of blocks and those of units are presented in Figure 3.2.

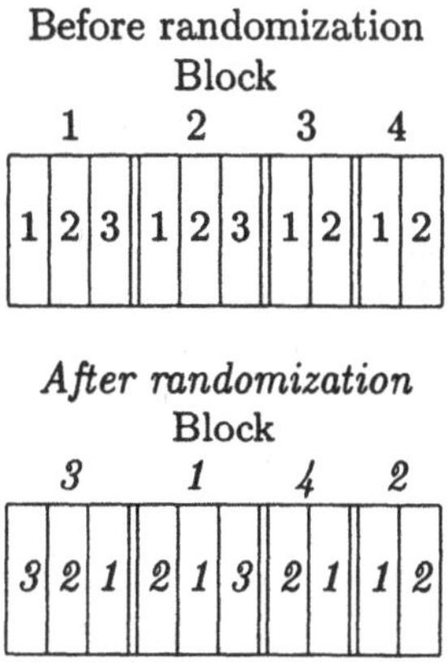
Before randomization
Block
1 2 3 4
1 2 3 1 2 3 1 2 1 2
After randomization
Block
3 1 4 2
3 2 1 2 1 3 2 1 1 2

Fig. 3.2. Randomization for an experiment in which at most four blocks with two units are to be used, unless an adjustment of the design is made to use all available units.

Because the smallest number of units in a block is two, the researcher may choose a design described by the incidence matrix

$$\boldsymbol{N} = \begin{bmatrix} 1 & 1 & 1 & 1 \\ 1 & 0 & 1 & 0 \\ 0 & 1 & 0 & 1 \end{bmatrix},$$

desirable for comparing treatments B and C with the standard treatment A, to which the first row of $\boldsymbol{N}$ corresponds. According to the randomized labels (Fig.

3.2), treatment A is to be assigned to all units with label 1, in all four blocks, treatment B is to be assigned to units with label 2 within blocks with labels 1 and 3, whereas treatment C is to be assigned to units with label 2 in blocks with labels 2 and 4. Evidently, when applying this design, two out of ten available units are not included in the experiment, which may be regarded as a waste of the already scarce experimental material. Therefore, to use all available units, the researcher may wish to choose a design given temporarily by the incidence matrix

$$\boldsymbol{N}_{\text{temp}} = \begin{bmatrix} 1 & 1 & 1 & 1 \\ 1 & 1 & 1 & 1 \\ 1 & 1 & 1 & 1 \end{bmatrix},$$

with the provision of being modified by deletion of one of the three treatments, B, say, in those randomized blocks in which room exists for two treatments only. Because, after the randomization, those blocks are labeled 2 and 4, the adjusted incidence matrix becomes

$$\boldsymbol{N}^* = \begin{bmatrix} 1 & 1 & 1 & 1 \\ 1 & 0 & 1 & 0 \\ 1 & 1 & 1 & 1 \end{bmatrix}.$$

Applying the assignment rule described above to this incidence matrix, it becomes clear that, in addition to the assignments made before in connection with the matrix $\boldsymbol{N}$, treatment C is now to be assigned also to units that after the randomization, are labeled 3 in blocks with labels 1 and 3. The reader may note that this modified design is suitable if the main interest is in comparing treatment A with treatment C. Why? Anyway, the choice of the treatment partially sacrificed in such design is to be made in advance. In general, one has to remember that an adjustment of the design to the availability of experimental units is justified only if performed independently of the unit responses to treatments (see White, 1975, p. 558).

With regard to the second question, concerning adjustment of the model (3.1.9), it can be answered by adopting the assumption of additivity, as mentioned at the beginning of the section, i.e., by assuming, implicitly, that the variances and covariances of the random variables $\{\beta_j\}$, $\{\eta_{\ell(j)}\}$ and $\{e_{\ell(j)}\}$ do not depend on the treatment applied (exactly as in Section 1.3). Then, the adjustment of the model (3.1.9) to a real situation of comparing several treatments in the same experiment can be made by changing the constant term only. Thus, the final model gets the form

$$y_{\ell(j)}(i) = \mu(i) + \beta_j + \eta_{\ell(j)} + e_{\ell(j)} \tag{3.1.12}$$

$(i = 1, 2, ..., v;\ j = 1, 2, ..., b;\ \ell = 1, 2, ..., k_j)$, with

$$\mathrm{E}[y_{\ell(j)}(i)] = \mu(i) = N_B^{-1} \sum_{\xi=1}^{N_B} K_\xi^{-1} \sum_{\pi(\xi)=1}^{K_\xi} \mu_{\pi(\xi)}(i), \tag{3.1.13}$$

where $\mu_{\pi(\xi)}(i)$ is the true response of unit π in block ξ to the treatment i, and with

$$\mathrm{Cov}[y_{\ell(j)}(i), y_{\ell'(j')}(i')] = \mathrm{Cov}(y_{\ell(j)}, y_{\ell'(j')}), \tag{3.1.14}$$

as given in (3.1.11).

Now, writing the observed variables $\{y_{\ell(j)}(i)\}$ in the form of an $n \times 1$ vector $\boldsymbol{y}$, and the corresponding unit-error and technical-error variables in the form of $n \times 1$ vectors $\boldsymbol{\eta}$ and $\boldsymbol{e}$, respectively, and writing the treatment parameters as $\boldsymbol{\tau} = [\tau_1, \tau_2, ..., \tau_v]'$, where $\tau_i = \mu(i)$, and the block variables as $\boldsymbol{\beta} = [\beta_1, \beta_2, ..., \beta_b]'$, one can express the model (3.1.12) as in (3.1.1), and the corresponding moments (3.1.13) and (3.1.14) in the form of the expectation vector

$$\mathrm{E}(\boldsymbol{y}) = \boldsymbol{\Delta}'\boldsymbol{\tau} \tag{3.1.15}$$

and the covariance (dispersion) matrix

$$\mathrm{Cov}(\boldsymbol{y}) = (\boldsymbol{D}'\boldsymbol{D} - N_B^{-1}\mathbf{1}_n\mathbf{1}_n')\sigma_B^2 + (\boldsymbol{I}_n - K_H^{-1}\boldsymbol{D}'\boldsymbol{D})\sigma_U^2 + \boldsymbol{I}_n\sigma_e^2, \tag{3.1.16}$$

respectively, where the matrices $\boldsymbol{\Delta}'$ and $\boldsymbol{D}'$ are as defined in Section 2.2.

The model (3.1.1), with properties (3.1.15) and (3.1.16), coincides with that of Patterson and Thompson (1971), when their matrix $\boldsymbol{\Gamma}$ in the formula $\mathrm{Cov}(\boldsymbol{y}) = (\boldsymbol{D}'\boldsymbol{\Gamma}\boldsymbol{D} + \boldsymbol{I}_n)\sigma^2$ is taken equal to $\boldsymbol{I}_b\gamma - N_B^{-1}\mathbf{1}_b\mathbf{1}_b'\sigma_B^2/\sigma^2$, where $\gamma = (\sigma_B^2 - K_H^{-1}\sigma_U^2)/\sigma^2$ and $\sigma^2 = \sigma_U^2 + \sigma_e^2$. In fact, they have considered a simplified model with $\boldsymbol{\Gamma} = \boldsymbol{I}_n\gamma$. Furthermore, if $N_B = b$, $k_1 = k_2 = \cdots = k_b = k$ (say) and $k = K_H$ (the latter implying the equality of all K_ξ), then the present model coincides essentially with that considered by Rao (1959) and by Shah (1992). In its general form, this model coincides completely with the model obtained by Kala (1991) under more general considerations.

Finally, it should be mentioned that the present model does not cover the situation in which the available blocks are subject to further grouping into some superblocks, i.e., form two strata of blocks. Then, a different randomization model is appropriate. It will be considered in Chapter 5.

3.1.2 Main estimation results

Under the present model, the following main results concerning the linear estimation of treatment parametric functions are obtainable.

Theorem 3.1.1. *Under the model* (3.1.1), *with properties* (3.1.15) *and* (3.1.16), *a function* $\boldsymbol{w}'\boldsymbol{y}$ *is uniformly the BLUE of* $\boldsymbol{c}'\boldsymbol{\tau}$ *if and only if* $\boldsymbol{w} = \boldsymbol{\Delta}'\boldsymbol{s}$, *where* $\boldsymbol{s} = \boldsymbol{r}^{-\delta}\boldsymbol{c}$ *satisfies the condition*

$$(\boldsymbol{k}^\delta - \boldsymbol{N}'\boldsymbol{r}^{-\delta}\boldsymbol{N})\boldsymbol{N}'\boldsymbol{s} = \mathbf{0}. \tag{3.1.17}$$

Proof. On account of Theorem 1.2.1, a function $\boldsymbol{w}'\boldsymbol{y}$ is, under the considered model, the BLUE of its expectation, i.e., of $\boldsymbol{w}'\boldsymbol{\Delta}'\boldsymbol{\tau}$, if and only if

$$\begin{aligned}(\boldsymbol{I}_n - \boldsymbol{P}_{\Delta'})[(\boldsymbol{D}'\boldsymbol{D} - N_B^{-1}\mathbf{1}_n\mathbf{1}_n')\sigma_B^2 \\ + (\boldsymbol{I}_n - K_H^{-1}\boldsymbol{D}'\boldsymbol{D})\sigma_U^2 + \boldsymbol{I}_n\sigma_e^2]\boldsymbol{w} = \mathbf{0},\end{aligned} \tag{3.1.18}$$

where $\boldsymbol{P}_{\Delta'} = \boldsymbol{\Delta}'\boldsymbol{r}^{-\delta}\boldsymbol{\Delta}$ denotes the orthogonal projector on $\mathcal{C}(\boldsymbol{\Delta}')$, the column space of $\boldsymbol{\Delta}'$ (see Appendix A.2). If this condition is to hold uniformly for any set of the variance components σ_B^2, σ_U^2 and σ_e^2, it is necessary and sufficient that $(\boldsymbol{I}_n - \boldsymbol{P}_{\Delta'})\boldsymbol{w} = \boldsymbol{0}$ and $(\boldsymbol{I}_n - \boldsymbol{P}_{\Delta'})\boldsymbol{D}'\boldsymbol{D}\boldsymbol{w} = \boldsymbol{0}$. But these equations hold simultaneously if and only if $\boldsymbol{w} = \boldsymbol{\Delta}'\boldsymbol{s}$ and $(\boldsymbol{I}_n - \boldsymbol{P}_{\Delta'})\boldsymbol{D}'\boldsymbol{D}\boldsymbol{\Delta}'\boldsymbol{s} = \boldsymbol{0}$ for some $\boldsymbol{s}$, the latter equation being equivalent to (3.1.17). □

Note that in view of this proof, the component $N_B^{-1}\boldsymbol{1}_n\boldsymbol{1}_n'\sigma_B^2$ in (3.1.16) does not play any role in establishing Theorem 3.1.1, as $(\boldsymbol{I}_n - \boldsymbol{P}_{\Delta'})\boldsymbol{1}_n = \boldsymbol{0}$ on account of (2.1.1). For the same reason, the simplification of $\boldsymbol{\Gamma}$ to the form $\boldsymbol{I}_n\gamma$, made by Patterson and Thompson (1971), does not affect the BLUE of $\boldsymbol{c}'\boldsymbol{\tau}$. (For more general considerations, see Kala, 1981, Theorem 6.2.)

Corollary 3.1.1. *For the estimation of* $\boldsymbol{c}'\boldsymbol{\tau} = \boldsymbol{s}'\boldsymbol{r}^{\delta}\boldsymbol{\tau}$ *under the model as in Theorem* 3.1.1 *the following applies:*

(a) *If* $\boldsymbol{N}'\boldsymbol{s} = \boldsymbol{0}$, *then* (3.1.17) *is satisfied and the estimated function is a contrast.*

(b) *If* $\boldsymbol{N}'\boldsymbol{s} \neq \boldsymbol{0}$, *then to satisfy* (3.1.17), *it is necessary and sufficient that the elements of* $\boldsymbol{N}'\boldsymbol{s}$ *obtained from the same connected subdesign are all equal, i.e., that* $\boldsymbol{N}'\boldsymbol{s} \in \mathcal{C}\{\text{diag}[\boldsymbol{1}_{b_1} : \boldsymbol{1}_{b_2} : \cdots : \boldsymbol{1}_{b_g}]\}$.

Proof. The result (a) is obvious, and $\boldsymbol{c}'\boldsymbol{1}_v = 0$ because of $\boldsymbol{N}\boldsymbol{1}_b = \boldsymbol{r}$. To prove (b), note that (3.1.17) is equivalent to the condition $\boldsymbol{N}'\boldsymbol{r}^{-\delta}\boldsymbol{N}\boldsymbol{N}'\boldsymbol{s} = \boldsymbol{k}^{\delta}\boldsymbol{N}'\boldsymbol{s}$, which in case of a connected design is satisfied if and only if $\boldsymbol{N}'\boldsymbol{s} \in \mathcal{C}(\boldsymbol{1}_b)$ [as it follows from Corollary 2.3.2(a) and Remark 2.3.1]. A straightforward use of Corollary 2.3.2(c) leads to the general result, applicable to any $\boldsymbol{N} = \text{diag}[\boldsymbol{N}_1 : \boldsymbol{N}_2 : \cdots : \boldsymbol{N}_g]$ with $g \geq 1$ (the notation being as in Definition 2.2.6a). □

Now, the question is, under which design conditions any function $\boldsymbol{s}'\boldsymbol{\Delta}\boldsymbol{y}$ is the BLUE of its expectation. An answer to this question can be given as follows.

Theorem 3.1.2. *Under the model as in Theorem* 3.1.1, *any function* $\boldsymbol{w}'\boldsymbol{y} = \boldsymbol{s}'\boldsymbol{\Delta}\boldsymbol{y}$, *i.e., with any* $\boldsymbol{s}$, *is uniformly the BLUE of* $\text{E}(\boldsymbol{w}'\boldsymbol{y}) = \boldsymbol{s}'\boldsymbol{r}^{\delta}\boldsymbol{\tau}$ *if and only if*

(i) *the design is orthogonal and*

(ii) *the block sizes of the design are constant within any of its connected subdesigns.*

Proof. From the proof of Theorem 3.1.1, it is evident that $\boldsymbol{s}'\boldsymbol{\Delta}\boldsymbol{y}$ is the BLUE of its expectation for any $\boldsymbol{s}$ if and only if

$$(\boldsymbol{I}_n - \boldsymbol{P}_{\Delta'})\boldsymbol{D}'\boldsymbol{D}\boldsymbol{\Delta}' = \mathbf{O}. \tag{3.1.19}$$

The equality (3.1.19), however, implies that $\boldsymbol{k}^{\delta}\boldsymbol{N}'\boldsymbol{1}_v = \boldsymbol{N}'\boldsymbol{r}^{-\delta}\boldsymbol{N}\boldsymbol{N}'\boldsymbol{1}_v$, i.e., that $\boldsymbol{k}$ is to be an eigenvector of $\boldsymbol{N}'\boldsymbol{r}^{-\delta}\boldsymbol{N}$ with respect to $\boldsymbol{k}^{\delta}$, corresponding to the eigenvalue 1. But, on account of Corollary 3.1.1(b), for this, it is necessary and sufficient that (ii) holds. On the other hand, if (ii) is satisfied, then the matrix (2.2.10) can be used to show that the equality (3.1.19) is equivalent to $(\boldsymbol{I}_n - \boldsymbol{P}_{\Delta'})\boldsymbol{D}'\boldsymbol{k}^{-\delta}\boldsymbol{D}\boldsymbol{\Delta}' = \mathbf{O}$, which on account of (2.3.13) and (2.2.13) is equivalent to (i). Thus, (3.1.19) implies (ii) and (i), subsequently, and, *vice versa*, (i) with (ii) implies (3.1.19). □

Remark 3.1.1. If (ii) of Theorem 3.1.2 holds, then from (2.2.14) and (2.2.9), the orthogonality condition can be written as

$$\boldsymbol{N} = \tilde{\boldsymbol{k}}^{-\delta}\boldsymbol{N}\boldsymbol{N}'\boldsymbol{r}^{-\delta}\boldsymbol{N}$$

in general, and on account of Corollary 2.3.4 as $\boldsymbol{N} = b^{-1}\boldsymbol{r}\boldsymbol{1}_b'$ for a connected design, because for such a design the condition (ii) reads that all design block sizes are equal. Also note that the two conditions of Theorem 3.1.2 coincide with condition (5.12) of Theorem 3 given by Kala (1991).

Remark 3.1.2. If the conditions (i) and (ii) stated in Theorem 3.1.2 are satisfied, i.e., if (3.1.19) holds, then

$$\text{Cov}(\boldsymbol{y})\boldsymbol{\Delta}' = \boldsymbol{\Delta}'[(\boldsymbol{r}^{-\delta}\boldsymbol{N}\boldsymbol{N}' - N_B^{-1}\boldsymbol{1}_v\boldsymbol{r}')\sigma_B^2 + (\boldsymbol{I}_v - K_H^{-1}\boldsymbol{r}^{-\delta}\boldsymbol{N}\boldsymbol{N}')\sigma_U^2 + \boldsymbol{I}_v\sigma_e^2],$$

which implies that both $\boldsymbol{\Delta}'\boldsymbol{s}$ and $\text{Cov}(\boldsymbol{y})\boldsymbol{\Delta}'\boldsymbol{s}$ belong to $\mathcal{C}(\boldsymbol{\Delta}')$ for any $\boldsymbol{s}$, and thus, by Theorem 1.2.2, the BLUEs obtainable under the model (3.1.1), with the moments (3.1.15) and (3.1.16), can equivalently be obtained under a simple alternative model in which the covariance matrix (3.1.16) is reduced to $\boldsymbol{I}_n\sigma^2$, i.e., as SLSEs. Moreover, it can be shown (applying, e.g., Theorem 2.3.2 of Rao and Mitra, 1971) that the equality (3.1.19) is not only a sufficient, but also a necessary condition for the BLUEs obtainable under the two alternative models to be the same. In other words, (3.1.19) is a necessary and sufficient condition for $\boldsymbol{s}'\boldsymbol{\Delta}\boldsymbol{y}$ to be both the SLSE and the BLUE of its expectation, $\boldsymbol{s}'\boldsymbol{r}^{\delta}\boldsymbol{\tau}$, whichever vector $\boldsymbol{s}$ is used. This result coincides with Theorem 3 of Kala (1991).

Remark 3.1.2 answers the question posed at the end of Section 1.3. It implies, in particular, that if the equality (3.1.19) holds, then the BLUE of the expectation vector (3.1.15) is obtainable by an SLSE procedure (see, e.g., Seber, 1980, Chapter 3), in the form $\boldsymbol{P}_{\Delta'}\boldsymbol{y}$. This implication can be checked (on account of Theorem 1.2.1) by noting that $\text{E}(\boldsymbol{P}_{\Delta'}\boldsymbol{y}) = \boldsymbol{P}_{\Delta'}\boldsymbol{\Delta}'\boldsymbol{\tau} = \boldsymbol{\Delta}'\boldsymbol{\tau} = \text{E}(\boldsymbol{y})$ and that $(\boldsymbol{I}_n - \boldsymbol{P}_{\Delta'})\text{Cov}(\boldsymbol{y})\boldsymbol{P}_{\Delta'} = \boldsymbol{O}$. In fact, the latter holds because under (3.1.19), i.e., under $\boldsymbol{D}'\boldsymbol{D}\boldsymbol{\Delta}' = \boldsymbol{P}_{\Delta'}\boldsymbol{D}'\boldsymbol{D}\boldsymbol{\Delta}'$, the equality

$$\text{Cov}(\boldsymbol{y})\boldsymbol{P}_{\Delta'} = \boldsymbol{P}_{\Delta'}[(\boldsymbol{D}'\boldsymbol{D} - N_B^{-1}\boldsymbol{1}_n\boldsymbol{1}_n')\sigma_B^2 + (\boldsymbol{I}_n - K_H^{-1}\boldsymbol{D}'\boldsymbol{D})\sigma_U^2 + \boldsymbol{I}_n\sigma_e^2]\boldsymbol{P}_{\Delta'}$$

holds. Thus, it follows that the vector $\boldsymbol{y}$ can be decomposed as

$$\boldsymbol{y} = \boldsymbol{P}_{\Delta'}\boldsymbol{y} + (\boldsymbol{I}_n - \boldsymbol{P}_{\Delta'})\boldsymbol{y},$$

where the components on the right-hand side are uncorrelated.

The above decomposition yields, then, the analysis of variance, of the form

$$\|\boldsymbol{y}\|^2 = \|\boldsymbol{P}_{\Delta'}\boldsymbol{y}\|^2 + \|(\boldsymbol{I}_n - \boldsymbol{P}_{\Delta'})\boldsymbol{y}\|^2,$$

applicable to any experiment in a block design satisfying the conditions of Theorem 3.1.2. In fact, under these conditions, it may be more convenient to present this analysis as

$$\begin{aligned} \boldsymbol{y}'(\boldsymbol{I}_n - n^{-1}\mathbf{1}_n\mathbf{1}_n')\boldsymbol{y} &= \boldsymbol{y}'\boldsymbol{\Delta}'(\boldsymbol{r}^{-\delta} - \boldsymbol{r}^{-\delta}\boldsymbol{N}\boldsymbol{k}^{-\delta}\boldsymbol{N}'\boldsymbol{r}^{-\delta})\boldsymbol{\Delta}\boldsymbol{y} \\ &\quad + \boldsymbol{y}'\boldsymbol{D}'(\boldsymbol{k}^{-\delta}\boldsymbol{N}'\boldsymbol{r}^{-\delta}\boldsymbol{N}\boldsymbol{k}^{-\delta} - n^{-1}\mathbf{1}_b\mathbf{1}_b')\boldsymbol{D}\boldsymbol{y} \\ &\quad + \boldsymbol{y}'(\boldsymbol{I}_n - \boldsymbol{\Delta}'\boldsymbol{r}^{-\delta}\boldsymbol{\Delta})\boldsymbol{y}, \end{aligned}$$

which if the design is connected reduces to

$$\boldsymbol{y}'(\boldsymbol{I}_n - n^{-1}\mathbf{1}_n\mathbf{1}_n')\boldsymbol{y} = \boldsymbol{y}'\boldsymbol{\Delta}'(\boldsymbol{r}^{-\delta} - n^{-1}\mathbf{1}_v\mathbf{1}_v')\boldsymbol{\Delta}\boldsymbol{y} + \boldsymbol{y}'(\boldsymbol{I}_n - \boldsymbol{\Delta}'\boldsymbol{r}^{-\delta}\boldsymbol{\Delta})\boldsymbol{y}.$$

Anyway, whether or not the design is connected, it is customary to write the residual sum of squares in the partitioned form

$$\boldsymbol{y}'(\boldsymbol{I}_n - \boldsymbol{\Delta}'\boldsymbol{r}^{-\delta}\boldsymbol{\Delta})\boldsymbol{y} = \boldsymbol{y}'\boldsymbol{\phi}_*\boldsymbol{D}'(\boldsymbol{D}\boldsymbol{\phi}_*\boldsymbol{D}')^-\boldsymbol{D}\boldsymbol{\phi}_*\boldsymbol{y} + \boldsymbol{y}'\boldsymbol{\psi}\boldsymbol{y}, \qquad (3.1.20)$$

where $\boldsymbol{\psi}$ $(= \boldsymbol{\psi}_*)$ and $\boldsymbol{\phi}_*$ are as defined in Sections 2.2 and 2.3.

3.2 Resolving into stratum submodels

Results of Section 3.1 sound discouraging, as in many designs, the BLUEs will exist under the model (3.1.1) for only a few parametric functions of interest, or for none of them. For example, in the class of a BIB design, for which from Definition 2.4.2 the equality $\boldsymbol{N}\boldsymbol{N}' = (r - \lambda)\boldsymbol{I}_v + \lambda\mathbf{1}_v\mathbf{1}_v'$ holds, none of the contrasts of treatment parameters will have the BLUE (on account of Corollary 3.1.1).

The apparent difficulty with the model (3.1.1) is usually evaded by resolving it into three submodels (two for the contrasts), in accordance with the stratification of the experimental units. In fact, the units of a block experiment can be seen as being grouped according to a nested classification with three "strata." Adopting the terminology used by Pearce (1983, p. 109), these strata may be specified as follows:

1st stratum—of units within blocks, called "intra-block,"
2nd stratum—of blocks within the total area, called "inter-block,"
3rd stratum—of the total experimental area.

Because of this stratification, the observed vector $\boldsymbol{y}$ can be decomposed as

$$\boldsymbol{y} = \boldsymbol{y}_1 + \boldsymbol{y}_2 + \boldsymbol{y}_3, \qquad (3.2.1)$$

where each of the three components is related to one of the strata. The component vectors $\boldsymbol{y}_\alpha, \alpha = 1, 2, 3$, are thus obtainable by projecting $\boldsymbol{y}$ orthogonally onto relevant mutually orthogonal subspaces. The first component in (3.2.1) can be written as

$$\boldsymbol{y}_1 = \boldsymbol{\phi}_1\boldsymbol{y}, \qquad (3.2.2)$$

where

$$\boldsymbol{\phi}_1 = \boldsymbol{I}_n - \boldsymbol{D}'\boldsymbol{k}^{-\delta}\boldsymbol{D} = \boldsymbol{P}_{(\boldsymbol{D}')^\perp} \qquad (3.2.3)$$

(so $\boldsymbol{\phi}_1 \equiv \boldsymbol{\phi}$ in the notation of Section 2.2), i.e., $\boldsymbol{y}_1$ is the orthogonal projection of $\boldsymbol{y}$ on $\mathcal{C}^{\perp}(\boldsymbol{D}')$, the orthogonal complement of $\mathcal{C}(\boldsymbol{D}')$. The second component is

$$\boldsymbol{y}_2 = \boldsymbol{\phi}_2 \boldsymbol{y}, \tag{3.2.4}$$

where

$$\boldsymbol{\phi}_2 = \boldsymbol{D}'\boldsymbol{k}^{-\delta}\boldsymbol{D} - n^{-1}\mathbf{1}_n\mathbf{1}_n' = \boldsymbol{P}_{D'} - \boldsymbol{P}_{\mathbf{1}_n}, \tag{3.2.5}$$

i.e., $\boldsymbol{y}_2$ is the orthogonal projection of $\boldsymbol{y}$ on $\mathcal{C}^{\perp}(\mathbf{1}_n) \cap \mathcal{C}(\boldsymbol{D}')$, the orthogonal complement of $\mathcal{C}(\mathbf{1}_n)$ in $\mathcal{C}(\boldsymbol{D}')$. The third is

$$\boldsymbol{y}_3 = \boldsymbol{\phi}_3 \boldsymbol{y}, \tag{3.2.6}$$

where

$$\boldsymbol{\phi}_3 = n^{-1}\mathbf{1}_n\mathbf{1}_n' = \boldsymbol{P}_{\mathbf{1}_n}, \tag{3.2.7}$$

i.e., $\boldsymbol{y}_3$ is the orthogonal projection of $\boldsymbol{y}$ on $\mathcal{C}(\mathbf{1}_n)$. Evidently, the three matrices (3.2.3), (3.2.5) and (3.2.7) satisfy the conditions

$$\boldsymbol{\phi}_\alpha = \boldsymbol{\phi}_\alpha', \quad \boldsymbol{\phi}_\alpha\boldsymbol{\phi}_\alpha = \boldsymbol{\phi}_\alpha, \quad \boldsymbol{\phi}_\alpha\boldsymbol{\phi}_{\alpha'} = \mathbf{O} \text{ for } \alpha \neq \alpha', \tag{3.2.8}$$

where $\alpha, \alpha' = 1, 2, 3$, and the condition

$$\boldsymbol{\phi}_1 + \boldsymbol{\phi}_2 + \boldsymbol{\phi}_3 = \boldsymbol{I}_n. \tag{3.2.9}$$

Note that the third equality in (3.2.8) implies, in particular, that

$$\boldsymbol{\phi}_1\boldsymbol{D}' = \mathbf{O} \quad \text{and} \quad \boldsymbol{\phi}_\alpha\mathbf{1}_n = \mathbf{0} \text{ for } \alpha = 1, 2, \tag{3.2.10}$$

whereas the first two equalities in (3.2.8) imply that

$$\text{rank}(\boldsymbol{\phi}_1) = n - b, \quad \text{rank}(\boldsymbol{\phi}_2) = b - 1 \quad \text{and} \quad \text{rank}(\boldsymbol{\phi}_3) = 1.$$

The resulting projections (3.2.2), (3.2.4) and (3.2.6) can be considered as submodels of the overall model (3.1.1). Their meaning can be seen by substituting the model (3.1.1) for $\boldsymbol{y}$ in (3.2.2), (3.2.4) and (3.2.6), respectively. The submodel (3.2.2) leads to the so-called intra-block analysis, resulting from the elimination of block effects, whereas (3.2.4) provides the so-called inter-block analysis, based on block totals (see, e.g., Pearce, Caliński and Marshall, 1974). The submodel (3.2.6) underlies the total-area analysis, suitable mainly for estimating the general parametric mean.

3.2.1 Intra-block submodel

The submodel (3.2.2) has the properties

$$\text{E}(\boldsymbol{y}_1) = \boldsymbol{\phi}_1\boldsymbol{\Delta}'\boldsymbol{\tau} = \boldsymbol{\Delta}'\boldsymbol{\tau} - \boldsymbol{D}'\boldsymbol{k}^{-\delta}\boldsymbol{D}\boldsymbol{\Delta}'\boldsymbol{\tau} \tag{3.2.11}$$

and

$$\text{Cov}(\boldsymbol{y}_1) = \boldsymbol{\phi}_1(\sigma_U^2 + \sigma_e^2). \tag{3.2.12}$$

From them, the main result concerning estimation under (3.2.2) can be stated as follows.

Theorem 3.2.1. *Under* (3.2.2), *a function* $\boldsymbol{w}'\boldsymbol{y}_1 = \boldsymbol{w}'\boldsymbol{\phi}_1\boldsymbol{y}$ *is uniformly the BLUE of* $\boldsymbol{c}'\boldsymbol{\tau}$ *if and only if* $\boldsymbol{\phi}_1\boldsymbol{w} = \boldsymbol{\phi}_1\boldsymbol{\Delta}'\boldsymbol{s}$, *where the vectors* $\boldsymbol{c}$ *and* $\boldsymbol{s}$ *are in the relation* $\boldsymbol{c} = \boldsymbol{\Delta}\boldsymbol{\phi}_1\boldsymbol{\Delta}'\boldsymbol{s}$ (*i.e.*, $\boldsymbol{c} = \boldsymbol{C}_1\boldsymbol{s}$, *where* $\boldsymbol{C}_1 = \boldsymbol{\Delta}\boldsymbol{\phi}_1\boldsymbol{\Delta}' \equiv \boldsymbol{C}$ *in the notation of Section* 2.2).

Proof. Under (3.2.2), with (3.2.11) and (3.2.12), the necessary and sufficient condition of Theorem 1.2.1 for a function $\boldsymbol{w}'\boldsymbol{y}_1 = \boldsymbol{w}'\boldsymbol{\phi}_1\boldsymbol{y}$ to be the BLUE of $\mathrm{E}(\boldsymbol{w}'\boldsymbol{y}_1) = \boldsymbol{w}'\boldsymbol{\phi}_1\boldsymbol{\Delta}'\boldsymbol{\tau}$ is the equality

$$(\boldsymbol{I}_n - \boldsymbol{P}_{\phi_1\Delta'})\boldsymbol{\phi}_1\boldsymbol{w} = \boldsymbol{0}.$$

It is satisfied if and only if $\boldsymbol{\phi}_1\boldsymbol{w} = \boldsymbol{\phi}_1\boldsymbol{\Delta}'\boldsymbol{s}$ for some $\boldsymbol{s}$. But $\mathrm{E}(\boldsymbol{s}'\boldsymbol{\Delta}\boldsymbol{y}_1) = \boldsymbol{s}'\boldsymbol{\Delta}\boldsymbol{\phi}_1\boldsymbol{\Delta}'\boldsymbol{\tau}$. Hence, the relation for $\boldsymbol{c}$ and $\boldsymbol{s}$ follows. □

Remark 3.2.1. Because $\boldsymbol{1}_v'\boldsymbol{\Delta}\boldsymbol{\phi}_1 = \boldsymbol{0}'$, on account of (2.2.1) and (3.2.10), the only parametric functions for which the BLUEs may exist under (3.2.2) are contrasts.

If $\boldsymbol{c}'\boldsymbol{\tau}$ is a contrast, and the condition of Theorem 3.2.1 is satisfied, then the variance of its BLUE under (3.2.2), i.e., of $\widehat{\boldsymbol{c}'\boldsymbol{\tau}} = \boldsymbol{s}'\boldsymbol{\Delta}\boldsymbol{y}_1 = \boldsymbol{c}'\boldsymbol{C}_1^-\boldsymbol{\Delta}\boldsymbol{\phi}_1\boldsymbol{y}$, is of the form

$$\mathrm{Var}(\widehat{\boldsymbol{c}'\boldsymbol{\tau}}) = \boldsymbol{s}'\boldsymbol{C}_1\boldsymbol{s}(\sigma_U^2 + \sigma_e^2) = \boldsymbol{c}'\boldsymbol{C}_1^-\boldsymbol{c}(\sigma_U^2 + \sigma_e^2), \tag{3.2.13}$$

where $\boldsymbol{C}_1^-$ is any g-inverse of the matrix $\boldsymbol{C}_1 = \boldsymbol{\Delta}\boldsymbol{\phi}_1\boldsymbol{\Delta}' = \boldsymbol{r}^\delta - \boldsymbol{N}\boldsymbol{k}^{-\delta}\boldsymbol{N}'$, as defined in (2.2.8) (remembering that $\boldsymbol{\phi}_1 \equiv \boldsymbol{\phi}$ and $\boldsymbol{C}_1 \equiv \boldsymbol{C}$).

Remark 3.2.2. Because $\mathrm{Cov}(\boldsymbol{y}_1)\boldsymbol{\phi}_1\boldsymbol{\Delta}' = \boldsymbol{\phi}_1\boldsymbol{\Delta}'(\sigma_U^2 + \sigma_e^2)$, which implies that both $\boldsymbol{\phi}_1\boldsymbol{\Delta}'\boldsymbol{s}$ and $\mathrm{Cov}(\boldsymbol{y}_1)\boldsymbol{\phi}_1\boldsymbol{\Delta}'\boldsymbol{s}$ belong to $\mathcal{C}(\boldsymbol{\phi}_1\boldsymbol{\Delta}')$ for any $\boldsymbol{s}$, it follows from Theorem 3.2.1 on account of Theorem 1.2.2 that the BLUEs under the submodel (3.2.2), with the moments (3.2.11) and (3.2.12), can equivalently be obtained under a simple alternative model in which the covariance matrix (3.2.12) is replaced by that of the form $(\sigma_U^2 + \sigma_e^2)\boldsymbol{I}_n$, i.e., that $\boldsymbol{s}'\boldsymbol{\Delta}\boldsymbol{\phi}_1\boldsymbol{y}$ is both the SLSE and the BLUE of its expectation, for any $\boldsymbol{s}$.

From Remark 3.2.2, it follows, in particular, that the BLUE of the expectation vector (3.2.11) can be obtained by a simple least-squares procedure, in the form $\widehat{\mathrm{E}(\boldsymbol{y}_1)} = \boldsymbol{P}_{\phi_1\Delta'}\boldsymbol{y}_1$, where $\boldsymbol{P}_{\phi_1\Delta'} = \boldsymbol{\phi}_1\boldsymbol{\Delta}'\boldsymbol{C}^-\boldsymbol{\Delta}\boldsymbol{\phi}_1$. To check it in view of Theorem 1.2.1, note that $\mathrm{E}(\boldsymbol{P}_{\phi_1\Delta'}\boldsymbol{y}_1) = \boldsymbol{\phi}_1\boldsymbol{\Delta}'\boldsymbol{\tau} = \mathrm{E}(\boldsymbol{y}_1)$ and that $(\boldsymbol{I}_n - \boldsymbol{P}_{\phi_1\Delta'})\mathrm{Cov}(\boldsymbol{y}_1)\boldsymbol{P}_{\phi_1\Delta'} = \boldsymbol{O}$, on account of (3.2.11) and (3.2.12). Furthermore, it follows that the vector $\boldsymbol{y}_1$ can be decomposed as

$$\begin{aligned}\boldsymbol{y}_1 &= \boldsymbol{P}_{\phi_1\Delta'}\boldsymbol{y}_1 + (\boldsymbol{I}_n - \boldsymbol{P}_{\phi_1\Delta'})\boldsymbol{y}_1 \quad \text{(in terms of } \boldsymbol{y}_1\text{)}\\ &= \boldsymbol{P}_{\phi_1\Delta'}\boldsymbol{y} + (\boldsymbol{\phi}_1 - \boldsymbol{P}_{\phi_1\Delta'})\boldsymbol{y} \quad \text{(in terms of } \boldsymbol{y}\text{)},\end{aligned}$$

where the second component on the right-hand side in each form of the decomposition is the residual vector providing the minimum norm quadratic unbiased estimator (MINQUE) of $\sigma_U^2 + \sigma_e^2$ (see Rao, 1974, Section 3).

More precisely, taking the squared norm on both sides of the above decomposition of $\boldsymbol{y}_1$, one can write

$$\|\boldsymbol{y}_1\|^2 = \|\boldsymbol{P}_{\phi_1\Delta'}\boldsymbol{y}_1\|^2 + \|(\boldsymbol{I}_n - \boldsymbol{P}_{\phi_1\Delta'})\boldsymbol{y}_1\|^2,$$

as $\boldsymbol{P}_{\phi_1\Delta'}(\boldsymbol{I}_n - \boldsymbol{P}_{\phi_1\Delta'}) = \boldsymbol{O}$ (see Appendix A.2). This equation provides the intra-block analysis of variance, which in terms of the observed vector $\boldsymbol{y}$ can be expressed in a more customary way as

$$\boldsymbol{y}'\boldsymbol{\phi}_1\boldsymbol{y} = \boldsymbol{y}'\boldsymbol{\phi}_1\boldsymbol{\Delta}'\boldsymbol{C}_1^-\boldsymbol{\Delta}\boldsymbol{\phi}_1\boldsymbol{y} + \boldsymbol{y}'(\boldsymbol{\phi}_1 - \boldsymbol{\phi}_1\boldsymbol{\Delta}'\boldsymbol{C}_1^-\boldsymbol{\Delta}\boldsymbol{\phi}_1)\boldsymbol{y} = \boldsymbol{Q}_1'\boldsymbol{C}_1^-\boldsymbol{Q}_1 + \boldsymbol{y}'\boldsymbol{\psi}_1\boldsymbol{y},$$

where $\boldsymbol{Q}_1 = \boldsymbol{\Delta}\boldsymbol{\phi}_1\boldsymbol{y}$ and $\boldsymbol{\psi}_1 \equiv \boldsymbol{\psi}$, as defined in (2.3.8). The quadratic form $\boldsymbol{y}'\boldsymbol{\phi}_1\boldsymbol{y}$ can be called the intra-block total sum of squares, and its components, $\boldsymbol{Q}_1'\boldsymbol{C}_1^-\boldsymbol{Q}_1$ and $\boldsymbol{y}'\boldsymbol{\psi}_1\boldsymbol{y}$, can be called the intra-block treatment sum of squares and the intra-block residual sum of squares, respectively. The corresponding degrees of freedom (d.f.) are $n - b = \text{rank}(\boldsymbol{\phi}_1)$ for the total, $h = \text{rank}(\boldsymbol{C}_1)$, by Lemma 2.3.5, for the treatment component and $n - b - h = \text{rank}(\boldsymbol{\psi}_1)$, on account of Lemma 2.3.6, for the residual component. The expectations of these component sums of squares are [according to formula (4a.1.7) in Rao, 1973]

$$\text{E}(\boldsymbol{Q}_1'\boldsymbol{C}_1^-\boldsymbol{Q}_1) = h(\sigma_U^2 + \sigma_e^2) + \boldsymbol{\tau}'\boldsymbol{C}_1\boldsymbol{\tau} \quad \text{and} \quad \text{E}(\boldsymbol{y}'\boldsymbol{\psi}_1\boldsymbol{y}) = (n - b - h)(\sigma_U^2 + \sigma_e^2).$$

It follows that the intra-block residual mean square $s_1^2 = \boldsymbol{y}'\boldsymbol{\psi}_1\boldsymbol{y}\ /(n - b - h)$ is an unbiased estimator of $\sigma_1^2 = \sigma_U^2 + \sigma_e^2$. Moreover, s_1^2 is the MINQUE of σ_1^2 under the submodel (3.2.2), as it may be seen from Theorem 3.4 of Rao (1974).

Thus, s_1^2 can be used to obtain an unbiased estimator of the variance (3.2.13), in the form

$$\widehat{\text{Var}(\boldsymbol{c}'\boldsymbol{\tau})} = \boldsymbol{s}'\boldsymbol{C}_1\boldsymbol{s}s_1^2 = \boldsymbol{c}'\boldsymbol{C}_1^-\boldsymbol{c}s_1^2.$$

Furthermore, because under the multivariate normal distribution of $\boldsymbol{y}$, and hence of $\boldsymbol{y}_1$, the quadratic functions $\boldsymbol{Q}_1'\boldsymbol{C}_1^-\boldsymbol{Q}_1/\sigma_1^2$ and $\boldsymbol{y}'\boldsymbol{\psi}_1\boldsymbol{y}/\sigma_1^2$ have independent χ^2 distributions, the first noncentral with h d.f. and the noncentrality parameter $\delta_1 = \boldsymbol{\tau}'\boldsymbol{C}_1\boldsymbol{\tau}/\sigma_1^2$, the second central with $n - b - h$ d.f. (as can be proved applying, e.g., Theorems 9.2.1 and 9.4.1 of Rao and Mitra, 1971), the hypothesis $\boldsymbol{\tau}'\boldsymbol{C}_1\boldsymbol{\tau} = 0$, equivalent to $\text{E}(\boldsymbol{y}_1) = \boldsymbol{0}$ [or $\text{E}(\boldsymbol{y}) \in \mathcal{C}(\boldsymbol{D}')$], can be tested by the variance ratio criterion

$$h^{-1}\boldsymbol{Q}_1'\boldsymbol{C}_1^-\boldsymbol{Q}_1/s_1^2,$$

which under the normality assumption has then the F distribution with h and $n - b - h$ d.f., central when the hypothesis is true (see, e.g., John, 1987, p. 13; note that he uses $\boldsymbol{q}$ for $\boldsymbol{Q}_1$).

Example 3.2.1. Ceranka (1975) has described the intra-block analysis of an agricultural field experiment with $v = 4$ treatments (varieties) of sunflower compared in $b = 10$ blocks of two different sizes, $k_1 = 2$ and $k_2 = 3$, according to the incidence matrix

$$N = \begin{bmatrix} 1 & 0 & 1 & 0 & 1 & 0 & 1 & 1 & 1 & 0 \\ 1 & 0 & 0 & 1 & 0 & 1 & 1 & 1 & 0 & 1 \\ 0 & 1 & 1 & 0 & 0 & 1 & 1 & 0 & 1 & 1 \\ 0 & 1 & 0 & 1 & 1 & 0 & 0 & 1 & 1 & 1 \end{bmatrix}.$$

It is assumed that the blocks and the units, i.e., plots, within blocks have been randomized according to the procedure described in Section 3.1.1. In the context of an agricultural field experiment, this means that the order in which the columns of the matrix N are assigned to the blocks of plots, to be formed in the experimental field, is chosen at random, and that for each block the order in which the treatments indicated by 1's in the assigned column of N are then assigned to the plots of the block is also chosen at random. Because, usually, the researcher has much freedom in forming blocks in an experimental field, the unequal block sizes should cause no concern here. The plot observations, concerning the average head diameter of the plant in centimeters, are as follows:

Block	1	1	2	2	3	3	4	4	5	5	6	6
Treatment	2	1	3	4	3	1	2	4	4	1	2	3
Observation	16.5	17.9	15.0	13.0	15.4	19.0	17.5	12.1	13.1	19.7	17.9	15.8

Block	7	7	7	8	8	8	9	9	9	10	10	10
Treatment	3	1	2	1	4	2	4	1	3	3	4	2
Observation	15.3	17.6	17.3	16.1	12.2	18.0	12.5	16.2	12.6	11.8	12.3	16.6

Because this design is equireplicate (see Definition 2.2.3) with two different block sizes and with the off-diagonal elements of its matrix $Nk^{-\delta}N'$ all equal, and, hence, proportional to the constant treatment replication (and its square), it can be seen as a design totally balanced in the sense of Jones (Definition 2.4.5; see also Ceranka, 1976). According to (2.4.8), its C-matrix is then of the form

$$C_1 = \varepsilon[6I_4 - (3/2)\mathbf{1}_4\mathbf{1}_4'] \quad \text{with} \quad \varepsilon = 7/9,$$

i.e.,

$$C_1 = (7/6)(4I_4 - \mathbf{1}_4\mathbf{1}_4').$$

A suitable choice of a g-inverse of C_1 will be here the matrix $(6\varepsilon)^{-1}[I_4 - (1/4)\mathbf{1}_4\mathbf{1}_4']$. With it, for any contrast $c'\tau$, its BLUE under the intra-block submodel is obtainable as

$$\widehat{c'\tau} = (6\varepsilon)^{-1}c'\Delta\phi_1 y = (3/14)c'Q_1,$$

where $Q_1 = \Delta\phi_1 y = [9.7667,\ 9.2167,\ -6.2167,\ -12.7667]'$, and its variance, according to (3.2.13), as

$$\mathrm{Var}(\widehat{c'\tau}) = (6\varepsilon)^{-1}c'c\sigma_1^2 = (3/14)c'c\sigma_1^2,$$

where $\sigma_1^2 = \sigma_U^2 + \sigma_e^2$. The resulting intra-block analysis of variance can be presented as

Source	Degrees of freedom	Sum of squares	Mean square	F
Treatments	$v-1=\ 3$	$Q_1'C_1^-Q_1 = 81.85060$	27.2835	31.26
Residuals	$n-b-v+1=11$	$y'\psi_1 y\ =\ \ 9.60107$	$s_1^2 = 0.872825$	
Total	$n-b=14$	$y'\phi_1 y\ = 91.45167$		

From it, s_1^2 can be used to obtain an unbiased estimate of the above $\text{Var}(\widehat{c'\tau})$ as $(3/14)c'cs_1^2 = 0.18703c'c$. Also note that the observed value of the variance ratio criterion F is highly significant, with the calculated value $P = 0.000011$. Thus, it can be concluded that there are significant intra-block differences between the strains of sunflower with regard to the analysed trait.

3.2.2 Inter-block submodel

As to the submodel (3.2.4), it has the properties

$$\text{E}(\boldsymbol{y}_2) = \boldsymbol{\phi}_2\boldsymbol{\Delta}'\boldsymbol{\tau} = \boldsymbol{D}'\boldsymbol{k}^{-\delta}\boldsymbol{D}\boldsymbol{\Delta}'\boldsymbol{\tau} - n^{-1}\mathbf{1}_n\boldsymbol{r}'\boldsymbol{\tau} \tag{3.2.14}$$

and

$$\text{Cov}(\boldsymbol{y}_2) = \boldsymbol{\phi}_2\boldsymbol{D}'\boldsymbol{D}\boldsymbol{\phi}_2(\sigma_B^2 - K_H^{-1}\sigma_U^2) + \boldsymbol{\phi}_2(\sigma_U^2 + \sigma_e^2). \tag{3.2.15}$$

The main estimation result under (3.2.4) can be expressed as follows.

Theorem 3.2.2. *Under* (3.2.4), *a function* $\boldsymbol{w}'\boldsymbol{y}_2 = \boldsymbol{w}'\boldsymbol{\phi}_2\boldsymbol{y}$ *is uniformly the BLUE of* $\boldsymbol{c}'\boldsymbol{\tau}$ *if and only if* $\boldsymbol{\phi}_2\boldsymbol{w} = \boldsymbol{\phi}_2\boldsymbol{\Delta}'\boldsymbol{s}$, *where the vectors* $\boldsymbol{c}$ *and* $\boldsymbol{s}$ *are in the relation* $\boldsymbol{c} = \boldsymbol{\Delta}\boldsymbol{\phi}_2\boldsymbol{\Delta}'\boldsymbol{s}$ (*i.e.,* $\boldsymbol{c} = \boldsymbol{C}_2\boldsymbol{s}$, *where* $\boldsymbol{C}_2 = \boldsymbol{\Delta}\boldsymbol{\phi}_2\boldsymbol{\Delta}'$, *as introduced in Section* 2.2), *and* $\boldsymbol{s}$ *satisfies the condition*

$$[\boldsymbol{K}_0 - \boldsymbol{N}_0'(\boldsymbol{N}_0\boldsymbol{k}^{-\delta}\boldsymbol{N}_0')^-\boldsymbol{N}_0]\boldsymbol{N}_0'\boldsymbol{s} = \mathbf{0}, \tag{3.2.16}$$

or the equivalent condition

$$[\boldsymbol{K}_0 - \boldsymbol{N}_0'\boldsymbol{r}^{-\delta}\boldsymbol{N}_0(\boldsymbol{N}_0'\boldsymbol{r}^{-\delta}\boldsymbol{N}_0)^-\boldsymbol{K}_0]\boldsymbol{N}_0'\boldsymbol{s} = \mathbf{0}, \tag{3.2.17}$$

where $\boldsymbol{K}_0 = \boldsymbol{k}^\delta - n^{-1}\boldsymbol{k}\boldsymbol{k}'$ *and* $\boldsymbol{N}_0 = \boldsymbol{N} - n^{-1}\boldsymbol{r}\boldsymbol{k}'$.

Proof. Under (3.2.4), with (3.2.14) and (3.2.15), the necessary and sufficient condition of Theorem 1.2.1 for a function $\boldsymbol{w}'\boldsymbol{y}_2 = \boldsymbol{w}'\boldsymbol{\phi}_2\boldsymbol{y}$ to be the BLUE of $\text{E}(\boldsymbol{w}'\boldsymbol{y}_2) = \boldsymbol{w}'\boldsymbol{\phi}_2\boldsymbol{\Delta}'\boldsymbol{\tau}$ is the equality $(\boldsymbol{I}_n - \boldsymbol{P}_{\phi_2\Delta'})[\boldsymbol{\phi}_2\boldsymbol{D}'\boldsymbol{D}\boldsymbol{\phi}_2(\sigma_B^2 - K_H^{-1}\sigma_U^2) + \boldsymbol{\phi}_2(\sigma_U^2 + \sigma_e^2)]\boldsymbol{w} = \mathbf{0}$. It holds uniformly if and only if the equalities $(\boldsymbol{I}_n - \boldsymbol{P}_{\phi_2\Delta'})\boldsymbol{\phi}_2\boldsymbol{w} = \mathbf{0}$ and $(\boldsymbol{I}_n - \boldsymbol{P}_{\phi_2\Delta'})\boldsymbol{\phi}_2\boldsymbol{D}'\boldsymbol{D}\boldsymbol{\phi}_2\boldsymbol{w} = \mathbf{0}$ hold simultaneously. The first equality holds if and only if $\boldsymbol{\phi}_2\boldsymbol{w} = \boldsymbol{\phi}_2\boldsymbol{\Delta}'\boldsymbol{s}$ for some $\boldsymbol{s}$, which holds if and only if $\boldsymbol{D}\boldsymbol{\phi}_2\boldsymbol{w} = \boldsymbol{D}\boldsymbol{\phi}_2\boldsymbol{\Delta}'\boldsymbol{s}$ for that $\boldsymbol{s}$. With this equality, the second equality reads $(\boldsymbol{I}_n - \boldsymbol{P}_{\phi_2\Delta'})\boldsymbol{\phi}_2\boldsymbol{D}'\boldsymbol{D}\boldsymbol{\phi}_2\boldsymbol{\Delta}'\boldsymbol{s} = \mathbf{0}$, which is equivalent to (3.2.16) because of the relations $\boldsymbol{D}'\boldsymbol{k}^{-\delta}\boldsymbol{D}\boldsymbol{\phi}_2 = \boldsymbol{\phi}_2$, $\boldsymbol{D}\boldsymbol{\phi}_2\boldsymbol{D}' = \boldsymbol{K}_0$, $\boldsymbol{\Delta}\boldsymbol{\phi}_2\boldsymbol{D}' = \boldsymbol{N}_0$ and $\boldsymbol{C}_2 = \boldsymbol{\Delta}\boldsymbol{\phi}_2\boldsymbol{\Delta}' = \boldsymbol{N}_0\boldsymbol{k}^{-\delta}\boldsymbol{N}_0'$. That the conditions (3.2.16) and (3.2.17) are equivalent can be checked by using the equality $\boldsymbol{N}_0'(\boldsymbol{N}_0\boldsymbol{k}^{-\delta}\boldsymbol{N}_0')^-\boldsymbol{N}_0\boldsymbol{k}^{-\delta}\boldsymbol{N}_0' = \boldsymbol{N}_0' =$

$\boldsymbol{N}_0'\boldsymbol{r}^{-\delta}\boldsymbol{N}_0(\boldsymbol{N}_0'\boldsymbol{r}^{-\delta}\boldsymbol{N}_0)^{-}\boldsymbol{N}_0'$, obtainable from Lemma 2.2.6(c) of Rao and Mitra (1971), and by noting that $\boldsymbol{N}_0\boldsymbol{k}^{-\delta}\boldsymbol{K}_0 = \boldsymbol{N}_0$. Finally, the relation between $\boldsymbol{c}$ and $\boldsymbol{s}$ follows from the fact that $\mathrm{E}(\boldsymbol{s}'\boldsymbol{\Delta y}_2) = \boldsymbol{s}'\boldsymbol{N}_0\boldsymbol{k}^{-\delta}\boldsymbol{N}_0'\boldsymbol{\tau}$. □

Corollary 3.2.1. *For the estimation of* $\boldsymbol{c}'\boldsymbol{\tau} = \boldsymbol{s}'\boldsymbol{C}_2\boldsymbol{\tau}$ *under the submodel* (3.2.4), *the following applies:*

(a) *The case* $\boldsymbol{N}_0'\boldsymbol{s} = \boldsymbol{0}$ *is to be excluded.*

(b) *If* $\boldsymbol{N}_0'\boldsymbol{s} \neq \boldsymbol{0}$, *then* $\boldsymbol{c}'\boldsymbol{\tau}$ *is a contrast, and to satisfy* (3.2.16) *or* (3.2.17) *by the vector* $\boldsymbol{s}$, *it is necessary and sufficient that* $\boldsymbol{K}_0\boldsymbol{N}_0'\boldsymbol{s} \in \mathcal{C}(\boldsymbol{N}_0'\boldsymbol{r}^{-\delta}\boldsymbol{N}_0) = \mathcal{C}(\boldsymbol{N}_0')$.

(c) *If* $\boldsymbol{s}$ *is such that* $\boldsymbol{r}'\boldsymbol{s} = 0$, *then the conditions* (3.2.16) *and* (3.2.17) *can be replaced by*

$$\boldsymbol{K}_0\boldsymbol{N}'\boldsymbol{s} = \boldsymbol{N}_0'(\boldsymbol{N}_0\boldsymbol{k}^{-\delta}\boldsymbol{N}_0')^{-}\boldsymbol{N}_0\boldsymbol{N}'\boldsymbol{s} \tag{3.2.18}$$

and

$$\boldsymbol{K}_0\boldsymbol{N}'\boldsymbol{s} = \boldsymbol{N}_0'\boldsymbol{r}^{-\delta}\boldsymbol{N}_0(\boldsymbol{N}_0'\boldsymbol{r}^{-\delta}\boldsymbol{N}_0)^{-}\boldsymbol{K}_0\boldsymbol{N}'\boldsymbol{s}, \tag{3.2.19}$$

respectively. To satisfy any of them, it is then necessary and sufficient that $\boldsymbol{K}_0\boldsymbol{N}'\boldsymbol{s} \in \mathcal{C}(\boldsymbol{N}_0')$.

(d) *The condition* (ii) *of Theorem* 3.1.2 *is sufficient for satisfying the equality* (3.2.19), *and hence* (3.2.18), *by any vector* $\boldsymbol{s}$, *thus being sufficient for the equalities* (3.2.16) *and* (3.2.17) *to be satisfied by any* $\boldsymbol{s}$.

Proof. The result (a) is obvious, as $\boldsymbol{N}_0'\boldsymbol{s} = \boldsymbol{0}$ implies $\boldsymbol{c} = \boldsymbol{0}$. To prove (b), note that $\boldsymbol{N}_0'\boldsymbol{1}_v = \boldsymbol{0}$ and that the equation $\boldsymbol{N}_0'\boldsymbol{r}^{-\delta}\boldsymbol{N}_0\boldsymbol{x} = \boldsymbol{K}_0\boldsymbol{N}_0'\boldsymbol{s}$ is consistent if and only if (3.2.17) holds [see Theorem 2.3.1(d) of Rao and Mitra, 1971]. Alternatively, note that the equation $\boldsymbol{N}_0'\boldsymbol{x} = \boldsymbol{K}_0\boldsymbol{N}_0'\boldsymbol{s}$ is consistent if and only if (3.2.16) holds, because $(\boldsymbol{N}_0\boldsymbol{k}^{-\delta}\boldsymbol{N}_0')^{-}\boldsymbol{N}_0\boldsymbol{k}^{-\delta}$ can be used as a g-inverse of $\boldsymbol{N}_0'$. The result (c) is obvious, as $\boldsymbol{N}_0'\boldsymbol{s} = \boldsymbol{N}'\boldsymbol{s}$ if $\boldsymbol{r}'\boldsymbol{s} = 0$. The result (d) can easily be checked by using Lemma 2.2.6(c) of Rao and Mitra (1971) and the relation $\boldsymbol{K}_0\boldsymbol{N}' = \boldsymbol{N}_0'\tilde{\boldsymbol{k}}^{\delta}$, held under the condition (ii) of Theorem 3.1.2, with the matrix $\tilde{\boldsymbol{k}}^{\delta}$ defined as in (2.2.10), and, further, by noting that $\boldsymbol{N}_0' = \boldsymbol{N}'(\boldsymbol{I}_v - n^{-1}\boldsymbol{1}_v\boldsymbol{r}')$. □

Now, it may be noted that if the conditions of Theorem 3.2.2 are satisfied, then $\widehat{\boldsymbol{c}'\boldsymbol{\tau}} = \boldsymbol{s}'\boldsymbol{\Delta y}_2 = \boldsymbol{c}'\boldsymbol{C}_2^{-}\boldsymbol{\Delta\phi}_2\boldsymbol{y}$ is the BLUE of the contrast $\boldsymbol{c}'\boldsymbol{\tau} = \boldsymbol{s}'\boldsymbol{C}_2\boldsymbol{\tau}$ under (3.2.4), and that its variance has the form

$$\begin{aligned}\mathrm{Var}(\widehat{\boldsymbol{c}'\boldsymbol{\tau}}) &= \boldsymbol{s}'\boldsymbol{N}_0\boldsymbol{N}_0'\boldsymbol{s}(\sigma_B^2 - K_H^{-1}\sigma_U^2) + \boldsymbol{s}'\boldsymbol{N}_0\boldsymbol{k}^{-\delta}\boldsymbol{N}_0'\boldsymbol{s}(\sigma_U^2 + \sigma_e^2)\\ &= \boldsymbol{c}'\boldsymbol{C}_2^{-}\boldsymbol{N}_0\boldsymbol{N}_0'\boldsymbol{C}_2^{-}\boldsymbol{c}(\sigma_B^2 - K_H^{-1}\sigma_U^2) + \boldsymbol{c}'\boldsymbol{C}_2^{-}\boldsymbol{c}(\sigma_U^2 + \sigma_e^2).\end{aligned} \tag{3.2.20}$$

Evidently, if $k_1 = k_2 = \cdots = k_b = k$, the variance (3.2.20) reduces to

$$\begin{aligned}\mathrm{Var}(\widehat{\boldsymbol{c}'\boldsymbol{\tau}}) &= k^{-1}\boldsymbol{s}'\boldsymbol{N}_0\boldsymbol{N}_0'\boldsymbol{s}[k\sigma_B^2 + (1 - K_H^{-1}k)\sigma_U^2 + \sigma_e^2]\\ &= \boldsymbol{c}'(k^{-1}\boldsymbol{N}_0\boldsymbol{N}_0')^{-}\boldsymbol{c}[k\sigma_B^2 + (1 - K_H^{-1}k)\sigma_U^2 + \sigma_e^2],\end{aligned} \tag{3.2.21}$$

which coincides with the formula given by Pearce (1983, Section 3.8) when $k = K_H$ and the technical error is ignored.

Remark 3.2.3. For a vector $\boldsymbol{s}$ such that $\boldsymbol{r}'\boldsymbol{s} = 0$, the condition of Theorem 3.2.2 is less restrictive than that of Theorem 3.1.1, as any such $\boldsymbol{s}$ satisfying (3.1.17) satisfies also the condition (3.2.19), whereas that satisfying (3.2.19) must not necessarily satisfy (3.1.17).

An answer to the question what is necessary and sufficient for the condition of Theorem 3.2.2 to be satisfied by any $\boldsymbol{s}$ can be given as follows.

Corollary 3.2.2. *The condition* (3.2.16) *holds for any* $\boldsymbol{s}$, *i.e., the equality*

$$\boldsymbol{K}_0\boldsymbol{N}_0' = \boldsymbol{N}_0'(\boldsymbol{N}_0\boldsymbol{k}^{-\delta}\boldsymbol{N}_0')^{-}\boldsymbol{N}_0\boldsymbol{N}_0' \tag{3.2.22}$$

holds, if and only if for any $b\times 1$ *nonzero vector* $\mathbf{t}$ *satisfying the equality* $\boldsymbol{N}_0\mathbf{t} = \mathbf{0}$, *the equality* $\boldsymbol{N}_0\boldsymbol{K}_0\mathbf{t} = \mathbf{0}$ *holds as well.*

Proof. From Corollary 3.2.1(b), the condition (3.2.16) is satisfied by any $\boldsymbol{s}$ if and only if $\mathcal{C}(\boldsymbol{K}_0\boldsymbol{N}_0') \subset \mathcal{C}(\boldsymbol{N}_0')$. But this inclusion holds if and only if $\mathcal{C}^\perp(\boldsymbol{N}_0') \subset \mathcal{C}^\perp(\boldsymbol{K}_0\boldsymbol{N}_0')$ or, equivalently, $\mathcal{N}(\boldsymbol{N}_0) \subset \mathcal{N}(\boldsymbol{N}_0\boldsymbol{K}_0)$ [as $\mathcal{C}^\perp(\boldsymbol{A}) = \mathcal{N}(\boldsymbol{A}')$ for any matrix $\boldsymbol{A}$; see Appendix A.1]. Evidently, the last inclusion means that for any vector $\boldsymbol{t}$ for which $\boldsymbol{N}_0\boldsymbol{t} = \mathbf{0}$, the equality $\boldsymbol{N}_0\boldsymbol{K}_0\boldsymbol{t} = \mathbf{0}$ also holds. □

To see the applicability of Corollary 3.2.2, it will be helpful to examine the following two examples taken from Pearce (1983, p. 102 and p. 117).

Example 3.2.2. Consider a design with the incidence matrix

$$\boldsymbol{N} = \begin{bmatrix} 1 & 1 & 1 & 1 & 1 & 1 \\ 1 & 1 & 1 & 1 & 1 & 1 \\ 1 & 1 & 1 & 1 & 1 & 1 \\ 1 & 1 & 1 & 1 & 1 & 1 \\ 0 & 1 & 1 & 0 & 1 & 0 \\ 0 & 0 & 1 & 1 & 1 & 0 \end{bmatrix}.$$

From it,

$$\boldsymbol{N}_0 = \frac{1}{10}\begin{bmatrix} 2 & 0 & -2 & 0 & -2 & 2 \\ 2 & 0 & -2 & 0 & -2 & 2 \\ 2 & 0 & -2 & 0 & -2 & 2 \\ 2 & 0 & -2 & 0 & -2 & 2 \\ -4 & 5 & 4 & -5 & 4 & -4 \\ -4 & -5 & 4 & 5 & 4 & -4 \end{bmatrix}$$

and

$$\boldsymbol{N}_0\boldsymbol{K}_0 = \frac{1}{150}\begin{bmatrix} 136 & 20 & -156 & 20 & -156 & 136 \\ 136 & 20 & -156 & 20 & -156 & 136 \\ 136 & 20 & -156 & 20 & -156 & 136 \\ 136 & 20 & -156 & 20 & -156 & 136 \\ -272 & 335 & 312 & -415 & 312 & -272 \\ -272 & 335 & 312 & -415 & 312 & -272 \end{bmatrix}.$$

Taking, e.g., $\boldsymbol{t} = [-1, 2, -1, 2, -1, -1]'$, it can be seen that $\boldsymbol{N}_0\boldsymbol{t} = \mathbf{0}$, whereas $\boldsymbol{N}_0\boldsymbol{K}_0\boldsymbol{t} = (10)^{-1}[8, 8, 8, 8, -16, -16]'$. Thus, this design does not satisfy the condition of Corollary 3.2.2, and hence, (3.2.16) does not hold for any $\boldsymbol{s}$.

Example 3.2.3. Consider a design with the incidence matrix

$$\boldsymbol{N} = \begin{bmatrix} 1 & 1 & 0 & 0 \\ 1 & 1 & 0 & 0 \\ 1 & 1 & 0 & 0 \\ 1 & 0 & 1 & 1 \\ 0 & 1 & 1 & 1 \\ 0 & 0 & 1 & 1 \end{bmatrix},$$

from which

$$\boldsymbol{N}_0 = \frac{1}{14}\begin{bmatrix} 6 & 6 & -6 & -6 \\ 6 & 6 & -6 & -6 \\ 6 & 6 & -6 & -6 \\ 2 & -12 & 5 & 5 \\ -12 & 2 & 5 & 5 \\ -8 & -8 & 8 & 8 \end{bmatrix}$$

and

$$\boldsymbol{N}_0\boldsymbol{K}_0 = \frac{2}{49}\begin{bmatrix} 36 & 36 & -36 & -36 \\ 36 & 36 & -36 & -36 \\ 36 & 36 & -36 & -36 \\ 19 & 79 & 30 & 30 \\ 79 & 19 & 30 & 30 \\ -48 & -48 & 48 & 48 \end{bmatrix}.$$

The matrix $\boldsymbol{N}_0$ above is of rank 2, and hence, the dimension of the null space of $\boldsymbol{N}_0$ is 2. From the definitions, $\boldsymbol{N}_0\mathbf{1}_b = \mathbf{0}$ and $\boldsymbol{N}_0\boldsymbol{K}_0\mathbf{1}_b = \mathbf{0}$. Also, taking $\boldsymbol{t} = [0, 0, -1, 1]'$, one obtains both $\boldsymbol{N}_0\boldsymbol{t} = \mathbf{0}$ and $\boldsymbol{N}_0\boldsymbol{K}_0\boldsymbol{t} = \mathbf{0}$, which shows that the condition of Corollary 3.2.2 is satisfied. Thus, in this example, the equality (3.2.16) holds for any $\boldsymbol{s}$, which implies that the design provides under the submodel (3.2.4) the BLUE for any contrast $\boldsymbol{c}'\boldsymbol{\tau} = \boldsymbol{s}'\boldsymbol{C}_2\boldsymbol{\tau}$, i.e., for any $\boldsymbol{c}'\boldsymbol{\tau}$ such that $\boldsymbol{c} \in \mathcal{C}(\boldsymbol{C}_2)$. Here

$$\boldsymbol{C}_2 = \boldsymbol{N}_0\boldsymbol{k}^{-\delta}\boldsymbol{N}_0' = \frac{1}{84}\begin{bmatrix} 18 & 18 & 18 & -15 & -15 & -24 \\ 18 & 18 & 18 & -15 & -15 & -24 \\ 18 & 18 & 18 & -15 & -15 & -24 \\ -15 & -15 & -15 & 23 & 2 & 20 \\ -15 & -15 & -15 & 2 & 23 & 20 \\ -24 & -24 & -24 & 20 & 20 & 32 \end{bmatrix},$$

and so the columns of this matrix span the subspace of all contrasts (of the vectors $\boldsymbol{c}$ representing them) for which the BLUEs under the inter-block submodel exist. The dimension of this subspace is 2. The reader is advised to check that the equivalent condition (3.2.17) also holds for any $\boldsymbol{s}$ in this example.

Remark 3.2.4. (a) Because the equalities $\boldsymbol{N}_0\mathbf{1}_b = \mathbf{0}$ and $\boldsymbol{N}_0\boldsymbol{K}_0\mathbf{1}_b = \mathbf{0}$ hold always, the necessary and sufficient condition for the equality (3.2.22), given in Corollary 3.2.2, can be replaced by the condition that $\boldsymbol{N}\boldsymbol{t}_0 = \mathbf{0}$ implies $\boldsymbol{N}_0\boldsymbol{k}^\delta\boldsymbol{t}_0 = \mathbf{0}$ for any vector $\boldsymbol{t}_0$ such that $\boldsymbol{k}'\boldsymbol{t}_0 = 0$.

(b) If $\text{rank}(\boldsymbol{N}_0) = b - 1$, i.e., the columns of the incidence matrix $\boldsymbol{N}$ are all linearly independent [as $\text{rank}(\boldsymbol{N}_0) = \text{rank}(\boldsymbol{N}) - 1$], then a vector $\boldsymbol{t}$ satisfying $\boldsymbol{N}_0\boldsymbol{t} = \mathbf{0}$ must be equal or proportional to $\mathbf{1}_b$ [i.e., $\boldsymbol{t} \in \mathcal{C}(\mathbf{1}_b)$], and so satisfy also the equality $\boldsymbol{N}_0\boldsymbol{K}_0\boldsymbol{t} = \mathbf{0}$. Thus, the condition of Corollary 3.2.2 is then satisfied automatically, whatever the k_j's are.

Example 3.2.4. Take as the incidence matrix $\boldsymbol{N}$ that denoted in Example 3.1.2 by $\boldsymbol{N}^*$, and apply to it Remark 3.2.4(a). For this case, first note that, as vectors orthogonal to the vector $\boldsymbol{k} = [3, 2, 3, 2]'$, it is sufficient to choose the vectors $\boldsymbol{t}_{01} = [-1, 0, 1, 0]'$, $\boldsymbol{t}_{02} = [0, -1, 0, -1]'$ and $\boldsymbol{t}_{03} = [2, -3, 2, -3]'$. Next find that for the incidence matrix taken here, one obtains

$$\boldsymbol{N}_0\boldsymbol{k}^\delta = \frac{1}{5}\begin{bmatrix} -3 & 2 & -3 & 2 \\ 6 & -4 & 6 & -4 \\ -3 & 2 & -3 & 2 \end{bmatrix}.$$

Now, note that the vectors $\boldsymbol{t}_{01}$ and $\boldsymbol{t}_{02}$ satisfy simultaneously the equalities $\boldsymbol{N}\boldsymbol{t}_{0j} = \mathbf{0}$ and $\boldsymbol{N}_0\boldsymbol{k}^\delta\boldsymbol{t}_{0j} = \mathbf{0}$, $j = 1, 2$, whereas $\boldsymbol{N}\boldsymbol{t}_{03} = [-2, 4, -2]'$. Thus, the condition of Remark 3.2.4(a) is satisfied, and so is the condition of Corollary 3.2.2.

The reader is also advised to apply Remark 3.2.4(a) to Examples 3.2.1 and 3.2.2.

Example 3.2.5. Consider a design (from Pearce, 1983, p. 225) for a 2^3 factorial structure of treatments, with the incidence matrix

$$\boldsymbol{N} = \begin{bmatrix} 0 & 1 & 1 & 1 & 0 & 0 & 0 & 0 \\ 0 & 0 & 0 & 0 & 2 & 1 & 1 & 1 \\ 0 & 0 & 0 & 0 & 1 & 2 & 1 & 1 \\ 0 & 0 & 0 & 0 & 1 & 1 & 2 & 1 \\ 1 & 1 & 1 & 0 & 0 & 0 & 0 & 0 \\ 1 & 1 & 0 & 1 & 0 & 0 & 0 & 0 \\ 1 & 0 & 1 & 1 & 0 & 0 & 0 & 0 \\ 0 & 0 & 0 & 0 & 1 & 1 & 1 & 2 \end{bmatrix}.$$

It is a nonbinary, nonproper, nonequireplicate disconnected block design. It can be seen that the rank of $\boldsymbol{N}$ is equal to the number of blocks (and of treatments, i.e., $\boldsymbol{N}$ is nonsingular), and so the condition of Corollary 3.2.2 is satisfied on account of Remark 3.2.4(b). The reader may note that the equality (3.2.22) holds here also on account of Corollary 3.2.1(d). Why?

Corollary 3.2.3. *Suppose that the incidence matrix $\boldsymbol{N}$, after an appropriate ordering of its columns, can be written as $\boldsymbol{N} = [\boldsymbol{N}_1 : \boldsymbol{N}_2 : \cdots : \boldsymbol{N}_a]$, where the $v \times b_h$ submatrix $\boldsymbol{N}_h$ has the properties $\boldsymbol{k}_h = \boldsymbol{N}'_h\boldsymbol{1}_v = k_h\boldsymbol{1}_{b_h}$ and $\boldsymbol{r}_h = \boldsymbol{N}_h\boldsymbol{1}_{b_h} = (n_h/n)\boldsymbol{r}$, with $n_h = \boldsymbol{1}'_v\boldsymbol{r}_h$, $\boldsymbol{r} = \boldsymbol{r}_1 + \boldsymbol{r}_2 + \cdots + \boldsymbol{r}_a$ and $n = \boldsymbol{1}'_v\boldsymbol{r}$, for $h = 1, 2, ..., a$. Then, the condition of Corollary 3.2.2 is satisfied, whether or not the $k_h, h = 1, 2, ..., a$, are equal.*

Proof. If the matrix $\boldsymbol{N}$ has the above form and properties, then the matrix $\boldsymbol{N}_0$ can be written as $\boldsymbol{N}_0 = [\boldsymbol{N}_{01} : \boldsymbol{N}_{02} : \cdots : \boldsymbol{N}_{0a}]$, where $\boldsymbol{N}_{0h} = \boldsymbol{N}_h - n_h^{-1}\boldsymbol{r}_h\boldsymbol{k}'_h$, because the equality $n^{-1}\boldsymbol{r}\boldsymbol{k}'_h = n_h^{-1}\boldsymbol{r}_h\boldsymbol{k}'_h$ holds for any h, and, with $\boldsymbol{k} = [k_1\boldsymbol{1}'_{b_1}, k_2\boldsymbol{1}'_{b_2}, ..., k_a\boldsymbol{1}'_{b_a}]'$, the matrix $\boldsymbol{N}_0\boldsymbol{K}_0$ can be written as $\boldsymbol{N}_0\boldsymbol{K}_0 = [k_1\boldsymbol{N}_{01} : k_2\boldsymbol{N}_{02} : \cdots : k_a\boldsymbol{N}_{0a}]$, because the equality $\boldsymbol{N}_{0h}\boldsymbol{k}_h = k_h\boldsymbol{N}_{0h}\boldsymbol{1}_{b_h} = \boldsymbol{0}$ holds for any h. Now, with $b = b_1 + b_2 + \cdots + b_a$, consider as a basis of the vector space $\mathcal{R}^b$ (see Appendix A.1) the following set of $\boldsymbol{k}^\delta$-orthogonal nonzero vectors:

$$
\begin{array}{lll}
a \text{ vectors of the type } \boldsymbol{t}_h = [t_{h1}\boldsymbol{1}'_{b_1}, t_{h2}\boldsymbol{1}'_{b_2}, ..., t_{ha}\boldsymbol{1}'_{b_a}]', & h = 1, 2, ..., a, \\
b_1 - 1 \text{ vectors of the type } \boldsymbol{t}_j^{(1)} = [\boldsymbol{t}'_{j1}, \boldsymbol{0}', ..., \boldsymbol{0}']', & j = 1, 2, ..., b_1 - 1, \\
b_2 - 1 \text{ vectors of the type } \boldsymbol{t}_j^{(2)} = [\boldsymbol{0}', \boldsymbol{t}'_{j2}, ..., \boldsymbol{0}']', & j = 1, 2, ..., b_2 - 1, \\
\quad\cdots \qquad\qquad \cdots & \cdots \\
b_a - 1 \text{ vectors of the type } \boldsymbol{t}_j^{(a)} = [\boldsymbol{0}', \boldsymbol{0}', ..., \boldsymbol{t}'_{ja}]', & j = 1, 2, ..., b_a - 1,
\end{array}
$$

where $n_1t_{h1}t_{h'1} + n_2t_{h2}t_{h'2} + \cdots + n_at_{ha}t_{h'a} = 0$ for any $h \neq h'$, $\boldsymbol{t}'_{jh}\boldsymbol{t}_{j'h} = 0$ for any h and $j \neq j'$ and $\boldsymbol{1}'_{b_h}\boldsymbol{t}_{jh} = 0$ for any h and j. It can then be seen that with any vector $\boldsymbol{t}_h$ ($h = 1, 2, ..., a$), the equalities $\boldsymbol{N}_0\boldsymbol{t}_h = \boldsymbol{0}$ and $\boldsymbol{N}_0\boldsymbol{K}_0\boldsymbol{t}_h = \boldsymbol{0}$ are both satisfied, as $\boldsymbol{N}_{0h}\boldsymbol{1}_{b_h} = \boldsymbol{0}$ for any h. Taking the remaining vectors of the basis, note that if the equality $\boldsymbol{N}_0\boldsymbol{t}_j^{(h)} = \boldsymbol{0}$ holds, the equality $\boldsymbol{N}_0\boldsymbol{K}_0\boldsymbol{t}_j^{(h)} = \boldsymbol{0}$ holds as well, for any h and j. Thus, the condition of Corollary 3.2.2 is satisfied, independently of the block sizes, k_h's. □

Corollary 3.2.4. *Suppose that the incidence matrix $\boldsymbol{N}$, after an appropriate ordering of its rows and columns, can be written as*

$$\boldsymbol{N} = \begin{bmatrix} \boldsymbol{1}_{v_0}\boldsymbol{1}'_{b_1} & \boldsymbol{O} \\ \boldsymbol{N}_1 & \boldsymbol{N}_2 \end{bmatrix},$$

where the $v \times b_h$ submatrix $\boldsymbol{N}_h$ has the properties $\boldsymbol{k}_h = \boldsymbol{N}'_h\boldsymbol{1}_v = k_h\boldsymbol{1}_{b_h}$ and $\boldsymbol{r}_h = \boldsymbol{N}_h\boldsymbol{1}_{b_h} = (n_h/n_*)\boldsymbol{r}_*$, with $n_h = \boldsymbol{1}'_v\boldsymbol{r}_h$, $\boldsymbol{r}_* = \boldsymbol{r}_1 + \boldsymbol{r}_2$ and $n_* = \boldsymbol{1}'_v\boldsymbol{r}_*$, for $h = 1, 2$. Then, the condition of Corollary 3.2.2 is satisfied, whether or not the $k_h, h = 1, 2$, are equal.

Proof. If the matrix $\boldsymbol{N}$ has the above form and properties, then

$$\boldsymbol{N}_0 = \begin{bmatrix} n^{-1}k_2b_2\boldsymbol{1}_{v_0}\boldsymbol{1}'_{b_1} & -n^{-1}k_2b_1\boldsymbol{1}_{v_0}\boldsymbol{1}'_{b_2} \\ \boldsymbol{N}_1 - n^{-1}(k_1 + v_0)\boldsymbol{r}_*\boldsymbol{1}'_{b_1} & \boldsymbol{N}_2 - n^{-1}k_2\boldsymbol{r}_*\boldsymbol{1}'_{b_2} \end{bmatrix}$$

and

$$\boldsymbol{N}_0\boldsymbol{k}^\delta = \begin{bmatrix} n^{-1}(k_1+v_0)k_2b_2\mathbf{1}_{v_0}\mathbf{1}'_{b_1} & -n^{-1}k_2^2b_1\mathbf{1}_{v_0}\mathbf{1}'_{b_2} \\ (k_1+v_0)\{\boldsymbol{N}_1 - n^{-1}(k_1+v_0)\boldsymbol{r}_*\mathbf{1}'_{b_1}\} & k_2(\boldsymbol{N}_2 - n^{-1}k_2\boldsymbol{r}_*\mathbf{1}'_{b_2}) \end{bmatrix},$$

where $n = n_* + v_0b_1$ and $\boldsymbol{k} = [(k_1+v_0)\mathbf{1}'_{b_1}, k_2\mathbf{1}'_{b_2}]'$. Now, let $b = b_1 + b_2$, and consider as a basis of the vector space $\mathcal{R}^b$, the following set of $\boldsymbol{k}^\delta$-orthogonal nonzero vectors:

the two vectors $\mathbf{1}_b$ and $\boldsymbol{t}_0 = [t_1\mathbf{1}'_{b_1}, t_2\mathbf{1}'_{b_2}]'$,
$b_1 - 1$ vectors of the type $\boldsymbol{t}_j^{(1)} = [\boldsymbol{t}'_{j1}, \mathbf{0}']'$, $\quad j = 1, 2, ..., b_1 - 1$,
$b_2 - 1$ vectors of the type $\boldsymbol{t}_j^{(2)} = [\mathbf{0}', \boldsymbol{t}'_{j2}]'$, $\quad j = 1, 2, ..., b_2 - 1$,

where $t_1(k_1+v_0)b_1 + t_2k_2b_2 = 0$, $\boldsymbol{t}'_{jh}\boldsymbol{t}_{j'h} = 0$ for any h and $j \neq j'$, and $\mathbf{1}'_{b_h}\boldsymbol{t}_{jh} = 0$ for any h and j. It can then be seen that

$$\boldsymbol{N}_0\boldsymbol{t}_0 = \boldsymbol{N}\boldsymbol{t}_0 = \begin{bmatrix} t_1b_1\mathbf{1}_{v_0} \\ t_1\boldsymbol{r}_1 + t_2\boldsymbol{r}_2 \end{bmatrix} \neq \mathbf{0}$$

for any $\boldsymbol{t}_0 \neq \mathbf{0}$. For the remaining vectors note that

$$\boldsymbol{N}_0\boldsymbol{t}_j^{(1)} = \boldsymbol{N}\boldsymbol{t}_j^{(1)} = \begin{bmatrix} \mathbf{0} \\ \boldsymbol{N}_1\boldsymbol{t}_{j1} \end{bmatrix}$$

and

$$\boldsymbol{N}_0\boldsymbol{K}_0\boldsymbol{t}_j^{(1)} = \boldsymbol{N}_0\boldsymbol{k}^\delta\boldsymbol{t}_j^{(1)} = \begin{bmatrix} \mathbf{0} \\ (k_1+v_0)\boldsymbol{N}_1\boldsymbol{t}_{j1} \end{bmatrix},$$

and that

$$\boldsymbol{N}_0\boldsymbol{t}_j^{(2)} = \boldsymbol{N}\boldsymbol{t}_j^{(2)} = \begin{bmatrix} \mathbf{0} \\ \boldsymbol{N}_2\boldsymbol{t}_{j2} \end{bmatrix} \text{ and } \boldsymbol{N}_0\boldsymbol{K}_0\boldsymbol{t}_j^{(2)} = \boldsymbol{N}_0\boldsymbol{k}^\delta\boldsymbol{t}_j^{(2)} = \begin{bmatrix} \mathbf{0} \\ k_2\boldsymbol{N}_2\boldsymbol{t}_{j2} \end{bmatrix},$$

which shows that if the equality $\boldsymbol{N}_0\boldsymbol{t}_j^{(h)} = \mathbf{0}$ holds, the equality $\boldsymbol{N}_0\boldsymbol{K}_0\boldsymbol{t}_j^{(h)} = \mathbf{0}$ holds as well, for any h and j. Thus, the condition of Corollary 3.2.2 is satisfied, independently of the block sizes, k_h's. □

Examples of the application of Corollaries 3.2.3 and 3.2.4 will be seen in Volume II of the book.

Remark 3.2.5. If the equation (3.2.16) holds for any $\boldsymbol{s}$, i.e., if (3.2.22) holds, then

$$\text{Cov}(\boldsymbol{y}_2)\boldsymbol{\phi}_2\boldsymbol{\Delta}' = \boldsymbol{\phi}_2\boldsymbol{\Delta}'[(\boldsymbol{N}'_0\boldsymbol{k}^{-\delta}\boldsymbol{N}'_0)^-\boldsymbol{N}_0\boldsymbol{N}'_0(\sigma_B^2 - K_H^{-1}\sigma_U^2) + \boldsymbol{I}_v(\sigma_U^2 + \sigma_e^2)],$$

which implies that both $\boldsymbol{\phi}_2\boldsymbol{\Delta}'\boldsymbol{s}$ and $\text{Cov}(\boldsymbol{y}_2)\boldsymbol{\phi}_2\boldsymbol{\Delta}'\boldsymbol{s}$ belong to $\mathcal{C}(\boldsymbol{\phi}_2\boldsymbol{\Delta}')$ for any $\boldsymbol{s}$, and thus, the conditions stated in Theorem 1.2.2, when applied to Theorem

3.2.2 above, are satisfied. This implication means that the BLUEs obtainable under the submodel (3.2.4), with the moments (3.2.14) and (3.2.15), and with the condition (3.2.22) satisfied, can equivalently be obtained under a simple alternative model, in which, instead of (3.2.15), the matrix $(\sigma_U^2 + \sigma_e^2)\boldsymbol{I}_n$ is used as the covariance matrix of $\boldsymbol{y}_2$. Moreover, it can be shown (applying, e.g., Theorem 2.3.2 of Rao and Mitra, 1971) that the equality (3.2.22) is not only a sufficient, but also a necessary condition for the SLSEs and the BLUEs to be the same. For this reason, it is helpful to note that (3.2.22) can equivalently be written as

$$\boldsymbol{\phi}_2\boldsymbol{D}'\boldsymbol{D}\boldsymbol{\phi}_2\boldsymbol{\Delta}' = \boldsymbol{\phi}_2\boldsymbol{\Delta}'\boldsymbol{C}_2^-\boldsymbol{\Delta}\boldsymbol{\phi}_2\boldsymbol{D}'\boldsymbol{D}\boldsymbol{\phi}_2\boldsymbol{\Delta}',$$

where $\boldsymbol{C}_2^-\boldsymbol{\Delta}$ can be regarded as a g-inverse of $\boldsymbol{\phi}_2\boldsymbol{\Delta}'$.

Remark 3.2.5 implies, in particular, that if (3.2.22) holds, then the BLUE of the expectation vector (3.2.14) is obtainable by a simple least-squares procedure, i.e., has the form $\widehat{\mathrm{E}(\boldsymbol{y}_2)} = \boldsymbol{P}_{\phi_2\Delta'}\boldsymbol{y}_2$, where $\boldsymbol{P}_{\phi_2\Delta'} = \boldsymbol{\phi}_2\boldsymbol{\Delta}'\boldsymbol{C}_2^-\boldsymbol{\Delta}\boldsymbol{\phi}_2$. This result can be checked by noting that

$$\mathrm{E}(\boldsymbol{P}_{\phi_2\Delta'}\boldsymbol{y}_2) = \boldsymbol{P}_{\phi_2\Delta'}\boldsymbol{\phi}_2\boldsymbol{\Delta}'\boldsymbol{\tau} = \boldsymbol{\phi}_2\boldsymbol{\Delta}'\boldsymbol{\tau} = \mathrm{E}(\boldsymbol{y}_2)$$

and that

$$(\boldsymbol{I}_n - \boldsymbol{P}_{\phi_2\Delta'})\mathrm{Cov}(\boldsymbol{y}_2)\boldsymbol{P}_{\phi_2\Delta'} = \mathbf{O}, \tag{3.2.23}$$

because under (3.2.22), which equivalently can be written as a relation

$$\boldsymbol{\phi}_2\boldsymbol{D}'\boldsymbol{D}\boldsymbol{\phi}_2\boldsymbol{\Delta}' = \boldsymbol{P}_{\phi_2\Delta'}\boldsymbol{D}'\boldsymbol{D}\boldsymbol{\phi}_2\boldsymbol{\Delta}',$$

the equality

$$\mathrm{Cov}(\boldsymbol{y}_2)\boldsymbol{P}_{\phi_2\Delta'} = \boldsymbol{P}_{\phi_2\Delta'}[\boldsymbol{D}'\boldsymbol{D}(\sigma_B^2 - K_H^{-1}\sigma_U^2) + \boldsymbol{I}_n(\sigma_U^2 + \sigma_e^2)]\boldsymbol{P}_{\phi_2\Delta'}$$

holds. Thus, the vector $\boldsymbol{y}_2$ can then be decomposed as

$$\begin{aligned} \boldsymbol{y}_2 &= \boldsymbol{P}_{\phi_2\Delta'}\boldsymbol{y}_2 + (\boldsymbol{I}_n - \boldsymbol{P}_{\phi_2\Delta'})\boldsymbol{y}_2 \quad \text{(in terms of } \boldsymbol{y}_2\text{)} \\ &= \boldsymbol{P}_{\phi_2\Delta'}\boldsymbol{y} + (\boldsymbol{\phi}_2 - \boldsymbol{P}_{\phi_2\Delta'})\boldsymbol{y} \quad \text{(in terms of } \boldsymbol{y}\text{)}, \end{aligned}$$

with $(\boldsymbol{I}_n - \boldsymbol{P}_{\phi_2\Delta'})\boldsymbol{y}_2 = (\boldsymbol{\phi}_2 - \boldsymbol{P}_{\phi_2\Delta'})\boldsymbol{y}$ as the residual vector.

The above decomposition yields the inter-block analysis of variance, of the form

$$\|\boldsymbol{y}_2\|^2 = \|\boldsymbol{P}_{\phi_2\Delta'}\boldsymbol{y}_2\|^2 + \|(\boldsymbol{I}_n - \boldsymbol{P}_{\phi_2\Delta'})\boldsymbol{y}_2\|^2,$$

expressible with the use of the observed vector $\boldsymbol{y}$ as

$$\begin{aligned} \boldsymbol{y}'\boldsymbol{\phi}_2\boldsymbol{y} &= \boldsymbol{y}'\boldsymbol{\phi}_2\boldsymbol{\Delta}'\boldsymbol{C}_2^-\boldsymbol{\Delta}\boldsymbol{\phi}_2\boldsymbol{y} + \boldsymbol{y}'(\boldsymbol{\phi}_2 - \boldsymbol{\phi}_2\boldsymbol{\Delta}'\boldsymbol{C}_2^-\boldsymbol{\Delta}\boldsymbol{\phi}_2)\boldsymbol{y} \\ &= \boldsymbol{Q}_2'\boldsymbol{C}_2{}^-\boldsymbol{Q}_2 + \boldsymbol{y}'\boldsymbol{\psi}_2\boldsymbol{y}, \end{aligned}$$

where $\boldsymbol{Q}_2 = \boldsymbol{\Delta}\boldsymbol{\phi}_2\boldsymbol{y}$, $\boldsymbol{C}_2 = \boldsymbol{\Delta}\boldsymbol{\phi}_2\boldsymbol{\Delta}'$ and $\boldsymbol{\psi}_2 = \boldsymbol{\phi}_2 - \boldsymbol{\phi}_2\boldsymbol{\Delta}'\boldsymbol{C}_2^-\boldsymbol{\Delta}\boldsymbol{\phi}_2 = \boldsymbol{\phi}_2(\boldsymbol{I}_n - \boldsymbol{\Delta}'\boldsymbol{C}_2^-\boldsymbol{\Delta})\boldsymbol{\phi}_2$, the matrix $\boldsymbol{\psi}_2$ having the readily seen properties

$$\boldsymbol{\psi}_2\boldsymbol{\phi}_2 = \boldsymbol{\psi}_2,\ \boldsymbol{\psi}_2\mathbf{1}_n = \mathbf{0},\ \boldsymbol{\psi}_2\boldsymbol{\Delta}' = \boldsymbol{O},\ \boldsymbol{\psi}_2' = \boldsymbol{\psi}_2, \text{ and } \boldsymbol{\psi}_2\boldsymbol{\psi}_2 = \boldsymbol{\psi}_2.$$

The quadratic form $y'\phi_2 y$ can be called the inter-block total sum of squares, whereas its components, $Q_2'C_2^-Q_2$ and $y'\psi_2 y = y'\phi_2(I_n - \Delta'C_2^-\Delta)\phi_2 y$, can be called the inter-block treatment sum of squares and the inter-block residual sum of squares, respectively. The corresponding d.f. are evidently $b - 1 = \text{rank}(\phi_2)$ for the total, $v - \rho - 1 = \text{rank}(C_2)$ (from Lemmas 2.3.5 and 2.3.3) for the treatment component and $b - v + \rho = \text{rank}(\psi_2) = \text{rank}(\phi_2) - \text{rank}(C_2)$ for the residual component. The expectations of these component sums of squares are

$$\text{E}(Q_2'C_2^-Q_2) = \text{tr}(N_0'C_2^-N_0)(\sigma_B^2 - K_H^{-1}\sigma_U^2) + (v - 1 - \rho)(\sigma_U^2 + \sigma_e^2) + \tau'C_2\tau$$

and

$$\text{E}(y'\psi_2 y) = [\text{tr}(K_0) - \text{tr}(N_0'C_2^-N_0)](\sigma_B^2 - K_H^{-1}\sigma_U^2) + (b - v + \rho)(\sigma_U^2 + \sigma_e^2).$$

It follows from the latter equality that the inter-block residual mean square $s_2^2 = y'\psi_2 y/(b - v + \rho)$ is an unbiased estimator of

$$\begin{aligned}\sigma_2^2 &= (b - v + \rho)^{-1}[\text{tr}(K_0) - \text{tr}(N_0'C_2^-N_0)](\sigma_B^2 - K_H^{-1}\sigma_U^2) \\ &\quad + \sigma_U^2 + \sigma_e^2.\end{aligned} \tag{3.2.24}$$

In the case of $k_1 = k_2 = \cdots = k_b = k$, the mean square s_2^2 is in fact the MINQUE of σ_2^2, which then reduces from (3.2.24) to

$$\sigma_2^2 = k\sigma_B^2 + (1 - K_H^{-1}k)\sigma_U^2 + \sigma_e^2,$$

further reducing to $k\sigma_B^2 + \sigma_e^2$ if $k = K_H$. It should be noted, however, that $b - v + \rho = 0$ if $b = v - \rho$ (that $b \geq v - \rho$ always, follows from Lemma 2.3.3). In that case, no estimator for σ_2^2 exists in the inter-block analysis.

Thus, in the case of equal k_j's and $b > v - \rho$, the mean square s_2^2 can be used to obtain an unbiased estimator of the variance (3.2.21), in the form

$$\widehat{\text{Var}(c'\tau)} = k^{-1}s'N_0N_0's s_2^2 = kc'(N_0N_0')^-c s_2^2.$$

In general, the estimation of (3.2.20) is not so simple.

Furthermore, if $k_1 = k_2 = \cdots = k_b = k$, then $\text{Cov}(y_2) = \phi_2\sigma_2^2$ and, as for the intra-block analysis, it can be shown that under the multivariate normality assumption the quadratic functions $Q_2'C_2^-Q_2/\sigma_2^2$ and $y'\psi_2 y/\sigma_2^2$ have independent χ^2 distributions, the first noncentral with $v-1-\rho$ d.f. and the noncentrality parameter $\delta_2 = \tau'C_2\tau/\sigma_2^2$, the second central with $b - v + \rho$ d.f. Hence, the hypothesis $\tau'C_2\tau = 0$, equivalent to $\text{E}(y_2) = \mathbf{0}$ [or $P_{D'}\text{E}(y) \in \mathcal{C}(1_n)$], can be tested by the variance ratio criterion

$$(v - 1 - \rho)^{-1}Q_2'C_2^-Q_2/s_2^2,$$

which under the assumed normality has then the F distribution with $v - 1 - \rho$ and $b - v + \rho$ d.f., central when the hypothesis is true. This result, however, does not apply to the general case, when k_j's are not all equal.

Example 3.2.6. Returning to Example 3.2.1, it may be interesting to see for which contrasts the BLUEs under the inter-block submodel exist there. This question requires the condition (3.2.16) to be checked. In fact, it is sufficient to note that because the incidence matrix of that design can be written as $\boldsymbol{N} = [\boldsymbol{N}_1 : \boldsymbol{N}_2]$, where the submatrices $\boldsymbol{N}_1$ and $\boldsymbol{N}_2$ are incidence matrices of proper and equireplicate subdesigns, Corollary 3.2.3 applies. It indicates that the condition of Corollary 3.2.2 is satisfied, i.e., the condition (3.2.16) holds for any $\boldsymbol{s}$. Hence, by Theorem 3.2.2, a contrast $\boldsymbol{c}'\boldsymbol{\tau} = \boldsymbol{s}'\boldsymbol{C}_2\boldsymbol{\tau}$, for any $\boldsymbol{s}$, has under the inter-block submodel the BLUE. Because of the equality $\boldsymbol{C}_2 = -\boldsymbol{C}_1 + \boldsymbol{r}^{\delta} - n^{-1}\boldsymbol{r}\boldsymbol{r}'$, one obtains for this example

$$\boldsymbol{C}_2 = (1-\varepsilon)[6\boldsymbol{I}_4 - (3/2)\mathbf{1}_4\mathbf{1}_4'] \quad \text{with} \quad \varepsilon = 7/9.$$

Note that as a g-inverse of it, the matrix $(1-\varepsilon)^{-1}[6^{-1}\boldsymbol{I}_4 - (1/24)\mathbf{1}_4\mathbf{1}_4']$ can be taken. With it, the BLUE of $\boldsymbol{c}'\boldsymbol{\tau}$ is obtainable under this submodel as

$$\widehat{\boldsymbol{c}'\boldsymbol{\tau}} = [6(1-\varepsilon)]^{-1}\boldsymbol{c}'\boldsymbol{Q}_2 = (3/4)\boldsymbol{c}'\boldsymbol{Q}_2,$$

where $\boldsymbol{Q}_2 = \boldsymbol{\Delta}\boldsymbol{\phi}_2\boldsymbol{y} = [3.8833,\ 1.7333,\ -0.7333,\ -4.8833]'$. To obtain its variance, one can use the formula (3.2.20), which gives here (as can easily be checked)

$$\operatorname{Var}(\widehat{\boldsymbol{c}'\boldsymbol{\tau}}) = (3/4)\boldsymbol{c}'\boldsymbol{c}\sigma^2_{(2)},$$

where

$$\sigma^2_{(2)} = (9/4)(\sigma^2_B - K_H^{-1}\sigma^2_U) + \sigma^2_U + \sigma^2_e.$$

[The symbol $\sigma^2_{(2)}$ is used here to distinguish from σ^2_2 defined in (3.2.24), but note that $\sigma^2_{(2)}$ and σ^2_2 coincide for a proper design.] Now, the inter-block analysis of variance can be obtained as

Source	Degrees of freedom	Sum of squares	Mean square
Treatments	$v-1=3$	$\boldsymbol{Q}_2'\boldsymbol{C}_2^-\boldsymbol{Q}_2 = 31.85208$	10.6174
Residuals	$b-v=6$	$\boldsymbol{y}'\boldsymbol{\psi}_2\boldsymbol{y} = 15.24125$	$s_2^2 = 2.540208$
Total	$b-1=9$	$\boldsymbol{y}'\boldsymbol{\phi}_2\boldsymbol{y} = 47.09333$	

As the design is not proper (i.e., the block sizes are not equal), the relevant variance ratio criterion $10.6174/s_2^2 = 4.180$ cannot be referred to the exact F distribution with 3 and 6 d.f. It gives only a rough indication of the significance of treatment differences. The mean square s_2^2 gives an unbiased estimate of σ_2^2 defined in (3.2.24), which here is $\sigma_2^2 = (14.75/6)(\sigma^2_B - K_H^{-1}\sigma^2_U) + \sigma^2_U + \sigma^2_e$. ¿From it and from the unbiased estimate of $\sigma_1^2 = \sigma_U^2 + \sigma_e^2$, i.e., $\hat{\sigma}_1^2 = s_1^2 = 0.872825$, an unbiased estimate of $\sigma_B^2 - K_H^{-1}\sigma_U^2$ can be obtained as $(6/14.75)(s_2^2 - s_1^2) = 0.678257$. With it, an unbiased estimate of $\sigma^2_{(2)}$ is obtainable as

$$\hat{\sigma}^2_{(2)} = (9/4)0.678257 + 0.872825 = 2.39890.$$

It provides an unbiased estimate of the above $\operatorname{Var}(\widehat{\boldsymbol{c}'\boldsymbol{\tau}})$ as

$$\widehat{\operatorname{Var}}(\widehat{\boldsymbol{c}'\boldsymbol{\tau}}) = (3/4)\boldsymbol{c}'\boldsymbol{c}\hat{\sigma}^2_{(2)} = 1.79918\boldsymbol{c}'\boldsymbol{c}.$$

3.2.3 Total-area submodel

Considering the third submodel, given in (3.2.6), note that its properties are

$$\mathrm{E}(\boldsymbol{y}_3) = \boldsymbol{\phi}_3\boldsymbol{\Delta}'\boldsymbol{\tau} = n^{-1}\mathbf{1}_n\boldsymbol{r}'\boldsymbol{\tau} \tag{3.2.25}$$

and

$$\mathrm{Cov}(\boldsymbol{y}_3) = \boldsymbol{\phi}_3[(n^{-1}\boldsymbol{k}'\boldsymbol{k} - N_B^{-1}n)\sigma_B^2 + (1 - K_H^{-1}n^{-1}\boldsymbol{k}'\boldsymbol{k})\sigma_U^2 + \sigma_e^2]. \tag{3.2.26}$$

They lead to the following main result concerning estimation under (3.2.6).

Theorem 3.2.3. *Under* (3.2.6), *a function* $\boldsymbol{w}'\boldsymbol{y}_3 = \boldsymbol{w}'\boldsymbol{\phi}_3\boldsymbol{y}$ *is uniformly the BLUE of* $\boldsymbol{c}'\boldsymbol{\tau}$ *if and only if* $\boldsymbol{\phi}_3\boldsymbol{w} = \boldsymbol{\phi}_3\boldsymbol{\Delta}'\boldsymbol{s}$, *where the vectors* $\mathbf{c}$ *and* $\boldsymbol{s}$ *are in the relation* $\mathbf{c} = \boldsymbol{\Delta}\boldsymbol{\phi}_3\boldsymbol{\Delta}'\boldsymbol{s}$ $(= n^{-1}\boldsymbol{r}\boldsymbol{r}'\boldsymbol{s}$, *as* $\boldsymbol{C}_3 = \boldsymbol{\Delta}\boldsymbol{\phi}_3\boldsymbol{\Delta}' = n^{-1}\boldsymbol{r}\boldsymbol{r}')$.

Proof. It follows exactly the same pattern as the proof of Theorem 3.2.1. □

Remark 3.2.6. (a) The only parametric functions for which the BLUEs under (3.2.6) exist are those defined as $\boldsymbol{c}'\boldsymbol{\tau} = (\boldsymbol{s}'\boldsymbol{r})n^{-1}\boldsymbol{r}'\boldsymbol{\tau}$, where $\boldsymbol{s}'\boldsymbol{r} \neq 0$, i.e., the general parametric mean and its multiplicities, contrasts being excluded *a fortiori* (as it follows that $\mathbf{1}_v'\boldsymbol{c} = \boldsymbol{r}'\boldsymbol{s}$ here).

(b) Because of the equality $\mathrm{Cov}(\boldsymbol{y}_3)\boldsymbol{\phi}_3\boldsymbol{\Delta}' = \boldsymbol{\phi}_3\boldsymbol{\Delta}'[(n^{-1}\boldsymbol{k}'\boldsymbol{k} - N_B^{-1}n)\sigma_B^2 + (1-K_H^{-1}n^{-1}\boldsymbol{k}'\boldsymbol{k})\sigma_U^2 + \sigma_e^2]$, the BLUEs under (3.2.6) and the SLSEs are the same (on account of Theorem 1.2.2 applied to Theorem 3.2.3).

If $\boldsymbol{c}'\boldsymbol{\tau} = (\boldsymbol{s}'\boldsymbol{r})n^{-1}\boldsymbol{r}'\boldsymbol{\tau} = (\boldsymbol{c}'\mathbf{1}_v)n^{-1}\boldsymbol{r}'\boldsymbol{\tau}$, then the variance of its BLUE under (3.2.6), i.e., of $\widehat{\boldsymbol{c}'\boldsymbol{\tau}} = \boldsymbol{s}'\boldsymbol{\Delta}\boldsymbol{y}_3 = (\boldsymbol{s}'\boldsymbol{r})n^{-1}\mathbf{1}_n'\boldsymbol{y} = (\boldsymbol{c}'\mathbf{1}_v)n^{-1}\mathbf{1}_n'\boldsymbol{y}$, is of the form

$$\begin{aligned}\mathrm{Var}(\widehat{\boldsymbol{c}'\boldsymbol{\tau}}) &= n^{-1}(\boldsymbol{s}'\boldsymbol{r})^2[(n^{-1}\boldsymbol{k}'\boldsymbol{k} - N_B^{-1}n)\sigma_B^2 + (1 - K_H^{-1}n^{-1}\boldsymbol{k}'\boldsymbol{k})\sigma_U^2 + \sigma_e^2] \\ &= n^{-1}(\boldsymbol{c}'\mathbf{1}_v)^2[(n^{-1}\boldsymbol{k}'\boldsymbol{k} - N_B^{-1}n)\sigma_B^2 \\ &\qquad + (1 - K_H^{-1}n^{-1}\boldsymbol{k}'\boldsymbol{k})\sigma_U^2 + \sigma_e^2]. \end{aligned} \tag{3.2.27}$$

Evidently, if all k_j are equal $(= k)$, the variance (3.2.27) reduces to

$$\mathrm{Var}(\widehat{\boldsymbol{c}'\boldsymbol{\tau}}) = n^{-1}(\boldsymbol{c}'\mathbf{1}_v)^2[(1 - N_B^{-1}b)k\sigma_B^2 + (1 - K_H^{-1}k)\sigma_U^2 + \sigma_e^2],$$

and if, in addition, $b = N_B$ and $K_H = k$, which may be considered as the usual case, then $\mathrm{Var}(\widehat{\boldsymbol{c}'\boldsymbol{\tau}}) = n^{-1}(\boldsymbol{c}'\mathbf{1}_v)^2\sigma_e^2$.

Finally, it may be noticed that because $\boldsymbol{P}_{\phi_3\Delta'} = \boldsymbol{\phi}_3$ (as $n^{-1}\mathbf{1}_v\mathbf{1}_v'$ is a g-inverse of $\boldsymbol{C}_3 = n^{-1}\boldsymbol{r}\boldsymbol{r}'$) and, hence, both $(\boldsymbol{I}_n - \boldsymbol{P}_{\phi_3\Delta'})\boldsymbol{y}_3 = \mathbf{0}$ and $(\boldsymbol{I}_n - \boldsymbol{P}_{\phi_3\Delta'})\mathrm{Cov}(\boldsymbol{y}_3) = \mathbf{O}$ hold simultaneously, the vector $\boldsymbol{P}_{\phi_3\Delta'}\boldsymbol{y}_3 = \boldsymbol{y}_3 = n^{-1}\mathbf{1}_n\mathbf{1}_n'\boldsymbol{y}$ is itself the BLUE of its expectation, $n^{-1}\mathbf{1}_n\boldsymbol{r}'\boldsymbol{\tau}$, leaving no residuals.

3.2.4 Some special cases

It follows from the considerations above that any function $\boldsymbol{s}'\boldsymbol{\Delta}\boldsymbol{y}$ can be resolved into three components in the form

$$\boldsymbol{s}'\boldsymbol{\Delta}\boldsymbol{y} = \boldsymbol{s}'\boldsymbol{\Delta}\boldsymbol{y}_1 + \boldsymbol{s}'\boldsymbol{\Delta}\boldsymbol{y}_2 + \boldsymbol{s}'\boldsymbol{\Delta}\boldsymbol{y}_3 = \boldsymbol{s}'\boldsymbol{\Delta}\boldsymbol{\phi}_1\boldsymbol{y} + \boldsymbol{s}'\boldsymbol{\Delta}\boldsymbol{\phi}_2\boldsymbol{y} + \boldsymbol{s}'\boldsymbol{\Delta}\boldsymbol{\phi}_3\boldsymbol{y},$$

which conveniently can be written as

$$\boldsymbol{s}'\boldsymbol{\Delta y} = \boldsymbol{s}'\boldsymbol{Q}_1 + \boldsymbol{s}'\boldsymbol{Q}_2 + \boldsymbol{s}'\boldsymbol{Q}_3, \tag{3.2.28}$$

with

$$\boldsymbol{Q}_1 = \boldsymbol{\Delta y}_1 = \boldsymbol{\Delta \phi}_1\boldsymbol{y},\ \boldsymbol{Q}_2 = \boldsymbol{\Delta y}_2 = \boldsymbol{\Delta \phi}_2\boldsymbol{y} \quad \text{and} \quad \boldsymbol{Q}_3 = \boldsymbol{\Delta y}_3 = \boldsymbol{\Delta \phi}_3\boldsymbol{y}.$$

Each of the components in (3.2.28) represents a contribution to $\boldsymbol{s}'\boldsymbol{\Delta y}$ from a different stratum. The component $\boldsymbol{s}'\boldsymbol{Q}_1$ may then be called the intra-block component, $\boldsymbol{s}'\boldsymbol{Q}_2$ the inter-block component, and $\boldsymbol{s}'\boldsymbol{Q}_3$ may be called the total-area component.

In connection with formula (3.2.28), it is interesting to consider three special cases of the vector $\boldsymbol{s}$ (and, hence, of the vector $\boldsymbol{c} = \boldsymbol{r}^\delta\boldsymbol{s}$). First suppose that $\boldsymbol{s}$ is such that $\boldsymbol{N}'\boldsymbol{s} = \boldsymbol{0}$, i.e., is orthogonal to the columns of $\boldsymbol{N}$, which also implies that $\boldsymbol{r}'\boldsymbol{s} = 0$ (i.e., $\boldsymbol{1}_v'\boldsymbol{c} = 0$). Then, $\boldsymbol{s}'\boldsymbol{Q}_2 = -\boldsymbol{s}'\boldsymbol{Q}_3$ and $\boldsymbol{s}'\boldsymbol{\Delta y} = \boldsymbol{s}'\boldsymbol{Q}_1$, which means that only the intra-block stratum contributes. As the second case, suppose that $\boldsymbol{s}$ is such that $\boldsymbol{N}'\boldsymbol{s} \neq \boldsymbol{0}$, but it satisfies the conditions $\boldsymbol{\phi}_1\boldsymbol{\Delta}'\boldsymbol{s} = \boldsymbol{0}$ and $\boldsymbol{s}'\boldsymbol{r} = 0$ (i.e., $\boldsymbol{c}'\boldsymbol{1}_v = 0$). Then, $\boldsymbol{s}'\boldsymbol{Q}_1 = \boldsymbol{s}'\boldsymbol{Q}_3 = 0$ and $\boldsymbol{s}'\boldsymbol{\Delta y} = \boldsymbol{s}'\boldsymbol{Q}_2$, which means that the contribution comes from the inter-block stratum only. As the third case, suppose that $\boldsymbol{s} \in \mathcal{C}(\boldsymbol{1}_v)$, i.e., that $\boldsymbol{s}$ is proportional to the vector $\boldsymbol{1}_v$. Then, on account of (3.2.10), $\boldsymbol{s}'\boldsymbol{Q}_1 = \boldsymbol{s}'\boldsymbol{Q}_2 = 0$ and $\boldsymbol{s}'\boldsymbol{\Delta y} = \boldsymbol{s}'\boldsymbol{Q}_3$, which means that the only contribution is from the total-area stratum.

Moreover, for the three cases, it is instructive to observe the following implications for estimating the function $\mathrm{E}(\boldsymbol{s}'\boldsymbol{\Delta y}) = \boldsymbol{s}'\boldsymbol{r}^\delta\boldsymbol{\tau}$ ($= \boldsymbol{c}'\boldsymbol{\tau}$, with $\boldsymbol{c} = \boldsymbol{r}^\delta\boldsymbol{s}$). If $\boldsymbol{N}'\boldsymbol{s} = \boldsymbol{0}$, then the condition (3.1.17) is satisfied. Also, if $\boldsymbol{N}'\boldsymbol{s} \neq \boldsymbol{0}$, but $\boldsymbol{\phi}_1\boldsymbol{\Delta}'\boldsymbol{s} = \boldsymbol{0}$, which is equivalent to $\boldsymbol{\Delta\phi}_1\boldsymbol{\Delta}'\boldsymbol{s} = \boldsymbol{0}$, the condition (3.1.17) holds, provided the condition (ii) of Theorem 3.1.2 holds. Finally, if $\boldsymbol{s} \in \mathcal{C}(\boldsymbol{1}_v)$, then $\boldsymbol{N}'\boldsymbol{s} \in \mathcal{C}(\boldsymbol{k})$ and (3.1.17) is satisfied, provided that

$$\mathcal{C}(\boldsymbol{k}) \subset \mathcal{C}\{\mathrm{diag}[\boldsymbol{1}_{b_1} : \boldsymbol{1}_{b_2} : \cdots : \boldsymbol{1}_{b_g}]\},$$

i.e., again under the condition (ii) of Theorem 3.1.2. The importance of these observations is that for the three discussed cases, the BLUE obtainable under the relevant submodel (3.2.2), (3.2.4) or (3.2.6), respectively (i.e., for the stratum concerned), is simultaneously the BLUE of $\boldsymbol{s}'\boldsymbol{r}^\delta\boldsymbol{\tau}$ ($= \boldsymbol{c}'\boldsymbol{\tau}$) under the overall model (3.1.1), for the second and the third case, however, provided that the design satisfies the condition (ii) of Theorem 3.1.2. (But see also Remark 3.2.8.)

The present discussion can be summarized as follows.

Corollary 3.2.5. *The function* $\boldsymbol{s}'\boldsymbol{\Delta y}$ *is the BLUE of* $\boldsymbol{c}'\boldsymbol{\tau} = \boldsymbol{s}'\boldsymbol{r}^\delta\boldsymbol{\tau}$ *under the overall model* (3.1.1) *in the following three cases:*

(a) *Let* $\boldsymbol{N}'\boldsymbol{s} = \boldsymbol{0}$ (*implying* $\boldsymbol{r}'\boldsymbol{s} = 0$); *the BLUE is then equal to* $\boldsymbol{s}'\boldsymbol{Q}_1$, *and its variance is of the form*

$$\mathrm{Var}(\widehat{\boldsymbol{c}'\boldsymbol{\tau}}) = \boldsymbol{s}'\boldsymbol{r}^\delta\boldsymbol{s}(\sigma_U^2 + \sigma_e^2) = \boldsymbol{c}'\boldsymbol{r}^{-\delta}\boldsymbol{c}(\sigma_U^2 + \sigma_e^2). \tag{3.2.29}$$

(b) *Let $N's \neq 0$, but $\phi_1 \Delta' s = 0$ and $r's = 0$, and assume that the condition* (ii) *of Theorem* 3.1.2 *holds; the BLUE is then equal to $s'Q_2$, and its variance is of the form*

$$\operatorname{Var}(\widehat{c'\tau}) = s'NN's(\sigma_B^2 - K_H^{-1}\sigma_U^2) + s'r^\delta s(\sigma_U^2 + \sigma_e^2). \tag{3.2.30}$$

(c) *Let $s = (n^{-1}c'1_v)1_v$, provided that the condition* (ii) *of Theorem* 3.1.2 *holds; the BLUE is then equal to $s'Q_3$, and its variance is of the form*

$$\operatorname{Var}(\widehat{c'\tau}) = (n^{-1}c'1_v)^2[(k'k - N_B^{-1}n^2)\sigma_B^2 + (n - K_H^{-1}k'k)\sigma_U^2 + n\sigma_e^2]. \tag{3.2.31}$$

Proof. These results follow from the discussion above, the formulae (3.2.29), (3.2.30) and (3.2.31) being obtainable directly from (3.1.16), but also from (3.2.13), (3.2.20) and (3.2.27), respectively. □

Remark 3.2.7. For case (a) of Corollary 3.2.5, note that if the design is connected and orthogonal, then $N's = 0$ for any s such that $r's = 0$, irrespectively of the vector k (see Corollary 2.3.4).

Remark 3.2.8. For case (b) of Corollary 3.2.5, it should be noted that the conditions $\phi_1 \Delta' s = 0$ and $r's = 0$ imply that the design is disconnected and, hence, on account of Corollary 2.3.2(b), that (in the notation of Definition 2.2.6)

$$N's = [s_1'N_1, s_2'N_2, ..., s_g'N_g]' = [s_1k_1', s_2k_2', ..., s_gk_g']',$$

where $s_\ell = s_\ell 1_{v_\ell}$ and $k_\ell = N_\ell' 1_{v_\ell}$. On the other hand, if the condition (3.1.17) is to be satisfied as well, then, on account of Corollary 3.1.1(b), it is necessary and sufficient that

$$k_\ell = k_\ell 1_{b_\ell} \quad \text{for any } \ell \text{ such that } s_\ell \neq 0, \tag{3.2.32}$$

which in turn holds if and only if the block sizes of the design are constant within any of its connected subdesigns to which the nonzero s_ℓ's correspond, a condition weaker than (ii) of Theorem 3.1.2.

If (3.2.32) holds, then

$$s'NN's = \sum_{\ell=1}^{g} s_\ell^2 k_\ell' k_\ell = \sum_{\ell=1}^{g} s_\ell^2 k_\ell^2 b_\ell \geq \frac{\left(\sum_{\ell=1}^{g} s_\ell^2 k_\ell b_\ell\right)^2}{\sum_{\ell=1}^{g} s_\ell^2 b_\ell}, \tag{3.2.33}$$

the equality in (3.2.33) evidently holding if and only if the k_ℓ's in (3.2.32) involved by the nonzero coefficients $s_\ell, \ell = 1, 2, ..., g$, are all equal ($= k$, say). In this extreme case, formula (3.2.30) is reduced to

$$\begin{aligned} \operatorname{Var}(\widehat{c'\tau}) &= s'r^\delta s[k\sigma_B^2 + (1 - K_H^{-1}k)\sigma_U^2 + \sigma_e^2] \\ &= c'r^{-\delta}c[k\sigma_B^2 + (1 - K_H^{-1}k)\sigma_U^2 + \sigma_e^2]. \end{aligned} \tag{3.2.34}$$

Remark 3.2.9. For case (c) of Corollary 3.2.5, it can be noted that if $\boldsymbol{s} = (n^{-1}\boldsymbol{c}'\mathbf{1}_v)\mathbf{1}_v$, then $\boldsymbol{N}'\boldsymbol{s} = (n^{-1}\boldsymbol{c}'\mathbf{1}_v)\boldsymbol{k}$. Hence, on account of Corollary 3.1.1(b), the condition (ii) of Theorem 3.1.2 is not only sufficient, but also necessary for the condition (3.1.17) to be satisfied in that case.

3.3 Estimating the same contrast under different submodels

The discussion in Section 3.2.4 has revealed that, under certain conditions, the function $\boldsymbol{s}'\boldsymbol{\Delta y}$ is the BLUE of its expectation $\boldsymbol{c}'\boldsymbol{\tau} = \boldsymbol{s}'\boldsymbol{r}^{\delta}\boldsymbol{\tau}$, the decomposition (3.2.28) being then reduced to one component only. In particular, $\boldsymbol{s}'\boldsymbol{\Delta y} = \boldsymbol{s}'\boldsymbol{Q}_3$ for any $\boldsymbol{s} \in \mathcal{C}(\mathbf{1}_v)$, as $\mathbf{1}_v'\boldsymbol{Q}_1 = \mathbf{1}_v'\boldsymbol{Q}_2 = 0$. On the other hand, for any $\boldsymbol{s}$ representing a contrast, i.e., for any $\boldsymbol{s}$ such that $\boldsymbol{s}'\boldsymbol{r} = 0$, the component $\boldsymbol{s}'\boldsymbol{Q}_3$ is equal to 0 and so (3.2.28) is reduced to

$$\boldsymbol{s}'\boldsymbol{\Delta y} = \boldsymbol{s}'\boldsymbol{Q}_1 + \boldsymbol{s}'\boldsymbol{Q}_2, \tag{3.3.1}$$

unless some additional conditions for $\boldsymbol{s}$ are met that reduce (3.3.1) further, to one component only (see Corollary 3.2.5).

The two components in (3.3.1) are uncorrelated, whatever the vector $\boldsymbol{s}$ and the matrix $\boldsymbol{\Delta}$ are, because the vectors $\boldsymbol{y}_1$ and $\boldsymbol{y}_2$ in (3.2.1) are not correlated, as can be seen from the equality

$$\boldsymbol{\phi}_1\text{Cov}(\boldsymbol{y})\boldsymbol{\phi}_2 = \mathbf{O},$$

holding on account of (3.1.16) and the properties of $\boldsymbol{\phi}_1$ and $\boldsymbol{\phi}_2$ shown in (3.2.8) and (3.2.10).

Now, one may ask, under which condition the intra-block component $\boldsymbol{s}'\boldsymbol{Q}_1$ and the inter-block component $\boldsymbol{s}'\boldsymbol{Q}_2$ estimate the same parametric function, the contrast $\boldsymbol{c}'\boldsymbol{\tau} = \boldsymbol{s}'\boldsymbol{r}^{\delta}\boldsymbol{\tau}$ (with the accuracy to a constant factor). To answer this question, note that

$$\text{E}(\boldsymbol{s}'\boldsymbol{Q}_1) = \boldsymbol{s}'\boldsymbol{\Delta}\boldsymbol{\phi}_1\boldsymbol{\Delta}'\boldsymbol{\tau} \tag{3.3.2}$$

and

$$\text{E}(\boldsymbol{s}'\boldsymbol{Q}_2) = \boldsymbol{s}'\boldsymbol{\Delta}\boldsymbol{\phi}_2\boldsymbol{\Delta}'\boldsymbol{\tau} = \boldsymbol{s}'\boldsymbol{\Delta}(\boldsymbol{I}_n - \boldsymbol{\phi}_1)\boldsymbol{\Delta}'\boldsymbol{\tau} \quad \text{if} \quad \boldsymbol{s}'\boldsymbol{r} = 0, \tag{3.3.3}$$

which imply the following result.

Lemma 3.3.1. *If $\boldsymbol{s}$ represents a contrast, then the equality $E(\boldsymbol{s}'\boldsymbol{Q}_1) = \kappa E(\boldsymbol{s}'\boldsymbol{Q}_2)$, where κ is a positive scalar, holds for any $\boldsymbol{\tau}$ if and only if*

$$\boldsymbol{\Delta}\boldsymbol{\phi}_1\boldsymbol{\Delta}'\boldsymbol{s} = \varepsilon\boldsymbol{r}^{\delta}\boldsymbol{s} \quad \textit{with} \quad \varepsilon = \kappa/(1+\kappa), \tag{3.3.4}$$

i.e., if and only if the vector $\boldsymbol{s}$ is an eigenvector of $\boldsymbol{C}_1 = \boldsymbol{\Delta}\boldsymbol{\phi}_1\boldsymbol{\Delta}'$ with respect to $\boldsymbol{r}^{\delta}$, corresponding to an eigenvalue ε such that $0 < \varepsilon < 1$.

This result was originally noticed by Jones (1959). Contrasts represented by eigenvectors of $\boldsymbol{C}_1$ with respect to $\boldsymbol{r}^\delta$ were later considered by Caliński (1971, 1977) and have been called basic contrasts by Pearce et al. (1974). The eigenvectors are usually $\boldsymbol{r}^\delta$-orthonormalized as in (2.3.2). (This will be formalized in Section 3.4.)

For any vector $\boldsymbol{s}$ satisfying the eigenvector condition (3.3.4), with $0 < \varepsilon < 1$ (which implies that $\boldsymbol{1}_v'\boldsymbol{s} = 0$), it follows from Theorem 3.2.1 that the component $\boldsymbol{s}'\boldsymbol{Q}_1$ is under (3.2.2), i.e., in the intra-block analysis, the BLUE of

$$\mathrm{E}(\boldsymbol{s}'\boldsymbol{Q}_1) = \varepsilon\boldsymbol{s}'\boldsymbol{r}^\delta\boldsymbol{\tau}$$

and has the variance

$$\mathrm{Var}(\boldsymbol{s}'\boldsymbol{Q}_1) = \varepsilon\boldsymbol{s}'\boldsymbol{r}^\delta\boldsymbol{s}(\sigma_U^2 + \sigma_e^2), \tag{3.3.5}$$

whereas from Theorem 3.2.2 and Corollary 3.2.1(c), when (3.2.18) or (3.2.19) holds, the component $\boldsymbol{s}'\boldsymbol{Q}_2$ is under (3.2.4), i.e., in the inter-block analysis, the BLUE of

$$\mathrm{E}(\boldsymbol{s}'\boldsymbol{Q}_2) = (1-\varepsilon)\boldsymbol{s}'\boldsymbol{r}^\delta\boldsymbol{\tau}$$

and has the variance

$$\begin{aligned}\mathrm{Var}(\boldsymbol{s}'\boldsymbol{Q}_2) &= \boldsymbol{s}'\boldsymbol{N}\boldsymbol{N}'\boldsymbol{s}(\sigma_B^2 - K_H^{-1}\sigma_U^2) + (1-\varepsilon)\boldsymbol{s}'\boldsymbol{r}^\delta\boldsymbol{s}(\sigma_U^2 + \sigma_e^2)\\ &= (1-\varepsilon)\boldsymbol{s}'\boldsymbol{r}^\delta\boldsymbol{s}[\boldsymbol{t}'\boldsymbol{k}^{2\delta}\boldsymbol{t}(\sigma_B^2 - K_H^{-1}\sigma_U^2) + \sigma_U^2 + \sigma_e^2],\end{aligned} \tag{3.3.6}$$

where $\boldsymbol{t}$ is a $\boldsymbol{k}^\delta$-normalized eigenvector of $\boldsymbol{N}'\boldsymbol{r}^{-\delta}\boldsymbol{N}$ with respect to $\boldsymbol{k}^\delta$ corresponding to the eigenvalue $\mu = 1-\varepsilon$ (as in the proof of Lemma 2.3.3) and related to the vector $\boldsymbol{s}$ by the formula

$$\boldsymbol{t} = (1-\varepsilon)^{-1/2}(\boldsymbol{s}'\boldsymbol{r}^\delta\boldsymbol{s})^{-1/2}\boldsymbol{k}^{-\delta}\boldsymbol{N}'\boldsymbol{s}, \tag{3.3.7}$$

and where $\boldsymbol{k}^{2\delta} = (\boldsymbol{k}^\delta)^2$. Thus, the following theorem is established.

Theorem 3.3.1. *For any* $\boldsymbol{c} = \boldsymbol{r}^\delta\boldsymbol{s}$, *where* $\boldsymbol{s}$ *is such that* $\boldsymbol{1}_v'\boldsymbol{c} = \boldsymbol{r}'\boldsymbol{s} = 0$ *and* (3.3.4) *is satisfied with* $0 < \varepsilon < 1$, *the BLUE of the contrast* $\boldsymbol{c}'\boldsymbol{\tau}$ *obtainable in the intra-block analysis, called the intra-block BLUE, can be written as*

$$(\widehat{\boldsymbol{c}'\boldsymbol{\tau}})_{\mathrm{intra}} = \varepsilon^{-1}\boldsymbol{s}'\boldsymbol{Q}_1 = \varepsilon^{-1}\boldsymbol{c}'\boldsymbol{r}^{-\delta}\boldsymbol{Q}_1, \tag{3.3.8}$$

with

$$\mathrm{Var}[(\widehat{\boldsymbol{c}'\boldsymbol{\tau}})_{\mathrm{intra}}] = \varepsilon^{-1}\boldsymbol{c}'\boldsymbol{r}^{-\delta}\boldsymbol{c}(\sigma_U^2 + \sigma_e^2), \tag{3.3.9}$$

and that obtainable in the inter-block analysis, under the condition (3.2.18) *or* (3.2.19), *called the inter-block BLUE, as*

$$(\widehat{\boldsymbol{c}'\boldsymbol{\tau}})_{\mathrm{inter}} = (1-\varepsilon)^{-1}\boldsymbol{s}'\boldsymbol{Q}_2 = (1-\varepsilon)^{-1}\boldsymbol{c}'\boldsymbol{r}^{-\delta}\boldsymbol{Q}_2, \tag{3.3.10}$$

with

$$\text{Var}[(\widehat{c'\tau})_{\text{inter}}] = (1-\varepsilon)^{-1}c'r^{-\delta}c[t'k^{2\delta}t(\sigma_B^2 - K_H^{-1}\sigma_U^2) + \sigma_U^2 + \sigma_e^2], \quad (3.3.11)$$

where t *is as in* (3.3.7). *If, in particular,* $k_1 = k_2 = \cdots = k_b = k$, *then the variance* (3.3.11) *gets the form*

$$\text{Var}[(\widehat{c'\tau})_{\text{inter}}] = (1-\varepsilon)^{-1}c'r^{-\delta}c[k\sigma_B^2 + (1 - K_H^{-1}k)\sigma_U^2 + \sigma_e^2] \quad (3.3.12)$$

or

$$\text{Var}[(\widehat{c'\tau})_{\text{inter}}] = (1-\varepsilon)^{-1}c'r^{-\delta}c(k\sigma_B^2 + \sigma_e^2),$$

if in addition $k = K_H$.

Proof. It follows from the discussion preceding the theorem. □

Example 3.3.1. In connection with Theorem 3.3.1, consider again the experiment analyzed in Examples 3.2.1 and 3.2.6, for which

$$C_1 = \varepsilon[6I_4 - (3/2)1_4 1_4'] \quad \text{with} \quad \varepsilon = 7/9.$$

This form of C_1 implies that any vector s representing a contrast satisfies the condition (3.3.4), i.e., is an eigenvector of C_1 with respect to $r^\delta = 6I_4$, corresponding to the unique nonzero eigenvalue of C_1 with that respect, of multiplicity $v - 1 = 3$, which here is $\varepsilon = 7/9$ (see Theorem 2.4.3). Hence, by Theorem 3.3.1, the intra-block BLUE of any contrast $c'\tau = s'r^\delta\tau$ is of the form (3.3.8), i.e.,

$$(\widehat{c'\tau})_{\text{intra}} = \varepsilon^{-1}s'Q_1 = \varepsilon^{-1}c'r^{-\delta}Q_1 = (3/14)c'Q_1,$$

where Q_1 is as given in Example 3.2.1, and its variance is given by (3.3.9), i.e.,

$$\text{Var}[(\widehat{c'\tau})_{\text{intra}}] = \varepsilon^{-1}c'r^{-\delta}c\sigma_1^2 = (3/14)c'c\sigma_1^2,$$

where $\sigma_1^2 = \sigma_U^2 + \sigma_e^2$. These results coincide with those in Example 3.2.1. On the other hand, by Theorem 3.3.1, the inter-block BLUE of any contrast $c'\tau = s'r^\delta\tau$ is of the form (3.3.10), i.e.,

$$(\widehat{c'\tau})_{\text{inter}} = (1-\varepsilon)^{-1}s'Q_2 = (1-\varepsilon)^{-1}c'r^{-\delta}Q_2 = (3/4)c'Q_2,$$

where Q_2 is as in Example 3.2.6, and its variance is given by (3.3.11), i.e.,

$$\text{Var}[(\widehat{c'\tau})_{\text{inter}}] = (1-\varepsilon)^{-1}c'r^{-\delta}c\sigma_{(2)}^2 = (3/4)c'c\sigma_{(2)}^2,$$

where

$$\sigma_{(2)}^2 = t'k^{2\delta}t(\sigma_B^2 - K_H^{-1}\sigma_U^2) + \sigma_1^2 = (9/4)(\sigma_B^2 - K_H^{-1}\sigma_U^2) + \sigma_1^2,$$

exactly as obtained in Example 3.2.6. As a particular application of these results, note that for any elementary contrast, i.e., of the type $\tau_i - \tau_{i'}$ $(i \neq i')$, the variance of its intra-block BLUE is equal to $(3/7)\sigma_1^2$ and the variance of its inter-block

BLUE is equal to $(3/2)\sigma_{(2)}^2$. In general, one can expect that $\sigma_{(2)}^2 \geq \sigma_1^2$, since usually $\sigma_B^2 \geq K_H^{-1}\sigma_U^2$ in a properly designed experiment (see the definitions in Section 3.1).

Remark 3.3.1. If ε is nonzero, but less than 1, it is said that the contrast represented by $\boldsymbol{s}$, i.e., $\boldsymbol{c}'\boldsymbol{\tau} = \boldsymbol{s}'\boldsymbol{r}^{\delta}\boldsymbol{\tau}$, is "partially confounded" with blocks. It would be said "totally confounded" in the case of $\varepsilon = 0$. In the opposite case, when $\boldsymbol{N}'\boldsymbol{s} = \boldsymbol{0}$, which corresponds to $\varepsilon = 1$, one can say that the contrast is not confounded with blocks at all.

It may also be noted that with (3.3.7) the component $\boldsymbol{s}'\boldsymbol{Q}_2$ can be written as

$$\boldsymbol{s}'\boldsymbol{Q}_2 = \boldsymbol{s}'\boldsymbol{N}\boldsymbol{k}^{-\delta}\boldsymbol{D}\boldsymbol{y} = (1-\varepsilon)^{1/2}(\boldsymbol{s}'\boldsymbol{r}^{\delta}\boldsymbol{s})^{1/2}\boldsymbol{t}'\boldsymbol{D}\boldsymbol{y}. \tag{3.3.13}$$

Now, as $\boldsymbol{D}\boldsymbol{y}$ is the vector of block totals (often denoted by $\boldsymbol{B}$, whereas the vector $\boldsymbol{\Delta}\boldsymbol{y}$ of treatment totals is denoted by $\boldsymbol{T}$), it is clearly seen from (3.3.13) that (3.3.10) is in fact an inter-block estimator of the contrast $\boldsymbol{c}'\boldsymbol{\tau}$. On the other hand, it is evident from (3.3.7) that if $\boldsymbol{N}'\boldsymbol{s} = \boldsymbol{0}$, then no inter-block estimator for the contrast $\boldsymbol{c}'\boldsymbol{\tau} = \boldsymbol{s}'\boldsymbol{r}^{\delta}\boldsymbol{\tau}$ exists.

Remark 3.3.2. If the vector $\boldsymbol{s}$ represents a contrast (i.e., $\boldsymbol{r}'\boldsymbol{s} = 0$), then the condition (3.2.18) can equivalently be written, in the notation of Lemma 2.3.3, as

$$\boldsymbol{K}_0\boldsymbol{N}'\boldsymbol{s} = \boldsymbol{k}^{\delta}\sum_{i=\rho+1}^{v-1}\boldsymbol{t}_i\boldsymbol{t}_i'\boldsymbol{k}^{\delta}\boldsymbol{N}'\boldsymbol{s}, \tag{3.3.14}$$

which holds if and only if

$$\sum_j \boldsymbol{t}_j\boldsymbol{t}_j'\boldsymbol{k}^{\delta}\boldsymbol{N}'\boldsymbol{s} = \boldsymbol{0},$$

where the vectors $\{\boldsymbol{t}_j\}$ form a complete set of $\boldsymbol{k}^{\delta}$-orthonormal eigenvectors of $\boldsymbol{N}'\boldsymbol{r}^{-\delta}\boldsymbol{N}$ corresponding to its zero eigenvalues (see the proof of Lemma 2.3.3).

It follows from Remark 3.3.2 that if the vector $\boldsymbol{s}$ satisfies the eigenvector condition (3.3.4) with $0 < \varepsilon < 1$, then (3.3.14), equivalent to the condition (3.2.18), holds if and only if $\boldsymbol{t}'\boldsymbol{k}^{2\delta}\boldsymbol{t}_j = 0$ for any $\boldsymbol{t}_j \in \{\boldsymbol{t}_j\}$ and the vector $\boldsymbol{t}$ given by (3.3.7). This result can also be written as follows. If $\boldsymbol{s}$ and $\boldsymbol{t}$ satisfy simultaneously the condition (3.3.4) and the equality $\boldsymbol{N}'\boldsymbol{r}^{-\delta}\boldsymbol{N}\boldsymbol{t} = (1-\varepsilon)\boldsymbol{k}^{\delta}\boldsymbol{t}$, respectively, with the same $0 < \varepsilon < 1$, then (3.2.18) holds if and only if $\boldsymbol{N}\boldsymbol{t}_0 = \boldsymbol{0}$ implies $\boldsymbol{t}'\boldsymbol{k}^{2\delta}\boldsymbol{t}_0 = 0$ for any conformable $\boldsymbol{t}_0$. [Compare this result with Remark 3.2.4(a).]

Example 3.3.2. In connection with Remark 3.3.2, consider again the design of Example 3.2.4, i.e., that with the incidence matrix

$$\boldsymbol{N} = \begin{bmatrix} 1 & 1 & 1 & 1 \\ 1 & 0 & 1 & 0 \\ 1 & 1 & 1 & 1 \end{bmatrix}$$

(denoted by $\boldsymbol{N}^*$ in Example 3.1.2). From it,

$$\boldsymbol{C}_1 = \boldsymbol{r}^{\delta} - \boldsymbol{N}\boldsymbol{k}^{-\delta}\boldsymbol{N}' = \frac{1}{3}\begin{bmatrix} 7 & -2 & -5 \\ -2 & 4 & -2 \\ -5 & -2 & 7 \end{bmatrix},$$

of rank 2, and

$$\boldsymbol{N}'\boldsymbol{r}^{-\delta}\boldsymbol{N} = \frac{1}{2}\begin{bmatrix} 2 & 1 & 2 & 1 \\ 1 & 1 & 1 & 1 \\ 2 & 1 & 2 & 1 \\ 1 & 1 & 1 & 1 \end{bmatrix},$$

of rank 2. Note that $\boldsymbol{C}_1\boldsymbol{s}_0 = \boldsymbol{r}^{\delta}\boldsymbol{s}_0$ for $\boldsymbol{s}_0 = [1, 0, -1]'$, as $\boldsymbol{N}'\boldsymbol{s}_0 = \boldsymbol{0}$, whereas $\boldsymbol{C}_1\boldsymbol{s}_1 = (5/6)\boldsymbol{r}^{\delta}\boldsymbol{s}_1$ for $\boldsymbol{s}_1 = [-1, 4, -1]'$. Thus, there is only one nonzero eigenvalue smaller than 1, $\varepsilon = 5/6$, of the matrix $\boldsymbol{C}_1$ with respect to $\boldsymbol{r}^{\delta}$. On the other hand, note that there is also only one nonzero eigenvalue smaller than 1, $\mu = 1 - \varepsilon = 1/6$, of the matrix $\boldsymbol{N}'\boldsymbol{r}^{-\delta}\boldsymbol{N}$ with respect to $\boldsymbol{k}^{\delta}$. Now, taking into account the two eigenvectors of $\boldsymbol{C}_1$ with respect to $\boldsymbol{r}^{\delta}$ corresponding to the nonzero eigenvalues, i.e., vectors $\boldsymbol{s}_0$ and $\boldsymbol{s}_1$, note that the condition (3.3.14), equivalent to (3.2.18), is satisfied automatically for $\boldsymbol{s}_0$. To see that it is also satisfied for $\boldsymbol{s}_1$, consider the corresponding vector $\boldsymbol{t}_1 = [-2, 3, -2, 3]'$ for which the equality $\boldsymbol{N}'\boldsymbol{r}^{-\delta}\boldsymbol{N}\boldsymbol{t}_1 = (1/6)\boldsymbol{k}^{\delta}\boldsymbol{t}_1$ holds, and note that it satisfies the condition $\boldsymbol{t}_1'\boldsymbol{k}^{2\delta}\boldsymbol{t}_0 = 0$ for any vector $\boldsymbol{t}_0$ for which $\boldsymbol{N}\boldsymbol{t}_0 = \boldsymbol{0}$. To check it, it is sufficient to take as $\boldsymbol{t}_0$ the vectors $\boldsymbol{t}_{01}$ and $\boldsymbol{t}_{02}$ from Example 3.2.3. Thus, the condition (3.3.14), and hence (3.2.18), holds for any vector $\boldsymbol{s}$ such that $\boldsymbol{r}'\boldsymbol{s} = 0$, as $\boldsymbol{s}_0$ and $\boldsymbol{s}_1$ span the subspace of all such vectors.

Remark 3.3.3. To give a statistical interpretation of the eigenvalue ε in (3.3.4), it will be instructive to suppose that, for the eigenvector $\boldsymbol{s}$ and the fixed vector $\boldsymbol{r}$ determining the contrast $\boldsymbol{c}'\boldsymbol{\tau} = \boldsymbol{s}'\boldsymbol{r}^{\delta}\boldsymbol{\tau}$ under study, a design exists such that its incidence matrix $\boldsymbol{N}$ satisfies the equality $\boldsymbol{N}'\boldsymbol{s} = \boldsymbol{0}$. Then, according to Corollary 3.2.5(a), $\boldsymbol{s}'\boldsymbol{Q}_1 = \boldsymbol{s}'\boldsymbol{\Delta y}$ $(= \boldsymbol{s}'\boldsymbol{T})$ is the BLUE of $\boldsymbol{c}'\boldsymbol{\tau} = \boldsymbol{s}'\boldsymbol{r}^{\delta}\boldsymbol{\tau}$ under the overall model (3.1.1), with the variance obtainable in the form (3.2.29). Comparing (3.2.29) with (3.3.9), it becomes evident that ε is the "efficiency factor" of the analyzed design for the contrast $\boldsymbol{c}'\boldsymbol{\tau}$ when it is estimated in the intra-block analysis, i.e., by (3.3.8). On the other hand, $1 - \varepsilon$ can be interpreted as the relative "loss of information," incurred in that analysis, due to partially confounding the contrast with blocks. This terminology is from Jones (1959, p. 176). The factor ε is also called, after James and Wilkinson (1971), the "canonical efficiency factor." In the extreme case of $\varepsilon = 0$, the whole information is lost when the analysis is confined to intra-block, as then $\boldsymbol{s}'\boldsymbol{Q}_1 = 0$.

It would usually be said in the case of $\varepsilon = 0$ that the contrast is "totally confounded" in the intra-block analysis but estimated with "full efficiency" in the inter-block analysis (see Pearce et al., 1974, p. 455). Thus, the question may be raised, whether it is justified to consider $1 - \varepsilon$ as the efficiency factor of the design for the contrast $\boldsymbol{c}'\boldsymbol{\tau} = \boldsymbol{s}'\boldsymbol{r}^{\delta}\boldsymbol{\tau}$ estimated in the inter-block analysis, i.e., by (3.3.10). This question can be answered as follows.

Remark 3.3.4. Suppose that there exists a disconnected design, which for the vector $\boldsymbol{s}$ and the fixed vector $\boldsymbol{r}$ satisfies the condition (3.1.17) of Theorem 3.1.1 and, simultaneously, the equality $\boldsymbol{\Delta}\boldsymbol{\phi}_1\boldsymbol{\Delta}'\boldsymbol{s} = \boldsymbol{0}$ (i.e. $\boldsymbol{\phi}_1\boldsymbol{\Delta}'\boldsymbol{s} = \boldsymbol{0}$). Then, on account of Corollary 3.2.5(b) and Remark 3.2.8, it will provide $\boldsymbol{s}'\boldsymbol{Q}_2 = \boldsymbol{s}'\boldsymbol{\Delta}\boldsymbol{y}$ as the BLUE of $\boldsymbol{c}'\boldsymbol{\tau}$ under the model (3.1.1), with the minimum variance of the form (3.2.34), attainable when the blocks involved by the vector $\boldsymbol{N}'\boldsymbol{s}$ are all of equal size k, say. Now, comparing the variance formulae (3.2.34) and (3.3.11), it becomes clear that the true inter-block efficiency factor is of the form

$$(1-\varepsilon)\frac{k\sigma_B^2 + (1 - K_H^{-1}k)\sigma_U^2 + \sigma_e^2}{\boldsymbol{t}'\boldsymbol{k}^{2\delta}\boldsymbol{t}(\sigma_B^2 - K_H^{-1}\sigma_U^2) + \sigma_U^2 + \sigma_e^2}.$$

So, the coefficient $1-\varepsilon$ becomes the efficiency factor of the design for estimating the contrast $\boldsymbol{c}'\boldsymbol{\tau} = \boldsymbol{s}'\boldsymbol{r}^{\delta}\boldsymbol{\tau}$ in the inter-block analysis if and only if $k = \boldsymbol{t}'\boldsymbol{k}^{2\delta}\boldsymbol{t}$, i.e., if and only if the involved blocks of the above minimum variance (optimal) design are all of equal size, equal to $\boldsymbol{t}'\boldsymbol{k}^{2\delta}\boldsymbol{t}$ of the design under cosideration.

It follows from Remark 3.3.4 that the coefficient $1-\varepsilon$, considered as the inter-block efficiency factor of the design for a contrast satisfying (3.3.4), is in general only notionally acceptable, as the optimal design to which comparison is made may not exist in reality if $\boldsymbol{t}'\boldsymbol{k}^{2\delta}\boldsymbol{t}$ of the considered design is not an integer. This condition may be one of the reasons (in addition to those following from the difficulties discussed in Section 3.2.2) why many authors have been reluctant to use the inter-block analysis to other than proper designs (see, e.g., Pearce, 1983, Section 3.8), and why it is really reasonable to use the term "proper design" for an equiblock-sized design (Definition 2.2.2). Some other reasons will become apparent later.

3.4 Basic contrasts and their intra-block analysis

In Section 3.3, it has been shown that certain contrasts of treatment parameters play a special role in the analysis of block designs. A relation between the design and these contrasts is given with the use of eigenvectors and corresponding eigenvalues of the C-matrix of the design with respect to the treatment replications. The eigenvectors represent the contrasts, and the eigenvalues express the efficiency factors of the design for the intra-block estimation of the contrasts. Thus, to each block design a set of contrasts corresponds, not necessarily unique, for which the efficiencies of estimation in the intra-block analysis are readily defined. This relation was originally noticed by Jones (1959).

In view of this relation, it appears that a block design can well be characterized by a set of contrasts for which the condition (3.3.4) is satisfied, and by the corresponding set of eigenvalues resulting from (3.3.4). Algebraically, this characterization is supplied by the spectral decomposition (2.3.2) of the C-

matrix of the design. Contrasts represented by vectors $\{\boldsymbol{s}_i\}$ appearing in that decomposition have been termed by Pearce et al. (1974) as follows.

Definition 3.4.1. For any block design, contrasts $\{\boldsymbol{c}_i'\boldsymbol{\tau} = \boldsymbol{s}_i'\boldsymbol{r}^\delta\boldsymbol{\tau}, \boldsymbol{s}_i'\boldsymbol{r} = 0, i = 1, 2, ..., v-1\}$ are said to be basic contrasts of the design if the vectors $\{\boldsymbol{s}_i\}$ are $\boldsymbol{r}^\delta$-orthonormal eigenvectors of the matrix $\boldsymbol{C}_1 = \boldsymbol{\Delta\phi}_1\boldsymbol{\Delta}'$ of the design with respect to $\boldsymbol{r}^\delta$.

Note, by referring to Remark 3.3.3, that the eigenvalues of $\boldsymbol{C}_1$ with respect to $\boldsymbol{r}^\delta$ are then the efficiency factors of the design for estimating the corresponding basic contrasts in the intra-block analysis. In this section, when the term "efficiency factor for a contrast" is used, it should be understood as the efficiency factor of the design for estimating the contrast in the intra-block analysis.

The following results give sense to the term "basic" used in Definition 3.4.1.

Theorem 3.4.1. *Let $\{\boldsymbol{c}_i'\boldsymbol{\tau} = \boldsymbol{s}_i'\boldsymbol{r}^\delta\boldsymbol{\tau}, i = 1, 2, ..., v-1\}$ be any set of basic contrasts of a block design, and let $\{\varepsilon_i, i = 1, 2, ..., v-1\}$ be the corresponding efficiency factors. Then*

(i) *the intra-block analysis provides the BLUEs, of the form*

$$(\widehat{\boldsymbol{c}_i'\boldsymbol{\tau}})_{\text{intra}} = \varepsilon_i^{-1}\boldsymbol{s}_i'\boldsymbol{Q}_1 = \varepsilon_i^{-1}\boldsymbol{c}_i'\boldsymbol{r}^{-\delta}\boldsymbol{Q}_1 \tag{3.4.1}$$

($\boldsymbol{Q}_1 = \boldsymbol{\Delta\phi}_1\boldsymbol{y}$), *with the corresponding variances*

$$\text{Var}[(\widehat{\boldsymbol{c}_i'\boldsymbol{\tau}})_{\text{intra}}] = \varepsilon_i^{-1}\sigma_1^2 \tag{3.4.2}$$

($\sigma_1^2 = \sigma_U^2 + \sigma_e^2$) *and covariances*

$$\text{Cov}[(\widehat{\boldsymbol{c}_i'\boldsymbol{\tau}})_{\text{intra}}, (\widehat{\boldsymbol{c}_{i'}'\boldsymbol{\tau}})_{\text{intra}}] = 0, \quad i \neq i', \tag{3.4.3}$$

for those of the basic contrasts for which the efficiency factors are nonzero (positive), and

(ii) *no BLUEs exist in the intra-block analysis for those basic contrasts for which the efficiency factors are zero.*

Proof. Part (i) follows from Theorem 3.3.1, formulae (3.4.1) and (3.4.2) following from (3.3.8) and (3.3.9), respectively, whereas (3.4.3) follows from the formula

$$\text{Cov}(\boldsymbol{Q}_1) = \boldsymbol{\Delta\phi}_1\boldsymbol{\Delta}'\sigma_1^2, \tag{3.4.4}$$

holding on account of (3.2.12), and from the fact that $\boldsymbol{s}_i'\boldsymbol{\Delta\phi}_1\boldsymbol{\Delta}'\boldsymbol{s}_{i'} = 0$ if $i \neq i'$, following from the properties of $\{\boldsymbol{s}_i\}$ given in Definition 3.4.1. Part (ii) is obvious, as $\boldsymbol{\Delta\phi}_1\boldsymbol{\Delta}'\boldsymbol{s} = \boldsymbol{0}$ implies that $\boldsymbol{s}'\boldsymbol{Q}_1 = \boldsymbol{0}$ (see also Remark 3.3.3). □

Theorem 3.4.2. *For any block design for which the vectors $\boldsymbol{s}_1, \boldsymbol{s}_2, ..., \boldsymbol{s}_h$ represent basic contrasts receiving nonzero (positive) efficiency factors, a set of contrasts $\boldsymbol{U}'\boldsymbol{\tau}$ obtains the BLUEs in the intra-block analysis if and only if the matrix $\boldsymbol{U}$ can be written as $\boldsymbol{U} = \boldsymbol{r}^\delta\boldsymbol{S}\boldsymbol{A}$, where $\boldsymbol{S} = [\boldsymbol{s}_1 : \boldsymbol{s}_2 : \cdots : \boldsymbol{s}_h]$ and*

$A = [a_1 : a_2 : \cdots : a_h]'$ is some matrix of h rows. If U is such, then the BLUEs provided by the intra-block analysis are of the form

$$(\widehat{U'\tau})_{\text{intra}} = A'\varepsilon^{-\delta}S'Q_1 = \sum_{i=1}^{h} \varepsilon_i{}^{-1} a_i s_i' Q_1, \tag{3.4.5}$$

and their covariance (dispersion) matrix is of the form

$$\text{Cov}[(\widehat{U'\tau})_{\text{intra}}] = A'\varepsilon^{-\delta}A\sigma_1^2 = \sum_{i=1}^{h} \varepsilon_i{}^{-1} a_i a_i' \sigma_1^2, \tag{3.4.6}$$

where $\varepsilon^{\delta} = \text{diag}[\varepsilon_1, \varepsilon_2, ..., \varepsilon_h]$ and $\varepsilon^{-\delta} = (\varepsilon^{\delta})^{-1}$.

Proof. On account of Theorem 3.2.1, the intra-block analysis provides the BLUEs for $U'\tau$ if and only if the columns of the matrix U are linear combinations of the matrix $\Delta\phi_1\Delta'$, i.e., if and only if a matrix A^* exists such that $U = \Delta\phi_1\Delta' A^*$. But, because from Lemma 2.3.3

$$C_1 = \Delta\phi_1\Delta' = r^{\delta}S\varepsilon^{\delta}S'r^{\delta}, \tag{3.4.7}$$

the condition for U can be written as $U = r^{\delta}S\varepsilon^{\delta}S'r^{\delta}A^* = r^{\delta}SA$, with $A = \varepsilon^{\delta}S'r^{\delta}A^*$. On the other hand, if $U = r^{\delta}SA$, then (as $S'r^{\delta}S = I_h$) one can write $U = r^{\delta}S\varepsilon^{\delta}S'r^{\delta}S\varepsilon^{-\delta}A = \Delta\phi_1\Delta'A^*$, with $A^* = S\varepsilon^{-\delta}A$. This argument proves the first part of the theorem. To prove the second, note that because $U = \Delta'\phi_1\Delta'S\varepsilon^{-\delta}A$, the functions $A'\varepsilon^{-\delta}S'\Delta y_1 = A'\varepsilon^{-\delta}S'Q_1$ are, again on account of Theorem 3.2.1, the BLUEs of the contrasts $U'\tau$. (That these are contrasts follows from the equality $1_v'\Delta\phi_1 = 0'$.) Thus, (3.4.5) is established, whereas (3.4.6) follows from (3.4.4) and (3.4.7).

Corollary 3.4.1. *For any block design for which the vectors $s_1, s_2, ..., s_\rho$ represent basic contrasts receiving the unit efficiency factors (in accordance with the notation of Lemma 2.3.3), a set of contrasts $U_0'\tau = A_0'S_0'r^{\delta}\tau$, where $S_0 = [s_1 : s_2 : \cdots : s_\rho]$, and $A_0 = [a_1 : a_2 : \cdots : a_\rho]'$ is some matrix of ρ rows, obtains the BLUEs under the overall model (3.1.1), in the form*

$$\widehat{U_0'\tau} = A_0'S_0'\Delta y = \sum_{i=1}^{\rho} a_i s_i' \Delta y, \tag{3.4.8}$$

for which the covariance matrix is

$$\text{Cov}(\widehat{U_0'\tau}) = A_0'A_0\sigma_1^2 = \sum_{i=1}^{\rho} a_i a_i' \sigma_1^2. \tag{3.4.9}$$

Proof. This result follows from Theorem 3.4.2 and Corollary 3.2.3(a), but it also follows directly from Theorem 3.1.1, after noting that as the equality $N_0'S_0A_0 = O$ holds, so does the equality in (3.1.17), and that

$$\text{Cov}(A_0'S_0'\Delta y) = A_0'S_0'\Delta\text{Cov}(y)\Delta'S_0A_0 = A_0'A_0(\sigma_U^2 + \sigma_e^2),$$

from (3.4.8) and (3.1.16), giving (3.4.9). □

Remark 3.4.1. (a) In the notation of Corollary 3.4.1, a block design for which $\rho \geq 1$ can be called orthogonal for the set of contrasts $\boldsymbol{U}_0'\boldsymbol{\tau} = \boldsymbol{A}_0'\boldsymbol{S}_0'\boldsymbol{r}^\delta\boldsymbol{\tau}$.

(b) It follows from Theorem 3.4.2 and Corollary 3.4.1 that the efficiency factor of a block design for a contrast $\boldsymbol{u}'\boldsymbol{\tau} = \boldsymbol{a}'\boldsymbol{S}'\boldsymbol{r}^\delta\boldsymbol{\tau}$ estimated in the intra-block analysis is (in the sense considered in Section 3.3) of the form

$$\varepsilon[(\widehat{\boldsymbol{u}'\boldsymbol{\tau}})_{\text{intra}}] = \boldsymbol{a}'\boldsymbol{a}/\boldsymbol{a}'\boldsymbol{\varepsilon}^{-\delta}\boldsymbol{a}$$

(see also Caliński, Ceranka and Mejza, 1980, p. 60).

(c) Replacing in (b) the vector $\boldsymbol{a}$ by $\mathbf{1}_h$ gives the contrast $\boldsymbol{u}'\boldsymbol{\tau} = \mathbf{1}_h'\boldsymbol{S}'\boldsymbol{r}^\delta\boldsymbol{\tau} = \sum_{i=1}^{h} \boldsymbol{s}_i'\boldsymbol{r}^\delta\boldsymbol{\tau}$, for which the BLUE in the intra-block analysis is of the form

$$(\mathbf{1}_h'\widehat{\boldsymbol{S}'\boldsymbol{r}^\delta\boldsymbol{\tau}})_{\text{intra}} = \sum_{i=1}^{h} \varepsilon_i^{-1}\boldsymbol{s}_i'\boldsymbol{Q}_1,$$

with the variance

$$\text{Var}[(\mathbf{1}_h'\widehat{\boldsymbol{S}'\boldsymbol{r}^\delta\boldsymbol{\tau}})_{\text{intra}}] = \sum_{i=1}^{h} \varepsilon_i^{-1}\sigma_1^2$$

and the efficiency factor

$$\varepsilon[(\mathbf{1}_h'\widehat{\boldsymbol{S}'\boldsymbol{r}^\delta\boldsymbol{\tau}})_{\text{intra}}] = h/\sum_{i=1}^{h} \varepsilon_i^{-1} = \varepsilon \quad \text{(say)},$$

the latter being called the "average" efficiency factor of a block design, with regard to the estimation in the intra-block analysis (see Caliński et al., 1980, p. 61; also John, 1987, p. 27).

From the results concerning basic contrasts, presented above with regard to their role in the intra-block analysis, it follows that three main subsets of basic contrasts can be distinguished. The first subset contains contrasts represented by eigenvectors $\boldsymbol{s}_1, \boldsymbol{s}_2, ..., \boldsymbol{s}_\rho$ (in accordance with the ordering used in Lemma 2.3.3). They are all estimated with full efficiency in the intra-block analysis, their BLUEs under the submodel $\boldsymbol{y}_1 = \boldsymbol{\phi}_1\boldsymbol{y}$ being simultaneously the BLUEs under the overall model (3.1.1). The second subset is that of contrasts represented by eigenvectors $\boldsymbol{s}_{\rho+1}, \boldsymbol{s}_{\rho+2}, ..., \boldsymbol{s}_h$ [$h = \text{rank}(\boldsymbol{C}_1)$]. For all of them, the intra-block analysis provides the BLUEs, under $\boldsymbol{y}_1 = \boldsymbol{\phi}_1\boldsymbol{y}$, with efficiencies less than one, because of some loss of information under this submodel due to partially confounding the contrasts with blocks. Finally, the third subset contains contrasts represented by $\boldsymbol{s}_{h+1}, \boldsymbol{s}_{h+2}, ..., \boldsymbol{s}_{v-1}$. The intra-block analysis provides BLUEs for none of them, because of the complete loss of information under $\boldsymbol{y}_1 = \boldsymbol{\phi}_1\boldsymbol{y}$ caused by totally confounding these contrasts with blocks. In a connected design, the last subset is empty [see Lemma 2.3.3 and Corollary 2.3.2(a)].

As to the second subset of basic contrasts, it might be worth noting that although the $h-\rho$ efficiency factors for contrasts belonging to this subset may all be different, the number of their distinct values will usually be smaller than $h-\rho$.

This fact gives rise to the following notation, in which the subscript i is replaced by a double subscript βj, β referring to the subset of contrasts corresponding to the same efficiency factor, and j to a contrast within the subset.

The spectral decomposition (2.3.2) can then be written in the form

$$\boldsymbol{C}_1 = \boldsymbol{\Delta}\boldsymbol{\phi}_1\boldsymbol{\Delta}' = \boldsymbol{r}^\delta \sum_{\beta=0}^{m-1} \varepsilon_\beta \boldsymbol{H}_\beta \boldsymbol{r}^\delta = \boldsymbol{r}^\delta \sum_{\beta=0}^{m-1} \varepsilon_\beta \boldsymbol{L}_\beta, \tag{3.4.10}$$

where

$$\boldsymbol{H}_\beta = \sum_{j=1}^{\rho_\beta} \boldsymbol{s}_{\beta j}\boldsymbol{s}'_{\beta j} \quad \text{and} \quad \boldsymbol{L}_\beta = \boldsymbol{H}_\beta \boldsymbol{r}^\delta,$$

$\boldsymbol{s}_{\beta 1}, \boldsymbol{s}_{\beta 2}, ..., \boldsymbol{s}_{\beta \rho_\beta}$ being $\boldsymbol{r}^\delta$-orthonormal eigenvectors of $\boldsymbol{C}_1$ with respect to $\boldsymbol{r}^\delta$, corresponding to a common eigenvalue ε_β of multiplicity ρ_β, for $\beta = 0, 1, ..., m-1$ (with $\rho_0 \equiv \rho$ in the original notation of Lemma 2.3.3), and where $m-1$ is the number of distinct, less than 1, nonzero (positive) eigenvalues of $\boldsymbol{C}_1$ with respect to $\boldsymbol{r}^\delta$.

With regard to (3.4.10), the following result is useful.

Corollary 3.4.2. *Let a subset of basic contrasts of a block design be represented by the eigenvectors $\boldsymbol{s}_{\beta 1}, \boldsymbol{s}_{\beta 2}, ..., \boldsymbol{s}_{\beta \rho_\beta}$ of $\boldsymbol{C}_1$ with respect to $\boldsymbol{r}^\delta$ corresponding to a common eigenvalue $\varepsilon_\beta > 0$ of multiplicity ρ_β. Then, for a set of contrasts $\boldsymbol{U}'_\beta \boldsymbol{\tau} = \boldsymbol{A}'_\beta \boldsymbol{S}'_\beta \boldsymbol{r}^\delta \boldsymbol{\tau}$, where*

$$\boldsymbol{S}_\beta = [\boldsymbol{s}_{\beta 1} : \boldsymbol{s}_{\beta 2} : \cdots : \boldsymbol{s}_{\beta \rho_\beta}],$$

and $\boldsymbol{A}_\beta$ is some matrix of ρ_β rows, the intra-block analysis provides the BLUEs of the form

$$(\widehat{\boldsymbol{U}'_\beta \boldsymbol{\tau}})_{\text{intra}} = \varepsilon_\beta^{-1} \boldsymbol{A}'_\beta \boldsymbol{S}'_\beta \boldsymbol{Q}_1, \tag{3.4.11}$$

and their covariance matrix of the form

$$\text{Cov}[(\widehat{\boldsymbol{U}'_\beta \boldsymbol{\tau}})_{\text{intra}}] = \varepsilon_\beta^{-1} \boldsymbol{A}'_\beta \boldsymbol{A}_\beta \sigma_1^2, \tag{3.4.12}$$

ε_β being the common efficiency factor of the design for all contrasts in the set, $\beta = 0, 1, ..., m-1$.

Proof. This result follows immediately from Theorem 3.4.2 and Remark 3.4.1(b). □

Remark 3.4.2. Formulae (3.4.11) and (3.4.12) of Corollary 3.4.2 show that the intra-block estimation of contrasts belonging to a subspace spanned by basic contrasts for which the design gives the same efficiency factor is simple, and the structure of the resulting covariance matrix of the obtained BLUEs is simple, everything being controlled by the common efficiency factor. Thus, an essential implication for planning of experiments is that if the experimental problem has its reflection in distinguishing certain subsets of contrasts, ordered according to their importance, a design should be chosen that gives for all members of a specified subset a common efficiency factor, of the higher value the more important the contrasts of the subset are. If possible, the design should allow the most important subset of contrasts to be estimated with full efficiency (i.e., of value 1).

In addition to Corollary 3.4.2, it may be noted that because of the decomposition (3.4.10), a possible choice of a g-inverse of the matrix $\boldsymbol{C}_1$ is the matrix

$$\sum_{\beta=0}^{m-1} \varepsilon_\beta^{-1} \boldsymbol{H}_\beta = \sum_{\beta=0}^{m-1} \varepsilon_\beta^{-1} \boldsymbol{L}_\beta \boldsymbol{r}^{-\delta}, \tag{3.4.13}$$

on account of Corollary 2.3.5 (see also Rao and Mitra, 1971, Section 3.3.1, Note 3). Formula (3.4.13) allows the projector $\boldsymbol{P}_{\phi_1\Delta'}$ used in Section 3.2.1 to be written in a decomposed form as

$$\boldsymbol{P}_{\phi_1\Delta'} = \sum_{\beta=0}^{m-1} \varepsilon_\beta^{-1} \boldsymbol{\phi}_1 \boldsymbol{\Delta}' \boldsymbol{H}_\beta \boldsymbol{\Delta} \boldsymbol{\phi}_1 = \sum_{\beta=0}^{m-1} \boldsymbol{P}_{\phi_1\Delta' S_\beta},$$

where

$$\boldsymbol{P}_{\phi_1\Delta' S_\beta} = \varepsilon_\beta^{-1} \boldsymbol{\phi}_1 \boldsymbol{\Delta}' \boldsymbol{H}_\beta \boldsymbol{\Delta} \boldsymbol{\phi}_1$$

is the orthogonal projector on the subspace $\mathcal{C}(\boldsymbol{\phi}_1 \boldsymbol{\Delta}' \boldsymbol{S}_\beta) \subset \mathcal{C}(\boldsymbol{\phi}_1 \boldsymbol{\Delta}')$. This decomposition further implies that the intra-block treatment sum of squares (defined in Section 3.2.1) can be decomposed in the form

$$\boldsymbol{y}_1' \boldsymbol{P}_{\phi_1\Delta'} \boldsymbol{y}_1 = \boldsymbol{y}' \boldsymbol{P}_{\phi_1\Delta'} \boldsymbol{y} = \sum_{\beta=0}^{m-1} \boldsymbol{y}' \boldsymbol{P}_{\phi_1\Delta' S_\beta} \boldsymbol{y}$$

or, in other equivalent notation, as

$$\boldsymbol{Q}_1' \boldsymbol{C}_1^- \boldsymbol{Q}_1 = \sum_{\beta=0}^{m-1} \varepsilon_\beta^{-1} \boldsymbol{Q}_1' \boldsymbol{H}_\beta \boldsymbol{Q}_1,$$

the βth component, with ρ_β d.f., being related to the βth subset of basic contrasts for which the common efficiency factor is ε_β (see also Pearce, 1983, p. 75).

In the discussion on basic contrasts conducted until now, only the estimation within the first stratum, i.e., intra-block, has been taken into account. This confinement in presenting the theory may be necessary if a general block design, possibly nonproper, is under consideration. In such a general case, as indicated in Section 3.2.2, the inter-block estimation of a contrast may fail in getting the BLUE of it in that stratum. It will be shown in the next section that only proper designs induce a block structure that allows the estimation in the inter-block stratum to remain straightforward.

3.5 Proper designs and the orthogonal block structure

In this section and the next, the decomposition (3.2.1) of the model (3.1.1) will be considered again, but now mainly with regard to proper block designs, for which some advantageous properties have already been indicated in the preceding sections. These properties are essential with regard to the inter-block submodel (3.2.4), for which the structure of the covariance matrix, shown in (3.2.15), is not satisfactory from the application point of view. In general, it obtains a simple form for the extreme case of $K_H\sigma_B^2 = \sigma_U^2$ only, i.e., when the grouping of units into blocks is not successful (see definitions of σ_B^2, σ_U^2 and K_H in Section 3.1.1). However, if the design is proper, with k as the constant block size, then

$$(\boldsymbol{I}_n - n^{-1}\boldsymbol{1}_n\boldsymbol{1}_n')\boldsymbol{D}'\boldsymbol{D}(\boldsymbol{I}_n - n^{-1}\boldsymbol{1}_n\boldsymbol{1}_n') = k\boldsymbol{\phi}_2 \tag{3.5.1}$$

and $\text{Cov}(\boldsymbol{y}_2)$ gets a manageable form. Therefore, further development of the theory connected with the decomposition (3.2.1) will be confined here to proper designs, i.e., designs with

$$k_1 = k_2 = \cdots = k_b = k. \tag{3.5.2}$$

This confinement will allow the theory to be presented in a unified form.

First note that, because under (3.5.2) not only (3.5.1), but also the equalities

$$n^{-1}\boldsymbol{k}'\boldsymbol{k} - N_B^{-1}n = k(1 - N_B^{-1}b) \quad \text{and} \quad 1 - K_H^{-1}n^{-1}\boldsymbol{k}'\boldsymbol{k} = 1 - K_H^{-1}k$$

hold, the covariance matrices (3.2.15) and (3.2.26) are simplified to

$$\text{Cov}(\boldsymbol{y}_2) = \boldsymbol{\phi}_2[k\sigma_B^2 + (1 - K_H^{-1}k)\sigma_U^2 + \sigma_e^2] \tag{3.5.3}$$

and

$$\text{Cov}(\boldsymbol{y}_3) = \boldsymbol{\phi}_3[(1 - N_B^{-1}b)k\sigma_B^2 + (1 - K_H^{-1}k)\sigma_U^2 + \sigma_e^2], \tag{3.5.4}$$

respectively, for any proper block design. Also note that for any such design, $\boldsymbol{D}'\boldsymbol{k}^{-\delta}\boldsymbol{D} = k^{-1}\boldsymbol{D}'\boldsymbol{D}$, simplifying the matrices (3.2.3) and (3.2.5), accordingly.

Thus, in the case of any proper block design, the expectation vector and the covariance matrix for each of the three submodels (3.2.2), (3.2.4) and (3.2.6) can be written, respectively, as

$$\text{E}(\boldsymbol{y}_\alpha) = \boldsymbol{\phi}_\alpha\boldsymbol{\Delta}'\boldsymbol{\tau} \tag{3.5.5}$$

and

$$\text{Cov}(\boldsymbol{y}_\alpha) = \boldsymbol{\phi}_\alpha\sigma_\alpha^2, \tag{3.5.6}$$

for $\alpha = 1, 2, 3$, where, from (3.2.12), (3.5.3) and (3.5.4),

$$\begin{aligned} \sigma_1^2 &= \sigma_U^2 + \sigma_e^2, \\ \sigma_2^2 &= k\sigma_B^2 + (1 - K_H^{-1}k)\sigma_U^2 + \sigma_e^2, \\ \sigma_3^2 &= (1 - N_B^{-1}b)k\sigma_B^2 + (1 - K_H^{-1}k)\sigma_U^2 + \sigma_e^2. \end{aligned} \tag{3.5.7}$$

These variances are called the "stratum variances" (see, e.g., John, 1987, p. 187). Furthermore, if $N_B = b$ and $k = K_H$ (the latter implying the equality of all K_ξ's), which can be considered as the most usual case, the variances σ_2^2 and σ_3^2 reduce to

$$\sigma_2^2 = k\sigma_B^2 + \sigma_e^2 \quad \text{and} \quad \sigma_3^2 = \sigma_e^2.$$

On the other hand, it should be noticed that, under the decomposition (3.2.1),

$$\text{Cov}(\boldsymbol{y}) = \sum_{\alpha=1}^{3} \text{Cov}(\boldsymbol{y}_\alpha) + \sum\sum_{\alpha \neq \alpha'} \text{Cov}(\boldsymbol{y}_\alpha, \boldsymbol{y}_{\alpha'}),$$

where the covariance matrix of the vectors $\boldsymbol{y}_\alpha$ and $\boldsymbol{y}_{\alpha'}$ has the form

$$\text{Cov}(\boldsymbol{y}_\alpha, \boldsymbol{y}_{\alpha'}) = \boldsymbol{\phi}_\alpha \text{Cov}(\boldsymbol{y}) \boldsymbol{\phi}_{\alpha'}.$$

On account of the formulae (3.1.16), (3.2.3), (3.2.5) and (3.2.7), it can easily be checked that, under (3.5.2),

$$\boldsymbol{\phi}_\alpha \text{Cov}(\boldsymbol{y}) \boldsymbol{\phi}_{\alpha'} = \mathbf{O} \tag{3.5.8}$$

for any $\alpha \neq \alpha'$. [In fact, the assumption (3.5.2) is necessary only for the pair $\alpha = 2$, $\alpha' = 3$.] Thus, for any proper block design, the decomposition (3.2.1) implies not only that

$$\text{E}(\boldsymbol{y}) = \text{E}(\boldsymbol{y}_1) + \text{E}(\boldsymbol{y}_2) + \text{E}(\boldsymbol{y}_3) = \boldsymbol{\phi}_1 \boldsymbol{\Delta}'\boldsymbol{\tau} + \boldsymbol{\phi}_2 \boldsymbol{\Delta}'\boldsymbol{\tau} + \boldsymbol{\phi}_3 \boldsymbol{\Delta}'\boldsymbol{\tau},$$

but also that

$$\text{Cov}(\boldsymbol{y}) = \text{Cov}(\boldsymbol{y}_1) + \text{Cov}(\boldsymbol{y}_2) + \text{Cov}(\boldsymbol{y}_3) = \boldsymbol{\phi}_1 \sigma_1^2 + \boldsymbol{\phi}_2 \sigma_2^2 + \boldsymbol{\phi}_3 \sigma_3^2, \tag{3.5.9}$$

where the matrices $\boldsymbol{\phi}_1, \boldsymbol{\phi}_2$ and $\boldsymbol{\phi}_3$ satisfy the conditions (3.2.8) and (3.2.9).

The representation (3.5.9) is a desirable property, as originally indicated for a more general class of designs by Nelder (1965a). After him, the following definition will be adopted (see also Houtman and Speed, 1983, Section 2.2).

Definition 3.5.1. An experiment is said to have the orthogonal block structure (OBS) if the covariance matrix of the random variables observed on the experimental units (plots) has a representation of the form (3.5.9), where the matrices $\{\boldsymbol{\phi}_\alpha\}$ are symmetric, idempotent and pairwise orthogonal, summing to the identity matrix, as in (3.2.8) and (3.2.9).

With this definition, it can now be said that any experiment of a proper block design has the orthogonal block structure, or that it has the OBS property. A natural question, appearing immediately, is whether the proper designs are the only block designs inducing the OBS property. The answer is as follows.

Lemma 3.5.1. *An experiment in a block design has the OBS property, corresponding to* (3.5.9), *if and only if the design is proper.*

Proof. Note that from (3.1.16), (3.2.8) and (3.2.10), in general, $\boldsymbol{\phi}_1 \text{Cov}(\boldsymbol{y}) \boldsymbol{\phi}_\alpha = \mathbf{O}$ for $\alpha = 2, 3$. Hence, the representation (3.5.9) holds if and only if

$$\boldsymbol{\phi}_2 \text{Cov}(\boldsymbol{y}) \boldsymbol{\phi}_3 = n^{-1}(\boldsymbol{D}'\boldsymbol{k}\mathbf{1}_n' - n^{-1}\boldsymbol{k}'\boldsymbol{k}\mathbf{1}_n\mathbf{1}_n')(\sigma_B^2 - K_H^{-1}\sigma_U^2) = \mathbf{O}.$$

But $D'k1'_n = n^{-1}k'k1_n1'_n$ if and only if $k = (k'k/n)1_n$, i.e., if and only if (3.5.2) holds. □

As it follows from Theorem 3.1.2, the condition (3.5.2) is, however, not sufficient to obtain for any s the BLUE of $s'r^{\delta}\tau$ under the overall model (3.1.1). For this purpose, the design needs to be not only proper, but also orthogonal, i.e., such that the condition (2.2.14), now of the form $k^{-1}NN'r^{-\delta}N = N$, holds. In particular, if the design is connected, it has to satisfy the condition $N = b^{-1}r1'_b$ (as it follows from Corollary 2.3.4). Thus, if a design is proper but not orthogonal, one cannot obtain the BLUEs for all parameteric functions of the type $s'r^{\delta}\tau = c'\tau$ under the model (3.1.1). It remains then to seek the estimators within each stratum separately, i.e., under the submodels

$$y_{\alpha} = \phi_{\alpha}y, \quad \alpha = 1, 2, 3, \tag{3.5.10}$$

which for any proper design have the properties (3.5.5) and (3.5.6), with the matrices ϕ_1 and ϕ_2 reduced from (3.2.3) and (3.2.5) to

$$\phi_1 = I_n - k^{-1}D'D \tag{3.5.11}$$

and

$$\phi_2 = k^{-1}D'D - n^{-1}1_n1'_n, \tag{3.5.12}$$

respectively, and with ϕ_3, as defined in (3.2.7). These matrices coincide with the projection matrices used in Example 1 of Bailey and Rowley (1987, Section 3), illustrating an OBS property.

It appears that the BLUEs of treatment parametric functions are obtainable under the submodels (3.5.10), according to the following theorem.

Theorem 3.5.1. *If the design is proper, then, under* (3.5.10), *a function* $w'y_{\alpha} = w'\phi_{\alpha}y$ *is uniformly the BLUE of* $c'\tau$ *if and only if* $\phi_{\alpha}w = \phi_{\alpha}\Delta's$, *where the vectors* c *and* s *are in the relation* $c = \Delta\phi_{\alpha}\Delta's$.

Proof. This can be proved exactly as Theorem 3.2.1, on account of (3.5.5) and (3.5.6). □

Remark 3.5.1. From (2.2.1) and (3.2.10), $1'_v\Delta\phi_{\alpha} = 0'$ for $\alpha = 1, 2$, and hence, the only parametric functions for which the BLUEs may exist under (3.5.10) for $\alpha = 1, 2$ are contrasts. On the other hand, because $\Delta\phi_3\Delta's = n^{-1}(r's)r$ for any s, no contrast will obtain a BLUE under (3.5.10) for $\alpha = 3$. In fact, a function for which the BLUE exists within the third stratum is $c'\tau = (s'r)n^{-1}r'\tau$, i.e., the overall total or any function proportional to that, the overall mean $n^{-1}r'\tau$ in particular. [See Remark 3.2.1, Corollary 3.2.1(b) and Remark 3.2.6(a).]

It follows from Theorem 3.5.1 that if for a given c ($\neq 0$), a vector s exists such that $c = \Delta\phi_{\alpha}\Delta's$, then the BLUE of $c'\tau$ in stratum α is obtainable as

$$\widehat{c'\tau} = s'\Delta y_{\alpha}, \tag{3.5.13}$$

with the variance of the form

$$\mathrm{Var}(\widehat{\boldsymbol{c}'\boldsymbol{\tau}}) = \boldsymbol{s}'\boldsymbol{\Delta}\boldsymbol{\phi}_\alpha\boldsymbol{\Delta}'\boldsymbol{s}\sigma_\alpha^2 = \boldsymbol{c}'(\boldsymbol{\Delta}\boldsymbol{\phi}_\alpha\boldsymbol{\Delta}')^-\boldsymbol{c}\sigma_\alpha^2, \tag{3.5.14}$$

where σ_α^2 is the appropriate stratum variance defined in (3.5.7) and $(\boldsymbol{\Delta}\boldsymbol{\phi}_\alpha\boldsymbol{\Delta}')^-$ is any g-inverse of $\boldsymbol{\Delta}\boldsymbol{\phi}_\alpha\boldsymbol{\Delta}'$.

Explicitly, from (3.5.11), (3.5.12) and (3.2.7), the matrices $\boldsymbol{\Delta}\boldsymbol{\phi}_\alpha\boldsymbol{\Delta}'$ in (3.5.14) are

$$\begin{aligned}
\boldsymbol{\Delta}\boldsymbol{\phi}_1\boldsymbol{\Delta}' &= \boldsymbol{r}^\delta - k^{-1}\boldsymbol{N}\boldsymbol{N}' = \boldsymbol{C}_1, && (3.5.15)\\
\boldsymbol{\Delta}\boldsymbol{\phi}_2\boldsymbol{\Delta}' &= k^{-1}\boldsymbol{N}\boldsymbol{N}' - n^{-1}\boldsymbol{r}\boldsymbol{r}' = \boldsymbol{C}_2, && (3.5.16)\\
\boldsymbol{\Delta}\boldsymbol{\phi}_3\boldsymbol{\Delta}' &= n^{-1}\boldsymbol{r}\boldsymbol{r}' = \boldsymbol{C}_3.
\end{aligned}$$

Now, returning to the decomposition (3.2.1), it implies, as stated at the beginning of Section 3.2.4, that any function $\boldsymbol{s}'\boldsymbol{\Delta}\boldsymbol{y}$, estimating unbiasedly the parametric function $\boldsymbol{s}'\boldsymbol{r}^\delta\boldsymbol{\tau} = \boldsymbol{c}'\boldsymbol{\tau}$, can be resolved into three components, in the form

$$\boldsymbol{s}'\boldsymbol{\Delta}\boldsymbol{y} = \boldsymbol{s}'\boldsymbol{Q}_1 + \boldsymbol{s}'\boldsymbol{Q}_2 + \boldsymbol{s}'\boldsymbol{Q}_3, \tag{3.5.17}$$

where, as in (3.2.28), $\boldsymbol{Q}_\alpha = \boldsymbol{\Delta}\boldsymbol{y}_\alpha = \boldsymbol{\Delta}\boldsymbol{\phi}_\alpha\boldsymbol{y}$ for $\alpha = 1, 2, 3$, i.e., each of the components is a contribution to the estimate from a different stratum.

In connection with formula (3.5.17), it is interesting to consider the three special cases discussed in Section 3.2.4. It should be noticed that when the design is proper, the following implications can be observed. If $\boldsymbol{N}'\boldsymbol{s} = \boldsymbol{0}$, then the condition (3.1.17) is satisfied. Also, if $\boldsymbol{N}'\boldsymbol{s} \neq \boldsymbol{0}$, but $\boldsymbol{\phi}_1\boldsymbol{\Delta}'\boldsymbol{s} = \boldsymbol{0}$, which is equivalent to $\boldsymbol{\Delta}\boldsymbol{\phi}_1\boldsymbol{\Delta}'\boldsymbol{s} = \boldsymbol{0}$, the condition $\boldsymbol{N}'\boldsymbol{r}^{-\delta}\boldsymbol{N}\boldsymbol{N}'\boldsymbol{s} = \boldsymbol{k}^\delta\boldsymbol{N}'\boldsymbol{s}$ holds, which is equivalent to (3.1.17). Finally, if $\boldsymbol{s} \in \mathcal{C}(\boldsymbol{1}_v)$, then $\boldsymbol{N}'\boldsymbol{s} \in \mathcal{C}(\boldsymbol{1}_b)$ and (3.1.17) is again satisfied. But it should be emphasized here that the above observations, except for the first case, may not be true when the condition (3.5.2) does not hold (see Corollary 3.2.5 and Remarks 3.2.8 and 3.2.9). The importance of this statement is that for the three discussed cases, the BLUE under the submodel (3.5.10) for the stratum concerned is simultaneously the BLUE of $\boldsymbol{s}'\boldsymbol{r}^\delta\boldsymbol{\tau}$ $(= \boldsymbol{c}'\boldsymbol{\tau})$ under the overall model (3.1.1), when the design is proper. Estimation of functions $\boldsymbol{s}'\boldsymbol{r}^\delta\boldsymbol{\tau}$ $(= \boldsymbol{c}'\boldsymbol{\tau})$ not restricted to the three special cases considered here will be discussed further after reconsidering the problem discussed in Section 3.3.

As stated in Remark 3.5.1, the only parametric functions for which the BLUEs exist under the submodels $\boldsymbol{y}_1 = \boldsymbol{\phi}_1\boldsymbol{y}$ and $\boldsymbol{y}_2 = \boldsymbol{\phi}_2\boldsymbol{y}$ are contrasts. The discussion held in connection with formula (3.5.17) has revealed that certain contrasts may obtain the BLUEs exclusively under one of these submodels, i.e., either in the intra-block analysis (within the first stratum) or in the inter-block analysis (within the second stratum). For other contrasts, the BLUEs may be provided under both of these submodels, i.e., in both of the analyses.

It has been indicated, in Lemma 3.3.1, that a necessary and sufficient condition for the intra-block and inter-block components to estimate the same contrast

$\boldsymbol{c}'\boldsymbol{\tau} = \boldsymbol{s}'\boldsymbol{r}^{\delta}\boldsymbol{\tau}$ (with the accuracy to a constant factor) is that the vector $\boldsymbol{s}$ satisfies the eigenvector condition (3.3.4). This condition, in particular, remains true for the components $\boldsymbol{s}'\boldsymbol{Q}_1 = \boldsymbol{s}'\boldsymbol{\Delta y}_1$ and $\boldsymbol{s}'\boldsymbol{Q}_2 = \boldsymbol{s}'\boldsymbol{\Delta y}_2$ of the resolution (3.5.17) in the case of a proper design. So, the necessary and sufficient condition for the equality $\mathrm{E}(\boldsymbol{s}'\boldsymbol{Q}_1) = \kappa\, \mathrm{E}(\boldsymbol{s}'\boldsymbol{Q}_2)$, when $\boldsymbol{s}'\boldsymbol{r} = 0$, is

$$\boldsymbol{C}_1\boldsymbol{s} = \varepsilon\boldsymbol{r}^{\delta}\boldsymbol{s} \quad \text{with} \quad 0 < \varepsilon < 1 \quad (\varepsilon = \frac{\kappa}{1+\kappa}), \tag{3.5.18}$$

or its equivalent

$$\boldsymbol{C}_2\boldsymbol{s} = (1-\varepsilon)\boldsymbol{r}^{\delta}\boldsymbol{s} \quad \text{with} \quad 0 < \varepsilon < 1. \tag{3.5.19}$$

Note that the equivalence of (3.5.18) and (3.5.19) holds for any block design, not necessarily proper, with any ε, provided that $\boldsymbol{s}'\boldsymbol{r} = 0$.

Comparing (3.5.18) and (3.5.19) with the condition of Theorem 3.5.1, one can write the following lemma.

Lemma 3.5.2. *If the design is proper, then for any* $\mathbf{c} = \boldsymbol{r}^{\delta}\boldsymbol{s}$ *such that* $\boldsymbol{s}$ *satisfies the equivalent eigenvector conditions* (3.5.18) *and* (3.5.19), *with* $0 < \varepsilon < 1$, *the BLUE of the contrast* $\boldsymbol{c}'\boldsymbol{\tau}$ *is obtainable in both of the analyses, in the intra-block analysis, i.e., under the submodel* $\boldsymbol{y}_1 = \boldsymbol{\phi}_1\boldsymbol{y}$, *and in the inter-block analysis, i.e., under the submodel* $\boldsymbol{y}_2 = \boldsymbol{\phi}_2\boldsymbol{y}$.

Proof. On account of Theorem 3.5.1, if (3.5.18) is satisfied, then, from (3.3.2), the function $\varepsilon^{-1}\boldsymbol{s}'\boldsymbol{Q}_1$ is under $\boldsymbol{y}_1 = \boldsymbol{\phi}_1\boldsymbol{y}$ the BLUE of $\boldsymbol{c}'\boldsymbol{\tau} = \boldsymbol{s}'\boldsymbol{r}^{\delta}\boldsymbol{\tau}$. Similarly, if (3.5.19) is satisfied, then, from (3.3.3), the function $(1-\varepsilon)^{-1}\boldsymbol{s}'\boldsymbol{Q}_2$ is under $\boldsymbol{y}_2 = \boldsymbol{\phi}_2\boldsymbol{y}$ the BLUE of $\boldsymbol{c}'\boldsymbol{\tau} = \boldsymbol{s}'\boldsymbol{r}^{\delta}\boldsymbol{\tau}$. Because, on account of (2.2.1) and (3.2.10), $\mathbf{1}_v'\boldsymbol{\Delta\phi}_1 = \mathbf{0}' = \mathbf{1}_v'\boldsymbol{\Delta\phi}_2$, it is easy to check that if any of the two conditions (3.5.18) and (3.5.19) holds, with $0 < \varepsilon < 1$, so does the other, any of them also implying the equality $\mathbf{1}_v'\boldsymbol{c} = \boldsymbol{r}'\boldsymbol{s} = 0$. □

It should, however, be noticed that because (in view of Lemma 2.3.3) in general the inequality $0 \leq \varepsilon \leq 1$ holds, the cases not considered in Lemma 3.5.2 are those of $\varepsilon = 0$ and $\varepsilon = 1$. To these cases, the following result applies.

Lemma 3.5.3. *If the design is proper, then for any* $\mathbf{c} = \boldsymbol{r}^{\delta}\boldsymbol{s}$ *such that* $\boldsymbol{s}$ *satisfies one of the eigenvector conditions*

$$\boldsymbol{C}_{\alpha}\boldsymbol{s} = \boldsymbol{r}^{\delta}\boldsymbol{s}, \quad \alpha = 1, 2 \text{ or } 3, \tag{3.5.20}$$

the BLUE of the function $\boldsymbol{c}'\boldsymbol{\tau}$ *is obtainable under the overall model* (3.1.1).

Proof. In case of a proper design, the condition (3.1.17) of Theorem 3.1.1 can equivalently be written as

$$\boldsymbol{N}'\boldsymbol{s} = k^{-1}\boldsymbol{N}'\boldsymbol{r}^{-\delta}\boldsymbol{N}\boldsymbol{N}'\boldsymbol{s}. \tag{3.5.21}$$

For $\alpha = 1$, the condition (3.5.20) holds if and only if $\boldsymbol{N}\boldsymbol{N}'\boldsymbol{s} = \mathbf{0}$, i.e., if and only if $\boldsymbol{N}'\boldsymbol{s} = \mathbf{0}$, and if the latter holds, (3.5.21) holds automatically. For $\alpha = 2$, the condition (3.5.20), implying $\boldsymbol{r}'\boldsymbol{s} = \boldsymbol{0}$, holds if and only if $k^{-1}\boldsymbol{N}\boldsymbol{N}'\boldsymbol{s} = \boldsymbol{r}^{\delta}\boldsymbol{s}$, and the latter implies (3.5.21). For $\alpha = 3$, it can easily be checked that the condition (3.5.20) is satisfied if and only if $\boldsymbol{s} \in \mathcal{C}(\mathbf{1}_v)$, i.e., when $\boldsymbol{s}$ is proportional to the vector $\mathbf{1}_v$, and for such $\boldsymbol{s}$ the equality (3.5.21) holds automatically. Thus, any of the three conditions (3.5.20), for $\alpha = 1, 2$ and 3, implies (3.1.17). □

It may be noticed that Lemma 3.5.3 formalizes the discussion following formula (3.5.17). Now a general result can be given.

Theorem 3.5.2. *In case of a proper block design, for any vector* $\mathbf{c} = \boldsymbol{r}^{\delta}\boldsymbol{s}$ *such that* $\boldsymbol{s}$ *satisfies the eigenvector condition*

$$\boldsymbol{C}_{\alpha}\boldsymbol{s} = \varepsilon_{\alpha}\boldsymbol{r}^{\delta}\boldsymbol{s} \quad \text{with} \quad 0 < \varepsilon_{\alpha} \leq 1, \quad \text{for some} \quad \alpha = 1, 2 \text{ or } 3, \tag{3.5.22}$$

where $\varepsilon_1 = \varepsilon, \varepsilon_2 = 1 - \varepsilon$ *and* $\varepsilon_3 = 1$, *the BLUE of the function* $\boldsymbol{c}'\boldsymbol{\tau}$ *is obtainable in the analysis within that stratum* α *[for which* (3.5.22) *is satisfied], i.e., under the corresponding submodel* $\boldsymbol{y}_{\alpha} = \boldsymbol{\phi}_{\alpha}\boldsymbol{y}$, *where it gets the form*

$$(\widehat{\boldsymbol{c}'\boldsymbol{\tau}})_{\alpha} = \varepsilon_{\alpha}^{-1}\boldsymbol{s}'\boldsymbol{Q}_{\alpha} = \varepsilon_{\alpha}^{-1}\boldsymbol{c}'\boldsymbol{r}^{-\delta}\boldsymbol{Q}_{\alpha}, \tag{3.5.23}$$

and its variance is

$$\text{Var}[(\widehat{\boldsymbol{c}'\boldsymbol{\tau}})_{\alpha}] = \varepsilon_{\alpha}^{-1}\boldsymbol{s}'\boldsymbol{r}^{\delta}\boldsymbol{s}\sigma_{\alpha}^{2} = \varepsilon_{\alpha}^{-1}\boldsymbol{c}'\boldsymbol{r}^{-\delta}\boldsymbol{c}\sigma_{\alpha}^{2}. \tag{3.5.24}$$

If (3.5.22) *is satisfied with* $0 < \varepsilon_{\alpha} < 1$, *then two BLUEs of* $\boldsymbol{c}'\boldsymbol{\tau}$ *are obtainable, one under the submodel* $\boldsymbol{y}_1 = \boldsymbol{\phi}_1\boldsymbol{y}$ *and another under* $\boldsymbol{y}_2 = \boldsymbol{\phi}_2\boldsymbol{y}$. *If* (3.5.22) *is satisfied with* $\varepsilon_{\alpha} = 1$, *then the unique BLUE is obtainable within stratum* α *only, being simultaneously the BLUE under the overall model* (3.1.1).

Proof. The existence of the BLUEs follows from Lemma 3.5.2 for $\alpha = 1, 2$ with $0 < \varepsilon_{\alpha} < 1$, and from Lemma 3.5.3 for that $\alpha = 1, 2$ or 3 for which $\varepsilon_{\alpha} = 1$. Formulae (3.5.23) and (3.5.24) follow from the definitions of $\boldsymbol{Q}_{\alpha}$ in (3.5.17) and from the properties (3.5.5) and (3.5.6), considered in view of (3.5.22). □

Remark 3.5.2. For $\alpha = 1$, i.e., for the intra-block analysis, Theorem 3.5.2 applies to proper as well as to nonproper block designs, as can be seen by taking $\boldsymbol{\phi}_1$ in the general form (3.2.3) and comparing formulae (3.5.22), (3.5.23) and (3.5.24) with (3.3.4), (3.3.8) and (3.3.9), respectively.

Remark 3.5.3. Formula (3.5.24) shows that the larger the coefficient ε_{α}, the smaller the variance of the BLUE of $\boldsymbol{c}'\boldsymbol{\tau}$ obtainable within stratum α, the minimum variance being attained when $\varepsilon_{\alpha} = 1$, i.e., when (3.5.23) is the BLUE under the overall model (3.1.1). Thus, for any proper design, ε_{α} can be interpreted as the efficiency factor of the analyzed design for the function $\boldsymbol{c}'\boldsymbol{\tau}$ when it is estimated in the analysis within stratum α. On the other hand, $1 - \varepsilon_{\alpha}$ can be regarded as the relative loss of information incurred when estimating $\boldsymbol{c}'\boldsymbol{\tau}$ in the within stratum α analysis.

Remark 3.5.4. Because $\text{E}(\varepsilon_{\alpha}^{-1}\boldsymbol{s}'\boldsymbol{Q}_{\alpha}) = \boldsymbol{s}'\boldsymbol{r}^{\delta}\boldsymbol{\tau}$ if and only if (3.5.22) holds, for a function $\boldsymbol{c}'\boldsymbol{\tau} = \boldsymbol{s}'\boldsymbol{r}^{\delta}\boldsymbol{\tau}$ to obtain the BLUE within stratum α in the form (3.5.23), the condition (3.5.22) is not only sufficient, but also necessary.

3.6 Proper designs and the notion of general balance

In Section 3.4, the notion of basic contrasts has been introduced, based on a simple algebraic relation between certain contrasts of treatment parameters and the efficiency factors with which the contrasts are estimated in the intra-block analysis of the designed experiment. If a contrast satisfies the eigenvector condition (3.3.4), then it is called a basic contrast of the design (Definition 3.4.1). It follows that the intra-block BLUE (i.e., obtainable in the intra-block analysis) of such contrast is then of a simple form, and also the form of its variance is simple then (see Theorem 3.4.1). On the other hand, it has been shown in Section 3.5 that the eigenvector condition (3.5.18), i.e., the condition (3.3.4), implies and is implied by a dual condition (3.5.19), which in a proper design offers the same kind of simplicity to the inter-block BLUE of the related contrast as the former condition does to the intra-block BLUE of the contrast (Theorem 3.5.2). This relation shows that in case of a proper block design the property determining basic contrasts of the design has its desirable effect not only on the estimation in the intra-block analysis, but also on that in the inter-block analysis. Thus, the results presented in Section 3.4 can now be suitably extended to cover the inter-block analysis as well, provided that this is confined to proper designs only.

First, the following remark can be made.

Remark 3.6.1. The efficiency factor ε_α in Theorem 3.5.2 attains the maximum value 1 for $\alpha = 1$ if and only if $\boldsymbol{N}'\boldsymbol{s} = \boldsymbol{0}$, and for $\alpha = 2$ if and only if $\boldsymbol{N}'\boldsymbol{s} \neq \boldsymbol{0}$, but $\boldsymbol{\phi}_1\boldsymbol{\Delta}'\boldsymbol{s} = \boldsymbol{0}$. On the other hand, a block design is called orthogonal if the condition (2.2.12) holds, i.e., if

$$\boldsymbol{\phi}_1\boldsymbol{\Delta}'\sum_{i=1}^{v-1}\boldsymbol{s}_i\boldsymbol{s}_i'\boldsymbol{N} = \mathbf{O},$$

where $\{\boldsymbol{s}_i\}$ represent basic contrasts (which can be seen from the representation $\boldsymbol{r}^{-\delta} = \sum_{i=1}^{v}\boldsymbol{s}_i\boldsymbol{s}_i'$ and the definition of $\boldsymbol{s}_v$ in Lemma 2.3.3). Thus, if the vector $\boldsymbol{s}$ representing a contrast $\boldsymbol{c}'\boldsymbol{\tau} = \boldsymbol{s}'\boldsymbol{r}^{\delta}\boldsymbol{\tau}$ satisfies (3.5.22) with $\varepsilon_\alpha = 1$ for $\alpha = 1$ or $\alpha = 2$, then the design can be called orthogonal for the contrast. If a proper block design is orthogonal for $\boldsymbol{c}'\boldsymbol{\tau}$, then $(\widehat{\boldsymbol{c}'\boldsymbol{\tau}})_\alpha$ given by (3.5.23), with $\varepsilon_\alpha = 1$, is the BLUE under the overall model (3.1.1), as stated in Theorem 3.5.2.

A relevant extension of Theorem 3.4.1 can be given as follows.

Theorem 3.6.1. *Let $\{\boldsymbol{c}_i'\boldsymbol{\tau} = \boldsymbol{s}_i'\boldsymbol{r}^{\delta}\boldsymbol{\tau}, i = 1, 2, ..., v-1\}$ be any set of basic contrasts of a proper block design, and let $\{\varepsilon_i, i = 1, 2, ..., v-1\}$ be the corresponding eigenvalues of the matrix $\boldsymbol{C}_1$ with respect to $\boldsymbol{r}^{\delta}$. Then, the analysis within stratum α $(= 1, 2)$ provides the BLUEs, of the form*

$$(\widehat{\boldsymbol{c}_i'\boldsymbol{\tau}})_\alpha = \varepsilon_{\alpha i}^{-1}\boldsymbol{s}_i'\boldsymbol{Q}_\alpha = \varepsilon_{\alpha i}^{-1}\boldsymbol{c}_i'\boldsymbol{r}^{-\delta}\boldsymbol{Q}_\alpha, \tag{3.6.1}$$

and their variances

$$\mathrm{Var}[(\widehat{\boldsymbol{c}_i'\boldsymbol{\tau}})_\alpha] = \varepsilon_{\alpha i}^{-1}\sigma_\alpha^2 \tag{3.6.2}$$

and covariances

$$\text{Cov}[(\widehat{c_i'\tau})_\alpha, (\widehat{c_{i'}'\tau})_\alpha] = 0, \quad i \neq i', \tag{3.6.3}$$

for those of the basic contrasts for which the efficiency factors in stratum α, $\varepsilon_{\alpha i} = \varepsilon_i$ *if* $\alpha = 1$ *and* $\varepsilon_{\alpha i} = 1 - \varepsilon_i$ *if* $\alpha = 2$, *are nonzero (positive). Also, the correlations between* $(\widehat{c_i'\tau})_\alpha$ *and* $(\widehat{c_{i'}'\tau})_{\alpha'}$ *are zero for* $\alpha \neq \alpha'$, *whether* $i = i'$ *or* $i \neq i'$.

Proof. On account of Theorem 3.5.2, formulae (3.6.1) and (3.6.2) follow immediately from (3.5.23) and (3.5.24), respectively, whereas (3.6.3) follows from the formula

$$\text{Cov}(\boldsymbol{Q}_\alpha) = \boldsymbol{\Delta}\boldsymbol{\phi}_\alpha\boldsymbol{\Delta}'\sigma_\alpha^2, \tag{3.6.4}$$

holding because of (3.5.6), and from the equality $\boldsymbol{s}_i'\boldsymbol{\Delta}\boldsymbol{\phi}_\alpha\boldsymbol{\Delta}'\boldsymbol{s}_{i'} = 0$, satisfied by any pair of vectors $\boldsymbol{s}_i, \boldsymbol{s}_{i'}$ $(i \neq i')$ in accordance with Definition 3.4.1 and the equivalence between (3.5.18) and (3.5.19). The last statement of the theorem follows from (3.5.8), from which the vectors $\boldsymbol{Q}_1$ and $\boldsymbol{Q}_2$ are uncorrelated. □

Next, an extension of Theorem 3.4.2 is possible.

Theorem 3.6.2. *For any proper block design for which the vectors*

$$\boldsymbol{s}_1, \boldsymbol{s}_2, ..., \boldsymbol{s}_\rho, \boldsymbol{s}_{\rho+1}, \boldsymbol{s}_{\rho+2}, ..., \boldsymbol{s}_h, \boldsymbol{s}_{h+1}, \boldsymbol{s}_{h+2}, ..., \boldsymbol{s}_{v-1}$$

represent basic contrasts and are ordered as in Lemma 2.3.3, a set of contrasts $\boldsymbol{U}'\boldsymbol{\tau}$ *obtains the BLUEs in the analysis within stratum* $\alpha(= 1, 2)$ *if and only if the matrix* $\boldsymbol{U}$ *can be written as* $\boldsymbol{U} = \boldsymbol{r}^\delta \boldsymbol{S}_{(\alpha)}\boldsymbol{A}_{(\alpha)}$, *where* $\boldsymbol{S}_{(1)} = [\boldsymbol{s}_1 : \boldsymbol{s}_2 : \cdots : \boldsymbol{s}_h]$ *and* $\boldsymbol{A}_{(1)} = [\boldsymbol{a}_{11} : \boldsymbol{a}_{12} : \cdots : \boldsymbol{a}_{1h}]'$ *is some matrix of h rows, and where* $\boldsymbol{S}_{(2)} = [\boldsymbol{s}_{\rho+1} : \boldsymbol{s}_{\rho+2} : \cdots : \boldsymbol{s}_{v-1}]$ *and* $\boldsymbol{A}_{(2)} = [\boldsymbol{a}_{2,\rho+1} : \boldsymbol{a}_{2,\rho+2} : \cdots : \boldsymbol{a}_{2,v-1}]'$ *is some matrix of* $v - \rho - 1$ *rows. If* $\boldsymbol{U}$ *is such, then the BLUEs provided by the analysis within stratum* α *are of the form*

$$(\widehat{\boldsymbol{U}'\boldsymbol{\tau}})_\alpha = \boldsymbol{A}_{(\alpha)}'\boldsymbol{\varepsilon}_{(\alpha)}^{-\delta}\boldsymbol{S}_{(\alpha)}'\boldsymbol{Q}_\alpha \tag{3.6.5}$$

and their covariance matrix is of the form

$$\text{Cov}[(\widehat{\boldsymbol{U}'\boldsymbol{\tau}})_\alpha] = \boldsymbol{A}_{(\alpha)}'\boldsymbol{\varepsilon}_{(\alpha)}^{-\delta}\boldsymbol{A}_{(\alpha)}\sigma_\alpha^2, \tag{3.6.6}$$

where $\boldsymbol{\varepsilon}_{(1)}^\delta = \text{diag}[\varepsilon_1, \varepsilon_2, ..., \varepsilon_h]$ *and* $\boldsymbol{\varepsilon}_{(2)}^\delta = \text{diag}[1 - \varepsilon_{\rho+1}, 1 - \varepsilon_{\rho+2}, ..., 1 - \varepsilon_{v-1}]$.

Proof. The proof follows the same pattern as that of Theorem 3.4.2, now on account of Theorem 3.5.1. Lemma 2.3.3 can be used here by noting that it allows the matrix $\boldsymbol{C}_\alpha = \boldsymbol{\Delta}\boldsymbol{\phi}_\alpha\boldsymbol{\Delta}'$ to be written as

$$\boldsymbol{C}_\alpha = \boldsymbol{r}^\delta\boldsymbol{S}_{(\alpha)}\boldsymbol{\varepsilon}_{(\alpha)}^\delta\boldsymbol{S}_{(\alpha)}'\boldsymbol{r}^\delta \quad \text{for } \alpha = 1, 2. \tag{3.6.7}$$

Formula (3.6.5) is then obtainable by writing $\boldsymbol{U} = \boldsymbol{C}_\alpha\boldsymbol{S}_{(\alpha)}\boldsymbol{\varepsilon}_{(\alpha)}^{-\delta}\boldsymbol{A}_{(\alpha)}$ and applying Theorem 3.5.1, whereas (3.6.6) results directly from (3.6.4) and (3.6.7). □

Corollary 3.6.1. *For any proper orthogonal block design a set of contrasts $U'\tau$ obtains the BLUEs under the overall model (3.1.1) if and only if the columns of the matrix U belong to $C(r^\delta)$. If the contrasts in the set are ordered so that U can be written as*

$$U = r^\delta[S_{(1)}A_{(1)} : S_{(2)}A_{(2)}], \tag{3.6.8}$$

where $S_{(1)} = [s_1 : s_2 : \cdots : s_h]$, $S_{(2)} = [s_{h+1} : s_{h+2} : \cdots : s_{v-1}]$ and $A_{(1)}, A_{(2)}$ are some matrices of conformable numbers of rows, then the BLUEs get the form

$$\widehat{U'\tau} = \begin{bmatrix} A'_{(1)}S'_{(1)} \\ A'_{(2)}S'_{(2)} \end{bmatrix} \Delta y, \tag{3.6.9}$$

and their covariance matrix is of the form

$$\mathrm{Cov}(\widehat{U'\tau}) = \begin{bmatrix} A'_{(1)}A_{(1)}\sigma_1^2 & O \\ O & A'_{(2)}A_{(2)}\sigma_2^2 \end{bmatrix}. \tag{3.6.10}$$

Proof. The first part of the corollary follows directly from Theorem 3.1.2. Formulae (3.6.9) and (3.6.10) are obtainable from Theorem 3.6.2 on account of Corollaries 2.3.3 and 3.2.3, the latter allowing the vectors Q_1 and Q_2 to be replaced by Δy. These formulae can also be obtained directly from Theorem 3.1.2 and formula (3.1.16), in view of (3.6.8). □

Remark 3.6.2. Corollary 3.6.1 is also true for a disconnected orthogonal block design, which instead of being proper satisfies the weaker condition (ii) of Theorem 3.1.2, but then (3.6.10) is to be replaced by

$$\mathrm{Cov}(\widehat{U'\tau}) =$$

$$\begin{bmatrix} A'_{(1)}A_{(1)}\sigma_1^2 & O \\ O & A'_{(2)}S'_{(2)}\tilde{k}^\delta r^\delta S_{(2)}A_{(2)}(\sigma_B^2 - K_H^{-1}\sigma_U^2) + A'_{(2)}A_{(2)}\sigma_1^2 \end{bmatrix},$$

with $\tilde{k}^\delta$ defined in (2.2.10).

An extension of Corollary 3.4.1 is the following corollary.

Corollary 3.6.2. *For any proper block design for which the vectors $s_1, ..., s_\rho$ and $s_{h+1}, ..., s_{v-1}$ represent basic contrasts, as in Theorem 3.6.2, the first ρ receiving the unit efficiency factors in the intra-block analysis and the last $v-h-1$ receiving such efficiency factors in the inter-block analysis (i.e., the zero efficiency factors in the intra-block analysis), a set of contrasts $U'\tau$ admits the BLUEs under the overall model (3.1.1) if and only if the matrix U can (possibly after reordering its columns) be written as in (3.6.8), where $S_{(1)} = S_0 = [s_1 : s_2 : \cdots : s_\rho]$, $S_{(2)} = S_m = [s_{h+1} : s_{h+2} : \cdots : s_{v-1}]$, and $A_{(1)} = A_0$, $A_{(2)} = A_m$ are some matrices of conformable numbers of rows. The BLUEs are then obtainable in the form (3.6.9) and their covariance matrix in the form (3.6.10).*

Proof. The first part of the corollary follows directly from Theorem 3.1.1, as it can be seen when applying (2.3.3) to the condition (3.1.17), which then becomes equivalent to the equality

$$\sum_{i=\rho+1}^{v} (1-\varepsilon_i)^{3/2} \boldsymbol{t}_i \boldsymbol{s}_i' \boldsymbol{r}^{\delta} \boldsymbol{s} = \sum_{i=\rho+1}^{v} (1-\varepsilon_i)^{1/2} \boldsymbol{t}_i \boldsymbol{s}_i' \boldsymbol{r}^{\delta} \boldsymbol{s},$$

and this, because any $\boldsymbol{s}$ can be written as $\boldsymbol{s} = \sum_{i=1}^{v-1} a_i \boldsymbol{s}_i$, where $\{a_i\}$ are some scalars, is equivalent to

$$\sum_{i=\rho+1}^{v} (1-\varepsilon_i)^{3/2} a_i \boldsymbol{t}_i = \sum_{i=\rho+1}^{v} (1-\varepsilon_i)^{1/2} a_i \boldsymbol{t}_i,$$

which in turn holds if and only if ε_i is either 0 or 1 for any i for which $a_i \neq 0$. The second part of the corollary is obtainable as in Corollary 3.6.1. □

Also, Remark 3.4.1(a) and (b) can be extended.

Remark 3.6.3. (a) In the notation of Corollary 3.6.2, a proper block design for which $\rho \geq 1$ and/or $v - h \geq 2$ can be called orthogonal for the set of contrasts $\boldsymbol{U}'\boldsymbol{\tau} = [\boldsymbol{S}_0\boldsymbol{A}_0 : \boldsymbol{S}_m\boldsymbol{A}_m]'\boldsymbol{r}^{\delta}\boldsymbol{\tau}$. (See again Remark 3.6.1.)

(b) It follows from Theorem 3.6.2 and Corollary 3.6.2 that the efficiency factor of a proper block design for a contrast $\boldsymbol{u}'\boldsymbol{\tau} = \boldsymbol{a}_\alpha' \boldsymbol{S}_\alpha' \boldsymbol{r}^{\delta}\boldsymbol{\tau}$ estimated in the stratum α analysis is of the form

$$\varepsilon[(\widehat{\boldsymbol{u}'\boldsymbol{\tau}})_\alpha)] = \boldsymbol{a}_{(\alpha)}'\boldsymbol{a}_{(\alpha)} / \boldsymbol{a}_{(\alpha)}' \boldsymbol{\varepsilon}_{(\alpha)}^{-\delta} \boldsymbol{a}_{(\alpha)} \quad \text{for } \alpha = 1, 2.$$

At this point, it will be useful to notice that when multiplicities of the eigenvalues of $\boldsymbol{C}_1$ with respect to $\boldsymbol{r}^{\delta}$ are taken into account, then appropriate spectral decomposition can be given not only for the matrix $\boldsymbol{C}_1$, as in (3.4.10), but also for the matrix $\boldsymbol{C}_2$ defined in (3.5.16). It can be written as

$$\boldsymbol{C}_2 = \boldsymbol{\Delta}\boldsymbol{\phi}_2\boldsymbol{\Delta}' = \boldsymbol{r}^{\delta} \sum_{\beta=1}^{m} (1-\varepsilon_\beta) \boldsymbol{H}_\beta \boldsymbol{r}^{\delta} = \boldsymbol{r}^{\delta} \sum_{\beta=1}^{m} (1-\varepsilon_\beta) \boldsymbol{L}_\beta, \tag{3.6.11}$$

where $\boldsymbol{H}_1, ..., \boldsymbol{H}_{m-1}$ and $\boldsymbol{L}_1, ..., \boldsymbol{L}_{m-1}$ are as in (3.4.10), and

$$\boldsymbol{H}_m = \sum_{j=1}^{\rho_m} \boldsymbol{s}_{mj}\boldsymbol{s}_{mj}' = \sum_{j=1}^{v-1-h} \boldsymbol{s}_{h+j}\boldsymbol{s}_{h+j}' \quad \text{and} \quad \boldsymbol{L}_m = \boldsymbol{H}_m \boldsymbol{r}^{\delta},$$

with $\boldsymbol{s}_{mj} = \boldsymbol{s}_{h+j}$ for $j = 1, 2, ..., \rho_m = v - 1 - h$, according to the notation in (2.3.1), and where $\varepsilon_m = 0$.

Now, the following extension of Corollary 3.4.2 can be given.

Corollary 3.6.3. *Consider a subset of basic contrasts of a proper block design represented by the eigenvectors $s_{\beta 1}, s_{\beta 2}, ..., s_{\beta \rho_\beta}$ of C_1 with respect to r^δ corresponding to a common eigenvalue ε_β (not necessarily positive) of multiplicity ρ_β. Then, for a set of contrasts $U'_\beta \tau = A'_\beta S'_\beta r^\delta \tau$, where $S_\beta = [s_{\beta 1} : s_{\beta 2} : \cdots : s_{\beta \rho_\beta}]$, and where A_β is some matrix of ρ_β rows, the stratum α $(= 1, 2)$ analysis provides the BLUEs of the form*

$$(\widehat{U'_\beta \tau})_\alpha = \varepsilon_{\alpha\beta}^{-1} A'_\beta S'_\beta Q_\alpha, \tag{3.6.12}$$

with the covariance matrix of the form

$$\mathrm{Cov}[(\widehat{U'_\beta \tau})_\alpha] = \varepsilon_{\alpha\beta}^{-1} A'_\beta A_\beta \sigma_\alpha^2, \tag{3.6.13}$$

provided that $\varepsilon_{\alpha\beta} > 0$, $\varepsilon_{1\beta} = \varepsilon_\beta$ being the common efficiency factor of the design for the contrasts estimated in the intra-block analysis and $\varepsilon_{2\beta} = 1 - \varepsilon_\beta$ being such a factor for the contrasts estimated in the inter-block analysis.

Proof. This result follows immediately from Theorem 3.6.2 and Remark 3.6.3(b). □

For completeness, it should also be noted that, in the spirit of Remark 3.6.1, one can say that any block design is orthogonal for a function of the type $c'\tau = as'_v r^\delta \tau$, where a is a nonzero scalar and $s_v = n^{-1/2} 1_v$ (as in Lemma 2.3.3). Furthermore, also the matrix $C_3 = \Delta \phi_3 \Delta'$ covered by Theorem 3.5.2 can formally be written in its "spectral decomposition" form as

$$C_3 = \Delta \phi_3 \Delta' = r^\delta s_v s'_v r^\delta \quad (= n^{-1} r r'). \tag{3.6.14}$$

The three representations (3.4.10), (3.6.11) and (3.6.14), together with the general results established in the present section for proper block designs, give rise to the following concept of balance.

Definition 3.6.1. A proper block design inducing the OBS property defined by $\{\phi_\alpha\}$ is said to be generally balanced (GB) with respect to a decomposition

$$\mathcal{C}(\Delta') = \oplus_\beta \mathcal{C}(\Delta' S_\beta) \tag{3.6.15}$$

(the symbol $\oplus_\beta$ denoting the direct sum of the subspaces taken over β), if there exist scalars $\{\varepsilon_{\alpha\beta}\}$ such that for all α (= 1, 2, 3)

$$C_\alpha = \Delta \phi_\alpha \Delta' = \sum_\beta \varepsilon_{\alpha\beta} r^\delta H_\beta r^\delta \tag{3.6.16}$$

[the summation being taken over all β that appear in (3.6.15), $\beta = 0, 1, ..., m, m+1$], where $H_\beta = S_\beta S'_\beta$ for $\beta = 0, 1, ..., m, m+1$ ($H_{m+1} = s_v s'_v$), and the matrices $\{S_\beta\}$ are such that

$$S'_\beta r^\delta S_\beta = I_{\rho_\beta} \text{ for any } \beta \text{ and } S'_\beta r^\delta S_{\beta'} = O \quad \text{for } \beta \neq \beta'.$$

It can easily be shown (see below) that Definition 3.6.1 is equivalent to the definition of GB given by Houtman and Speed (1983, Section 4.1) when applied to a proper block design, and so it coincides with the notion of general balance introduced by Nelder (1965b).

The following result (from Caliński, 1993b, p. 32) explains the meaning of Definition 3.6.1, relating it to the notion of basic contrasts and the theory established for them.

Lemma 3.6.1. *A proper block design is GB with respect to the decomposition* (3.6.15) *if and only if the matrices* $\{\boldsymbol{S}_\beta\}$ *in Definition* 3.6.1 *satisfy the conditions*

$$\boldsymbol{\Delta}\boldsymbol{\phi}_\alpha\boldsymbol{\Delta}'\boldsymbol{S}_\beta = \varepsilon_{\alpha\beta}\boldsymbol{r}^\delta\boldsymbol{S}_\beta \qquad (3.6.17)$$

for all α *and* β.

Proof. The implication from (3.6.17) to (3.6.16) can be shown as for the representations of $\boldsymbol{\Delta}\boldsymbol{\phi}_\alpha\boldsymbol{\Delta}'$ given in (3.4.10), (3.6.11) and (3.6.14), i.e. by noting that $\boldsymbol{r}^{-\delta} = \sum_{\beta=0}^{m+1}\boldsymbol{S}_\beta\boldsymbol{S}_\beta'$, with $\boldsymbol{S}_{m+1} = \boldsymbol{s}_v$, which also implies that

$$\boldsymbol{\Delta}' = \boldsymbol{\Delta}'\sum_{\beta=0}^{m}\boldsymbol{S}_\beta\boldsymbol{S}_\beta'\boldsymbol{r}^\delta + \boldsymbol{\Delta}'\boldsymbol{s}_v\boldsymbol{s}_v'\boldsymbol{r}^\delta. \qquad (3.6.18)$$

The reverse implication is immediate, because of the assumed $\boldsymbol{r}^\delta$-orthonormality of the columns of $\{\boldsymbol{S}_\beta\}$, within and between the matrices. □

Remark 3.6.4. It follows from Lemma 3.6.1, on account of Definition 3.4.1, that any proper block design is GB with respect to the decomposition

$$\mathcal{C}(\boldsymbol{\Delta}') = \mathcal{C}(\boldsymbol{\Delta}'\boldsymbol{S}_0) \oplus \mathcal{C}(\boldsymbol{\Delta}'\boldsymbol{S}_1) \oplus \cdots \oplus \mathcal{C}(\boldsymbol{\Delta}'\boldsymbol{S}_m) \oplus \mathcal{C}(\boldsymbol{\Delta}'\boldsymbol{s}_v), \qquad (3.6.19)$$

where the matrices $\boldsymbol{S}_0, \boldsymbol{S}_1, ..., \boldsymbol{S}_m$ represent basic contrasts of the design, those represented by the columns of $\boldsymbol{S}_\beta$ receiving in the intra-block analysis a common efficiency factor $\varepsilon_{1\beta} = \varepsilon_\beta$ and in the inter-block analysis a common efficiency factor $\varepsilon_{2\beta} = 1 - \varepsilon_\beta$, and $\boldsymbol{s}_v = n^{-1/2}\boldsymbol{1}_v$. Thus, for the estimation of contrasts in the subspaces $\mathcal{C}(\boldsymbol{\Delta}'\boldsymbol{S}_\beta)$, $\beta = 0, 1, ..., m$, Corollary 3.6.3, with (3.6.12) and (3.6.13), is applicable.

Note that the equality (3.6.16) above can equivalently be written as

$$\boldsymbol{\Delta}'\boldsymbol{r}^{-\delta}\boldsymbol{\Delta}\boldsymbol{\phi}_\alpha\boldsymbol{\Delta}'\boldsymbol{r}^{-\delta}\boldsymbol{\Delta} = \sum_\beta \varepsilon_{\alpha\beta}\boldsymbol{\Delta}'\boldsymbol{H}_\beta\boldsymbol{\Delta} = \sum_\beta \varepsilon_{\alpha\beta}\boldsymbol{P}_{\boldsymbol{\Delta}'\boldsymbol{S}_\beta},$$

which is exactly the condition of Houtman and Speed (1983, Section 4.1).

Also, it should be mentioned that the notion of GB stems back to the early work by Jones (1959), as it will be indicated in Section 4.3, together with some illustration of the origins and the sense of the concept of GB (Section 4.3.1). For more general considerations concerning certain constructional conditions that imply GB of the design, see Bailey and Rowley (1990). Further discussion about the pros and cons of GB can be found in Mejza (1992) and Bailey (1994).

3.7 Recovery of inter-block information in a general block design

The theory of estimation developed until now has shown that unless the function $c'\tau$ satisfies one of the conditions (a), (b) or (c) of Corollary 3.2.5 [or the weaker condition given in Remark 3.2.8 for (b)], i.e., satisfies the equality (3.1.17), the BLUE of it does not exist under the overall model (3.1.1). Instead, BLUEs obtainable under submodels (3.2.2), (3.2.4) or (3.2.6) can be used. In particular, if $c'\tau$ is a contrast that does not satisfy (3.1.17), its BLUE is available under (3.2.2) and under (3.2.4), provided that the condition (3.2.16), or its equivalence (3.2.17), is satisfied. This statement means that in many cases, the estimation of a contrast can be based on information available in two strata of the experiment, the intra-block and the inter-block strata. But each of them gives a separate estimate of the contrast, the best within the relevant stratum. These estimates are usually different in their actual values, sometimes even quite diverse. This is a serious disadvantage and, therefore, a natural question that may be asked in this context is whether and how it is possible to combine the two different BLUEs of the same contrast into one estimator that would provide a single estimate in a somehow optimal way. Methods dealing with this problem, originated by Yates (1939, 1940a, 1940b), are known in the literature as methods of the recovery of inter-block information. In this section, a general theory underlying the recovery of inter-block information will be presented in a way that is applicable to any block design, proper or not. Application of the theory to proper designs will be discussed in Section 3.8.

3.7.1 BLUEs under known variance components

In an attempt to solve the problem of combining BLUEs from different strata on the basis of the theory established earlier in the present chapter, it will be instructive to make provisionally an unrealistic assumption that the variance components appearing in the covariance matrix (3.1.16) are known, and then to look for the BLUE under the so-assumed model (3.1.1). For this reason, the following results are essential.

Lemma 3.7.1. *Let the model be as in* (3.1.1), *with the expectation vector* (3.1.15) *and the covariance matrix* (3.1.16), *which can also be written as*

$$\mathrm{Cov}(\boldsymbol{y}) = \sigma_1^2(\boldsymbol{D}'\boldsymbol{\Gamma}\boldsymbol{D} + \boldsymbol{I}_n), \tag{3.7.1}$$

where

$$\boldsymbol{\Gamma} = \gamma \boldsymbol{I}_b - N_B^{-1}(\sigma_B^2/\sigma_1^2)\mathbf{1}_b\mathbf{1}_b', \tag{3.7.2}$$

with

$$\sigma_1^2 = \sigma_U^2 + \sigma_e^2 \quad \textit{and} \quad \gamma = (\sigma_B^2 - K_H^{-1}\sigma_U^2)/\sigma_1^2.$$

Further, suppose that the true value of γ *is known. Then, under this model,*

(a) *any function* $\boldsymbol{w}'\boldsymbol{y}$ *that is the BLUE of its expectation* $\boldsymbol{w}'\boldsymbol{\Delta}'\boldsymbol{\tau}$,

(b) *a vector that is the BLUE of the expectation vector* $\boldsymbol{\Delta}'\boldsymbol{\tau}$,
(c) *a vector that gives the residuals,*
all remain the same under an alternative model resulting from the replacement of (3.7.2) *by* $\boldsymbol{\Gamma} = \gamma \boldsymbol{I}_b$, *i.e., a model with the covariance matrix*

$$\text{Cov}(\boldsymbol{y}) = \sigma_1^2(\gamma \boldsymbol{D}'\boldsymbol{D} + \boldsymbol{I}_n) = \sigma_1^2 \boldsymbol{T}, \tag{3.7.3}$$

where $\boldsymbol{T} = \gamma \boldsymbol{D}'\boldsymbol{D} + \boldsymbol{I}_n$. *The matrix* $\boldsymbol{T}$ *is positive definite (p.d.) if* $\gamma > -1/k_{\max}$, *where* $k_{\max} = \max_j k_j$.

Proof. These results follow directly from the fact that $\boldsymbol{1}_n'(\boldsymbol{I}_n - \boldsymbol{P}_{\Delta'}) = \boldsymbol{0}'$, which implies the equality

$$(\boldsymbol{D}'\boldsymbol{\Gamma}\boldsymbol{D} + \boldsymbol{I}_n)(\boldsymbol{I}_n - \boldsymbol{P}_{\Delta'}) = (\gamma \boldsymbol{D}'\boldsymbol{D} + \boldsymbol{I}_n)(\boldsymbol{I}_n - \boldsymbol{P}_{\Delta'}). \tag{3.7.4}$$

This equality, in turn, implies that (a) the condition for $\boldsymbol{w}$, given in Theorem 1.2.1, is satisfied under the original model if and only if it is satisfied under the alternative model, and that (b) the BLUE of $\boldsymbol{\Delta}'\boldsymbol{\tau}$, as given by Rao (1974, Theorem 3.2), remains unchanged when (3.7.1) is replaced by (3.7.3), so that (c) the residual vector is also unchanged then. [To see this, refer to formulae (3.5) and (3.8) of Rao, 1974, in view of (3.7.4) above.] The matrix $\boldsymbol{T}$ in (3.7.3) is p.d. if, for any vector $\boldsymbol{x}$, $\gamma > -\boldsymbol{x}'\boldsymbol{x}/\boldsymbol{x}'\boldsymbol{D}'\boldsymbol{D}\boldsymbol{x}$, and this holds if $\gamma > -1/\kappa_{\max}$, where $\kappa_{\max}$ is the maximum eigenvalue of $\boldsymbol{D}'\boldsymbol{D}$ (and, hence, of $\boldsymbol{D}\boldsymbol{D}'$), this being $k_{\max}$. (This assumption is reasonable, as noticed by Patterson and Thompson, 1971, p. 546.) □

The matrix $\boldsymbol{T}$ used here, and henceforth, should not be confused with $\boldsymbol{T} = \boldsymbol{\Delta}\boldsymbol{y}$ used in Chapter 1.

A result more general than Lemma 3.7.1 has been given by Kala (1981, Theorem 6.2). Now the following can be established.

Theorem 3.7.1. *Under the model and assumptions as those adopted in Lemma* 3.7.1, *including the assumption that* γ *is known and exceeds* $-1/k_{\max}$, *a function of* $\boldsymbol{w}'\boldsymbol{y}$ *is the BLUE of* $\boldsymbol{c}'\boldsymbol{\tau}$ *if and only if*

$$\boldsymbol{w} = \boldsymbol{T}^{-1}\boldsymbol{\Delta}'\boldsymbol{s} = [\boldsymbol{\phi}_1 + \boldsymbol{D}'\boldsymbol{k}^{-\delta}(\boldsymbol{k}^{-\delta} + \gamma \boldsymbol{I}_b)^{-1}\boldsymbol{k}^{-\delta}\boldsymbol{D}]\boldsymbol{\Delta}'\boldsymbol{s}, \tag{3.7.5}$$

where the vectors $\mathbf{c}$ *and* $\boldsymbol{s}$ *are in the relation*

$$\mathbf{c} = \boldsymbol{\Delta}\boldsymbol{T}^{-1}\boldsymbol{\Delta}'\boldsymbol{s} = \boldsymbol{\Delta}[\boldsymbol{\phi}_1 + \boldsymbol{D}'\boldsymbol{k}^{-\delta}(\boldsymbol{k}^{-\delta} + \gamma \boldsymbol{I}_b)^{-1}\boldsymbol{k}^{-\delta}\boldsymbol{D}]\boldsymbol{\Delta}'\boldsymbol{s}, \tag{3.7.6}$$

with $\boldsymbol{\phi}_1$ *as defined in* (3.2.3).

Proof. On account of Theorem 1.2.1, under the considered model a function $\boldsymbol{w}'\boldsymbol{y}$ is the BLUE of its expectation if and only if the condition (3.1.18) is satisfied. But, by (3.7.4), the condition is reduced to $(\boldsymbol{I}_n - \boldsymbol{P}_{\Delta'})\boldsymbol{T}\boldsymbol{w} = \boldsymbol{0}$, where $\boldsymbol{T}$ is as in (3.7.3), or equivalently to

$$\boldsymbol{T}\boldsymbol{w} = \boldsymbol{\Delta}'\boldsymbol{s} \text{ for some } \boldsymbol{s}. \tag{3.7.7}$$

Furthermore, from Lemma 3.7.1, if $\gamma > -1/k_{\max}$, which is assumed here, the matrix $\boldsymbol{T}$ in (3.7.7) is p.d., and so has the inverse, which can be written as

$$\begin{aligned}\boldsymbol{T}^{-1} = (\gamma\boldsymbol{D}'\boldsymbol{D}+\boldsymbol{I}_n)^{-1} &= \boldsymbol{I}_n - \boldsymbol{D}'(\boldsymbol{k}^{\delta}+\gamma^{-1}\boldsymbol{I}_b)^{-1}\boldsymbol{D} \\ &= \boldsymbol{\phi}_1 + \boldsymbol{D}'\boldsymbol{k}^{-\delta}(\boldsymbol{k}^{-\delta}+\gamma\boldsymbol{I}_b)^{-1}\boldsymbol{k}^{-\delta}\boldsymbol{D} \quad (3.7.8)\end{aligned}$$

(see, e.g., Rao, 1973, Section 1f, Complement 34). Hence, the result (3.7.5). The relation (3.7.6) follows from (3.7.5) on account of the equality $\boldsymbol{w}'\boldsymbol{\Delta}'\boldsymbol{\tau} = \boldsymbol{c}'\boldsymbol{\tau}$ to be true for any $\boldsymbol{\tau}$. □

On the other hand, taking into account Lemma 3.7.1 and applying a general theory given by Rao (1974), one obtains the following results.

Theorem 3.7.2. *Under the model and assumptions as those adopted in Theorem* 3.7.1,

(a) *the BLUE of* $\boldsymbol{\tau}$ *is of the form*

$$\hat{\boldsymbol{\tau}} = (\boldsymbol{\Delta}\boldsymbol{T}^{-1}\boldsymbol{\Delta}')^{-1}\boldsymbol{\Delta}\boldsymbol{T}^{-1}\boldsymbol{y}, \quad (3.7.9)$$

with $\boldsymbol{T}^{-1}$ *as in* (3.7.8);

(b) *the covariance matrix of* $\hat{\boldsymbol{\tau}}$ *is then*

$$\text{Cov}(\hat{\boldsymbol{\tau}}) = \sigma_1^2(\boldsymbol{\Delta}\boldsymbol{T}^{-1}\boldsymbol{\Delta}')^{-1} - N_B^{-1}\sigma_B^2\mathbf{1}_v\mathbf{1}_v'; \quad (3.7.10)$$

(c) *the BLUE of* $\boldsymbol{c}'\boldsymbol{\tau}$ *for any* $\boldsymbol{c}$ *is* $\boldsymbol{c}'\hat{\boldsymbol{\tau}}$, *with the variance* $\boldsymbol{c}'\text{Cov}(\hat{\boldsymbol{\tau}})\boldsymbol{c}$, *which reduces to*

$$\text{Var}(\widehat{\boldsymbol{c}'\boldsymbol{\tau}}) = \sigma_1^2\boldsymbol{c}'(\boldsymbol{\Delta}\boldsymbol{T}^{-1}\boldsymbol{\Delta}')^{-1}\boldsymbol{c}, \quad (3.7.11)$$

if $\boldsymbol{c}'\boldsymbol{\tau}$ *is a contrast;*

(d) *the MINQUE of* σ_1^2 *is*

$$\begin{aligned}\hat{\sigma}_1^2 &= (n-v)^{-1}\|\boldsymbol{y}-\boldsymbol{\Delta}'\hat{\boldsymbol{\tau}}\|^2_{\boldsymbol{T}^{-1}} \\ &= (n-v)^{-1}(\boldsymbol{y}-\boldsymbol{\Delta}'\hat{\boldsymbol{\tau}})'\boldsymbol{T}^{-1}(\boldsymbol{y}-\boldsymbol{\Delta}'\hat{\boldsymbol{\tau}}). \quad (3.7.12)\end{aligned}$$

Proof. From Theorem 3.2(c) of Rao (1974), the BLUE of $\boldsymbol{\Delta}'\boldsymbol{\tau}$ is

$$\widehat{\boldsymbol{\Delta}'\boldsymbol{\tau}} = \boldsymbol{P}_{\Delta'(T^{-1})}\boldsymbol{y},$$

on account of Lemma 3.7.1. Because of the equalities

$$\boldsymbol{P}_{\Delta'(T^{-1})} = \boldsymbol{\Delta}'(\boldsymbol{\Delta}\boldsymbol{T}^{-1}\boldsymbol{\Delta}')^{-1}\boldsymbol{\Delta}\boldsymbol{T}^{-1}, \quad (3.7.13)$$

and $\boldsymbol{\tau} = (\boldsymbol{\Delta}\boldsymbol{\Delta}')^{-1}\boldsymbol{\Delta}\boldsymbol{\Delta}'\boldsymbol{\tau}$, the result (3.7.9) follows, which then implies (3.7.10), when noting that (3.7.1) can be written as

$$\text{Cov}(\boldsymbol{y}) = \sigma_1^2\boldsymbol{T} - N_B^{-1}\sigma_B^2\mathbf{1}_n\mathbf{1}_n'. \quad (3.7.14)$$

Formula (3.7.11) follows from (3.7.10) directly, as $c'\mathbf{1}_v = 0$ for a contrast. Formula (3.7.12) follows from Theorem 3.4(c) of Rao (1974) by noting that the residual vector

$$y - \widehat{\Delta'\tau} = (I_n - P_{\Delta'(T^{-1})})y,$$

when using the relevant T^{-1}-norm (see Appendix A.1), gives the residual sum of squares of the form

$$\|(I_n - P_{\Delta'(T^{-1})})y\|^2_{T^{-1}} = y'(I_n - P_{\Delta'(T^{-1})})'T^{-1}(I_n - P_{\Delta'(T^{-1})})y, \quad (3.7.15)$$

equivalent to (3.13) of Rao (1974), which provides the MINQUE of $d\sigma_1^2$, where $d = \text{rank}(T : \Delta') - \text{rank}(\Delta') = n - v$, because T is p.d. [Note, from (3.7.14), that the matrix T used here can be treated exactly as the matrix T used in Rao's (1974) Theorem 3.2, because the present T satisfies the conditions required there, as can easily be checked.] □

Remark 3.7.1. (a) Using as c the vector given in (3.7.6), one obtains from (3.7.9) the BLUE

$$\widehat{c'\tau} = c'\hat{\tau} = s'\Delta T^{-1}y = s'\Delta[\phi_1 + D'k^{-\delta}(k^{-\delta} + \gamma I_b)^{-1}k^{-\delta}D]y,$$

which agrees with $w'y$ given by (3.7.5).

(b) The result (3.7.9) can be seen as the solution of the normal equations

$$\Delta T^{-1}\Delta'\hat{\tau} = \Delta T^{-1}y,$$

which can also be written as

$$\Delta[\phi_1 + D'k^{-\delta}(k^{-\delta} + \gamma I_b)^{-1}k^{-\delta}D]\Delta'\hat{\tau} = \Delta[\phi_1 + D'k^{-\delta}(k^{-\delta} + \gamma I_b)^{-1}k^{-\delta}D]y.$$

Remark 3.7.2. Comparing the results in Remark 3.7.1 with those given in Section 3.2.1, it can easily be seen that the recovery of the inter-block information is controlled by the matrix $D'k^{-\delta}(k^{-\delta} + \gamma I_b)^{-1}k^{-\delta}D$, a component of T^{-1}. Its value depends very much on the true value of γ. Because, for $\gamma > -1/k_{\max}$ (as assumed),

$$(k^{-\delta} + \gamma I_b)^{-1} = \begin{cases} k^{\delta} - k^{\delta}(k^{\delta} + \gamma^{-1}I_b)^{-1}k^{\delta} & \text{if } \gamma \neq 0, \\ k^{\delta} & \text{if } \gamma = 0, \end{cases}$$

it is permissible to write

$$T^{-1} = \phi_1 + D'k^{-\delta}(k^{-\delta} + \gamma I_b)^{-1}k^{-\delta}D = I_n - D'k_*^{-\delta}D \quad (3.7.16)$$

in any case, where $k_*^{-\delta}$ is a diagonal matrix with its jth diagonal element equal to

$$\begin{aligned} k_{*j}^{-1} &= \gamma/(k_j\gamma + 1) \\ &= 1/(k_j + \gamma^{-1}) \quad \text{if } \gamma \neq 0. \end{aligned} \quad (3.7.17)$$

With this notation, formulae (3.7.9), (3.7.10) and (3.7.11) can be written, respectively, as

$$\begin{aligned}\hat{\boldsymbol{\tau}} &= (\boldsymbol{r}^{\delta} - \boldsymbol{N}\boldsymbol{k}_{*}^{-\delta}\boldsymbol{N}')^{-1}(\boldsymbol{\Delta} - \boldsymbol{N}\boldsymbol{k}_{*}^{-\delta}\boldsymbol{D})\boldsymbol{y} = \boldsymbol{C}_c^{-1}\boldsymbol{Q}_c,\\ \mathrm{Cov}(\hat{\boldsymbol{\tau}}) &= \sigma_1^2(\boldsymbol{r}^{\delta} - \boldsymbol{N}\boldsymbol{k}_{*}^{-\delta}\boldsymbol{N}')^{-1} - N_B^{-1}\sigma_B^2\mathbf{1}_v\mathbf{1}_v' = \sigma_1^2\boldsymbol{C}_c^{-1} - N_B^{-1}\sigma_B^2\mathbf{1}_v\mathbf{1}_v'\end{aligned}$$

and, for $\boldsymbol{c}'\mathbf{1}_v = 0$,

$$\mathrm{Var}(\widehat{\boldsymbol{c}'\boldsymbol{\tau}}) = \sigma_1^2\boldsymbol{c}'(\boldsymbol{r}^{\delta} - \boldsymbol{N}\boldsymbol{k}_{*}^{-\delta}\boldsymbol{N}')^{-1}\boldsymbol{c} = \sigma_1^2\boldsymbol{c}'\boldsymbol{C}_c^{-1}\boldsymbol{c},$$

where

$$\begin{aligned}\boldsymbol{C}_c &= \boldsymbol{\Delta}\boldsymbol{T}^{-1}\boldsymbol{\Delta}' = \boldsymbol{r}^{\delta} - \boldsymbol{N}\boldsymbol{k}_{*}^{-\delta}\boldsymbol{N}',\\ \boldsymbol{Q}_c &= \boldsymbol{\Delta}\boldsymbol{T}^{-1}\boldsymbol{y} = (\boldsymbol{\Delta} - \boldsymbol{N}\boldsymbol{k}_{*}^{-\delta}\boldsymbol{D})\boldsymbol{y}.\end{aligned} \qquad (3.7.18)$$

Note that the matrix $\boldsymbol{C}_c$ and the vector $\boldsymbol{Q}_c$ in (3.7.18) are similar, respectively, to $\boldsymbol{C}_1$ ($\equiv \boldsymbol{C}$) and $\boldsymbol{Q}_1$ used in the intra-block analysis (Section 3.2.1), but they are more general in the sense of combining the intra-block and inter-block information (hence, the subscript c). From (3.7.16) and (3.7.17), it is evident that a maximum recovery of the inter-block information is achieved in the case of $-1/k_{\max} < \gamma \leq 0$, i.e., when the blocking is completely unsuccessful, whereas no recovery of that information occurs in the case of $\gamma \to \infty$, i.e., when the formation of blocks is extremely successful, in the sense of eliminating the intra-block variation of experimental units, measured by σ_1^2. In the latter case, $\boldsymbol{C}_c \to \boldsymbol{C}_1$ and $\boldsymbol{Q}_c \to \boldsymbol{Q}_1$.

Also, note that if the design is proper, i.e., if $k_j = k$ for all j, then

$$\boldsymbol{C}_c = \boldsymbol{r}^{\delta} - k_{*}^{-1}\boldsymbol{N}\boldsymbol{N}' \quad \text{and} \quad \boldsymbol{Q}_c = (\boldsymbol{\Delta} - k_{*}^{-1}\boldsymbol{N}\boldsymbol{D})\boldsymbol{y}, \qquad (3.7.19)$$

where $k_{*}^{-1} = k^{-1}(1 - \sigma_1^2/\sigma_2^2)$, with σ_2^2 defined as in (3.5.7) and, hence, $\sigma_2^2/\sigma_1^2 = k\gamma + 1$, i.e.,

$$\boldsymbol{C}_c = \boldsymbol{r}^{\delta} - (1 - \sigma_1^2/\sigma_2^2)k^{-1}\boldsymbol{N}\boldsymbol{N}' = \boldsymbol{\Delta}[\boldsymbol{\phi}_1 + (\sigma_1^2/\sigma_2^2)(\boldsymbol{I}_n - \boldsymbol{\phi}_1)]\boldsymbol{\Delta}'$$

and

$$\boldsymbol{Q}_c = \boldsymbol{\Delta}[\boldsymbol{\phi}_1 + (\sigma_1^2/\sigma_2^2)(\boldsymbol{I}_n - \boldsymbol{\phi}_1)]\boldsymbol{y}.$$

Moreover, note that for a proper block design the matrix $\boldsymbol{C}_c$ and the vector $\boldsymbol{Q}_c$ in (3.7.19) are comparable with the matrix $\boldsymbol{A}_c$ and the vector $\boldsymbol{q}_c$, respectively, used by John (1987, p. 193), a full equivalence holding if his variances σ^2 and σ_1^2 are defined in the present notation as $\sigma^2 = \sigma_U^2 + \sigma_e^2$ and $\sigma_1^2 = \sigma_B^2 - K_H^{-1}\sigma_U^2$ (the latter measuring how much the inter-block variation exceeds the intra-block variation of the experimental units) and, hence, his $\phi = k\sigma_1^2/\sigma^2$ ($= \eta$ in the second edition, John and Williams, 1995, p. 142) is defined as $k\gamma$. Note, however, that John (1987) implicitly assumes that $N_B^{-1}b = 0$ (i.e., $N_B \to \infty$), which in the present notation, introduced in Section 3.5, would mean that the stratum variances σ_2^2 and σ_3^2 are equal, i.e., in his notation, that $\xi_0 = \xi_1$.

3.7.2 Estimation of unknown variance components

All of the results established until now in this section are based on the assumption that the ratio γ (defined in Lemma 3.7.1) is known. In practice, however, this is usually not the case. Therefore, to make the theory applicable, estimators of both, σ_1^2 and γ, are needed. Various approaches may be adopted for finding these estimators. To choose a practically suitable one, it may help first to write the residual sum of squares (3.7.15) as

$$\|(\boldsymbol{I}_n - \boldsymbol{P}_{\Delta'(T^{-1})})\boldsymbol{y}\|^2_{T^{-1}} = \boldsymbol{y}'(\boldsymbol{I}_n - \boldsymbol{P}_{\Delta'(T^{-1})})'\boldsymbol{T}^{-1}(\boldsymbol{I}_n - \boldsymbol{P}_{\Delta'(T^{-1})})\boldsymbol{y}$$

$$= \boldsymbol{y}'(\boldsymbol{T}^{-1} - \boldsymbol{T}^{-1}\boldsymbol{P}_{\Delta'(T^{-1})})'\boldsymbol{T}(\boldsymbol{T}^{-1} - \boldsymbol{T}^{-1}\boldsymbol{P}_{\Delta'(T^{-1})})\boldsymbol{y} = \boldsymbol{y}'\boldsymbol{RTRy} = \boldsymbol{y}'\boldsymbol{Ry},$$

where [see also Rao, 1974, formula (3.13)]

$$\begin{aligned} \boldsymbol{R} &= \boldsymbol{T}^{-1}(\boldsymbol{I}_n - \boldsymbol{P}_{\Delta'(T^{-1})}) = \boldsymbol{T}^{-1} - \boldsymbol{T}^{-1}\boldsymbol{\Delta}'(\boldsymbol{\Delta T}^{-1}\boldsymbol{\Delta}')^{-1}\boldsymbol{\Delta T}^{-1} \\ &= (\boldsymbol{I}_n - \boldsymbol{P}_{\Delta'(T^{-1})})'\boldsymbol{T}^{-1}. \end{aligned} \tag{3.7.20}$$

Thus, in view of the definition of $\boldsymbol{T}$ [see (3.7.3)],

$$\|(\boldsymbol{I}_n - \boldsymbol{P}_{\Delta'(T^{-1})})\boldsymbol{y}\|^2_{T^{-1}} = \boldsymbol{y}'\boldsymbol{RTRy} = \gamma\boldsymbol{y}'\boldsymbol{RD}'\boldsymbol{DRy} + \boldsymbol{y}'\boldsymbol{RRy}. \tag{3.7.21}$$

Following the approach used by Nelder (1968), and generalized by Caliński and Kageyama (1996a), the simultaneous estimators of σ_1^2 and γ can be obtained by equating the partial sums of squares in (3.7.21) to their expectations. These expectations are

$$\begin{aligned} \mathrm{E}(\boldsymbol{y}'\boldsymbol{RD}'\boldsymbol{DRy}) &= \sigma_1^2\mathrm{tr}(\boldsymbol{RD}'\boldsymbol{DRT}) + \boldsymbol{\tau}'\boldsymbol{\Delta RD}'\boldsymbol{DR\Delta}'\boldsymbol{\tau} = \sigma_1^2\mathrm{tr}(\boldsymbol{RD}'\boldsymbol{DRT}) \\ &= \sigma_1^2[\gamma\mathrm{tr}(\boldsymbol{RD}'\boldsymbol{DRD}'\boldsymbol{D}) + \mathrm{tr}(\boldsymbol{RD}'\boldsymbol{DR})] \end{aligned} \tag{3.7.22}$$

and

$$\begin{aligned} \mathrm{E}(\boldsymbol{y}'\boldsymbol{RRy}) &= \sigma_1^2\mathrm{tr}(\boldsymbol{RRT}) + \boldsymbol{\tau}'\boldsymbol{\Delta RR\Delta}'\boldsymbol{\tau} = \sigma_1^2\mathrm{tr}(\boldsymbol{RRT}) \\ &= \sigma_1^2[\gamma\mathrm{tr}(\boldsymbol{RRD}'\boldsymbol{D}) + \mathrm{tr}(\boldsymbol{RR})], \end{aligned} \tag{3.7.23}$$

because $\boldsymbol{R}\mathrm{Cov}(\boldsymbol{y}) = \sigma_1^2\boldsymbol{RT}$, as $\boldsymbol{R}\mathbf{1}_n = \mathbf{0}$ from $\boldsymbol{\Delta}'\mathbf{1}_v = \mathbf{1}_n$ [see (2.2.1)], and $\boldsymbol{\Delta R} = \mathbf{O}$, obviously. Hence, when equating $\boldsymbol{y}'\boldsymbol{RD}'\boldsymbol{DRy}$ and $\boldsymbol{y}'\boldsymbol{RRy}$ to their expectations, one obtains (in view of $\boldsymbol{T}$) the equations

$$\begin{aligned} \boldsymbol{y}'\boldsymbol{RD}'\boldsymbol{DRy} &= \sigma_1^2\gamma\mathrm{tr}(\boldsymbol{RD}'\boldsymbol{DRD}'\boldsymbol{D}) + \sigma_1^2\mathrm{tr}(\boldsymbol{RD}'\boldsymbol{DR}), \\ \boldsymbol{y}'\boldsymbol{RRy} &= \sigma_1^2\gamma\mathrm{tr}(\boldsymbol{RRD}'\boldsymbol{D}) + \sigma_1^2\mathrm{tr}(\boldsymbol{RR}), \end{aligned}$$

which simultaneously can be written as

$$\begin{bmatrix} \mathrm{tr}(\boldsymbol{RD}'\boldsymbol{DRD}'\boldsymbol{D}) & \mathrm{tr}(\boldsymbol{RD}'\boldsymbol{DR}) \\ \mathrm{tr}(\boldsymbol{RD}'\boldsymbol{DR}) & \mathrm{tr}(\boldsymbol{RR}) \end{bmatrix} \begin{bmatrix} \sigma_1^2\gamma \\ \sigma_1^2 \end{bmatrix} = \begin{bmatrix} \boldsymbol{y}'\boldsymbol{RD}'\boldsymbol{DRy} \\ \boldsymbol{y}'\boldsymbol{RRy} \end{bmatrix}. \tag{3.7.24}$$

The solution of (3.7.24) for the vector $[\sigma_1^2\gamma, \sigma_1^2]'$ will provide estimators of its elements, $\sigma_1^2\gamma$ and σ_1^2, and hence of γ.

The equations (3.7.24) clearly have no direct analytic solution, as both, the coefficient matrix on the left-hand side and the vector of the quadratic forms in $\boldsymbol{y}$ on the right, contain the unknown parameter γ. Therefore, the equations have to be solved numerically by an iterative procedure. Before considering it in detail, it will be useful to relate the equations (3.7.24) to those resulting from other approaches commonly used.

First, it can easily be seen that exactly the same equations result from the MINQUE approach of Rao (1970, 1971a, 1972). In fact, writting $\boldsymbol{V}_1 = \boldsymbol{D}'\boldsymbol{D}$ and $\boldsymbol{V}_2 = \boldsymbol{I}_n$, one will find complete equivalence between (3.7.24) and Rao's (1972) equation (4.4).

It may also be seen from (3.7.22) and (3.7.23) that the equations obtained from equating the sums of squares $\boldsymbol{y}'\boldsymbol{R}\boldsymbol{D}'\boldsymbol{D}\boldsymbol{R}\boldsymbol{y}$ and $\boldsymbol{y}'\boldsymbol{R}\boldsymbol{R}\boldsymbol{y}$ to their expectations can equivalently be written as

$$\boldsymbol{y}'\boldsymbol{R}\boldsymbol{D}'\boldsymbol{D}\boldsymbol{R}\boldsymbol{y} = \sigma_1^2 \text{tr}(\boldsymbol{R}\boldsymbol{D}'\boldsymbol{D}) \tag{3.7.25}$$

and

$$\boldsymbol{y}'\boldsymbol{R}\boldsymbol{R}\boldsymbol{y} = \sigma_1^2 \text{tr}(\boldsymbol{R}), \tag{3.7.26}$$

because $\text{tr}(\boldsymbol{R}\boldsymbol{D}'\boldsymbol{D}\boldsymbol{R}\boldsymbol{T}) = \text{tr}(\boldsymbol{R}\boldsymbol{D}'\boldsymbol{D})$ and $\text{tr}(\boldsymbol{R}\boldsymbol{R}\boldsymbol{T}) = \text{tr}(\boldsymbol{R})$, as can easily be checked. Now, as $(\sigma_1^2)^{-1}\boldsymbol{R}$ can be shown to be the Moore–Penrose inverse of the matrix

$$\boldsymbol{\phi}_* \text{Cov}(\boldsymbol{y}) \boldsymbol{\phi}_* = \sigma_1^2 \boldsymbol{\phi}_* \boldsymbol{T} \boldsymbol{\phi}_*,$$

where $\boldsymbol{\phi}_* = \boldsymbol{I}_n - \boldsymbol{\Delta}'\boldsymbol{r}^{-\delta}\boldsymbol{\Delta}$ [as in (2.3.13)], the equations (3.7.25) and (3.7.26) coincide with equations (5.1) of Patterson and Thompson (1975), on which the so-called modified maximum likelihood (MML) estimation method is based. The method is also known as the restricted maximum likelihood (REML) estimation method (see, e.g., Corbeil and Searle, 1976). Thus not only the original approach of Nelder (1968) devised for generally balanced designs, but also its generalization presented here must, in principle, give the same results as those obtainable from the MML (REML) equations derived under the multivariate normality assumption. This equivalence has already been indicated by Patterson and Thompson (1971, p. 552) for proper block designs and by Patterson and Thompson (1975, p. 206) for any block design (in fact, even for a more general design case).

As far as the methods of solving the equations (3.7.24), or their equivalences, are concerned, note that in both, the approach of Nelder (1968) and that of Patterson and Thompson (1971, 1975), an iterative procedure is to be used (Fisher's iterative method of scoring). According to the original MINQUE principle of Rao (1970, 1971a, 1972), a solution of (3.7.24) would be obtained under some *a priori* value of γ. Thus, the results of the latter approach will coincide with those obtained from the other two approaches when the computational procedures used there are restricted to a single iteration, provided that in all three methods the same *a priori* or preliminary estimate of γ is used (see Patterson and Thompson, 1975, p. 204). But, as indicated by Rao (1972, p. 113), the MINQUE method can also be used iteratively, and then the three approaches will lead to the same results, the exactness of their coincidence depending only on numerical details of the computational procedures involved. (See also Rao, 1979, p. 151.)

In view of the above comparative comments, it will be useful to give now a suitable computational procedure for obtaining a practical solution of equations (3.7.24). Following Patterson and Thompson (1971, Section 6), let the procedure start with a preliminary estimate γ_0 of γ, usually with $\gamma_0 \geq 0$. Incorporating it into the coefficients of (3.7.24), the equations can be written as

$$\begin{bmatrix} \operatorname{tr}(\boldsymbol{R}_0\boldsymbol{D}'\boldsymbol{D}\boldsymbol{R}_0\boldsymbol{D}'\boldsymbol{D}) & \operatorname{tr}(\boldsymbol{R}_0\boldsymbol{D}'\boldsymbol{D}\boldsymbol{R}_0) \\ \operatorname{tr}(\boldsymbol{R}_0\boldsymbol{D}'\boldsymbol{D}\boldsymbol{R}_0) & \operatorname{tr}(\boldsymbol{R}_0\boldsymbol{R}_0) \end{bmatrix} \begin{bmatrix} \sigma_1^2\gamma \\ \sigma_1^2 \end{bmatrix} = \begin{bmatrix} \boldsymbol{y}'\boldsymbol{R}_0\boldsymbol{D}'\boldsymbol{D}\boldsymbol{R}_0\boldsymbol{y} \\ \boldsymbol{y}'\boldsymbol{R}_0\boldsymbol{R}_0\boldsymbol{y} \end{bmatrix}, \tag{3.7.27}$$

where $\boldsymbol{R}_0$ is defined as in (3.7.20), but with $\boldsymbol{T}^{-1}$ replaced by $\boldsymbol{T}_0^{-1}$ obtained after replacing γ by γ_0 in (3.7.8), i.e., with

$$\begin{aligned} \boldsymbol{T}_0^{-1} &= \boldsymbol{I}_n - \boldsymbol{D}'(\boldsymbol{k}^{\delta} + \gamma_0^{-1}\boldsymbol{I}_b)^{-1}\boldsymbol{D} \\ &= \boldsymbol{\phi}_1 + \boldsymbol{D}'\boldsymbol{k}^{-\delta}(\boldsymbol{k}^{-\delta} + \gamma_0\boldsymbol{I}_b)^{-1}\boldsymbol{k}^{-\delta}\boldsymbol{D} \end{aligned} \tag{3.7.28}$$

(with $\boldsymbol{T}_0^{-1} \to \boldsymbol{\phi}_1$ if $\gamma_0 \to \infty$). However, instead of the equations (3.7.27), it will be more convenient to consider their equivalent transformed form obtained as follows. First, premultiply both sides of (3.7.27) by the matrix

$$\begin{bmatrix} 1 & 0 \\ \gamma_0 & 1 \end{bmatrix}.$$

This operation will give

$$\begin{bmatrix} \operatorname{tr}(\boldsymbol{R}_0\boldsymbol{D}'\boldsymbol{D}\boldsymbol{R}_0\boldsymbol{D}'\boldsymbol{D}) & \operatorname{tr}(\boldsymbol{R}_0\boldsymbol{D}'\boldsymbol{D}\boldsymbol{R}_0) \\ \operatorname{tr}(\boldsymbol{R}_0\boldsymbol{D}'\boldsymbol{D}) & \operatorname{tr}(\boldsymbol{R}_0) \end{bmatrix} \begin{bmatrix} \sigma_1^2\gamma \\ \sigma_1^2 \end{bmatrix} = \begin{bmatrix} \boldsymbol{y}'\boldsymbol{R}_0\boldsymbol{D}'\boldsymbol{D}\boldsymbol{R}_0\boldsymbol{y} \\ \boldsymbol{y}'\boldsymbol{R}_0\boldsymbol{y} \end{bmatrix},$$

on account of the equalities

$$\begin{aligned} \gamma_0\operatorname{tr}(\boldsymbol{R}_0\boldsymbol{D}'\boldsymbol{D}\boldsymbol{R}_0\boldsymbol{D}'\boldsymbol{D}) + \operatorname{tr}(\boldsymbol{R}_0\boldsymbol{D}'\boldsymbol{D}\boldsymbol{R}_0) &= \operatorname{tr}(\boldsymbol{R}_0\boldsymbol{D}'\boldsymbol{D}\boldsymbol{R}_0\boldsymbol{T}_0) = \operatorname{tr}(\boldsymbol{R}_0\boldsymbol{D}'\boldsymbol{D}), \\ \gamma_0\operatorname{tr}(\boldsymbol{R}_0\boldsymbol{D}'\boldsymbol{D}\boldsymbol{R}_0) + \operatorname{tr}(\boldsymbol{R}_0\boldsymbol{R}_0) &= \operatorname{tr}(\boldsymbol{R}_0\boldsymbol{T}_0\boldsymbol{R}_0) = \operatorname{tr}(\boldsymbol{R}_0), \\ \gamma_0\boldsymbol{y}'\boldsymbol{R}_0\boldsymbol{D}'\boldsymbol{D}\boldsymbol{R}_0\boldsymbol{y} + \boldsymbol{y}'\boldsymbol{R}_0\boldsymbol{R}_0\boldsymbol{y} &= \boldsymbol{y}'\boldsymbol{R}_0\boldsymbol{T}_0\boldsymbol{R}_0\boldsymbol{y} = \boldsymbol{y}'\boldsymbol{R}_0\boldsymbol{y}. \end{aligned}$$

Then, make an additional transformation, by writing

$$\begin{bmatrix} \operatorname{tr}(\boldsymbol{R}_0\boldsymbol{D}'\boldsymbol{D}\boldsymbol{R}_0\boldsymbol{D}'\boldsymbol{D}) & \operatorname{tr}(\boldsymbol{R}_0\boldsymbol{D}'\boldsymbol{D}\boldsymbol{R}_0) \\ \operatorname{tr}(\boldsymbol{R}_0\boldsymbol{D}'\boldsymbol{D}) & \operatorname{tr}(\boldsymbol{R}_0) \end{bmatrix} \begin{bmatrix} 1 & \gamma_0 \\ 0 & 1 \end{bmatrix} \begin{bmatrix} 1 & -\gamma_0 \\ 0 & 1 \end{bmatrix} \begin{bmatrix} \sigma_1^2\gamma \\ \sigma_1^2 \end{bmatrix} = \begin{bmatrix} \boldsymbol{y}'\boldsymbol{R}_0\boldsymbol{D}'\boldsymbol{D}\boldsymbol{R}_0\boldsymbol{y} \\ \boldsymbol{y}'\boldsymbol{R}_0\boldsymbol{y} \end{bmatrix},$$

which gives

$$\begin{bmatrix} \mathrm{tr}(\boldsymbol{R}_0\boldsymbol{D}'\boldsymbol{D}\boldsymbol{R}_0\boldsymbol{D}'\boldsymbol{D}) & \mathrm{tr}(\boldsymbol{R}_0\boldsymbol{D}'\boldsymbol{D}) \\ \mathrm{tr}(\boldsymbol{R}_0\boldsymbol{D}'\boldsymbol{D}) & n-v \end{bmatrix} \begin{bmatrix} \sigma_1^2(\gamma-\gamma_0) \\ \sigma_1^2 \end{bmatrix} = \begin{bmatrix} \boldsymbol{y}'\boldsymbol{R}_0\boldsymbol{D}'\boldsymbol{D}\boldsymbol{R}_0\boldsymbol{y} \\ \boldsymbol{y}'\boldsymbol{R}_0\boldsymbol{y} \end{bmatrix}, \tag{3.7.29}$$

because of the equality

$$\gamma_0\mathrm{tr}(\boldsymbol{R}_0\boldsymbol{D}'\boldsymbol{D}) + \mathrm{tr}(\boldsymbol{R}_0) = n - v.$$

Now, the solution of (3.7.29) gives a revised estimate of γ in the form

$$\hat{\gamma} = \gamma_0 + \frac{(n-v)\boldsymbol{y}'\boldsymbol{R}_0\boldsymbol{D}'\boldsymbol{D}\boldsymbol{R}_0\boldsymbol{y} - \mathrm{tr}(\boldsymbol{D}\boldsymbol{R}_0\boldsymbol{D}')\boldsymbol{y}'\boldsymbol{R}_0\boldsymbol{y}}{\mathrm{tr}[(\boldsymbol{D}\boldsymbol{R}_0\boldsymbol{D}')^2]\boldsymbol{y}'\boldsymbol{R}_0\boldsymbol{y} - \mathrm{tr}(\boldsymbol{D}\boldsymbol{R}_0\boldsymbol{D}')\boldsymbol{y}'\boldsymbol{R}_0\boldsymbol{D}'\boldsymbol{D}\boldsymbol{R}_0\boldsymbol{y}}. \tag{3.7.30}$$

Thus, a single iteration of Fisher's iterative method of scoring, suggested by Patterson and Thompson (1971, p. 550), consists here of the following two steps:

(0) One starts with a preliminary estimate γ_0 ($> -1/k_{\max}$) of γ to obtain the equation (3.7.29).

(1) By solving (3.7.29), a revised estimate $\hat{\gamma}$ of γ is obtained, of the form (3.7.30), and this is then used as a new preliminary estimate in step (0) of the next iteration.

It should, however, be examined whether γ_0 is always above the lower bound $-1/k_{\max}$, as required. If $\hat{\gamma} \leq -1/k_{\max}$, it cannot be used as a new γ_0 in step (0). In such a case, formula (3.7.30) is to be replaced by

$$\hat{\gamma} = \gamma_0 + \alpha\frac{(n-v)\boldsymbol{y}'\boldsymbol{R}_0\boldsymbol{D}'\boldsymbol{D}\boldsymbol{R}_0\boldsymbol{y} - \mathrm{tr}(\boldsymbol{D}\boldsymbol{R}_0\boldsymbol{D}')\boldsymbol{y}'\boldsymbol{R}_0\boldsymbol{y}}{\mathrm{tr}[(\boldsymbol{D}\boldsymbol{R}_0\boldsymbol{D}')^2]\boldsymbol{y}'\boldsymbol{R}_0\boldsymbol{y} - \mathrm{tr}(\boldsymbol{D}\boldsymbol{R}_0\boldsymbol{D}')\boldsymbol{y}'\boldsymbol{R}_0\boldsymbol{D}'\boldsymbol{D}\boldsymbol{R}_0\boldsymbol{y}},$$

with $\alpha \in (0,1)$ chosen so that $\hat{\gamma} > -1/k_{\max}$ (as suggested by Rao and Kleffe, 1988, p. 237). (See also the discussion in Verdooren, 1980, Section 6.)

The above iteration is to be repeated until convergence, i.e., until $\hat{\gamma}$ so obtained coincides with the latest γ_0 or, equivalently, until the equality

$$\frac{\boldsymbol{y}'\boldsymbol{R}_0\boldsymbol{D}'\boldsymbol{D}\boldsymbol{R}_0\boldsymbol{y}}{\mathrm{tr}(\boldsymbol{D}\boldsymbol{R}_0\boldsymbol{D}')} = \frac{\boldsymbol{y}'\boldsymbol{R}_0\boldsymbol{y}}{n-v} \tag{3.7.31}$$

is reached. The solution of (3.7.24) can then be written as $[\widehat{\sigma_1^2}\hat{\gamma}, \widehat{\sigma_1^2}]'$, where $\hat{\gamma}$ is the final estimate of γ obtained after the convergence of the iterative process, and

$$\widehat{\sigma_1^2} = \frac{\boldsymbol{y}'\hat{\boldsymbol{R}}\boldsymbol{y}}{n-v}, \tag{3.7.32}$$

where $\hat{\boldsymbol{R}}$ is defined according to (3.7.20) in the same way as $\boldsymbol{R}_0$ has been, but now with γ in (3.7.8) replaced by the final $\hat{\gamma}$. Although the convergence of this process has not been proved, experiences indicate that the procedure works well in practice (see, e.g., Nelder, 1968, p. 308; Patterson and Thompson, 1971, p. 553; Rao and Kleffe, 1988, p. 226).

Note that the equality (3.7.31) holds if and only if the MML equations (15) and (16) of Patterson and Thompson (1971) yield, with γ fixed at γ_0, the same estimate of σ_1^2. In the present notation, these equations can equivalently be written as

$$\boldsymbol{y}'\boldsymbol{R}\boldsymbol{D}'\boldsymbol{D}\boldsymbol{R}\boldsymbol{y} = \sigma_1^2 \text{tr}(\boldsymbol{D}\boldsymbol{R}\boldsymbol{D}') \quad \text{and} \quad \boldsymbol{y}'\boldsymbol{R}\boldsymbol{y} = \sigma_1^2(n-v).$$

Also note that (3.7.32) is the MINQUE estimate of σ_1^2 when $\gamma = \hat{\gamma}$, as it follows from Theorem 3.4 of Rao (1974).

As to the numerical aspects of the considered procedure of solving the equations (3.7.24) or, rather, its equivalence (3.7.29), the main task is to compute the elements of the matrix appearing on the left and those of the vector on the right-hand side of (3.7.29), i.e., elements involving the matrix $\boldsymbol{R}_0$. From (3.7.20), it can be seen that to compute $\boldsymbol{R}_0$, one has to invert the matrices $\boldsymbol{T}_0$ and $\boldsymbol{\Delta}\boldsymbol{T}_0^{-1}\boldsymbol{\Delta}'$. The inverse $\boldsymbol{T}_0^{-1}$ is readily available from (3.7.28), but to obtain $(\boldsymbol{\Delta}\boldsymbol{T}_0^{-1}\boldsymbol{\Delta}')^{-1}$, a problem may occur if v is large, as is often the case, e.g., in plant breeding experiments. A suitable computational procedure that provides all elements needed is the sweep operation. It can be described, after Dempster (1969, Section 4.3.2), as follows.

A $p \times p$ matrix $\boldsymbol{A}$ is said to have been swept on row h and column h if $\boldsymbol{A}$ is replaced by another $p \times p$ matrix $\boldsymbol{B}$ whose elements $\{b_{ij}\}$ are related to the elements $\{a_{ij}\}$ of $\boldsymbol{A}$ by the equalities

$$b_{hh} = -1/a_{hh}, \quad b_{ih} = a_{ih}/a_{hh}, \quad b_{hj} = a_{hj}/a_{hh}$$

and

$$b_{ij} = a_{ij} - a_{ih}a_{hj}/a_{hh}, \quad \text{for } i \neq h \text{ and } j \neq h.$$

It will be convenient to denote the above sweep operation by SWP[h] and the resulting matrix $\boldsymbol{B}$ by SWP[h]$\boldsymbol{A}$. Accordingly, the result of successively applying the operations SWP[1], SWP[2], ... , SWP[m] to $\boldsymbol{A}$ will conveniently be denoted by SWP[1, 2, ... , m]$\boldsymbol{A}$.

To embed the above sweep operation in the procedure of solving the equations (3.7.24), i.e., to use it for obtaining the equations (3.7.29), it will be convenient to proceed as follows (partially following Giesbrecht, 1986).

With a preliminary estimate of γ, i.e., with γ_0, compute the matrix $\boldsymbol{T}_0^{-1}$ by a suitable version of formula (3.7.28). Then, obtain the matrix

$$\boldsymbol{A} = \begin{bmatrix} \boldsymbol{\Delta}\boldsymbol{T}_0^{-1}\boldsymbol{\Delta}' & \boldsymbol{\Delta}\boldsymbol{T}_0^{-1}\boldsymbol{y} & \boldsymbol{\Delta}\boldsymbol{T}_0^{-1}\boldsymbol{D}' \\ \boldsymbol{y}'\boldsymbol{T}_0^{-1}\boldsymbol{\Delta}' & \boldsymbol{y}'\boldsymbol{T}_0^{-1}\boldsymbol{y} & \boldsymbol{y}'\boldsymbol{T}_0^{-1}\boldsymbol{D}' \\ \boldsymbol{D}\boldsymbol{T}_0^{-1}\boldsymbol{\Delta}' & \boldsymbol{D}\boldsymbol{T}_0^{-1}\boldsymbol{y} & \boldsymbol{D}\boldsymbol{T}_0^{-1}\boldsymbol{D}' \end{bmatrix}.$$

Sweeping the matrix $\boldsymbol{A}$ on its first v rows and the first v columns successively, i.e., on the submatrix $\boldsymbol{\Delta}\boldsymbol{T}_0^{-1}\boldsymbol{\Delta}'$ (in other words, pivoting on the diagonal elements of this submatrix), one obtains a matrix

$$\boldsymbol{B} = \mathrm{SWP}[1, 2, ..., v]\boldsymbol{A}, \tag{3.7.33}$$

of the form

$$\begin{bmatrix} -(\boldsymbol{\Delta T}_0^{-1}\boldsymbol{\Delta}')^{-1} & (\boldsymbol{\Delta T}_0^{-1}\boldsymbol{\Delta}')^{-1}\boldsymbol{\Delta T}_0^{-1}\boldsymbol{y} & (\boldsymbol{\Delta T}_0^{-1}\boldsymbol{\Delta}')^{-1}\boldsymbol{\Delta T}_0^{-1}\boldsymbol{D}' \\ \boldsymbol{y}'\boldsymbol{T}_0^{-1}\boldsymbol{\Delta}'(\boldsymbol{\Delta T}_0^{-1}\boldsymbol{\Delta}')^{-1} & \boldsymbol{y}'\boldsymbol{R}_0\boldsymbol{y} & \boldsymbol{y}'\boldsymbol{R}_0\boldsymbol{D}' \\ \boldsymbol{D T}_0^{-1}\boldsymbol{\Delta}'(\boldsymbol{\Delta T}_0^{-1}\boldsymbol{\Delta}')^{-1} & \boldsymbol{D R}_0\boldsymbol{y} & \boldsymbol{D R}_0\boldsymbol{D}' \end{bmatrix}.$$

From it, using the submatrices $\boldsymbol{y}'\boldsymbol{R}_0\boldsymbol{y}$, $\boldsymbol{y}'\boldsymbol{R}_0\boldsymbol{D}'$, $\boldsymbol{D R}_0\boldsymbol{y}$, and $\boldsymbol{D R}_0\boldsymbol{D}'$, the equations (3.7.29) can easily be obtained, or directly their solution for γ, as written in (3.7.30). This operation is to be repeated for each cycle of the iterative process, until a final convergence, i.e., until $\hat{\gamma} = \gamma_0$ and, hence, $\hat{\boldsymbol{R}} = \boldsymbol{R}_0$ is reached. Then, (3.7.32) is also obtained directly.

It may also be noted that the sweep operation (3.7.33) provides an estimator

$$\boldsymbol{\tau}_0 = (\boldsymbol{\Delta T}_0^{-1}\boldsymbol{\Delta}')^{-1}\boldsymbol{\Delta T}_0^{-1}\boldsymbol{y}$$

of $\boldsymbol{\tau}$, according to formula (3.7.9) for the BLUE of $\boldsymbol{\tau}$, but now with $\boldsymbol{T}$ replaced by $\boldsymbol{T}_0$. When the iterative process is terminated at convergence, i.e., when $\hat{\gamma} = \gamma_0$, then $\boldsymbol{T}_0^{-1}$ becomes

$$\hat{\boldsymbol{T}}^{-1} = \boldsymbol{I}_n - \boldsymbol{D}'(\boldsymbol{k}^{\delta} + \hat{\gamma}^{-1}\boldsymbol{I}_b)^{-1}\boldsymbol{D} = \boldsymbol{\phi}_1 + \boldsymbol{D}'\boldsymbol{k}^{-\delta}(\boldsymbol{k}^{-\delta} + \hat{\gamma}\boldsymbol{I}_b)^{-1}\boldsymbol{k}^{-\delta}\boldsymbol{D}$$

(provided that $\hat{\gamma} \neq 0$, otherwise reducing to $\hat{\boldsymbol{T}}^{-1} = \boldsymbol{I}_n$), and the so-obtained estimator of $\boldsymbol{\tau}$ can be written as

$$\tilde{\boldsymbol{\tau}} = (\boldsymbol{\Delta}\hat{\boldsymbol{T}}^{-1}\boldsymbol{\Delta}')^{-1}\boldsymbol{\Delta}\hat{\boldsymbol{T}}^{-1}\boldsymbol{y}. \tag{3.7.34}$$

The vector (3.7.34) may be called (after Rao and Kleffe, 1988, p. 274) an "empirical" estimator of $\boldsymbol{\tau}$, which of course is not the same as the BLUE obtainable with the exact value of γ. Hence, the properties of (3.7.34) are not the same as those of (3.7.9).

Example 3.7.1. Consider the data originally used by Cunningham and Henderson (1968) and reanalyzed by Patterson and Thompson (1971, Section 9) in the framework of the MML (REML) method. The data are supposed to result from an experiment designed for two treatments in three blocks according to the incidence matrix

$$\boldsymbol{N} = \begin{bmatrix} 2 & 5 & 1 \\ 2 & 3 & 5 \end{bmatrix}.$$

It is assumed that the experimental conditions have made the randomizations described in Section 3.1.1 applicable. The data observed on the 18 experimental units can be presented as follows:

Block	1	1	1	1	2	2	2	2	2	2	2	2	3	3	3	3	3
Treatment	1	1	2	2	1	1	1	1	1	2	2	2	1	2	2	2	2
Observation	3	2	2	3	2	3	5	6	7	8	8	9	3	4	4	3	2

For the intra-block analysis, the matrix $\boldsymbol{C}_1$ and the vector $\boldsymbol{Q}_1$ are needed (see Section 3.2.1). In the present example, they are

$$\boldsymbol{C}_1 = \frac{89}{24}\begin{bmatrix} 1 & -1 \\ -1 & 1 \end{bmatrix} \quad \text{and} \quad \boldsymbol{Q}_1 = \frac{15}{2}\begin{bmatrix} -1 \\ 1 \end{bmatrix}.$$

The unique basic contrast of the considered design is represented by the vector

$$\boldsymbol{s} = (1/3)\sqrt{10}[-1/4,\ 1/5]' = 1.05409[-1/4,\ 1/5]',$$

as it satisfies the condition (see Definition 3.4.1)

$$\boldsymbol{C}_1\boldsymbol{s} = \varepsilon \boldsymbol{r}^{\delta}\boldsymbol{s}, \quad \text{with } \varepsilon = 267/320 = 0.8344.$$

The corresponding eigenvalue of $\boldsymbol{C}_1$ with respect to $\boldsymbol{r}^{\delta}$, ε, is the efficiency factor of the analyzed design for the contrast $\boldsymbol{c}'\boldsymbol{\tau} = \boldsymbol{s}'\boldsymbol{r}^{\delta}\boldsymbol{\tau}$ when estimated in the intra-block analysis (see Remark 3.3.3). Hence, a possible choice of a g-inverse of $\boldsymbol{C}_1$ is the matrix (3.4.13), here equal to

$$\varepsilon^{-1}\boldsymbol{s}\boldsymbol{s}' = \frac{8}{2403}\begin{bmatrix} 25 & -20 \\ -20 & 16 \end{bmatrix}.$$

With these preliminary results, the intra-block BLUE of the contrast $\boldsymbol{c}'\boldsymbol{\tau} = \boldsymbol{s}'\boldsymbol{r}^{\delta}\boldsymbol{\tau} = (2/3)\sqrt{10}(\tau_2 - \tau_1)$ is, by Theorem 3.4.1, equal to

$$(\widehat{\boldsymbol{c}'\boldsymbol{\tau}})_{\text{intra}} = \varepsilon^{-1}\boldsymbol{s}'\boldsymbol{Q}_1 = \frac{120}{89}\sqrt{10} = 4.264,$$

and its variance is

$$\text{Var}[(\widehat{\boldsymbol{c}'\boldsymbol{\tau}})_{\text{intra}}] = \frac{320}{267}\sigma_1^2 = 1.1985\sigma_1^2,$$

where $\sigma_1^2 = \sigma_U^2 + \sigma_e^2$. This result means that the intra-block BLUE of $\tau_2 - \tau_1$ is equal to

$$(\widehat{\tau_2 - \tau_1})_{\text{intra}} = \frac{3}{2\sqrt{10}}(\widehat{\boldsymbol{c}'\boldsymbol{\tau}})_{\text{intra}} = \frac{180}{89} = 2.0225,$$

and its variance is

$$\text{Var}[(\widehat{\tau_2 - \tau_1})_{\text{intra}}] = \frac{9}{40}\text{Var}[(\widehat{\boldsymbol{c}'\boldsymbol{\tau}})_{\text{intra}}] = \frac{24}{89}\sigma_1^2 = 0.26966\sigma_1^2.$$

Note that these results could also be obtained directly from Section 3.2.1, i.e., as

$$(\widehat{\tau_2 - \tau_1})_{\text{intra}} = [-1,\ 1]\boldsymbol{C}_1^{-}\boldsymbol{Q}_1 = \frac{180}{89} = 2.0225,$$

and its variance, from (3.2.13), as

$$\mathrm{Var}[\widehat{(\tau_2 - \tau_1)}_{\mathrm{intra}}] = [-1,\ 1]C_1^- \begin{bmatrix} -1 \\ 1 \end{bmatrix} = \frac{24}{89}\sigma_1^2 = 0.26966\sigma_1^2,$$

easily obtainable when, alternatively, as a g-inverse of $\boldsymbol{C}_1$ the matrix

$$\frac{24}{89}\begin{bmatrix} 1 & 0 \\ 0 & 0 \end{bmatrix}$$

is used. To complete the intra-block analysis, one needs the relevant sums of squares, the intra-block treatment sum of squares

$$\boldsymbol{Q}_1'\boldsymbol{C}_1^-\boldsymbol{Q}_1 = \left(\frac{15}{2}\right)^2 \frac{24}{89}[-1,\ 1]\begin{bmatrix} 1 & 0 \\ 0 & 0 \end{bmatrix}\begin{bmatrix} -1 \\ 1 \end{bmatrix} = \frac{1350}{89} = 15.16854,$$

with $v - 1 = 1$ d.f., and the intra-block residual sum of squares

$$\boldsymbol{y}'\boldsymbol{\psi}_1\boldsymbol{y} = \boldsymbol{y}'\boldsymbol{y} - \boldsymbol{y}'\boldsymbol{D}'\boldsymbol{k}^{-\delta}\boldsymbol{D}\boldsymbol{y} - \boldsymbol{Q}_1'\boldsymbol{C}_1^-\boldsymbol{Q}_1 = 437 - 386.5 - 15.16854 = 35.33146,$$

with $n - b - v + 1 = 14$ d.f. The latter sum of squares gives the intra-block MINQUE of σ_1^2 as $s_1^2 = \boldsymbol{y}'\boldsymbol{\psi}_1\boldsymbol{y}/(n - b - v + 1) = 2.52368$ (see Cunningham and Henderson, 1968, Table 3). Hence, the intra-block estimate of the variance of $\widehat{(\tau_2 - \tau_1)}_{\mathrm{intra}}$ is $(24/89)s_1^2 = 0.269663s_1^2 = 0.68054$.

So far, for estimating the considered contrast only, the intra-block information has been used. To make use also of the inter-block information, the matrix $\boldsymbol{C}_1$ and the vector $\boldsymbol{Q}_1$ are to be modified as described in (3.7.18). The required change concerns the diagonal matrix $\boldsymbol{k}^{-\delta}$, which is to be replaced by $\boldsymbol{k}_*^{-\delta}$ defined in (3.7.17), i.e., instead of

$$\boldsymbol{k}^{-\delta} = \mathrm{diag}[4^{-1},\ 8^{-1},\ 6^{-1}],$$

the diagonal matrix

$$\gamma \mathrm{diag}[(4\gamma + 1)^{-1},\ (8\gamma + 1)^{-1},\ (6\gamma + 1)^{-1}],$$

or

$$\mathrm{diag}[(4 + \gamma^{-1})^{-1},\ (8 + \gamma^{-1})^{-1},\ (6 + \gamma^{-1})^{-1}] \quad \text{if } \gamma \neq 0,$$

is to be used. However, this formula becomes applicable if the parameter γ (defined in Lemma 3.7.1) is known. Otherwise, it is to be estimated, as described in Section 3.7.2, which involves some iterative computational procedure that can be tedious. Fortunately, this procedure is included in the statistical software packages GENSTAT and SAS (see the comment in John and Williams, 1995, p. 140). Here the "REML directive" in GENSTAT 5 (see Genstat 5 Committee, 1996) has been applied to obtain the estimates of γ and σ_1^2. It has provided the following iterative solutions of equations (3.7.29), obtainable through formula (3.7.30) with $\gamma_0 = 1$ as a preliminary estimate of γ :

Iteration cycle	1	2	3	4
$\hat{\gamma}$	1.60064	1.57083	1.57180	the same
$\hat{\sigma}_1^2$	2.51229	2.51863	2.51846	the same

These results are exactly as those of Patterson and Thompson (1971, Table 2).

The final estimate of γ, i.e., $\hat{\gamma} = 1.57180$, when used instead of γ, allows the matrix $\boldsymbol{T}^{-1}$ to be estimated. This estimation is done by replacing $\boldsymbol{k}_*^{-\delta}$ in (3.7.16) by its estimate, here,

$$\hat{\boldsymbol{k}}_*^{-\delta} = \text{diag}[(4+\hat{\gamma}^{-1})^{-1},\ (8+\hat{\gamma}^{-1})^{-1},\ (6+\hat{\gamma}^{-1})^{-1}]$$

$$= \text{diag}[0.215693,\ 0.115791,\ 0.150688].$$

Replacing now $\boldsymbol{T}^{-1}$ in (3.7.20) by its estimate $\hat{\boldsymbol{T}}^{-1}$, one obtains the matrix $\hat{\boldsymbol{R}}$, which when used in (3.7.32) gives the final estimate of σ_1^2. It should coincide with the estimate of σ_1^2 obtained at the convergence of the above iterative procedure. In fact, in this example, the result obtained from (3.7.32) is $\hat{\sigma}_1^2 = 2.51846$, which is equal to that obtained at the third (final) iteration cycle. Furthermore, the matrix $\hat{\boldsymbol{k}}_*^{-\delta}$ can be used to estimate the matrix $\boldsymbol{C}_c$ and the vector $\boldsymbol{Q}_c$ defined in (3.7.18), which then provide the empirical estimate of $\boldsymbol{\tau}$, according to (3.7.34). Here, the results are

$$\hat{\boldsymbol{C}}_c = \boldsymbol{\Delta}\hat{\boldsymbol{T}}^{-1}\boldsymbol{\Delta}' = \begin{bmatrix} 4.091765 & -3.353077 \\ -3.353077 & 4.327909 \end{bmatrix}$$

and

$$\hat{\boldsymbol{Q}}_c = \boldsymbol{\Delta}\hat{\boldsymbol{T}}^{-1}\boldsymbol{y} = \begin{bmatrix} -4.268148 \\ 11.189996 \end{bmatrix},$$

from which

$$\tilde{\boldsymbol{\tau}} = [2.946,\ 4.868]',$$

exactly as obtained directly from GENSTAT 5. Now, the empirical estimate of $\tau_2 - \tau_1$ is

$$\widetilde{(\tau_2 - \tau_1)} = 4.868 - 2.946 = 1.922.$$

If $\hat{\gamma}$ used in $\hat{\boldsymbol{T}}^{-1}$ were the true value of the parameter γ, then, by Theorem 3.7.2, the resulting estimate could be considered as a value of the BLUE of the contrast $\tau_2 - \tau_1$, having the variance as in (3.7.11), i.e.,

$$\sigma_1^2[-1,\ 1]\begin{bmatrix} 4.091765 & -3.353077 \\ -3.353077 & 4.327909 \end{bmatrix}^{-1}\begin{bmatrix} -1 \\ 1 \end{bmatrix} = 0.26502\sigma_1^2.$$

Comparing this result with the variance of the intra-block BLUE of $\tau_2 - \tau_1$, note that the inclusion of inter-block information reduces the variance, though in this example the reduction is not substantial (from $0.2697\sigma_1^2$ to $0.2650\sigma_1^2$).

3.7.3 Properties of the empirical estimators

The question of interest now is, what can be said about the properties of the estimators of $\boldsymbol{\tau}$ and $\boldsymbol{c}'\boldsymbol{\tau}$ considered in Theorem 3.7.2 when the unknown value of the ratio γ appearing in the formulae (3.7.3) and (3.7.8) is replaced by $\hat{\gamma}$ obtained from the solution of the equations (3.7.24), or their equivalence, as described above.

To get some insight into the properties of the empirical estimators $\tilde{\boldsymbol{\tau}}$ and $\boldsymbol{c}'\tilde{\boldsymbol{\tau}}$, where $\tilde{\boldsymbol{\tau}}$ is as in (3.7.34), it will be useful to represent them in terms of some simple linear functions of the observed vector $\boldsymbol{y}$. For this reason, let the matrix $\boldsymbol{D\phi_*D}' = \boldsymbol{k}^\delta - \boldsymbol{N}'\boldsymbol{r}^{-\delta}\boldsymbol{N} = \boldsymbol{C}_*$ (which may be seen as a dual of the C-matrix in the same sense as $\boldsymbol{\phi}_*$ is a dual of $\boldsymbol{\phi}$; see Section 2.2) be written in the form of the spectral decomposition

$$\boldsymbol{D\phi_*D}' = \sum_{j=1}^{h_*} \lambda_j \boldsymbol{u}_j \boldsymbol{u}_j', \tag{3.7.35}$$

where $\boldsymbol{u}_j'\boldsymbol{u}_j = \delta_{jj'}$ (the Kronecker delta) and $h_* = \text{rank}(\boldsymbol{D\phi_*D}') = b - v + h$, as it follows from Lemma 2.3.3 and Remark 2.3.1 [with $h = \text{rank}(\boldsymbol{C}_1)$]. From (3.7.35), assuming that $h_* > 0$, it is permissible to write

$$\boldsymbol{\phi_*D}'\boldsymbol{D\phi_*} = \sum_{j=1}^{h_*} \lambda_j \boldsymbol{v}_j \boldsymbol{v}_j' \tag{3.7.36}$$

[as in formula (22) of Patterson and Thompson, 1971], where

$$\boldsymbol{v}_j = \lambda_j^{-1/2}\boldsymbol{\phi_*D}'\boldsymbol{u}_j \quad \text{for } j = 1, 2, ..., h_*. \tag{3.7.37}$$

The vectors $\{\boldsymbol{v}_j\}$ can then be completed by additional $n - h_*$ orthonormal vectors chosen so to satisfy the equation

$$\boldsymbol{\phi}_*\boldsymbol{v}_j = \boldsymbol{v}_j \quad \text{for } j = h_* + 1, h_* + 2, ..., n - v \tag{3.7.38}$$

or

$$\boldsymbol{\phi}_*\boldsymbol{v}_j = \boldsymbol{0} \quad \text{for } j = n - v + 1, n - v + 2, ..., n,$$

as the matrix $\boldsymbol{\phi}_* = \boldsymbol{I}_n - \boldsymbol{\Delta}'\boldsymbol{r}^{-\delta}\boldsymbol{\Delta}$ is idempotent of rank $n - v$. In fact, it can now be written in its spectral form as $\boldsymbol{\phi}_* = \sum_{j=1}^{n-v} \boldsymbol{v}_j\boldsymbol{v}_j'$, because $\{\boldsymbol{v}_j\}$ defined in (3.7.37) are also orthonormal eigenvectors of $\boldsymbol{\phi}_*$. Hence, it is permissible to write

$$\begin{aligned}\boldsymbol{\phi_*T\phi_*} &= \boldsymbol{\phi}_*(\gamma\boldsymbol{D}'\boldsymbol{D} + \boldsymbol{I}_n)\boldsymbol{\phi}_* = \gamma\boldsymbol{\phi_*D}'\boldsymbol{D\phi_*} + \boldsymbol{\phi}_* \\ &= \sum_{j=1}^{h_*}(\gamma\lambda_j + 1)\boldsymbol{v}_j\boldsymbol{v}_j' + \sum_{j=h_*+1}^{n-v} \boldsymbol{v}_j\boldsymbol{v}_j'\end{aligned}$$

[as in formula (11) of Patterson and Thompson, 1971], from which the representation

$$R = (\phi_* T \phi_*)^+ = \sum_{j=1}^{h_*} \frac{1}{\gamma\lambda_j + 1} v_j v_j' + \sum_{j=h_*+1}^{n-v} v_j v_j' \tag{3.7.39}$$

is obtainable. [See the comment following formula (3.7.26).]

Now, noting that $(\Delta T^{-1}\Delta')^{-1}\Delta T^{-1} = r^{-\delta}\Delta(I_n - TR)$ and that $Dv_j = 0$ for $j = h_* + 1, h_* + 2, ..., n - v$ as well as $\Delta v_j = 0$ for $j = 1, 2, ..., n - v$, the representation (3.7.39) can be used to write the estimator (3.7.9) as

$$\hat{\tau} = r^{-\delta}\Delta\left(I_n - D'D\sum_{j=1}^{h_*} \frac{\gamma}{\gamma\lambda_j + 1} v_j v_j'\right) y$$

and, similarly, the empirical estimator (3.7.34) as

$$\tilde{\tau} = r^{-\delta}\Delta\left(I_n - D'D\sum_{j=1}^{h_*} \frac{\hat{\gamma}}{\hat{\gamma}\lambda_j + 1} v_j v_j'\right) y,$$

provided that not only the value of γ, but also of $\hat{\gamma}$ is above the lower limit $-1/k_{\max}$, which secures that both $\gamma\lambda_j + 1$ and $\hat{\gamma}\lambda_j + 1$ are positive for all j. The above formulae for $\hat{\tau}$ and $\tilde{\tau}$ allow the latter to be written in the form

$$\tilde{\tau} = \hat{\tau} + r^{-\delta} N D \sum_{j=1}^{h_*} \frac{z_j}{\gamma\lambda_j + 1} v_j, \tag{3.7.40}$$

and hence, the estimator $\widetilde{c'\tau} = c'\tilde{\tau}$ of a function $c'\tau$ to be written in the form

$$\widetilde{c'\tau} = \widehat{c'\tau} + c' r^{-\delta} N D \sum_{j=1}^{h_*} \frac{z_j}{\gamma\lambda_j + 1} v_j, \tag{3.7.41}$$

where

$$z_j = \frac{\gamma - \hat{\gamma}}{\hat{\gamma}\lambda_j + 1} v_j' y \quad \text{for } j = 1, 2, ..., h_*. \tag{3.7.42}$$

The representation (3.7.40) leads to the following result, which can be traced back to Roy and Shah (1962, Lemma 6.1).

Lemma 3.7.2. *Let the model be as in Lemma* 3.7.1, *and suppose that the value of* γ *is unknown, except that it is above the lower limit* $-1/k_{\max}$, *as is its estimate* $\hat{\gamma}$. *Further, let the random variables* $\{z_j\}$ *defined in* (3.7.42) *satisfy the conditions*

$$\mathrm{E}(z_j) = 0, \quad \mathrm{E}(z_j z_{j'}) = 0 \ \ \textit{and} \ \ \mathrm{Var}(z_j) < \infty \tag{3.7.43}$$

for all j *and* $j' \neq j$ $(= 1, 2, ..., h_*)$ *and for all admissible values of* γ. *Then, the estimator* $\tilde{\tau}$ *given in* (3.7.34) *has the properties*

$$\mathrm{E}(\tilde{\tau}) = \mathrm{E}(\hat{\tau}) = \tau \tag{3.7.44}$$

and

$$\mathrm{Cov}(\tilde{\boldsymbol{\tau}}) = \mathrm{Cov}(\hat{\boldsymbol{\tau}}) + \boldsymbol{r}^{-\delta}\boldsymbol{N}\boldsymbol{D}\sum_{j=1}^{h_*}\frac{\mathrm{Var}(z_j)}{(\gamma\lambda_j+1)^2}\boldsymbol{v}_j\boldsymbol{v}_j'\boldsymbol{D}'\boldsymbol{N}'\boldsymbol{r}^{-\delta}, \tag{3.7.45}$$

and hence, those of $\widetilde{\boldsymbol{c}'\boldsymbol{\tau}}$ *are*

$$\mathrm{E}(\widetilde{\boldsymbol{c}'\boldsymbol{\tau}}) = \boldsymbol{c}'\boldsymbol{\tau}, \quad \mathrm{Var}(\widetilde{\boldsymbol{c}'\boldsymbol{\tau}}) = \mathrm{Var}(\widehat{\boldsymbol{c}'\boldsymbol{\tau}}) + \sum_{j=1}^{h_*}(\boldsymbol{c}'\boldsymbol{r}^{-\delta}\boldsymbol{N}\boldsymbol{D}\boldsymbol{v}_j)^2\frac{\mathrm{Var}(z_j)}{(\gamma\lambda_j+1)^2}. \tag{3.7.46}$$

Proof. The unbiasedness given in (3.7.44) follows directly from the first of the assumptions (3.7.43), on account of (3.7.40) and the fact that $\hat{\boldsymbol{\tau}}$ is the BLUE of $\boldsymbol{\tau}$. The formula (3.7.45) is obtainable from (3.7.40) and the assumptions (3.7.43), which by Theorem 5a.2(i) of Rao (1973) imply that $\mathrm{Cov}(\hat{\boldsymbol{\tau}}, z_j) = 0$ for all j, as $\hat{\boldsymbol{\tau}}$ is the BLUE of $\boldsymbol{\tau}$. The properties (3.7.46) are immediate consequences of (3.7.44) and (3.7.45). □

To see when the variables $\{z_j\}$, as functions of the statistic $\hat{\gamma}$ and the observed vector $\boldsymbol{y}$, satisfy the conditions of Lemma 3.7.2, it will be helpful first to note that the quadratic forms in $\boldsymbol{y}$ appearing on the right-hand side of the equations (3.7.24) can be expressed as

$$\boldsymbol{y}'\boldsymbol{R}\boldsymbol{D}'\boldsymbol{D}\boldsymbol{R}\boldsymbol{y} = \sum_{j=1}^{h_*}\frac{\lambda_j}{(\gamma\lambda_j+1)^2}(\boldsymbol{v}_j'\boldsymbol{y})^2, \tag{3.7.47}$$

$$\boldsymbol{y}'\boldsymbol{R}\boldsymbol{R}\boldsymbol{y} = \sum_{j=1}^{h_*}\frac{1}{(\gamma\lambda_j+1)^2}(\boldsymbol{v}_j'\boldsymbol{y})^2 + \sum_{j=h_*+1}^{n-v}(\boldsymbol{v}_j'\boldsymbol{y})^2, \tag{3.7.48}$$

on account of the formula (3.7.39). Thus, the solution of (3.7.24) and, hence, the estimator $\hat{\gamma}$, depends on $\boldsymbol{y}$ through the linear forms $\boldsymbol{v}_1'\boldsymbol{y}, \boldsymbol{v}_2'\boldsymbol{y}, ..., \boldsymbol{v}_{n-v}'\boldsymbol{y}$ only. This fact implies that the random variables $\{z_j\}$, defined in (3.7.42), are functions of $\boldsymbol{y}$ through these forms only. Denote them by

$$x_j = \boldsymbol{v}_j'\boldsymbol{y} \quad \text{for} \quad j = 1, 2, ..., n - v. \tag{3.7.49}$$

Now, it will be helpful to recall the following result, used earlier by Graybill and Weeks (1959), Roy and Shah (1962) and, in some other form, by Kackar and Harville (1981).

Lemma 3.7.3. *Let* $\boldsymbol{x} = [x_1, x_2, ..., x_p]'$ *be a random vector such that all its* p *elements have mutually independent symmetric distributions around zero, in the sense that* x_i *and* $-x_i$ *are distributed identically and for each* i $(= 1, 2, ..., p)$ *independently. Further, let each* $g_j(\boldsymbol{x}) = g_j(x_1, x_2, ..., x_p)$, *for* $j = 1, 2, ..., p_1$ $(< p)$, *be an even function of any* x_i, *in the sense that* $g_j(\boldsymbol{x})$ *is invariant under the change of* x_i *to* $-x_i$ *for any* i *and* j. *Then, the distribution of the random vector*

$$\boldsymbol{z} = [g_1(\boldsymbol{x})x_1, g_2(\boldsymbol{x})x_2, ..., g_{p_1}(\boldsymbol{x})x_{p_1}]',$$

i.e., the joint distribution of the random variables $z_j = g_j(\boldsymbol{x})x_j, j = 1, 2, ..., p_1$, is symmetric around zero with regard to each z_j, in the sense that the distribution is invariant under the change of z_j to $-z_j$ for any j.

Proof. For convenience of this proof, let the random variables and their vectors be denoted here by capital letters, and let the corresponding small letters be used to denote any relevant real numbers or vectors. Then, the distribution function of $\boldsymbol{Z} = [Z_1, Z_2, ..., Z_{p_1}]'$ can be written as

$$\begin{aligned}
\Pr(\boldsymbol{Z} \leq z) &= \Pr(Z_1 \leq z_1, Z_2 \leq z_2, ..., Z_{p_1} \leq z_{p_1}) \\
&= \Pr[g_1(\boldsymbol{X})X_1 \leq z_1, g_2(\boldsymbol{X})X_2 \leq z_2, ..., g_{p_1}(\boldsymbol{X})X_{p_1} \leq z_{p_1}] \\
&= \Pr[-g_1(\boldsymbol{X})X_1 \geq -z_1, g_2(\boldsymbol{X})X_2 \leq z_2, ..., g_{p_1}(\boldsymbol{X})X_{p_1} \leq z_{p_1}].
\end{aligned}$$

But, because X_1 and $-X_1$ are distributed identically and independently of X_2, ..., X_{p_1}, and each $g_i(\boldsymbol{X})$ is an even function of x_1, it is justified to write

$$\begin{aligned}
&\Pr[-g_1(\boldsymbol{X})X_1 \geq -z_1, g_2(\boldsymbol{X})X_2 \leq z_2, ..., g_{p_1}(\boldsymbol{X})X_{p_1} \leq z_{p_1}] \\
&= \Pr[g_1(\boldsymbol{X})X_1 \geq -z_1, g_2(\boldsymbol{X})X_2 \leq z_2, ..., g_{p_1}(\boldsymbol{X})X_{p_1} \leq z_{p_1}] \\
&= \Pr[-g_1(\boldsymbol{X})X_1 \leq z_1, g_2(\boldsymbol{X})X_2 \leq z_2, ..., g_{p_1}(\boldsymbol{X})X_{p_1} \leq z_{p_1}] \\
&= \Pr(-Z_1 \leq z_1, Z_2 \leq z_2, ..., Z_{p_1} \leq z_{p_1}).
\end{aligned}$$

This derivation shows that the distribution function of the random vector $\boldsymbol{Z}$ is invariant under the change of its element Z_1 to $-Z_1$. The same can similarly be shown for the remaining Z_j variables, i.e., for $j = 2, 3, ..., p_1$. Thus, the distribution of $\boldsymbol{Z}$ is invariant under the change of the sign of any of its elements. □

This result leads to the following corollary.

Corollary 3.7.1. *Under the assumptions of Lemma* 3.7.3, $\mathrm{E}(z_j) = 0$ *for* $j = 1, 2, ..., p_1$ *and* $\mathrm{E}(z_j z_{j'}) = 0$ *for all* $j \neq j'$, *provided that these expectations exist.*

Proof. Derivations of these expectation results are straightforward. For example, suppose that the random variables $z_1, z_2, ..., z_{p_1}$ have a density function $f(\boldsymbol{z}) = f(z_1, z_2, ..., z_{p_1})$. Then,

$$\begin{aligned}
\mathrm{E}(z_1) &= \int_{-\infty}^{\infty}\int_{-\infty}^{\infty}\cdots\int_{-\infty}^{\infty} z_1 f(\boldsymbol{z}) d\boldsymbol{z} \\
&= \int_{-\infty}^{\infty}\cdots\int_{-\infty}^{\infty}\left[\int_{-\infty}^{0} z_1 f(\boldsymbol{z}) dz_1 + \int_{0}^{\infty} z_1 f(\boldsymbol{z}) dz_1\right] dz_2 \cdots dz_{p_1} \\
&= \int_{-\infty}^{\infty}\cdots\int_{-\infty}^{\infty}\left[-\int_{0}^{\infty} z_1 f(\boldsymbol{z}) dz_1 + \int_{0}^{\infty} z_1 f(\boldsymbol{z}) dz_1\right] dz_2 \cdots dz_{p_1} = 0,
\end{aligned}$$

as $f(-z_1, z_2, ..., z_{p_1}) = f(z_1, z_2, ..., z_{p_1}) = f(\boldsymbol{z})$ by Lemma 3.7.3. Similarly,

$$\begin{aligned}
\mathrm{E}(z_1z_2) &= \int_{-\infty}^{\infty}\int_{-\infty}^{\infty}\cdots\int_{-\infty}^{\infty} z_1z_2 f(\boldsymbol{z})d\boldsymbol{z} \\
&= \int_{-\infty}^{\infty}\cdots\int_{-\infty}^{\infty}\left[\int_{-\infty}^{0} z_1z_2f(\boldsymbol{z})dz_1 + \int_{0}^{\infty} z_1z_2f(\boldsymbol{z})dz_1\right]dz_2\cdots dz_{p_1} \\
&= \int_{-\infty}^{\infty}\cdots\int_{-\infty}^{\infty}\left[-\int_{0}^{\infty} z_1z_2f(\boldsymbol{z})dz_1 + \int_{0}^{\infty} z_1z_2f(\boldsymbol{z})dz_1\right]dz_2\cdots dz_{p_1} \\
&= 0
\end{aligned}$$

by the same argument as before. □

With the results obtained until now, the following main theorem (similar though more general than Theorem 6.1 of Roy and Shah, 1962) can be proved.

Theorem 3.7.3. *Let for a general block design the observed vector* $\boldsymbol{y}$ *have the model* (3.1.1) *with the properties* (3.1.15) *and* (3.1.16), *and suppose that the value of the ratio* γ *appearing in the representation* (3.7.1) *of the covariance matrix of* $\boldsymbol{y}$ *is unknown. Further, let the distribution of* $\boldsymbol{y}$ *be such that it induces the random variables* $\{x_j\}$ *defined in* (3.7.49) *to have mutually independent symmetric distributions around zero. Under these assumptions, if the statistic* $\hat{\gamma}$ *used to estimate* γ *is completely expressible by even functions of* $\{x_j\}$, *i.e., depends on* $\boldsymbol{y}$ *through such functions only, then, for all values of* γ *and* $\hat{\gamma}$ *exceeding the lower limit* $-1/k_{\max}$, *the estimator* $\tilde{\boldsymbol{\tau}}$ *defined in* (3.7.34) *has the properties given in* (3.7.44) *and* (3.7.45), *and hence, the properties of the estimator* $\widetilde{\boldsymbol{c}'\boldsymbol{\tau}}$ *are as in* (3.7.46) *for any* $\boldsymbol{c}$.

Proof. Under its assumptions, the theorem follows from Lemma 3.7.2 on account of Lemma 3.7.3 and Corollary 3.7.1, provided that the expectations in (3.7.43) exist and $\mathrm{Var}(z_j) < \infty$ for all j (= $1, 2, ..., h_*$). To check this, note that $\{\lambda_j\}$, as the eigenvalues of $\boldsymbol{C}_* = \boldsymbol{k}^{\delta} - \boldsymbol{N}'\boldsymbol{r}^{-\delta}\boldsymbol{N}$, satisfy the condition $\lambda_j \le k_{\max}$ for $j = 1, 2, ..., h_*$. On the other hand, because of the assumption that both γ and $\hat{\gamma}$ exceed $-1/k_{\max}$, the statistic $\hat{\gamma}$ is to be truncated at a low value (say) $\gamma_L > -1/k_{\max}$. Hence, as can easily be shown, the ratio $(\gamma - \hat{\gamma})/(\hat{\gamma}\lambda_j + 1)$ appearing in the definition (3.7.42) of z_j is a monotonically decreasing function of $\hat{\gamma}$ in the interval (γ_L, ∞), such that

$$-\frac{1}{\lambda_j} < \frac{\gamma - \hat{\gamma}}{\hat{\gamma}\lambda_j + 1} \le \frac{\gamma - \gamma_L}{\gamma_L\lambda_j + 1} \left(< \frac{k_{\max}\gamma + 1}{k_{\max} - \lambda_j} \right).$$

These inequalities imply that

$$\left|\frac{\gamma - \hat{\gamma}}{\hat{\gamma}\lambda_j + 1}\right| \le \max\left\{\frac{1}{\lambda_j}, \frac{\gamma - \gamma_L}{\gamma_L\lambda_j + 1}\right\} = M_j \quad \text{(say)},$$

where the equality holds only if either $\hat{\gamma} = \gamma_L$ and $\gamma \geq 2\gamma_L + 1/\lambda_j$ or $\hat{\gamma}$ approaches infinity and $\gamma \leq 2\gamma_L + 1/\lambda_j$. From this inequality then,

$$\mathrm{E}(|z_j|) \leq M_j \mathrm{E}(|x_j|) < \infty \quad \text{for } j = 1, 2, ..., h_*,$$

because $\mathrm{E}(x_j)$ exists [as $\mathrm{E}(\boldsymbol{y})$ exists from the assumptions]. Similarly, it follows that

$$\mathrm{E}(|z_j z_{j'}|) \leq M_j M_{j'} \mathrm{E}(|x_j x_{j'}|) < \infty \quad \text{for } j, j' = 1, 2, ..., h_*,$$

as $\mathrm{E}(x_j x_{j'})$ exists for any j, j' [as $\mathrm{Cov}(\boldsymbol{y})$ exists from the assumptions]. Thus, it has been shown that $\mathrm{E}(z_j)$ and $\mathrm{E}(z_j z_{j'})$ exist for all $j, j' = 1, 2, ..., h_*$. They are then

$$\mathrm{E}(z_j) = 0 \text{ and } \mathrm{E}(z_j z_{j'}) = \begin{cases} \mathrm{Var}(z_j) < \infty & \text{if } j = j', \\ 0 & \text{if } j \neq j', \end{cases} \tag{3.7.50}$$

on account of Corollary 3.7.1. This result, finally, verifies the assumptions of Lemma 3.7.2. □

It should be noted that Theorem 3.7.3 is general in the sense that it applies, under its assumptions, to any estimator of γ, which depends on the observed vector $\boldsymbol{y}$ through even functions of $\{x_j\}$ only. The estimator $\hat{\gamma}$ obtainable by solving the equations (3.7.24) is a particular case to which the theorem is applicable. Also note the following remark.

Remark 3.7.2. Under the assumptions of Theorem 3.7.3:

(a) The random variables $\{x_j\}$ defined in (3.7.49) have the properties

$$\mathrm{E}(x_j) = 0 \tag{3.7.51}$$

and

$$\mathrm{E}(x_j x_{j'}) = \begin{cases} \sigma_1^2(\gamma\lambda_j + 1) & \text{if } j = j', \text{ with } \lambda_j = 0 \text{ for } j > h_*, \\ 0 & \text{if } j \neq j', \end{cases} \tag{3.7.52}$$

resulting from the properties (3.1.15) and (3.1.16) of $\boldsymbol{y}$ and the definitions (3.7.37) and (3.7.38) of the vectors $\{\boldsymbol{v}_j\}$.

(b) If $\boldsymbol{y}$ has an n-variate normal distribution, then also the vector $\boldsymbol{x} = [x_1, x_2, ..., x_{n-v}]'$, with $x_j = \boldsymbol{v}_j'\boldsymbol{y}$, has an $(n - v)$-variate normal distribution, which automatically implies that its elements have mutually independent symmetric distributions around zero, on account of (3.7.51) and (3.7.52).

(c) As the functions of the vector $\boldsymbol{y}$ appearing in the equations (3.7.24) are, as shown in (3.7.47) and (3.7.48), completely expressible by the squares $x_j^2 = (\boldsymbol{v}_j'\boldsymbol{y})^2$, $j = 1, 2, ..., n - v$, the statistic $\hat{\gamma}$ obtained from the solution of the equations (3.7.24) is an even function of any of the x_j's, and as such satisfies the condition of Theorem 3.7.3, provided that its value exceeds $-1/k_{\max}$.

Remark 3.7.3. If it can be assumed that $\gamma \geq \gamma^*$, an *a priori* known value exceeding $-1/k_{\max}$, then it is better to truncate $\hat{\gamma}$ at γ^* than at any lower value

γ_L ($> -1/k_{\max}$). This recommendation follows from the fact that if $\hat{\gamma}^*$ denotes the estimator so truncated, then, under $\gamma \geq \gamma^*$,

$$\mathrm{Var}[z_j(\hat{\gamma}^*)] < \mathrm{Var}[z_j(\hat{\gamma})],$$

where $z_j(\hat{\gamma}^*)$ and $z_j(\hat{\gamma})$ are as defined in (3.7.42), with

$$\hat{\gamma}^* = \begin{cases} \hat{\gamma} & \text{if } \hat{\gamma} \geq \gamma^*, \\ \gamma^* & \text{otherwise.} \end{cases}$$

This result can easily be proved by noting that $\hat{\gamma}^* \geq \hat{\gamma}$ implies $z_j(\hat{\gamma}^*) \leq z_j(\hat{\gamma})$ for all $\gamma \geq \gamma^*$, the equality $z_j(\hat{\gamma}^*) = z_j(\hat{\gamma})$ holding only if $\hat{\gamma} \geq \gamma^*$ ($> \gamma_L$). (See also Shah, 1971.) Usually, it can be assumed that $\gamma \geq 0$, because, from the definition in Lemma 3.7.1, $\gamma < 0$ if and only if $K_H\sigma_B^2 < \sigma_U^2$, i.e., if the inter-block variation is smaller than the intra-block variation of experimental units, a case that would be in contradiction with the way the blocks are usually formed.

It follows from Theorem 3.7.3 that if the unknown ratio γ appearing in the matrix $\boldsymbol{T}$ defined in (3.7.3) is replaced by its estimator $\hat{\gamma}$ obtained according to the conditions of the theorem, then the unbiasedness of the estimators of $\boldsymbol{\tau}$ and $\widetilde{\boldsymbol{c}'\boldsymbol{\tau}}$ established in Theorem 3.7.2 is not violated. However, the formula for $\mathrm{Var}(\widetilde{\boldsymbol{c}'\boldsymbol{\tau}})$ in (3.7.46) shows that the replacement of γ by $\hat{\gamma}$ will cause the increase of the variance by an amount equal to

$$\sum_{j=1}^{h_*}(\boldsymbol{c}'\boldsymbol{r}^{-\delta}\boldsymbol{N}\boldsymbol{D}\boldsymbol{v}_j)^2\frac{\mathrm{Var}(z_j)}{(\gamma\lambda_j+1)^2},$$

where the random variables $\{z_j\}$ are as defined in (3.7.42). Unfortunately, the exact formula of $\mathrm{Var}(z_j) = \mathrm{E}(z_j^2)$ is in general intractable. It can, however, be approximated. For this note that, because z_j as a function of $\hat{\gamma}$ is $z_j(\hat{\gamma}) = (\hat{\gamma}\lambda_j+1)^{-1}(\gamma-\hat{\gamma})x_j$, its derivative with respect to $\hat{\gamma}$ is

$$\frac{dz_j}{d\hat{\gamma}} = -\frac{\gamma\lambda_j+1}{(\hat{\gamma}\lambda_j+1)^2}x_j,$$

from which (by Taylor's series expansion)

$$z_j \equiv z_j(\hat{\gamma}) = z_j(\gamma) + (\hat{\gamma}-\gamma)\left.\frac{dz_j}{d\hat{\gamma}}\right|_{\hat{\gamma}=\gamma} + \cdots = 0 - \frac{\hat{\gamma}-\gamma}{\gamma\lambda_j+1}x_j + \cdots,$$

giving the approximations

$$z_j \cong -(\gamma\lambda_j+1)^{-1}(\hat{\gamma}-\gamma)x_j$$

and, hence,

$$z_j^2 \cong \left(\frac{\hat{\gamma}-\gamma}{\gamma\lambda_j+1}\right)^2 x_j^2. \tag{3.7.53}$$

This derivation corresponds to the first stage of the approximation suggested by Kackar and Harville (1984). Applying to the expectation of (3.7.53) Theorem 2b.3(i) and (iii) of Rao (1973), one obtains

$$\mathrm{E}(z_j^2) \cong \mathrm{E}\left[\left(\frac{\hat{\gamma}-\gamma}{\gamma\lambda_j+1}\right)^2 \mathrm{E}(x_j^2|\hat{\gamma})\right]. \tag{3.7.54}$$

The conditional expectation $\mathrm{E}(x_j^2|\hat{\gamma})$ is usually not readily available, unless $\hat{\gamma}$ does not involve x_j defined in (3.7.49), which is not the case when $\hat{\gamma}$ is obtained by solving the equations (3.7.24), as can be seen from (3.7.47) and (3.7.48). However, if the number of x_j's from which the statistic $\hat{\gamma}$ is calculated is large, i.e., in the case of solving the equations (3.7.24) the number $n-v$ is large, then the statistical dependence between individual x_j^2 and $\hat{\gamma}$ can be ignored and instead of $\mathrm{E}(x_j^2|\hat{\gamma})$ the unconditional expectation $\mathrm{E}(x_j^2)$ can be used. This decision would allow the approximation (3.7.54) to be replaced by

$$\mathrm{E}(z_j^2) \cong \mathrm{E}\left[\left(\frac{\hat{\gamma}-\gamma}{\gamma\lambda_j+1}\right)^2 \mathrm{E}(x_j^2)\right] = \frac{\sigma_1^2}{\gamma\lambda_j+1}\mathrm{E}[(\hat{\gamma}-\gamma)^2] \tag{3.7.55}$$

[from (3.7.52)], where the mean squared error $\mathrm{E}[(\hat{\gamma}-\gamma)^2]$ (the MSE of $\hat{\gamma}$) becomes $\mathrm{Var}(\hat{\gamma})$ if $\hat{\gamma}$ is an unbiased estimator of γ. With the approximation (3.7.55), the variance of the empirical estimator $\widetilde{\boldsymbol{c}'\boldsymbol{\tau}} = \boldsymbol{c}'\tilde{\boldsymbol{\tau}}$, given in (3.7.46), can be approximated as

$$\mathrm{Var}(\widetilde{\boldsymbol{c}'\boldsymbol{\tau}}) \cong \mathrm{Var}(\widehat{\boldsymbol{c}'\boldsymbol{\tau}}) + \sigma_1^2\mathrm{E}[(\hat{\gamma}-\gamma)^2]\sum_{j=1}^{h_*}\frac{(\boldsymbol{c}'\boldsymbol{r}^{-\delta}\boldsymbol{N}\boldsymbol{D}\boldsymbol{v}_j)^2}{(\gamma\lambda_j+1)^3}. \tag{3.7.56}$$

This formula is equivalent to the two-stage approximation suggested by Kackar and Harville (1984).

Now, to make the approximation (3.7.56) applicable, the MSE of $\hat{\gamma}$ needs evaluation or approximation. If the distribution of $\boldsymbol{y}$ is assumed to be normal and $\hat{\gamma}$ is obtainable by solving (3.7.24), i.e., by the MML (REML) method, then $\mathrm{E}[(\hat{\gamma}-\gamma)^2]$, the MSE of $\hat{\gamma}$, can be replaced by the asymptotic variance of $\hat{\gamma}$ obtained from the inverse of the appropriate information matrix. As given by Patterson and Thompson (1971, p. 548), also by Harville (1977, p. 326), the information matrix associated with the MML (REML) estimation of γ and σ_1^2 is

$$\boldsymbol{I}(\gamma,\sigma_1^2) = \frac{1}{2}\begin{bmatrix} \mathrm{tr}[(\boldsymbol{R}\boldsymbol{D}'\boldsymbol{D})^2] & \sigma_1^{-2}\mathrm{tr}(\boldsymbol{R}\boldsymbol{D}'\boldsymbol{D}) \\ \sigma_1^{-2}\mathrm{tr}(\boldsymbol{R}\boldsymbol{D}'\boldsymbol{D}) & \sigma_1^{-4}(n-v) \end{bmatrix},$$

and its inverse is

$$[\boldsymbol{I}(\gamma,\sigma_1^2)]^{-1} = c\begin{bmatrix} \sigma_1^{-4}(n-v) & -\sigma_1^{-2}\mathrm{tr}(\boldsymbol{R}\boldsymbol{D}'\boldsymbol{D}) \\ -\sigma_1^{-2}\mathrm{tr}(\boldsymbol{R}\boldsymbol{D}'\boldsymbol{D}) & \mathrm{tr}[(\boldsymbol{R}\boldsymbol{D}'\boldsymbol{D})^2] \end{bmatrix},$$

where

$$c = \frac{2}{\sigma_1^{-4}(n-v)\mathrm{tr}[(\boldsymbol{RD}'\boldsymbol{D})^2] - \sigma_1^{-4}[\mathrm{tr}(\boldsymbol{RD}'\boldsymbol{D})]^2}.$$

Hence, the asymptotic variance of $\hat{\gamma}$ is

$$\mathrm{Var}_{\mathbf{as}}(\hat{\gamma}) = \frac{2(n-v)}{(n-v)\mathrm{tr}[(\boldsymbol{RD}'\boldsymbol{D})^2] - [\mathrm{tr}(\boldsymbol{RD}'\boldsymbol{D})]^2}, \tag{3.7.57}$$

which, on account of (3.7.39) and the property that $\boldsymbol{Dv}_j = \boldsymbol{0}$ for $j = h_* + 1, h_* + 2, ..., n - v$, can be written as a function of the eigenvalues of $\boldsymbol{D\phi}_*\boldsymbol{D}'$, shown in (3.7.35), i.e., as

$$\mathrm{Var}_{\mathbf{as}}(\hat{\gamma}) = \frac{2}{\sum_{j=1}^{h_*}\left(\frac{\lambda_j}{\gamma\lambda_j+1}\right)^2 - \frac{1}{n-v}\left(\sum_{j=1}^{h_*}\frac{\lambda_j}{\gamma\lambda_j+1}\right)^2}.$$

Thus, when under the normality assumption the ratio γ is estimated by the MML (REML) method, then $\mathrm{Var}(\widetilde{\boldsymbol{c}'\boldsymbol{\tau}})$ can further be approximated as

$$\mathrm{Var}(\widetilde{\boldsymbol{c}'\boldsymbol{\tau}}) \cong \mathrm{Var}(\widehat{\boldsymbol{c}'\boldsymbol{\tau}}) + C, \tag{3.7.58}$$

where

$$C = \frac{2\sigma_1^2}{\mathrm{tr}[(\boldsymbol{RD}'\boldsymbol{D})^2] - (n-v)^{-1}[\mathrm{tr}(\boldsymbol{RD}'\boldsymbol{D})]^2}\sum_{j=1}^{h_*}\frac{(\boldsymbol{c}'\boldsymbol{r}^{-\delta}\boldsymbol{NDv}_j)^2}{(\gamma\lambda_j+1)^3}$$

$$= \frac{2\sigma_1^2}{\sum_{j=1}^{h_*}\left(\frac{\lambda_j}{\gamma\lambda_j+1}\right)^2 - \frac{1}{n-v}\left(\sum_{j=1}^{h_*}\frac{\lambda_j}{\gamma\lambda_j+1}\right)^2}\sum_{j=1}^{h_*}\frac{\lambda_j(\boldsymbol{c}'\boldsymbol{r}^{-\delta}\boldsymbol{Nu}_j)^2}{(\gamma\lambda_j+1)^3},$$

as $\boldsymbol{Dv}_j = \lambda_j^{1/2}\boldsymbol{u}_j$ from (3.7.37) and (3.7.35).

The approximation (3.7.58) is equivalent to that which has been suggested by Kackar and Harville (1984), with all their three stages of approximation, here applied to the variance of the empirical estimator $\widetilde{\boldsymbol{c}'\boldsymbol{\tau}}$, as given in (3.7.46). This approximation would become exact if

(a) all $\{z_j\}$ defined in (3.7.42) were linear in $\hat{\gamma}$,

(b) the statistic $\hat{\gamma}$ were independent of the random variables $x_j = \boldsymbol{v}_j'\boldsymbol{y}$, $j = 1, 2, ..., h_*$, and

(c) the right-hand side of (3.7.57) were the exact MSE of $\hat{\gamma}$ rather than a mere approximation of it.

Considering now the asymptotic properties of the MML (REML) estimator of γ, or equivalently those of the iterated MINQUE of γ (see, e.g., Brown, 1976; Rao and Kleffe, 1988, Chapter 10), it can be seen that under the normality assumption the estimator $\hat{\gamma}$ obtained by solving the equations (3.7.24) is asymptotically unbiased and efficient (i.e., with the smallest possible limiting variance). Thus, it can be concluded in view of the above three conditions, (a), (b) and (c), that the approximation (3.7.58) approaches the exact value of $\text{Var}(\widetilde{\boldsymbol{c}'\boldsymbol{\tau}})$ as $n - v$ tends to infinity. However, because in practical applications the increase of n over v is possible only within some limits, the formula (3.7.58) will always remain only an approximation. Except for some special cases (see Kackar and Harville, 1984, Section 5), not much is known about the closeness of this approximation.

Example 3.7.2. To illustrate the use of the above approximation, it will be helpful to return to Example 3.7.1, where it has been shown that the inclusion of the inter-block information reduces the variance of the BLUE of a contrast. In connection with this, one should, however, be aware that when the unknown value of γ is replaced by its estimate, and so instead of the BLUE of a contrast the empirical estimator based on (3.7.34) is used, its variance exceeds that of the BLUE, as shown in (3.7.46). This expansion of the variance cannot be evaluated exactly, but an approximate formula is available, as given in (3.7.58). To apply it, one has first to obtain the matrix (3.7.35), which here is

$$\boldsymbol{D}\boldsymbol{\phi}_*\boldsymbol{D}' = \begin{bmatrix} 3.100 & -1.850 & -1.250 \\ -1.850 & 3.975 & 62.125 \\ -1.250 & -2.125 & 3.375 \end{bmatrix}, \quad \text{of rank } h^* = b - 1 = 2,$$

with the eigenvalues $\lambda_1 = 6.00$ and $\lambda_2 = 4.45$ and the corresponding eigenvectors

$$\boldsymbol{u}_1 = \begin{bmatrix} 0.2933 \\ -0.8066 \\ 0.5133 \end{bmatrix} \quad \text{and} \quad \boldsymbol{u}_2 = \begin{bmatrix} -0.7620 \\ 0.1270 \\ 0.6350 \end{bmatrix}.$$

They give, for the contrast $\tau_2 - \tau_1$,

$$\sum_{j=1}^{2} \frac{\lambda_j(\boldsymbol{c}'\boldsymbol{r}^{-\delta}\boldsymbol{N}\boldsymbol{u}_j)^2}{(\hat{\gamma}\lambda_j + 1)^3} = 0.001504$$

and

$$\sum_{j=1}^{2} \left(\frac{\lambda_j}{\hat{\gamma}\lambda_j + 1}\right)^2 - \frac{1}{n-v}\left(\sum_{j=1}^{2} \frac{\lambda_j}{\hat{\gamma}\lambda_j + 1}\right)^2 = 0.560647,$$

with which, by (3.7.58),

$$\begin{aligned} \text{Var}(\widetilde{\tau_2 - \tau_1}) &\cong \text{Var}(\widehat{\tau_2 - \tau_1}) + 2\frac{0.001504}{0.560647}\sigma_1^2 \\ &= (0.265019 + 0.005365)\sigma_1^2 = 0.270384\sigma_1^2. \end{aligned}$$

Inserting for σ_1^2 its estimate $\hat{\sigma}_1^2 = 2.51846$, one obtains finally the estimate of $\text{Var}(\widetilde{\tau_2 - \tau_1})$ as $0.270384\hat{\sigma}_1^2 = 0.68095$, which is slightly larger than that obtained in the intra-block analysis, viz., 0.68054. Thus, this example shows that although the recovery of inter-block information can improve the precision of estimating a contrast under known γ, it may happen that this improvement vanishes when γ is replaced by its estimate $\hat{\gamma}$.

In the special case of a BIB design, formula (3.7.58) reduces to

$$\mathrm{Var}(\widetilde{c'\tau}) \cong \mathrm{Var}(\widehat{c'\tau})\left\{1+\frac{2vw(1-w)(r-1)}{v(r-1)[(v-1)w^2+b-v]-[w(v-1)+b-v]^2}\right\}, \quad (3.7.59)$$

where

$$w=\frac{v(k-1)(k\gamma+1)}{v(k-1)(k\gamma+1)+v-k}=\frac{v(k-1)\delta}{v(k-1)\delta+v-k},$$

with $\delta=\sigma_2^2/\sigma_1^2$. Formula (3.7.59) is exactly equal to formula (5.4) of Kackar and Harville (1984). More general results applicable to proper block designs, from which (3.7.59) follows directly, will be presented in Section 3.8.

3.7.4 Testing hypotheses in a general block design

In the analysis of a general block design, a researcher may be interested not only in the estimation of treatment parameters and their linear functions, contrasts in particular, but also in testing hypotheses concerning such functions. Let these functions be written as $\boldsymbol{U}'\boldsymbol{\tau}$ (following the notation in Section 3.4), where $\boldsymbol{U}$ is a $v\times\ell$ matrix of rank ℓ satisfying the condition $\boldsymbol{U}'\mathbf{1}_v=\mathbf{0}$, i.e., let $\boldsymbol{U}'\boldsymbol{\tau}$ denote ℓ linearly independent contrasts of treatment parameters. Suppose that the researcher wants to test a hypothesis that can be defined by a consistent equation

$$H_0: \boldsymbol{U}'\boldsymbol{\tau}=\boldsymbol{q}. \quad (3.7.60)$$

To devise an appropriate test for H_0, it will be convenient first to assume that the true value of γ, defined in (3.7.2), is known. This will allow the BLUEs of the set of contrasts $\boldsymbol{U}'\boldsymbol{\tau}$ to be obtained in the form $\boldsymbol{U}'\hat{\boldsymbol{\tau}}$, where $\hat{\boldsymbol{\tau}}$ is given in (3.7.9), and their covariance matrix as $\mathrm{Cov}(\boldsymbol{U}'\hat{\boldsymbol{\tau}})=\sigma_1^2\boldsymbol{U}'(\boldsymbol{\Delta}\boldsymbol{T}^{-1}\boldsymbol{\Delta}')^{-1}\boldsymbol{U}=\sigma_1^2\boldsymbol{U}'\boldsymbol{C}_c^{-1}\boldsymbol{U}$, on account of (3.7.10) and (3.7.18). Referring now to Rao (1973, Section 4b), it can be seen that under the multivariate normal distribution of $\boldsymbol{y}$, the quadratic forms

$$\sigma_1^{-2}(\boldsymbol{U}'\hat{\boldsymbol{\tau}}-\boldsymbol{q})'(\boldsymbol{U}'\boldsymbol{C}_c^{-1}\boldsymbol{U})^{-1}(\boldsymbol{U}'\hat{\boldsymbol{\tau}}-\boldsymbol{q}) \quad \text{and} \quad \sigma_1^{-2}\boldsymbol{y}'\boldsymbol{R}\boldsymbol{y} \quad (3.7.61)$$

have independent χ^2 distributions. More precisely, the first in (3.7.61) has the χ^2 distribution with ℓ d.f. and the noncentrality parameter

$$\delta=\sigma_1^{-2}(\boldsymbol{U}'\boldsymbol{\tau}-\boldsymbol{q})'(\boldsymbol{U}'\boldsymbol{C}_c^{-1}\boldsymbol{U})^{-1}(\boldsymbol{U}'\boldsymbol{\tau}-\boldsymbol{q}),$$

equal to zero under H_0, and the second has the central χ^2 distribution with $n-v$ d.f. This can be checked applying Theorem 9.2.1 of Rao and Mitra (1971), and the independence of these quadratic forms can be proved using their Theorem 9.4.2(b). From the latter theorem, the necessary and sufficient conditions for the quadratic forms in (3.7.61) to be distributed independently are

$$\boldsymbol{T}^{-1}\boldsymbol{\Delta}'\boldsymbol{C}_c^{-1}\boldsymbol{U}(\boldsymbol{U}'\boldsymbol{C}_c^{-1}\boldsymbol{U})^{-1}\boldsymbol{U}'\boldsymbol{C}_c^{-1}\boldsymbol{\Delta}\boldsymbol{T}^{-1}\boldsymbol{T}\boldsymbol{R} = \mathbf{O}$$

and

$$\boldsymbol{R}\boldsymbol{T}\boldsymbol{T}^{-1}\boldsymbol{\Delta}'\boldsymbol{C}_c^{-1}\boldsymbol{U}(\boldsymbol{U}'\boldsymbol{C}_c^{-1}\boldsymbol{U})^{-1}\boldsymbol{q} = \mathbf{0}.$$

Evidently, they are satisfied because $\boldsymbol{\Delta}\boldsymbol{T}^{-1}\boldsymbol{T}\boldsymbol{R} = \boldsymbol{\Delta}\boldsymbol{R} = \mathbf{O}$, as it follows from (3.7.20). On account of these properties, the hypothesis (3.7.60) can be tested by the variance ratio criterion

$$\frac{n-v}{\ell}\,\frac{(\boldsymbol{U}'\hat{\boldsymbol{\tau}} - \boldsymbol{q})'(\boldsymbol{U}'\boldsymbol{C}_c^{-1}\boldsymbol{U})^{-1}(\boldsymbol{U}'\hat{\boldsymbol{\tau}} - \boldsymbol{q})}{\boldsymbol{y}'\boldsymbol{R}\boldsymbol{y}}, \tag{3.7.62}$$

which under the normality of $\boldsymbol{y}$ has then the F distribution with ℓ and $n - v$ d.f., central when the hypothesis is true (see, e.g., Rao, 1973, Section 4b.2).

In practice, however, the true value of γ is usually unknown and is subject to estimation, e.g., as described in Section 3.7.2. Then, instead of $\hat{\boldsymbol{\tau}}$, i.e., the BLUE of $\boldsymbol{\tau}$, the empirical estimator $\tilde{\boldsymbol{\tau}}$, given in (3.7.34), is to be used, and instead of the first quadratic form in (3.7.61), the quadratic form

$$(\boldsymbol{U}'\tilde{\boldsymbol{\tau}} - \boldsymbol{q})'[\boldsymbol{U}'\text{Cov}(\tilde{\boldsymbol{\tau}})\boldsymbol{U}]^{-1}(\boldsymbol{U}'\tilde{\boldsymbol{\tau}} - \boldsymbol{q}) \tag{3.7.63}$$

might be considered. But, as the covariance matrix of $\tilde{\boldsymbol{\tau}}$, of the form given in (3.7.45), is usually not available, the quadratic form (3.7.63) cannot in general be used directly. However, because of the result (3.7.55), $\text{Cov}(\tilde{\boldsymbol{\tau}})$ can be approximated by the formula

$$\begin{aligned}\text{Cov}(\tilde{\boldsymbol{\tau}}) &\cong \text{Cov}(\hat{\boldsymbol{\tau}}) + \sigma_1^2\boldsymbol{r}^{-\delta}\boldsymbol{N}\boldsymbol{D}\sum_{j=1}^{h_*}\frac{\text{E}[(\hat{\gamma}-\gamma)^2]}{(\gamma\lambda_j+1)^3}\boldsymbol{v}_j\boldsymbol{v}_j'\boldsymbol{D}'\boldsymbol{N}'\boldsymbol{r}^{-\delta} \\ &= \sigma_1^2\boldsymbol{C}_{\text{ap}}^{-1} \quad \text{(say)},\end{aligned} \tag{3.7.64}$$

applicable in practice after replacing the unknown value of γ in the denominators by its estimate. Thus, it can be suggested that, instead of (3.7.63), the quadratic form

$$\sigma_1^{-2}(\boldsymbol{U}'\tilde{\boldsymbol{\tau}} - \boldsymbol{q})'(\boldsymbol{U}'\hat{\boldsymbol{C}}_{\text{ap}}^{-1}\boldsymbol{U})^{-1}(\boldsymbol{U}'\tilde{\boldsymbol{\tau}} - \boldsymbol{q}) = Y \quad \text{(say)} \tag{3.7.65}$$

is used, where

$$\hat{\boldsymbol{C}}_{\text{ap}}^{-1} = \hat{\boldsymbol{C}}_c^{-1} + \boldsymbol{r}^{-\delta}\boldsymbol{N}\boldsymbol{D}\sum_{j=1}^{h_*}\frac{\text{E}[(\hat{\gamma}-\gamma)^2]}{(\hat{\gamma}\lambda_j+1)^3}\boldsymbol{v}_j\boldsymbol{v}_j'\boldsymbol{D}'\boldsymbol{N}'\boldsymbol{r}^{-\delta}, \tag{3.7.66}$$

with $\hat{\boldsymbol{C}}_c^{-1} = (\boldsymbol{\Delta}\hat{\boldsymbol{T}}^{-1}\boldsymbol{\Delta}')^{-1}$, obtainable according to (3.7.18) but with $\{k_{*j}^{-1}\}$, defined in (3.7.17), replaced by $\{\hat{\gamma}/(k_j\hat{\gamma}+1)\}$, and with $\text{E}[(\hat{\gamma}-\gamma)^2]$ (the MSE of $\hat{\gamma}$), if unknown, replaced by the asymptotic variance of $\hat{\gamma}$ defined in (3.7.57). Unfortunately, the distribution of Y is not of a χ^2 type, its exact form being in general intractable. An attempt, however, can be made to approximate it by the distribution of an appropriately scaled χ^2 variable having approximately the same expectation and variance as those of Y.

To approximate the distribution of the random variable Y defined in (3.7.65), consider first its expectation and variance conditional at a given $\hat{\gamma}$, assuming as above that the distribution of $\boldsymbol{y}$ is multivariate normal. Referring, e.g., to Rao and Kleffe (1988, Section 2.1), it can be seen that they are

$$\begin{aligned} \mathrm{E}(Y|\hat{\gamma}) &= \sigma_1^{-2}\{\boldsymbol{\tau}'\boldsymbol{U}\boldsymbol{H}\boldsymbol{U}'\boldsymbol{\tau} - 2\boldsymbol{q}'\boldsymbol{H}\boldsymbol{U}'\boldsymbol{\tau} + \boldsymbol{q}'\boldsymbol{H}\boldsymbol{q} + \mathrm{tr}[\boldsymbol{U}\boldsymbol{H}\boldsymbol{U}'\mathrm{Cov}(\tilde{\boldsymbol{\tau}})]\}, \\ \mathrm{Var}(Y|\hat{\gamma}) &= \sigma_1^{-4}\{4(\boldsymbol{U}\boldsymbol{H}\boldsymbol{U}'\boldsymbol{\tau} - \boldsymbol{U}\boldsymbol{H}\boldsymbol{q})'\mathrm{Cov}(\tilde{\boldsymbol{\tau}})(\boldsymbol{U}\boldsymbol{H}\boldsymbol{U}'\boldsymbol{\tau} - \boldsymbol{U}\boldsymbol{H}\boldsymbol{q}) \\ &\quad + 2\mathrm{tr}[\boldsymbol{U}\boldsymbol{H}\boldsymbol{U}'\mathrm{Cov}(\tilde{\boldsymbol{\tau}})\boldsymbol{U}\boldsymbol{H}\boldsymbol{U}'\mathrm{Cov}(\tilde{\boldsymbol{\tau}})]\}, \end{aligned}$$

where $\boldsymbol{H} = (\boldsymbol{U}'\hat{\boldsymbol{C}}_{\mathrm{ap}}^{-1}\boldsymbol{U})^{-1}$. Under H_0, they reduce to

$$\mathrm{E}(Y|\hat{\gamma}) = \sigma_1^{-2}\mathrm{tr}[\boldsymbol{H}\boldsymbol{U}'\mathrm{Cov}(\tilde{\boldsymbol{\tau}})\boldsymbol{U}], \tag{3.7.67}$$

$$\mathrm{Var}(Y|\hat{\gamma}) = 2\sigma_1^{-4}\mathrm{tr}[\boldsymbol{H}\boldsymbol{U}'\mathrm{Cov}(\tilde{\boldsymbol{\tau}})\boldsymbol{U}\boldsymbol{H}\boldsymbol{U}'\mathrm{Cov}(\tilde{\boldsymbol{\tau}})\boldsymbol{U}], \tag{3.7.68}$$

respectively. To obtain these moments unconditional, the formulae

$$\mathrm{E}(Y) = \mathrm{E}_{\hat{\gamma}}[\mathrm{E}(Y|\hat{\gamma})] \quad \text{and} \quad \mathrm{Var}(Y) = \mathrm{E}_{\hat{\gamma}}[\mathrm{Var}(Y|\hat{\gamma})] + \mathrm{Var}_{\hat{\gamma}}[\mathrm{E}(Y|\hat{\gamma})] \tag{3.7.69}$$

are to be used (see, e.g., Rao, 1973, Section 2b.3; also Kenward and Roger, 1997, Section 3). Their application here is complicated, as the involved distributions of (3.7.67) and (3.7.68) depend on the usually unknown distribution of $\hat{\gamma}$. To make the application more straightforward, some approximations of the matrices $\mathrm{Cov}(\tilde{\boldsymbol{\tau}})$ and $\boldsymbol{H} = (\boldsymbol{U}'\hat{\boldsymbol{C}}_{\mathrm{ap}}^{-1}\boldsymbol{U})^{-1}$ are needed. For the first, the approximation (3.7.64) can be used. For the second, Taylor's series expansion can be employed. Taking only its lower order terms, the approximation

$$(\boldsymbol{U}'\hat{\boldsymbol{C}}_{\mathrm{ap}}^{-1}\boldsymbol{U})^{-1} \cong (\boldsymbol{U}'\boldsymbol{C}_{\mathrm{ap}}^{-1}\boldsymbol{U})^{-1} + (\hat{\gamma} - \gamma)\left.\frac{d(\boldsymbol{U}'\hat{\boldsymbol{C}}_{\mathrm{ap}}^{-1}\boldsymbol{U})^{-1}}{d\hat{\gamma}}\right|_{\hat{\gamma}=\gamma}$$

$$= (\boldsymbol{U}'\boldsymbol{C}_{\mathrm{ap}}^{-1}\boldsymbol{U})^{-1} - (\hat{\gamma} - \gamma)(\boldsymbol{U}'\boldsymbol{C}_{\mathrm{ap}}^{-1}\boldsymbol{U})^{-1}\left.\frac{d(\boldsymbol{U}'\hat{\boldsymbol{C}}_{\mathrm{ap}}^{-1}\boldsymbol{U})}{d\hat{\gamma}}\right|_{\hat{\gamma}=\gamma}(\boldsymbol{U}'\boldsymbol{C}_{\mathrm{ap}}^{-1}\boldsymbol{U})^{-1}$$

can be obtained. To calculate $d(\boldsymbol{U}'\hat{\boldsymbol{C}}_{ap}^{-1}\boldsymbol{U})/d\hat{\gamma}$, it is helpful first to note that, using the relation $\boldsymbol{C}_c^{-1} = \boldsymbol{r}^{-\delta}\boldsymbol{\Delta}(\boldsymbol{T} - \boldsymbol{TRT})\boldsymbol{\Delta}'\boldsymbol{r}^{-\delta}$, following from (3.7.18) and (3.7.20), and those in (3.7.35), (3.7.36), (3.7.37) and (3.7.39), it is possible to write (3.7.66) with the use of the vectors $\{\boldsymbol{u}_j\}$ as

$$\hat{C}_{\mathrm{ap}}^{-1} = r^{-\delta} + r^{-\delta}N\left\{\sum_{j=1}^{h_*}\frac{\hat{\gamma}(\hat{\gamma}\lambda_j+1)^2+\lambda_j\mathrm{E}[(\hat{\gamma}-\gamma)^2]}{(\hat{\gamma}\lambda_j+1)^3}u_ju_j' + \hat{\gamma}\sum_{j=h_*+1}^{b}u_ju_j'\right\}N'r^{-\delta}.$$

Using this representation, it is easy to obtain

$$\frac{d\hat{C}_{\mathrm{ap}}^{-1}}{d\hat{\gamma}} = r^{-\delta}N\left\{\sum_{j=1}^{h_*}\frac{(\hat{\gamma}\lambda_j+1)^2-3\lambda_j^2\mathrm{E}[(\hat{\gamma}-\gamma)^2]}{(\hat{\gamma}\lambda_j+1)^4}u_ju_j' + \sum_{j=h_*+1}^{b}u_ju_j'\right\}N'r^{-\delta},$$

from which the considered approximation can be written as

$$\begin{aligned} H &= (U'\hat{C}_{\mathrm{ap}}^{-1}U)^{-1} \cong (U'C_{\mathrm{ap}}^{-1}U)^{-1} \\ &\quad - (\hat{\gamma}-\gamma)(U'C_{\mathrm{ap}}^{-1}U)^{-1}U'r^{-\delta}N\left\{\sum_{j=1}^{h_*}\frac{(\gamma\lambda_j+1)^2-3\lambda_j^2\mathrm{E}[(\hat{\gamma}-\gamma)^2]}{(\gamma\lambda_j+1)^4}u_ju_j'\right. \\ &\qquad \left. + \sum_{j=h_*+1}^{b}u_ju_j'\right\}N'r^{-\delta}U(U'C_{\mathrm{ap}}^{-1}U)^{-1}. \end{aligned}$$

Hence, the results (3.7.67) and (3.7.68) obtain, on account of (3.7.64), more suitable approximate forms

$$\mathrm{E}(Y|\hat{\gamma}) \cong \ell - (\hat{\gamma}-\gamma)\mathrm{tr}(A), \tag{3.7.70}$$
$$\mathrm{Var}(Y|\hat{\gamma}) \cong 2[\ell - 2(\hat{\gamma}-\gamma)\mathrm{tr}(A) + (\hat{\gamma}-\gamma)^2\mathrm{tr}(AA)], \tag{3.7.71}$$

respectively, where

$$\begin{aligned} A &= (U'C_{\mathrm{ap}}^{-1}U)^{-1}U'\left.\frac{d\hat{C}_{\mathrm{ap}}^{-1}}{d\hat{\gamma}}\right|_{\hat{\gamma}=\gamma} U \\ &= (U'C_{\mathrm{ap}}^{-1}U)^{-1}U'r^{-\delta}N\left\{\sum_{j=1}^{h_*}\frac{(\gamma\lambda_j+1)^2-3\lambda_j^2\mathrm{E}[(\hat{\gamma}-\gamma)^2]}{(\gamma\lambda_j+1)^4}u_ju_j'\right. \\ &\qquad \left. + \sum_{j=h_*+1}^{b}u_ju_j'\right\}N'r^{-\delta}U. \end{aligned} \tag{3.7.72}$$

From (3.7.69), (3.7.70) and (3.7.71), taking $\mathrm{E}(\hat{\gamma}) = \gamma$ and $\mathrm{Var}(\hat{\gamma}) = \mathrm{E}[(\hat{\gamma}-\gamma)^2] = \mathrm{Var}_{\mathrm{as}}(\hat{\gamma})$, which is justified asymptotically (as can be seen from Brown, 1976, Section 5), one can approximate the required moments, respectively, as follows:

$$\mathrm{E}(Y) \cong \ell = \mathrm{E}_{\mathrm{ap}}(Y) \quad \text{(say)}$$

and

$$\mathrm{Var}(Y) \cong 2\ell + \mathrm{Var}_{\mathrm{as}}(\hat{\gamma})\{2\mathrm{tr}(AA) + [\mathrm{tr}(A)]^2\} = \mathrm{Var}_{\mathrm{ap}}(Y) \quad \text{(say)}.$$

With them, to approximate the distribution of Y by that of $g\chi^2_d$, where g is a scale factor and χ^2_d represents a random variable having the central χ^2 distribution with d d.f., the following two equations,

$$\mathrm{E_{ap}}(Y) = \mathrm{E}(g\chi^2_d) = gd \quad \text{and} \quad \mathrm{Var_{ap}}(Y) = \mathrm{Var}(g\chi^2_d) = 2g^2 d,$$

are to be solved. The unique solution is supplied (similarly as in Box, 1954, Section 3) by

$$g = \frac{\mathrm{Var_{ap}}(Y)}{2\mathrm{E_{ap}}(Y)} = \frac{2\ell + \mathrm{Var_{as}}(\hat{\gamma})\{2\mathrm{tr}(\boldsymbol{AA}) + [\mathrm{tr}(\boldsymbol{A})]^2\}}{2\ell} \tag{3.7.73}$$

and

$$d = \frac{2[\mathrm{E_{ap}}(Y)]^2}{\mathrm{Var_{ap}}(Y)} = \frac{2\ell^2}{2\ell + \mathrm{Var_{as}}(\hat{\gamma})\{2\mathrm{tr}(\boldsymbol{AA}) + [\mathrm{tr}(\boldsymbol{A})]^2\}}. \tag{3.7.74}$$

Thus, it has been shown that to approximate the distribution, under H_0, of the quadratic form Y given in (3.7.65) by the distribution of $g\chi^2_d$, the factor g is to be taken as in (3.7.73) and d as in (3.7.74). Hence, as a candidate for the numerator of a possible test criterion similar to (3.7.62), the statistic $Y/(gd)$ can be considered. Because of $gd = \ell$, it appears that the quadratic form (3.7.65) divided by ℓ may be treated in this approach as having approximately the distribution of χ^2_d/d, if H_0 is true.

For the denominator of the criterion analogous to (3.7.62), the quadratic form $\boldsymbol{y}'\hat{\boldsymbol{R}}\boldsymbol{y}$, with $n - v$ d.f., might be considered as a possible candidate. It should, however, be recalled that the independence of the distributions of the two quadratic forms in (3.7.62) is conditioned on $\boldsymbol{\Delta T}^{-1}\boldsymbol{TR} = \mathbf{O}$, which holds because $\boldsymbol{\Delta R} = \mathbf{O}$ from the definition (3.7.20) of $\boldsymbol{R}$. If in the first quadratic form the vector $\hat{\boldsymbol{\tau}}$ is replaced by $\tilde{\boldsymbol{\tau}}$, and in the second the matrix $\boldsymbol{R}$ is replaced by $\hat{\boldsymbol{R}}$, i.e., if in both of these quadratic forms the parameter γ is replaced by its estimate $\hat{\gamma}$, then the above condition becomes $\boldsymbol{\Delta}\hat{\boldsymbol{T}}^{-1}\boldsymbol{T}\hat{\boldsymbol{R}} = \mathbf{O}$, which does not hold in general, unless $\hat{\boldsymbol{T}}^{-1}\boldsymbol{T} = \boldsymbol{I}_n$, usually not attainable for a finite n. But, if the matrix

$$\hat{\boldsymbol{R}} = \sum_{j=1}^{h_*} \frac{1}{\hat{\gamma}\lambda_j + 1}\boldsymbol{v}_j\boldsymbol{v}_j' + \sum_{j=h_*+1}^{n-v} \boldsymbol{v}_j\boldsymbol{v}_j'$$

is reduced to

$$\boldsymbol{R}_{\mathrm{red}} = \sum_{j=h_*+1}^{n-v} \boldsymbol{v}_j\boldsymbol{v}_j',$$

then, on account of (3.7.3),

$$\boldsymbol{TR}_{\mathrm{red}} = (\gamma\boldsymbol{D}'\boldsymbol{D} + \boldsymbol{I}_n) \sum_{j=h_*+1}^{n-v} \boldsymbol{v}_j\boldsymbol{v}_j' = \boldsymbol{R}_{\mathrm{red}},$$

because $\boldsymbol{D}\boldsymbol{v}_j = \mathbf{0}$ for $j = h_* + 1, h_* + 2, ..., n - v$, and hence, on account of (3.7.8),

$$\boldsymbol{\Delta}\hat{\boldsymbol{T}}^{-1}\boldsymbol{T}\boldsymbol{R}_{\text{red}} = \boldsymbol{\Delta}\hat{\boldsymbol{T}}^{-1}\boldsymbol{R}_{\text{red}} = \boldsymbol{\Delta}[\boldsymbol{I}_n - \boldsymbol{D}'(\boldsymbol{k}^{\delta} + \hat{\gamma}^{-1}\boldsymbol{I}_b)^{-1}\boldsymbol{D}] \sum_{j=h_*+1}^{n-v} \boldsymbol{v}_j\boldsymbol{v}_j' = \mathbf{O},$$

as $\boldsymbol{\Delta}\boldsymbol{v}_j = \mathbf{0}$ for $j = 1, 2, ..., n - v$. This result shows that $\boldsymbol{\Delta}\hat{\boldsymbol{T}}^{-1}\boldsymbol{T}\boldsymbol{R}_{\text{red}} = \mathbf{O}$ at any value of $\hat{\gamma}$, which makes the conditional distribution of the quadratic form Y, given a fixed value of $\hat{\gamma}$, to be independent of the distribution of $\sigma_1^{-2}\boldsymbol{y}'\boldsymbol{R}_{\text{red}}\boldsymbol{y} = X$ (say) for any of that value [see again Rao and Mitra, 1971, Theorem 9.4.2(b)]. As to the distribution of X, it can easily be seen from (3.7.49) and Remark 3.7.2 that, under the normality assumption of $\boldsymbol{y}$, X has the central χ^2 distribution with $n-v-h_* = n-b-h$ d.f. Now, taking into account the above approximation of the distribution of Y, it can be concluded that if the hypothesis (3.7.60) is true, then the statistic

$$F = \frac{n-b-h}{\ell} \frac{(\boldsymbol{U}'\tilde{\boldsymbol{\tau}} - \boldsymbol{q})'(\boldsymbol{U}'\hat{\boldsymbol{C}}_{\text{ap}}^{-1}\boldsymbol{U})^{-1}(\boldsymbol{U}'\tilde{\boldsymbol{\tau}} - \boldsymbol{q})}{\boldsymbol{y}'\boldsymbol{R}_{\text{red}}\boldsymbol{y}} \tag{3.7.75}$$

has a distribution that can be, justifiably, approximated by the distribution of the ratio

$$\frac{\chi_d^2/d}{\chi_{n-b-h}^2/(n-b-h)},$$

where χ_d^2 and χ_{n-b-h}^2 are independent random variables having central χ^2 distributions with d and $n-b-h$ d.f., respectively, i.e., by the central F distribution with d and $n-b-h$ d.f., where d is given in (3.7.74).

Considering the d.f. d, note that in general $d \leq \ell$, this being the closer to $d = \ell$ the smaller the coefficient $\text{Var}_{\text{as}}(\hat{\gamma})\{2\text{tr}(\boldsymbol{A}\boldsymbol{A}) + [\text{tr}(\boldsymbol{A})]^2\}$. It becomes zero if $\boldsymbol{A} = \mathbf{O}$, which will be attained in particular when $\boldsymbol{N}'\boldsymbol{r}^{-\delta}\boldsymbol{U} = \mathbf{O}$, as in the case considered in Corollary 3.2.5(a). Otherwise, the coefficient depends on $\text{Var}_{\text{as}}(\hat{\gamma})$ and on the structure of the matrix $\boldsymbol{A}$ defined in (3.7.72), where usually $\text{E}[(\hat{\gamma} - \gamma)^2]$ will also be replaced by $\text{Var}_{\text{as}}(\hat{\gamma})$. Furthermore, note that because $\boldsymbol{u}_b = b^{-1/2}\mathbf{1}_b$, giving $\boldsymbol{r}^{-\delta}\boldsymbol{N}\boldsymbol{u}_b = b^{-1/2}\mathbf{1}_v$, and $\boldsymbol{U}'\mathbf{1}_v = \mathbf{0}$, the sum $\sum_{j=h_*+1}^{b} \boldsymbol{u}_j\boldsymbol{u}_j'$ in the formula of $\boldsymbol{A}$ can be reduced to $\sum_{j=h_*+1}^{b-1} \boldsymbol{u}_j\boldsymbol{u}_j'$, which in turn disappears when $h_* = b - 1$, i.e., when the design is connected. Then, (3.7.72) can be written as

$$\boldsymbol{A} = (\boldsymbol{U}'\boldsymbol{C}_{\text{ap}}^{-1}\boldsymbol{U})^{-1}\boldsymbol{U}'\boldsymbol{r}^{-\delta}\boldsymbol{N}\left\{\sum_{j=1}^{h_*} \frac{(\gamma\lambda_j + 1)^2 - 3\lambda_j^2\text{E}[(\hat{\gamma} - \gamma)^2]}{(\gamma\lambda_j + 1)^4}\boldsymbol{u}_j\boldsymbol{u}_j'\right\}\boldsymbol{N}'\boldsymbol{r}^{-\delta}\boldsymbol{U}. \tag{3.7.76}$$

In addition, note that, on account of the formulae (3.7.35)–(3.7.38), the matrix $\boldsymbol{R}_{\text{red}}$ appearing in (3.7.75) can be presented as

$$\boldsymbol{R}_{\text{red}} = \boldsymbol{\phi}_* - \boldsymbol{\phi}_*\boldsymbol{D}'(\boldsymbol{D}\boldsymbol{\phi}_*\boldsymbol{D}')^+\boldsymbol{D}\boldsymbol{\phi}_* = \boldsymbol{\psi}_*,$$

which, as shown in Section 2.3, coincides with $\boldsymbol{\psi}_1$ ($\equiv \boldsymbol{\psi}$) defined in (2.3.8), and equivalently in (2.3.14). Thus, $\boldsymbol{y}'\boldsymbol{R}_{\text{red}}\boldsymbol{y} = \boldsymbol{y}'\boldsymbol{\psi}_1\boldsymbol{y}$, i.e., is the intra-block residual sum of squares considered in Section 3.2.1.

To conclude the discussion on testing hypotheses, note that the above derivation of the approximate F test of the hypothesis stated in (3.7.60), though based on the same approximation techniques as those used by Kenward and Roger (1997), differs at some points from that suggested in their paper. In the procedure devised here, the approximation by Taylor's series expansion is performed with regard to only one estimated parameter, $\hat{\gamma}$, which makes its application much simpler. This result has become possible because of the use, as the denominator sum of squares in the F statistic (3.7.75), the intra-block residual sum of squares instead of the residual sum of squares $\boldsymbol{y}'\hat{\boldsymbol{R}}\boldsymbol{y}$ obtained from the MML (REML) method to estimate σ_1^2, as shown in (3.7.32). This replacement seems to be justified by the fact that the intra-block mean square $s_1^2 = \boldsymbol{y}'\boldsymbol{\psi}_1\boldsymbol{y}/(n-b-h)$, considered in Section 3.2.1, is under the overall model (3.1.1) with the covariance matrix (3.7.1) admissible among all invariant quadratic and unbiased estimators of σ_1^2 [see Olsen, Seely and Birkes, 1976, Proposition 6.2(a); also Caliński, Gnot and Michalski, 1998, Lemma 4.1]. A particular advantage of using $\boldsymbol{y}'\boldsymbol{R}_{\text{red}}\boldsymbol{y} = \boldsymbol{y}'\boldsymbol{\psi}_1\boldsymbol{y}$ in (3.7.75) is its known exact distribution under normality of $\boldsymbol{y}$. This is also the reason why not its d.f. but that of the numerator sum of squares is estimated in the present procedure, unlike in the approach adopted by Kenward and Roger (1997, Section 3), where the nominal d.f. of the numerator sum of squares, i.e., ℓ [$=$ rank($\boldsymbol{U}$)], is kept fixed. This requirement results in a scaled form of the F statistic to be used, with the scale factor possibly smaller than 1. Such modification is avoided in the present approach. How much is gained by this is difficult to predict. A comparative study is in progress.

Example 3.7.3. Let the results given in Examples 3.2.1 and 3.2.6 be now supplemented by use of the procedures presented above. For this reason, note that to combine the results from both analyses, the intra-block analysis (Example 3.2.1) and the inter-block analysis (Example 3.2.6), one has first to refer to Section 3.7.1, particularly to definitions of σ_1^2 and γ given in Lemma 3.7.1 and to the explanation of their role given in Remark 3.7.2. Next, one has to obtain the estimates of σ_1^2 and γ. This can be done using the method described in Section 3.7.2. Employing the GENSTAT 5 (1996) software for the REML estimation procedure, the following iterative estimates of these parameters have been obtained:

Iteration cycle	1	2	3	4	5
$\hat{\gamma}$	0.571347	0.600761	0.597936	0.598200	the same
$\hat{\sigma}_1^2$	0.861422	0.851515	0.852540	0.852444	the same

With the final estimate of γ, obtained here at the fourth iteration cycle, the estimate of the matrix $\boldsymbol{T}^{-1}$, defined in (3.7.8), and with it that of $\boldsymbol{R}$, defined in (3.7.20), can be obtained. Using then the formula (3.7.32), the convergence can be checked. In fact, (3.7.32) gives here $\hat{\sigma}_1^2 = 0.852444$, which is exactly the value obtained at the fourth cycle of the iterative REML procedure. Also, with

$$\hat{k}_{*j}^{-1} = \begin{cases} 0.272355 & \text{for } j = 1,2,3,4,5,6, \\ 0.214056 & \text{for } j = 7,8,9,10, \end{cases}$$

obtained according to (3.7.17) but with γ replaced by its estimate $\hat{\gamma} = 0.598200$, the matrix $\boldsymbol{C}_c$ and the vector $\boldsymbol{Q}_c$, defined in (3.7.18), can be estimated. The results are

$$\hat{\boldsymbol{C}}_c = 5.241234\,\boldsymbol{I}_4 - 0.700467\,\mathbf{1}_4\mathbf{1}_4' \quad \text{and} \quad \hat{\boldsymbol{Q}}_c = [49.33,\ 47.82,\ 31.43,\ 23.12]'.$$

From them, using (3.7.34), the empirical estimate of $\boldsymbol{\tau}$ is obtained as

$$\tilde{\boldsymbol{\tau}} = [17.723,\ 17.435,\ 14.307,\ 12.721]'.$$

These estimates of treatment parameters coincide exactly with the "predicted means for treatments" obtained directly from GENSTAT 5 (1996). With them one can obtain the empirical estimate of any contrast $\boldsymbol{c}'\boldsymbol{\tau}$, i.e., $\widetilde{\boldsymbol{c}'\boldsymbol{\tau}} = \boldsymbol{c}'\tilde{\boldsymbol{\tau}}$, of the elementary contrasts in particular. When considering the variance of any such empirical estimator, it is instructive to take into account two possibilities. The first possibility is to ignore the fact that instead of the unknown true value of γ, its estimate has been used in obtaining the estimate of $\boldsymbol{\tau}$ and, hence, of any $\boldsymbol{c}'\boldsymbol{\tau}$. This approach means to regard $\hat{\gamma}$ as if it were the true value of γ. Then, the variance to be used is that given by the formula (3.7.11), which can also be written as $\text{Var}(\widehat{\boldsymbol{c}'\boldsymbol{\tau}}) = \sigma_1^2\boldsymbol{c}'\boldsymbol{C}_c^{-1}\boldsymbol{c}$, with $\boldsymbol{C}_c$ replaced by $\hat{\boldsymbol{C}}_c$. Because here

$$\hat{\boldsymbol{C}}_c^{-1} = (5.241234)^{-1}(\boldsymbol{I}_4 + 0.287151\,\mathbf{1}_4\mathbf{1}_4'),$$

the above variance becomes equal to

$$\text{Var}(\widehat{\boldsymbol{c}'\boldsymbol{\tau}}) = (5.241234)^{-1}\boldsymbol{c}'\boldsymbol{c}\sigma_1^2 = 0.190795\boldsymbol{c}'\boldsymbol{c}\sigma_1^2.$$

In particular, for any elementary contrast, it is equal to $0.381590\sigma_1^2$. Noting, from the results above, that

$$\text{Var}[(\widehat{\boldsymbol{c}'\boldsymbol{\tau}})_{\text{intra}}]/\text{Var}(\widehat{\boldsymbol{c}'\boldsymbol{\tau}}) = 1.123,$$

it could be said that a 12.3% increase in efficiency has been obtained from the recovery of inter-block information. However, as it follows from Lemma 3.7.2, this increase will be smaller when taking into account that $\hat{\gamma}$ is not the true value of γ but merely its estimate. To approximate the real variance of the above empirical estimate $\widetilde{\boldsymbol{c}'\boldsymbol{\tau}}$, the formula (3.7.58) can be used. For it, eigenvalues and eigenvectors of the matrix $\boldsymbol{C}_* = \boldsymbol{D}\boldsymbol{\phi}_*\boldsymbol{D}'$ are needed. It has been found that there are four distinct nonzero eigenvalues of $\boldsymbol{C}_*$,

$$\begin{aligned} \lambda_1 &= 2.8792 \quad \text{with multiplicity 3,} \\ \lambda_2 &= 2.5000 \quad \text{with multiplicity 1,} \\ \lambda_3 &= 2.0000 \quad \text{with multiplicity 2,} \\ \lambda_4 &= 1.6208 \quad \text{with multiplicity 3.} \end{aligned}$$

These multiplicities sum to $h_* = b - 1 = 9$, because the design is connected. With them and the corresponding eigenvectors, the following intermediate results applicable for any elementary contrast can be obtained:

$$\sum_{j=1}^{9} \frac{\lambda_j (\boldsymbol{c}'\boldsymbol{r}^{-\delta}\boldsymbol{N}\boldsymbol{u}_j)^2}{(\hat{\gamma}\lambda_j + 1)^3} = 0.033402$$

and

$$\sum_{j=1}^{9}\left(\frac{\lambda_j}{\hat{\gamma}\lambda_j + 1}\right)^2 - \frac{1}{20}\left(\sum_{j=1}^{9}\frac{\lambda_j}{\hat{\gamma}\lambda_j + 1}\right)^2 = 4.466749.$$

Using these results in (3.7.58), one obtains for any elementary contrast $\boldsymbol{c}'\boldsymbol{\tau}$ (for which $\boldsymbol{c}'\boldsymbol{c} = 2$) the approximate variance of its empirical estimator, $\widetilde{\boldsymbol{c}'\boldsymbol{\tau}} = \boldsymbol{c}'\tilde{\boldsymbol{\tau}}$, in the form

$$\text{Var}(\widetilde{\boldsymbol{c}'\boldsymbol{\tau}}) \cong \text{Var}(\widehat{\boldsymbol{c}'\boldsymbol{\tau}}) + 2\frac{0.033402}{4.466749}\sigma_1^2 = (0.381590 + 0.014956)\sigma_1^2 = 0.396546\sigma_1^2.$$

Note that this variance, though slightly larger than $\text{Var}(\widehat{\boldsymbol{c}'\boldsymbol{\tau}})$, is still smaller than that obtained from the intra-block analysis, for any elementary contrast. The increase in efficiency resulting from the recovery of inter-block information is now of 8.1%. Finally, inserting for σ_1^2 its estimate $\hat{\sigma}_1^2 = 0.852444$, the estimate of the above variance is $0.396546\hat{\sigma}_1^2 = 0.33803$. For comparison, note that the corresponding variances from the intra-block and the inter-block analyses are 0.37407 and 3.59836, respectively.

Results of estimating the elementary contrasts in the intra-block, the inter-block and the combined analysis, together with the estimates of the relevant standard errors, are presented in Table 3.1.

Table 3.1
Intra-block, inter-block and combined estimates of the elementary contrasts, obtained in Example 3.7.3

Contrast	Intra-block estimate	Inter-block estimate	Combined estimate
$\tau_1 - \tau_2$	0.12	1.61	0.29
$\tau_1 - \tau_3$	3.43	3.46	3.42
$\tau_1 - \tau_4$	4.83	6.57	5.00
$\tau_2 - \tau_3$	3.31	1.85	3.13
$\tau_2 - \tau_4$	4.71	4.96	4.71
$\tau_3 - \tau_4$	1.40	3.11	1.59
Estimated standard error	0.612	1.897	0.581

Now, to test the hypothesis $H_0 : \tau_1 = \tau_2 = \tau_3 = \tau_4$, equivalent to $H_0 : \boldsymbol{U}'\boldsymbol{\tau} = \boldsymbol{0}$, where $\boldsymbol{U}$ is an 4×3 matrix of rank 3 such that $\boldsymbol{U}'\boldsymbol{1}_4 = \boldsymbol{0}$, so that $\boldsymbol{U}'\boldsymbol{\tau}$ represents three linearly independent contrasts of treatment parameters, refer to the theory presented above. It has been suggested there to use the statistic F defined in (3.7.75), where $\boldsymbol{y}'\boldsymbol{R}_{\text{red}}\boldsymbol{y}$ is equivalent to the intra-block residual sum of squares $\boldsymbol{y}'\boldsymbol{\psi}_1\boldsymbol{y}$. Also, the degrees of freedom, here $n-b-h = 11$ and $\ell = 3$, are the same as in the intra-block analysis. However, the treatment sum of squares in the numerator of this F statistic differs from that in the intra-block analysis. As here $\boldsymbol{q} = \boldsymbol{0}$, it is of the form

$$(\boldsymbol{U}'\tilde{\boldsymbol{\tau}})'(\boldsymbol{U}'\hat{\boldsymbol{C}}_{\text{ap}}^{-1}\boldsymbol{U})^{-1}(\boldsymbol{U}'\tilde{\boldsymbol{\tau}}) = \text{TSS} \quad (\text{say}),$$

where $\hat{\boldsymbol{C}}_{\text{ap}}^{-1}$ is obtainable from (3.7.66), with $\text{E}[(\hat{\gamma}-\gamma)^2]$ replaced by the estimate of the asymptotic variance of $\hat{\gamma}$ given in (3.7.57). For the present example, the computational results are

$$\widehat{\text{Var}_{\text{as}}(\hat{\gamma})} = \frac{2}{4.466749}$$

and

$$\hat{\boldsymbol{C}}_{\text{ap}}^{-1} = \begin{bmatrix} 0.251491 & 0.053218 & 0.053218 & 0.053218 \\ 0.053218 & 0.251491 & 0.053218 & 0.053218 \\ 0.053218 & 0.053218 & 0.251491 & 0.053218 \\ 0.053218 & 0.053218 & 0.053218 & 0.251491 \end{bmatrix},$$

from which TSS = 89.89 and, by (3.7.75), $F = 34.33$. To relate this result to the approximating F distribution, one has to evaluate the d.f. d, given in (3.7.74). For this reason, it is necessary to estimate the matrix $\boldsymbol{A}$ defined in (3.7.72), which here reduces to (3.7.76) because the design is connected. Replacing there $\boldsymbol{C}_{\text{ap}}^{-1}$ by $\hat{\boldsymbol{C}}_{\text{ap}}^{-1}$, $\text{E}[(\hat{\gamma} - \gamma)^2]$ by $\widehat{\text{Var}_{\text{as}}(\hat{\gamma})}$ and γ by $\hat{\gamma}$ elsewhere, one obtains the estimated matrix $\hat{\boldsymbol{A}}$ with elements each within the interval $(-0.005, 0.005)$, such that

$$\widehat{\text{Var}_{\text{as}}(\hat{\gamma})}\{2\text{tr}(\hat{\boldsymbol{A}}\hat{\boldsymbol{A}}) + [\text{tr}(\hat{\boldsymbol{A}})]^2\} = 0.00007404,$$

from which $\hat{d} = 2.99996$, which means that one can take $d = 3$, exactly as for ℓ. Thus, here, the statistic F can be considered as having, under H_0, approximately the central F distribution with 3 and 11 d.f. From this distribution, the critical level of significance at which H_0 can be rejected is obtained equal to $P = 0.000007$, which fully confirms the result of the intra-block F test.

A final remark worth mentioning is that if the contrasts $\boldsymbol{U}'\boldsymbol{\tau}$ in (3.7.60) obtain the BLUEs in the intra-block analysis with full efficiency, then the test based on (3.7.75) coincides with the exact F test in the intra-block analysis.

Acknowledgement. Certain parts of Sections 3.1 and 3.7 have been reprinted, with some changes and additions, from the article by Caliński and Kageyama (1996a), under kind permission from Elsevier Science – NL, Sara Burgerhartstraat 25, 1055 KV Amsterdam, The Netherlands.

3.8 Proper designs and the recovery of inter-block information

3.8.1 BLUEs under known stratum variances

The results of Section 3.7 are applicable to any block design, proper or not. However, they can be simplified considerably if the design is proper, i.e., if $k_j = k$ for all j. This simplification has already been indicated in connection with the formulae in (3.7.19), from which it follows that if for a proper block design the stratum variances σ_1^2, σ_2^2 and σ_3^2, defined in (3.5.7), are known, or at least the ratio σ_1^2/σ_2^2 and the difference $\sigma_2^2 - \sigma_3^2 = N_B^{-1} n\sigma_B^2$ are known, then the BLUE of $\boldsymbol{\tau}$ can be written as

$$\hat{\boldsymbol{\tau}} = \boldsymbol{C}_c^{-1}\boldsymbol{Q}_c, \tag{3.8.1}$$

where

$$\boldsymbol{C}_c = \boldsymbol{\Delta}[\boldsymbol{\phi}_1 + (\sigma_1^2/\sigma_2^2)(\boldsymbol{I}_n - \boldsymbol{\phi}_1)]\boldsymbol{\Delta}' \quad \text{and} \quad \boldsymbol{Q}_c = \boldsymbol{\Delta}[\boldsymbol{\phi}_1 + (\sigma_1^2/\sigma_2^2)(\boldsymbol{I}_n - \boldsymbol{\phi}_1)]\boldsymbol{y},$$

and its covariance matrix as

$$\text{Cov}(\hat{\boldsymbol{\tau}}) = \sigma_1^2\boldsymbol{C}_c^{-1} - N_B^{-1}\sigma_B^2\mathbf{1}_v\mathbf{1}_v' = \sigma_1^2\boldsymbol{C}_c^{-1} - (\sigma_2^2 - \sigma_3^2)n^{-1}\mathbf{1}_v\mathbf{1}_v'. \tag{3.8.2}$$

Thus, under this knowledge of stratum variances, for any proper block design, the BLUE of a function $\boldsymbol{c}'\boldsymbol{\tau}$ is

$$\widehat{\boldsymbol{c}'\boldsymbol{\tau}} = \boldsymbol{c}'\boldsymbol{C}_c^{-1}\boldsymbol{Q}_c, \tag{3.8.3}$$

with the variance

$$\text{Var}(\widehat{\boldsymbol{c}'\boldsymbol{\tau}}) = \sigma_1^2\boldsymbol{c}'\boldsymbol{C}_c^{-1}\boldsymbol{c} - (\sigma_2^2 - \sigma_3^2)n^{-1}\boldsymbol{c}'\mathbf{1}_v\mathbf{1}_v'\boldsymbol{c}. \tag{3.8.4}$$

Evidently, if $\boldsymbol{c}'\boldsymbol{\tau}$ is a contrast, then the variance (3.8.4) reduces to

$$\text{Var}(\widehat{\boldsymbol{c}'\boldsymbol{\tau}}) = \sigma_1^2\boldsymbol{c}'\boldsymbol{C}_c^{-1}\boldsymbol{c}.$$

These results follow from Theorem 3.7.2 on account of Remark 3.7.2 and formula (3.7.19). But they can also be obtained directly from Theorem 3.7.1, which for proper block designs can be restated as follows.

Theorem 3.8.1. *If the design is proper, then under the model* (3.1.1), *with properties* (3.1.15) *and* (3.5.9), *assuming that the variance components defined in* (3.5.7) *are known and positive* (*or at least the ratio* σ_1^2/σ_2^2 *is known and positive*), *a function* $\boldsymbol{w}'\boldsymbol{y}$ *is the BLUE of a function* $\boldsymbol{c}'\boldsymbol{\tau}$ *if and only if*

$$\boldsymbol{w} = [\boldsymbol{\phi}_1 + (\sigma_1^2/\sigma_2^2)(\boldsymbol{I}_n - \boldsymbol{\phi}_1)]\boldsymbol{\Delta}'\boldsymbol{s} = [\boldsymbol{\phi}_1 + (\sigma_1^2/\sigma_2^2)(\boldsymbol{\phi}_2 + \boldsymbol{\phi}_3)]\boldsymbol{\Delta}'\boldsymbol{s}, \tag{3.8.5}$$

where the vectors $\boldsymbol{c}$ *and* $\boldsymbol{s}$ *are in the relation*

$$\boldsymbol{c} = \boldsymbol{\Delta}[\boldsymbol{\phi}_1 + (\sigma_1^2/\sigma_2^2)(\boldsymbol{I}_n - \boldsymbol{\phi}_1)]\boldsymbol{\Delta}'\boldsymbol{s} = \boldsymbol{\Delta}[\boldsymbol{\phi}_1 + (\sigma_1^2/\sigma_2^2)(\boldsymbol{\phi}_2 + \boldsymbol{\phi}_3)]\boldsymbol{\Delta}'\boldsymbol{s}. \tag{3.8.6}$$

Proof. This result follows immediately from Theorem 3.7.1, when taking into account the equality $k_1 = k_2 = \cdots = k_b = k$ and the relation $1 + k\gamma = \sigma_2^2/\sigma_1^2$, holding for any proper block design. □

Note that the equalities (3.8.5) and (3.8.6) give jointly the result (3.8.3). That the matrix $\boldsymbol{\Delta}[\boldsymbol{\phi}_1 + (\sigma_1^2/\sigma_2^2)(\boldsymbol{I}_n - \boldsymbol{\phi}_1)]\boldsymbol{\Delta}'$ is p.d., follows from the fact that $\boldsymbol{\phi}_1 + (\sigma_1^2/\sigma_2^2)(\boldsymbol{I}_n - \boldsymbol{\phi}_1)$ is p.d. (see Lemma 3.7.1, noting that the relation $\gamma > -1/k_{\max}$ is for a proper block design equivalent to $\sigma_2^2/\sigma_1^2 > 0$) and that the matrix $\boldsymbol{\Delta}'$ is of rank v. As (3.8.3) is true for any conformable vector $\mathbf{c}$, formula (3.8.1) follows immediately. It can also be seen as the solution of the normal equations $\boldsymbol{C}_c\hat{\boldsymbol{\tau}} = \boldsymbol{Q}_c$, i.e., the equations

$$\boldsymbol{\Delta}[\boldsymbol{\phi}_1 + (\sigma_1^2/\sigma_2^2)(\boldsymbol{I}_n - \boldsymbol{\phi}_1)]\boldsymbol{\Delta}'\boldsymbol{\tau} = \boldsymbol{\Delta}[\boldsymbol{\phi}_1 + (\sigma_1^2/\sigma_2^2)(\boldsymbol{I}_n - \boldsymbol{\phi}_1)]\boldsymbol{y}$$

(similar to those considered by Roy and Shah, 1962, and Shah, 1964a).

Now, referring to Remark 3.7.2, it is interesting to compare the results in (3.8.3) and (3.8.4) with the corresponding results in Section 3.2.1. It becomes then evident that for the recovery of inter-block information in a proper block design the matrix $(\sigma_1^2/\sigma_2^2)(\boldsymbol{I}_n - \boldsymbol{\phi}_1) = (\sigma_1^2/\sigma_2^2)k^{-1}\boldsymbol{D}'\boldsymbol{D}$ is responsible. More precisely, the amount of the recovery depends very much on the ratio σ_1^2/σ_2^2, the gain from the recovery increasing with the increase of the value of this ratio (or the decrease of the value of σ_2^2/σ_1^2). However, to get a better insight into the process of the recovery of inter-block information in a proper block design, it may be interesting to look at the problem through the estimators of basic contrasts. For this reason, it will be useful to consider first. the following corollary.

Corollary 3.8.1. *If the design is proper, then for any* $\mathbf{c} = \boldsymbol{r}^{\delta}\boldsymbol{s}$ *such that* $\boldsymbol{s}$ *satisfies the equivalent eigenvector conditions* (3.5.18) *and* (3.5.19), *with* $0 < \varepsilon < 1$, *the BLUE of the contrast* $\mathbf{c}'\boldsymbol{\tau}$ *is, under the assumptions of Theorem* 3.8.1, *of the form*

$$\begin{aligned} \widehat{\mathbf{c}'\boldsymbol{\tau}} = \widehat{\boldsymbol{s}'\boldsymbol{r}^{\delta}\boldsymbol{\tau}} &= w_1\varepsilon^{-1}\boldsymbol{s}'\boldsymbol{Q}_1 + w_2(1-\varepsilon)^{-1}\boldsymbol{s}'\boldsymbol{Q}_2 \\ &= w_1(\widehat{\mathbf{c}'\boldsymbol{\tau}})_1 + w_2(\widehat{\mathbf{c}'\boldsymbol{\tau}})_2, \end{aligned} \tag{3.8.7}$$

where $(\widehat{\mathbf{c}'\boldsymbol{\tau}})_\alpha$, $\alpha = 1, 2$, *defined in* (3.5.23), *are uncorrelated, and their "weights" are*

$$w_1 = \frac{\varepsilon\sigma_2^2}{\varepsilon\sigma_2^2 + (1-\varepsilon)\sigma_1^2} \quad \text{and} \quad w_2 = \frac{(1-\varepsilon)\sigma_1^2}{\varepsilon\sigma_2^2 + (1-\varepsilon)\sigma_1^2}. \tag{3.8.8}$$

Proof. It follows from Theorem 3.8.1 that, with $\boldsymbol{w}$ as in (3.8.5),

$$\boldsymbol{w}'\boldsymbol{y} = \boldsymbol{s}'\boldsymbol{Q}_1 + (\sigma_1^2/\sigma_2^2)\boldsymbol{s}'\boldsymbol{Q}_2 + (\sigma_1^2/\sigma_2^2)\boldsymbol{s}'\boldsymbol{Q}_3 \tag{3.8.9}$$

is the BLUE of $[\boldsymbol{s}'\boldsymbol{\Delta}\boldsymbol{\phi}_1\boldsymbol{\Delta}' + (\sigma_1^2/\sigma_2^2)\boldsymbol{s}'\boldsymbol{\Delta}\boldsymbol{\phi}_2\boldsymbol{\Delta}' + (\sigma_1^2/\sigma_2^2)\boldsymbol{s}'\boldsymbol{\Delta}\boldsymbol{\phi}_3\boldsymbol{\Delta}']\boldsymbol{\tau}$. Hence, for any $\boldsymbol{s}$ satisfying (3.5.18) or (3.5.19) with $0 < \varepsilon < 1$, which also implies that $\boldsymbol{s}'\boldsymbol{r} = 0$ (see the proof of Lemma 3.5.2), and hence, $\boldsymbol{s}'\boldsymbol{\Delta}\boldsymbol{\phi}_3 = \mathbf{0}'$,

$$\boldsymbol{s}'\boldsymbol{Q}_1 + (\sigma_1^2/\sigma_2^2)\boldsymbol{s}'\boldsymbol{Q}_2 \quad \text{is the BLUE of} \quad [\varepsilon + (1-\varepsilon)\sigma_1^2/\sigma_2^2]\boldsymbol{s}'\boldsymbol{r}^{\delta}\boldsymbol{\tau}$$

or, equivalently,

$$[\varepsilon\sigma_2^2 + (1-\varepsilon)\sigma_1^2]^{-1}(\sigma_2^2 \boldsymbol{s}'\boldsymbol{Q}_1 + \sigma_1^2 \boldsymbol{s}'\boldsymbol{Q}_2) \quad \text{is the BLUE of} \quad \boldsymbol{c}'\boldsymbol{\tau} = \boldsymbol{s}'\boldsymbol{r}^{\delta}\boldsymbol{\tau}.$$

Now, applying the notation of Theorem 3.5.2, with $\boldsymbol{Q}_\alpha$ defined in (3.5.17), the result (3.8.7) follows. The components in (3.8.7) are uncorrelated on account of (3.5.8). □

It may be seen that the result of Corollary 3.8.1 coincides with that given by Houtman and Speed (1983, p.1078) for BIB designs, except that their model is different, with (in our notation) $\sigma_1^2 = \sigma_e^2$ and $\sigma_2^2 = k\sigma_B^2 + \sigma_e^2$, i.e., with σ_U^2 ignored.

In general, a function $\boldsymbol{c}'\boldsymbol{\tau}$ for which the BLUE is of the combined form (3.8.7) can be said, after Martin and Zyskind (1966), to be best combinable by simple weighting of intra-block and inter-block BLUEs, or shortly, to be "simply combinable." Accordingly, the result (3.8.7), obtainable for a contrast $\boldsymbol{c}'\boldsymbol{\tau}$ when the stratum variances are known, can be called the property of "simple combinability" (see also Houtman and Speed, 1983, p. 1073).

Remark 3.8.1. It follows from Lemma 3.3.1 that $\boldsymbol{s}'\boldsymbol{Q}_1$ and $\boldsymbol{s}'\boldsymbol{Q}_2$ appearing in (3.8.9) are the intra-block and inter-block BLUEs of the same contrast (with the accuracy to a constant factor) if and only if the vector $\boldsymbol{s}$ satisfies (3.3.4) with $0 < \varepsilon < 1$, implying $\boldsymbol{s}'\boldsymbol{Q}_3 = 0$ then. Thus, the equivalent conditions (3.5.18) and (3.5.19) are not only sufficient (as stated in Corollary 3.8.1), but also necessary for the contrast $\boldsymbol{c}'\boldsymbol{\tau} = \boldsymbol{s}'\boldsymbol{r}^{\delta}\boldsymbol{\tau}$ to be simply combinable (a result originally established by Martin and Zyskind, 1966, Theorem 2.6). (See also Remark 3.5.4.)

Remark 3.8.2. It follows from (3.5.24) and Corollary 3.8.1 that, in the case of a proper block design and under the assumptions of Theorem 3.8.1, for any contrast $\boldsymbol{c}'\boldsymbol{\tau} = \boldsymbol{s}'\boldsymbol{r}^{\delta}\boldsymbol{\tau}$ such that $\boldsymbol{s}$ satisfies the equivalent conditions (3.5.18) and (3.5.19), the variance of the BLUE (3.8.7) is

$$\operatorname{Var}(\widehat{\boldsymbol{c}'\boldsymbol{\tau}}) = \frac{\sigma_1^2\sigma_2^2}{\varepsilon\sigma_2^2 + (1-\varepsilon)\sigma_1^2}\boldsymbol{c}'\boldsymbol{r}^{-\delta}\boldsymbol{c}, \tag{3.8.10}$$

for which the inequalities

$$\operatorname{Var}(\widehat{\boldsymbol{c}'\boldsymbol{\tau}}) \leq \operatorname{Var}[(\widehat{\boldsymbol{c}'\boldsymbol{\tau}})_1], \tag{3.8.11}$$

$$\operatorname{Var}(\widehat{\boldsymbol{c}'\boldsymbol{\tau}}) \leq \operatorname{Var}[(\widehat{\boldsymbol{c}'\boldsymbol{\tau}})_2] \tag{3.8.12}$$

hold, the equality in (3.8.11) holding if and only if $\varepsilon = 1$, and that in (3.8.12) if and only if $\varepsilon = 0$.

The results obtained above show that a combination of the intra-block and inter-block BLUEs of a contrast for which the conditions of Corollary 3.8.1 are satisfied (with $0 < \varepsilon < 1$) may lead to an improved estimator. This improvement will certainly be achieved when the stratum variances σ_1^2 and σ_2^2 are known (or their ratio is known). The gain obtainable from using the BLUE (3.8.7) instead of the intra-block BLUE (3.3.8) can be measured by considering the ratio of precisions

$$\frac{\mathrm{Var}[(\widehat{c'\tau})_1]}{\mathrm{Var}(\widehat{c'\tau})} = 1 + \frac{(1-\varepsilon)\sigma_1^2}{\varepsilon\sigma_2^2} \qquad \text{(for a proper design).}$$

Thus, the gain, $\varepsilon^{-1}(1-\varepsilon)(\sigma_1^2/\sigma_2^2) = w_2/w_1$, depends on the efficiency factor, ε, and the variance ratio

$$\frac{\sigma_1^2}{\sigma_2^2} = \frac{\sigma_U^2 + \sigma_e^2}{k\sigma_B^2 + (1 - K_H^{-1}k)\sigma_U^2 + \sigma_e^2} \qquad \left(= \frac{\sigma_U^2 + \sigma_e^2}{k\sigma_B^2 + \sigma_e^2} \text{ if } k = K_H\right).$$

This observation means that the recovery of inter-block information (as it is called after Yates, 1939, 1940b) is worth undertaking, particularly when the efficiency factor in the intra-block analysis is small and the inter-block variation does not exceed much the variation of units within blocks. Thus, one can expect to obtain the larger gain, the less one is successful (i) in choosing a design with high intra-block efficiency for the contrast of interest and (ii) in blocking the available experimental units aimed at removing their intra-block variation.

Now, to proceed further, it will be useful to find, still under the assumptions of Theorem 3.8.1, the BLUE of $\mathrm{E}(\boldsymbol{y}) = \boldsymbol{\Delta}'\boldsymbol{\tau}$ and, hence, the residuals $\boldsymbol{y} - \widehat{\boldsymbol{\Delta}'\boldsymbol{\tau}}$. As shown by Rao (1974, Theorem 3.2), the BLUE of $\boldsymbol{\Delta}'\boldsymbol{\tau}$ can be expressed as

$$\widehat{\boldsymbol{\Delta}'\boldsymbol{\tau}} = \boldsymbol{P}_{\Delta'(V^{-1})}\boldsymbol{y}, \tag{3.8.13}$$

where $\boldsymbol{V} = \mathrm{Cov}(\boldsymbol{y})$, and

$$\boldsymbol{P}_{\Delta'(V^{-1})} = \boldsymbol{\Delta}'(\boldsymbol{\Delta}\boldsymbol{V}^{-1}\boldsymbol{\Delta}')^{-1}\boldsymbol{\Delta}\boldsymbol{V}^{-1}$$

(see Appendix A.2). On account of (3.5.9), (3.6.16) and the properties of the matrices $\boldsymbol{H}_\beta = \boldsymbol{S}_\beta\boldsymbol{S}_\beta'$, $\beta = 0, 1, ..., m, m+1$, appearing in Definition 3.6.1, it is justified to write

$$(\boldsymbol{\Delta}\boldsymbol{V}^{-1}\boldsymbol{\Delta}')^{-1} = \sigma_1^2\boldsymbol{H}_0 + \sum_{\beta=1}^{m-1} \frac{\sigma_1^2\sigma_2^2}{\varepsilon_{1\beta}\sigma_2^2 + \varepsilon_{2\beta}\sigma_1^2}\boldsymbol{H}_\beta + \sigma_2^2\boldsymbol{H}_m + \sigma_3^2\boldsymbol{H}_{m+1}. \tag{3.8.14}$$

Hence, for any proper block design, the projector $\boldsymbol{P}_{\Delta'(V^{-1})}$ can be written in a decomposed form as

$$\boldsymbol{P}_{\Delta'(V^{-1})} = \sum_{\beta=0}^{m-1} w_{1\beta}\varepsilon_{1\beta}^{-1}\boldsymbol{P}_{\Delta' S_\beta}\boldsymbol{\phi}_1 + \sum_{\beta=1}^{m} w_{2\beta}\varepsilon_{2\beta}^{-1}\boldsymbol{P}_{\Delta' S_\beta}\boldsymbol{\phi}_2 + \boldsymbol{\phi}_3, \tag{3.8.15}$$

where the weights are defined in accordance with (3.8.8), i.e., as

$$w_{1\beta} = \frac{\varepsilon_{1\beta}\sigma_2^2}{\varepsilon_{1\beta}\sigma_2^2 + \varepsilon_{2\beta}\sigma_1^2} \quad \text{and} \quad w_{2\beta} = \frac{\varepsilon_{2\beta}\sigma_1^2}{\varepsilon_{1\beta}\sigma_2^2 + \varepsilon_{2\beta}\sigma_1^2}, \tag{3.8.16}$$

with $\varepsilon_{1\beta} = \varepsilon_\beta$ and $\varepsilon_{2\beta} = 1 - \varepsilon_\beta$, ε_β being the eigenvalue of $\boldsymbol{C}_1$ with respect to $\boldsymbol{r}^\delta$ corresponding to all $\boldsymbol{s}_{\beta j}$, $j = 1, 2, ..., \rho_\beta$, and where $\boldsymbol{P}_{\Delta' S_\beta} = \boldsymbol{\Delta}'\boldsymbol{S}_\beta\boldsymbol{S}_\beta'\boldsymbol{\Delta}$ is the orthogonal projector on the subspace $\mathcal{C}(\boldsymbol{\Delta}'\boldsymbol{S}_\beta)$, for $\beta = 0, 1, ..., m$ (where the matrices $\boldsymbol{S}_0, \boldsymbol{S}_1, ..., \boldsymbol{S}_m$ represent basic contrasts of the design, as defined in Corollary 3.6.3). Also note that formula (3.8.15) coincides with formula (4.1) of Houtman and Speed (1983) given for the general case of GB designs (see Definition 3.6.1).

From (3.8.13) and (3.8.15), the BLUE of $\boldsymbol{\Delta}'\boldsymbol{\tau}$ can be written for any proper block design in the form

$$\widehat{\boldsymbol{\Delta}'\boldsymbol{\tau}} = (\sum_{\beta=0}^{m-1} w_{1\beta}\varepsilon_{1\beta}^{-1}\boldsymbol{P}_{\Delta' S_\beta}\boldsymbol{\phi}_1 + \sum_{\beta=1}^{m} w_{2\beta}\varepsilon_{2\beta}^{-1}\boldsymbol{P}_{\Delta' S_\beta}\boldsymbol{\phi}_2 + \boldsymbol{\phi}_3)\boldsymbol{y}. \tag{3.8.17}$$

Hence, as $\boldsymbol{\tau} = \boldsymbol{r}^{-\delta}\boldsymbol{\Delta}\boldsymbol{\Delta}'\boldsymbol{\tau}$, the BLUE of $\boldsymbol{\tau}$ can be written, in terms of the estimators of basic contrasts, as

$$\begin{aligned}
\hat{\boldsymbol{\tau}} &= \boldsymbol{r}^{-\delta}\boldsymbol{\Delta}(\widehat{\boldsymbol{\Delta}'\boldsymbol{\tau}}) \\
&= \sum_{\beta=0}^{m-1}\sum_{j=1}^{\rho_\beta} w_{1\beta}\varepsilon_{1\beta}^{-1}\boldsymbol{s}_{\beta j}\boldsymbol{s}_{\beta j}'\boldsymbol{Q}_1 + \sum_{\beta=1}^{m}\sum_{j=1}^{\rho_\beta} w_{2\beta}\varepsilon_{2\beta}^{-1}\boldsymbol{s}_{\beta j}\boldsymbol{s}_{\beta j}'\boldsymbol{Q}_2 + \boldsymbol{s}_v\boldsymbol{s}_v'\boldsymbol{Q}_3 \\
&= \sum_{\beta=0}^{m}\sum_{j=1}^{\rho_\beta} \boldsymbol{s}_{\beta j}(\widehat{\boldsymbol{s}_{\beta j}'\boldsymbol{r}^\delta\boldsymbol{\tau}}) + \boldsymbol{s}_v(\widehat{\boldsymbol{s}_v'\boldsymbol{r}^\delta\boldsymbol{\tau}}) \\
&= \boldsymbol{r}^{-\delta}[\sum_{\beta=0}^{m}\sum_{j=1}^{\rho_\beta} \boldsymbol{c}_{\beta j}(\widehat{\boldsymbol{c}_{\beta j}'\boldsymbol{\tau}}) + \boldsymbol{c}_v(\widehat{\boldsymbol{c}_v'\boldsymbol{\tau}})],
\end{aligned} \tag{3.8.18}$$

where

$$\widehat{\boldsymbol{c}_{\beta j}'\boldsymbol{\tau}} = w_{1\beta}(\widehat{\boldsymbol{c}_{\beta j}'\boldsymbol{\tau}})_1 + w_{2\beta}(\widehat{\boldsymbol{c}_{\beta j}'\boldsymbol{\tau}})_2, \tag{3.8.19}$$

with $(\widehat{\boldsymbol{c}_{\beta j}'\boldsymbol{\tau}})_\alpha$, $\alpha = 1, 2$, given as in (3.6.1), and where $\widehat{\boldsymbol{c}_v'\boldsymbol{\tau}} = \widehat{\boldsymbol{s}_v'\boldsymbol{r}^\delta\boldsymbol{\tau}} = \boldsymbol{s}_v'\boldsymbol{Q}_3$, as in Corollary 3.2.5(c). [It should always be remembered that $\rho_0 \equiv \rho$, which is the multiplicity of $\varepsilon_{10} = \varepsilon_0 = 1$, as introduced in Section 3.4.]

That $\hat{\boldsymbol{\tau}}$, given in the form (3.8.18), is the BLUE of $\boldsymbol{\tau}$ follows, in view of (3.8.17), from a general theory (see, e.g., the proof of Theorem 3.3 of Rao, 1974), but it can also be checked by Theorem 3.8.1. For this reason, note that from (3.4.10), (3.6.11) and (3.6.14), one can write

$$\begin{aligned}
\boldsymbol{\Delta}\left[\boldsymbol{\phi}_1 + \frac{\sigma_1^2}{\sigma_2^2}(\boldsymbol{I}_n - \boldsymbol{\phi}_1)\right]\boldsymbol{\Delta}' &= \boldsymbol{r}^\delta\left(\boldsymbol{H}_0 + \sum_{\beta=1}^{m-1}\frac{\varepsilon_{1\beta}}{w_{1\beta}}\boldsymbol{H}_\beta + \frac{\sigma_1^2}{\sigma_2^2}\boldsymbol{H}_m + \frac{\sigma_1^2}{\sigma_2^2}\boldsymbol{H}_{m+1}\right)\boldsymbol{r}^\delta \\
&= \boldsymbol{r}^\delta\left[\sum_{\beta=0}^{m-1}\sum_{j=1}^{\rho_\beta}\varepsilon_{1\beta}\boldsymbol{s}_{\beta j}\boldsymbol{s}_{\beta j}' + \frac{\sigma_1^2}{\sigma_2^2}(\sum_{\beta=1}^{m}\sum_{j=1}^{\rho_\beta}\varepsilon_{2\beta}\boldsymbol{s}_{\beta j}\boldsymbol{s}_{\beta j}' + \boldsymbol{s}_v\boldsymbol{s}_v')\right]\boldsymbol{r}^\delta.
\end{aligned}$$

This formula allows (3.8.18) to be obtained from (3.8.1), which, as already shown, follows from Theorem 3.8.1. In fact, the representation (3.8.1) shows that the condition (3.8.5) is satisfied by the vector $\hat{\boldsymbol{\tau}}$ and any function $\boldsymbol{c}'\hat{\boldsymbol{\tau}}$, i.e., that $\hat{\boldsymbol{\tau}}$ and any $\boldsymbol{c}'\hat{\boldsymbol{\tau}}$ are really the BLUEs of their expectations, $\boldsymbol{\tau}$ and $\boldsymbol{c}'\boldsymbol{\tau}$, respectively.

Thus, the following results can be established (see also Caliński, 1996a, Theorem 3.1).

Corollary 3.8.2. *If the design is proper, then under the assumptions of Theorem 3.8.1, the following applies:*

(a) *The BLUE of* $\boldsymbol{\tau}$ *is* $\hat{\boldsymbol{\tau}}$ *given in* (3.8.18) *or, equivalently, in* (3.8.1), *with*

$$\mathrm{Cov}(\hat{\boldsymbol{\tau}}) = \boldsymbol{r}^{-\delta}\boldsymbol{\Delta}\boldsymbol{P}_{\Delta'(V^{-1})}\boldsymbol{V}(\boldsymbol{P}_{\Delta'(V^{-1})})'\boldsymbol{\Delta}'\boldsymbol{r}^{-\delta} = (\boldsymbol{\Delta}\boldsymbol{V}^{-1}\boldsymbol{\Delta}')^{-1}, \quad (3.8.20)$$

which can be expressed as in (3.8.14), *i.e., as*

$$\begin{aligned} \mathrm{Cov}(\hat{\boldsymbol{\tau}}) &= \sigma_1^2\left[\boldsymbol{H}_0 + \sum_{\beta=1}^{m-1}\frac{w_{1\beta}}{\varepsilon_{1\beta}}\boldsymbol{H}_\beta + \frac{\sigma_2^2}{\sigma_1^2}\boldsymbol{H}_m + \frac{\sigma_3^2}{\sigma_1^2}\boldsymbol{H}_{m+1}\right] \\ &= \sigma_1^2\boldsymbol{r}^{-\delta}\left[\sum_{j=1}^{\rho_0}\boldsymbol{c}_{0j}\boldsymbol{c}_{0j}' + \sum_{\beta=1}^{m-1}\frac{w_{1\beta}}{\varepsilon_{1\beta}}\sum_{j=1}^{\rho_\beta}\boldsymbol{c}_{\beta j}\boldsymbol{c}_{\beta j}' \right. \\ &\quad \left. + \frac{\sigma_2^2}{\sigma_1^2}\sum_{j=1}^{\rho_m}\boldsymbol{c}_{mj}\boldsymbol{c}_{mj}' + \frac{\sigma_3^2}{\sigma_1^2}\boldsymbol{c}_v\boldsymbol{c}_v'\right]\boldsymbol{r}^{-\delta} \end{aligned} \quad (3.8.21)$$

or, equivalently, as in (3.8.2).

(b) *For any* $\boldsymbol{c} = \boldsymbol{r}^\delta\boldsymbol{s}$, *the BLUE of* $\boldsymbol{c}'\boldsymbol{\tau}$ *is*

$$\begin{aligned} \widehat{\boldsymbol{c}'\boldsymbol{\tau}} = \boldsymbol{c}'\hat{\boldsymbol{\tau}} &= \sum_{\beta=0}^{m}\sum_{j=1}^{\rho_\beta}\boldsymbol{s}'\boldsymbol{r}^\delta\boldsymbol{s}_{\beta j}(\widehat{\boldsymbol{s}_{\beta j}'\boldsymbol{r}^\delta\boldsymbol{\tau}}) + \boldsymbol{s}'\boldsymbol{r}^\delta\boldsymbol{s}_v(\widehat{\boldsymbol{s}_v'\boldsymbol{r}^\delta\boldsymbol{\tau}}) \\ &= \sum_{\beta=0}^{m}\sum_{j=1}^{\rho_\beta}\boldsymbol{c}'\boldsymbol{r}^{-\delta}\boldsymbol{c}_{\beta j}(\widehat{\boldsymbol{c}_{\beta j}'\boldsymbol{\tau}}) + \boldsymbol{c}'\boldsymbol{r}^{-\delta}\boldsymbol{c}_v(\widehat{\boldsymbol{c}_v'\boldsymbol{\tau}}), \end{aligned} \quad (3.8.22)$$

where $\widehat{\boldsymbol{c}_{\beta j}'\boldsymbol{\tau}}$ *is as in* (3.8.19) *and* $\widehat{\boldsymbol{c}_v'\boldsymbol{\tau}} = n^{-1/2}\boldsymbol{1}_n'\boldsymbol{y}$, *and the variance of the BLUE is*

$$\begin{aligned} \mathrm{Var}(\widehat{\boldsymbol{c}'\boldsymbol{\tau}}) &= \boldsymbol{s}'\boldsymbol{r}^\delta\mathrm{Cov}(\hat{\boldsymbol{\tau}})\boldsymbol{r}^\delta\boldsymbol{s} = \boldsymbol{c}'\mathrm{Cov}(\hat{\boldsymbol{\tau}})\boldsymbol{c} \\ &= \sigma_1^2\left[\sum_{j=1}^{\rho_0}(\boldsymbol{c}'\boldsymbol{r}^{-\delta}\boldsymbol{c}_{0j})^2 + \sum_{\beta=1}^{m-1}\frac{w_{1\beta}}{\varepsilon_{1\beta}}\sum_{j=1}^{\rho_\beta}(\boldsymbol{c}'\boldsymbol{r}^{-\delta}\boldsymbol{c}_{\beta j})^2 \right. \\ &\quad \left. + \frac{\sigma_2^2}{\sigma_1^2}\sum_{j=1}^{\rho_m}(\boldsymbol{c}'\boldsymbol{r}^{-\delta}\boldsymbol{c}_{mj})^2 + \frac{\sigma_3^2}{\sigma_1^2}(\boldsymbol{c}'\boldsymbol{r}^{-\delta}\boldsymbol{c}_v)^2\right], \end{aligned} \quad (3.8.23)$$

the term $(\sigma_3^2/\sigma_1^2)(\boldsymbol{c}'\boldsymbol{r}^{-\delta}\boldsymbol{c}_v)^2$ *disappearing when* $\boldsymbol{c}'\boldsymbol{\tau}$ *is a contrast. Equivalently,* (3.8.22) *can also be expressed as in* (3.8.3), *and* (3.8.23) *as in* (3.8.4).

(c) *For any basic contrast* $c_i'\tau = s_i' r^{\delta}\tau$, *the BLUE is*

$$\widehat{c_i'\tau} = w_{1i}(\widehat{c_i'\tau})_1 + w_{2i}(\widehat{c_i'\tau})_2 \quad \text{[see (3.8.19)]},$$

with

$$\operatorname{Var}(\widehat{c_i'\tau}) = \begin{cases} \sigma_1^2 & \textit{for } i = 1, 2, ..., \rho_0 \ (\equiv \rho), \\ \sigma_1^2 w_{1i}/\varepsilon_{1i} & \textit{for } i = \rho_0 + 1, \rho_0 + 2, ..., h, \\ \sigma_2^2 & \textit{for } i = h+1, h+2, ..., v-1, \end{cases} \tag{3.8.24}$$

where $w_{1i}/\varepsilon_{1i} = \sigma_2^2/(\varepsilon_{1i}\sigma_2^2 + \varepsilon_{2i}\sigma_1^2)$, *and with*

$$\operatorname{Cov}(\widehat{c_i'\tau}, \widehat{c_{i'}'\tau}) = 0 \quad \textit{for} \quad i \neq i'. \tag{3.8.25}$$

(d) *For* $c_v'\tau = s_v' r^{\delta}\tau = n^{-1/2} r'\tau$, *the variance of its BLUE, i.e., of* $\widehat{c_v'\tau} = n^{-1/2}\mathbf{1}_n' y$, *is*

$$\operatorname{Var}(\widehat{c_v'\tau}) = \sigma_3^2 (s_v' r^{\delta} s_v)^2 = \sigma_3^2.$$

Proof. The best linear unbiasedness property of the estimators follows from Theorem 3.8.1, as already shown at the beginning of this section. Formula (3.8.20) follows from (3.8.13) on account of the relation $\tau = r^{-\delta}\Delta\Delta'\tau$, and formula (3.8.21) from (3.8.14) and (3.8.16). Formula (3.8.22) is obtainable directly from (3.8.18). Other results are readily obtainable from (3.8.21), (3.8.22) and Theorem 3.6.1, taking into account the properties of the vectors $c_i = r^{\delta} s_i$, $i = 1, 2, ..., v-1$, representing basic contrasts, and of the vector $c_v = r^{\delta} s_v$, where $s_v = n^{-1/2}\mathbf{1}_v$. □

Remark 3.8.3. In the case of $N_B^{-1} b \to 0$, with K_H remaining finite, σ_3^2 tends to σ_2^2 and, hence, (3.8.23) becomes

$$\operatorname{Var}(\widehat{c'\tau}) = \sigma_1^2 \left[\sum_{\beta=0}^{m-1} \frac{w_{1\beta}}{\varepsilon_{1\beta}} \sum_{j=1}^{\rho_\beta} (c' r^{-\delta} c_{\beta j})^2 + \frac{\sigma_2^2}{\sigma_1^2} \sum_{\beta=m}^{m+1} \sum_{j=1}^{\rho_\beta} (c' r^{-\delta} c_{mj})^2 \right].$$

This case corresponds to the assumptions adopted by Houtman and Speed (1983, p. 1071) for an experiment in a proper block design.

Finally, note that the vector y can be decomposed as

$$y = P_{\Delta'(V^{-1})} y + (I_n - P_{\Delta'(V^{-1})}) y,$$

where the second term is the residual vector giving, in accordance with the V^{-1}-norm (see Appendix A.1), the residual sum of squares of the form

$$\begin{aligned} \|(I_n - P_{\Delta'(V^{-1})}) y\|_{V^{-1}}^2 &= y'(I_n - P_{\Delta'(V^{-1})})' V^{-1} (I_n - P_{\Delta'(V^{-1})}) y \\ &= \sigma_1^{-2} y'(I_n - P_{\Delta'(V^{-1})})' \phi_1 (I_n - P_{\Delta'(V^{-1})}) y \\ &\quad + \sigma_2^{-2} y'(I_n - P_{\Delta'(V^{-1})})' \phi_2 (I_n - P_{\Delta'(V^{-1})}) y, \end{aligned} \tag{3.8.26}$$

as it follows from (3.5.9), which also implies that

$$\phi_3(\boldsymbol{I}_n - \boldsymbol{P}_{\Delta'(V^{-1})}) = \sigma_3^2\phi_3\boldsymbol{V}^{-1}(\boldsymbol{I}_n - \boldsymbol{P}_{\Delta'(V^{-1})}) = \mathbf{O},$$

on account of (3.2.7), (2.2.1) and the form of $\boldsymbol{P}_{\Delta'(V^{-1})}$. The residual degrees of freedom corresponding to (3.8.26) are given by the rank difference $d = \text{rank}(\boldsymbol{V} : \boldsymbol{\Delta}') - \text{rank}(\boldsymbol{\Delta}') = n - v$ (see Rao, 1974, Theorem 3.4).

3.8.2 Estimation of unknown stratum variances

The results of Section 3.8.1 have been obtained under the assumptions of Theorem 3.8.1, which in particular imply that σ_1^2, σ_2^2 and, hence, all weights $w_{1\beta}$, $w_{2\beta}$, for $\beta = 0, 1, ..., m$, are known. In real applications, however, these quantities are usually not known and need to be estimated. In fact, it is sufficient to estimate the stratum variances σ_1^2 and σ_2^2 only, as the weights are simple functions of them, defined in (3.8.16). So, the problem is how to estimate σ_1^2 and σ_2^2 so that the formulae appearing in Theorem 3.8.1 and Corollaries 3.8.1 and 3.8.2, i.e., those of the type (3.8.1), (3.8.18) and all other involving these stratum variances, could be applied in the analysis of real experimental data.

To solve the problem of estimating σ_1^2 and σ_2^2, first note that, from (3.8.26),

$$\begin{aligned}
&\mathrm{E}\{\|(\boldsymbol{I}_n - \boldsymbol{P}_{\Delta'(V^{-1})})\boldsymbol{y}\|^2_{V^{-1}}\} \\
&= \sigma_1^{-2}\mathrm{E}\{\|\phi_1(\boldsymbol{I}_n - \boldsymbol{P}_{\Delta'(V^{-1})})\boldsymbol{y}\|^2\} + \sigma_2^{-2}\mathrm{E}\{\|\phi_2(\boldsymbol{I}_n - \boldsymbol{P}_{\Delta'(V^{-1})})\boldsymbol{y}\|^2\} \\
&= d_1 + \sum_{\beta=0}^{m-1}(1 - w_{1\beta})\rho_\beta + d_2 + \sum_{\beta=1}^{m}(1 - w_{2\beta})\rho_\beta \\
&= d_1 + d_2 + h - \rho = n - v,
\end{aligned}$$

because, as can be shown,

$$\mathrm{E}\{\|\phi_1(\boldsymbol{I}_n - \boldsymbol{P}_{\Delta'(V^{-1})})\boldsymbol{y}\|^2\} = \sigma_1^2 d_1',$$

where

$$\begin{aligned}
d_1' &= \mathrm{tr}[\phi_1(\boldsymbol{I}_n - \boldsymbol{P}_{\Delta'(V^{-1})})] \\
&= n - b - \sum_{\beta=0}^{m-1} w_{1\beta}\rho_\beta = d_1 + \sum_{\beta=0}^{m-1}(1 - w_{1\beta})\rho_\beta, \qquad (3.8.27)
\end{aligned}$$

with $d_1 = n - b - h$, the residual d.f. in the intra-block analysis of variance (see Section 3.2.1), and

$$\mathrm{E}\{\|\phi_2(\boldsymbol{I}_n - \boldsymbol{P}_{\Delta'(V^{-1})})\boldsymbol{y}\|^2\} = \sigma_2^2 d_2',$$

where

$$\begin{aligned}
d_2' &= \mathrm{tr}[\phi_2(\boldsymbol{I}_n - \boldsymbol{P}_{\Delta'(V^{-1})})] \\
&= b - 1 - \sum_{\beta=1}^{m} w_{2\beta}\rho_\beta = d_2 + \sum_{\beta=1}^{m}(1 - w_{2\beta})\rho_\beta, \qquad (3.8.28)
\end{aligned}$$

with $d_2 = b - v + \rho$, the residual d.f. in the inter-block analysis of variance (see Section 3.2.2), the results in (3.8.27) and (3.8.28) following from (3.8.15).

With these results, it is natural to consider as estimators of σ_1^2 and σ_2^2 the solutions of the equations

$$\|\boldsymbol{\phi}_1(\boldsymbol{I}_n - \boldsymbol{P}_{\Delta'(V^{-1})})\boldsymbol{y}\|^2 = \sigma_1^2 d_1', \quad (3.8.29)$$

$$\|\boldsymbol{\phi}_2(\boldsymbol{I}_n - \boldsymbol{P}_{\Delta'(V^{-1})})\boldsymbol{y}\|^2 = \sigma_2^2 d_2', \quad (3.8.30)$$

respectively (as suggested by Nelder, 1968). However, it should be noted that the explicit forms of the sums of squares in (3.8.29) and (3.8.30) are

$$\|\boldsymbol{\phi}_1(\boldsymbol{I}_n - \boldsymbol{P}_{\Delta'(V^{-1})})\boldsymbol{y}\|^2 = \boldsymbol{y}'(\boldsymbol{I}_n - \boldsymbol{P}_{\Delta'(V^{-1})})'\boldsymbol{\phi}_1(\boldsymbol{I}_n - \boldsymbol{P}_{\Delta'(V^{-1})})\boldsymbol{y}$$
$$= \|\boldsymbol{\psi}_1\boldsymbol{y}\|^2 + \sum_{\beta=1}^{m-1} w_{2\beta}^2\varepsilon_{1\beta}\|\varepsilon_{1\beta}^{-1}\boldsymbol{P}_{\Delta' S_\beta}\boldsymbol{y}_1 - \varepsilon_{2\beta}^{-1}\boldsymbol{P}_{\Delta' S_\beta}\boldsymbol{y}_2\|^2 \quad (3.8.31)$$

and

$$\|\boldsymbol{\phi}_2(\boldsymbol{I}_n - \boldsymbol{P}_{\Delta'(V^{-1})})\boldsymbol{y}\|^2 = \boldsymbol{y}'(\boldsymbol{I}_n - \boldsymbol{P}_{\Delta'(V^{-1})})'\boldsymbol{\phi}_2(\boldsymbol{I}_n - \boldsymbol{P}_{\Delta'(V^{-1})})\boldsymbol{y}$$
$$= \|\boldsymbol{\psi}_2\boldsymbol{y}\|^2 + \sum_{\beta=1}^{m-1} w_{1\beta}^2\varepsilon_{2\beta}\|\varepsilon_{2\beta}^{-1}\boldsymbol{P}_{\Delta' S_\beta}\boldsymbol{y}_2 - \varepsilon_{1\beta}^{-1}\boldsymbol{P}_{\Delta' S_\beta}\boldsymbol{y}_1\|^2, \quad (3.8.32)$$

respectively, as it follows from (3.8.15) and from the definitions $\boldsymbol{\psi}_\alpha = \boldsymbol{\phi}_\alpha - \boldsymbol{P}_{\phi_\alpha\Delta'}$ and $\boldsymbol{y}_\alpha = \boldsymbol{\phi}_\alpha\boldsymbol{y}$, for $\alpha = 1, 2$, used in Section 3.2 (see also Caliński, 1996a, Section 3.3).

Evidently, unlike the original intra-block and inter-block residual sums of squares, $\|\boldsymbol{\psi}_1\boldsymbol{y}\|^2$ and $\|\boldsymbol{\psi}_2\boldsymbol{y}\|^2$, introduced in Sections 3.2.1 and 3.2.2, respectively, their augmented forms (3.8.31) and (3.8.32) depend on the weights $\{w_{2\beta}\}$ or $\{w_{1\beta}\}$. The weights in turn depend on the stratum variances, σ_1^2 and σ_2^2, as shown in (3.8.16). Thus, the solutions of (3.8.29) and (3.8.30) for σ_1^2 and σ_2^2 are not directly obtainable, unless the weights are known, which is usually not the case. Therefore, it has been suggested by Nelder (1968, Section 3.4) to use an iterative procedure, with a set of initial values $\{w_{\alpha\beta}^{(0)}\}$ for the unknown values of $\{w_{\alpha\beta}\}$, in (3.8.31) and (3.8.32), from which the "working" estimates of σ_1^2 and σ_2^2 can be obtained by solving (3.8.29) and (3.8.30), respectively, giving then a revised set of values of $\{w_{\alpha\beta}\}$ by (3.8.16), and so on, until the process converges. The final values of the weights so obtained can then be used in place of their unknown true values appearing in formulae like (3.8.18) and (3.8.21), to obtain some approximations to the unavailable BLUEs.

Alternatively, one can start the iterative procedure with a pair of initial values $\sigma_1^{2(0)}$ and $\sigma_2^{2(0)}$ for the unknown stratum variances, from which the "working" estimates of the weights $\{w_{\alpha\beta}\}$ can be obtained by (3.8.16), giving then a revised pair of values of σ_1^2 and σ_2^2 from the sums of squares (3.8.31) and (3.8.32) divided by d_1' and d_2', respectively, and so on (according to Fisher's method of scoring).

The above approach, adopted from Nelder (1968), which can be considered as a generalized alternative to that introduced by Yates (1939, 1940b), is also advocated by Houtman and Speed (1983, Sections 4.5 and 5.1). Following the terminology used by these authors, the augmented quadratic forms (3.8.31) and (3.8.32) may be called the "actual" residual sums of squares in the intra-block and in the inter-block strata, respectively, to distinguish them from the "apparent" residual sums of squares $\|\boldsymbol{\psi}_1\boldsymbol{y}\|^2$ and $\|\boldsymbol{\psi}_2\boldsymbol{y}\|^2$ in those strata. The former differ from the latter by a term measuring the difference between the estimates of contrasts of treatment parameters in the two strata. This difference can be better seen when writing (3.8.31) and (3.8.32) with the use of estimated basic contrasts, i.e., as

$$\|\boldsymbol{\phi}_1(\boldsymbol{I}_n - \boldsymbol{P}_{\Delta'(V^{-1})})\boldsymbol{y}\|^2 = \boldsymbol{y}'\boldsymbol{\psi}_1\boldsymbol{y} + \sum_{i=\rho+1}^{h} w_{2i}^2\varepsilon_{1i}[(\widehat{\boldsymbol{c}_i'\boldsymbol{\tau}})_1 - (\widehat{\boldsymbol{c}_i'\boldsymbol{\tau}})_2]^2 \quad (3.8.33)$$

for the intra-block stratum and as

$$\|\boldsymbol{\phi}_2(\boldsymbol{I}_n - \boldsymbol{P}_{\Delta'(V^{-1})})\boldsymbol{y}\|^2 = \boldsymbol{y}'\boldsymbol{\psi}_2\boldsymbol{y} + \sum_{i=\rho+1}^{h} w_{1i}^2\varepsilon_{2i}[(\widehat{\boldsymbol{c}_i'\boldsymbol{\tau}})_2 - (\widehat{\boldsymbol{c}_i'\boldsymbol{\tau}})_1]^2 \quad (3.8.34)$$

for the inter-block stratum, where $(\widehat{\boldsymbol{c}_i'\boldsymbol{\tau}})_\alpha$ is the BLUE of the ith basic contrast ($i = \rho+1, \rho+2, ..., h$; with $\rho \equiv \rho_0$, as recalled in Section 3.8.1) obtainable from the analysis within stratum α (= 1 for intra-block and = 2 for inter-block), on account of Theorem 3.6.1. Thus, the adopted Nelder estimators of the stratum variances σ_1^2 and σ_2^2 can formally be written for any proper block design as

$$\hat{\sigma_1^2} \equiv \hat{\sigma}_{1(N)}^2 = (d_1')^{-1}\{\boldsymbol{y}'\boldsymbol{\psi}_1\boldsymbol{y} + \sum_{i=\rho+1}^{h} (\hat{w}_{2i})^2\varepsilon_{1i}[(\widehat{\boldsymbol{c}_i'\boldsymbol{\tau}})_1 - (\widehat{\boldsymbol{c}_i'\boldsymbol{\tau}})_2]^2\}, \quad (3.8.35)$$

$$\hat{\sigma_2^2} \equiv \hat{\sigma}_{2(N)}^2 = (d_2')^{-1}\{\boldsymbol{y}'\boldsymbol{\psi}_2\boldsymbol{y} + \sum_{i=\rho+1}^{h} (\hat{w}_{1i})^2\varepsilon_{2i}[(\widehat{\boldsymbol{c}_i'\boldsymbol{\tau}})_2 - (\widehat{\boldsymbol{c}_i'\boldsymbol{\tau}})_1]^2\}, \quad (3.8.36)$$

where $\hat{w}_{1i}$ and $\hat{w}_{2i}$ are the estimated weights,

$$\hat{w}_{1i} = \frac{\varepsilon_{1i}\hat{\sigma_2^2}}{\varepsilon_{1i}\hat{\sigma_2^2} + \varepsilon_{2i}\hat{\sigma_1^2}} \quad \text{and} \quad \hat{w}_{2i} = \frac{\varepsilon_{2i}\hat{\sigma_1^2}}{\varepsilon_{1i}\hat{\sigma_2^2} + \varepsilon_{2i}\hat{\sigma_1^2}},$$

for $i = \rho+1, \rho+2, ..., h$.

Now, to compare the Nelder estimators (3.8.35) and (3.8.36) with other estimators of the stratum variances σ_1^2 and σ_2^2 suggested in the literature, it is useful to note first that the equations (3.8.29) and (3.8.30) can equivalently be written as

$$\boldsymbol{y}'\bar{\boldsymbol{R}}\boldsymbol{\phi}_1\bar{\boldsymbol{R}}\boldsymbol{y} = \text{tr}(\bar{\boldsymbol{R}}\boldsymbol{\phi}_1), \quad (3.8.37)$$

$$\boldsymbol{y}'\bar{\boldsymbol{R}}\boldsymbol{\phi}_2\bar{\boldsymbol{R}}\boldsymbol{y} = \text{tr}(\bar{\boldsymbol{R}}\boldsymbol{\phi}_2), \quad (3.8.38)$$

where

$$\begin{aligned}\bar{\boldsymbol{R}} &= \boldsymbol{V}^{-1}(\boldsymbol{I}_n - \boldsymbol{P}_{\Delta'(V^{-1})}) = \boldsymbol{V}^{-1} - \boldsymbol{V}^{-1}\boldsymbol{\Delta}'(\boldsymbol{\Delta}\boldsymbol{V}^{-1}\boldsymbol{\Delta}')^{-1}\boldsymbol{\Delta}\boldsymbol{V}^{-1} \\ &= (\boldsymbol{I}_n - \boldsymbol{P}_{\Delta'(V^{-1})})'\boldsymbol{V}^{-1}, \qquad (3.8.39)\end{aligned}$$

with $\boldsymbol{V} = \text{Cov}(\boldsymbol{y})$ of the form (3.5.9). [The symbol $\bar{\boldsymbol{R}}$ in (3.8.39) is to be distinguished from $\boldsymbol{R}$ defined in (3.7.20).] It can easily be shown that the matrix $\bar{\boldsymbol{R}}$ defined in (3.8.39) is the Moore–Penrose inverse of the matrix $\boldsymbol{\phi}_*\boldsymbol{V}\boldsymbol{\phi}_*$, where $\boldsymbol{\phi}_*$ is as in (2.2.5). So, the equations (3.8.37) and (3.8.38) coincide with equations (5.1) of Patterson and Thompson (1975), on which the MML estimation procedure is based, a method proposed by these authors for more general designs (see also Patterson and Thompson, 1971, and Section 3.7.2 in this book). Therefore, the Nelder estimators obtainable by solving the equations (3.8.29) and (3.8.30) are exactly the same as the MML estimators of Patterson and Thompson (1971, 1975) derived under the multivariate normality assumption. Both of these methods are iterative in their numerical applications. It should also be mentioned that the MML method is virtually the same as the marginal maximum likelihood estimation method (for which the same abbreviation MML is used), applied to block designs by El-Shaarawi, Prentice and Shah (1975). Also, as noted in Section 3.7, the MML method of Patterson and Thompson (1971, 1975) is often called the REML estimation method.

Another observation that can be made with regard to the equations (3.8.37) and (3.8.38) is that, with $\boldsymbol{V}$ of the form (3.5.9),

$$\text{tr}(\bar{\boldsymbol{R}}\boldsymbol{\phi}_\alpha) = \text{tr}(\bar{\boldsymbol{R}}\boldsymbol{V}\bar{\boldsymbol{R}}\boldsymbol{\phi}_\alpha) = \sigma_1^2\text{tr}(\bar{\boldsymbol{R}}\boldsymbol{\phi}_1\bar{\boldsymbol{R}}\boldsymbol{\phi}_\alpha) + \sigma_2^2\text{tr}(\bar{\boldsymbol{R}}\boldsymbol{\phi}_2\bar{\boldsymbol{R}}\boldsymbol{\phi}_\alpha) \quad \text{for } \alpha = 1, 2,$$

as $\bar{\boldsymbol{R}}\boldsymbol{V}\bar{\boldsymbol{R}} = \bar{\boldsymbol{R}}$ and $\bar{\boldsymbol{R}}\boldsymbol{\phi}_3 = \mathbf{O}$. This formula allows the equations to be written equivalently as

$$\sigma_1^2\text{tr}(\bar{\boldsymbol{R}}\boldsymbol{\phi}_1\bar{\boldsymbol{R}}\boldsymbol{\phi}_1) + \sigma_2^2\text{tr}(\bar{\boldsymbol{R}}\boldsymbol{\phi}_1\bar{\boldsymbol{R}}\boldsymbol{\phi}_2) = \boldsymbol{y}'\bar{\boldsymbol{R}}\boldsymbol{\phi}_1\bar{\boldsymbol{R}}\boldsymbol{y}, \qquad (3.8.40)$$

$$\sigma_1^2\text{tr}(\bar{\boldsymbol{R}}\boldsymbol{\phi}_2\bar{\boldsymbol{R}}\boldsymbol{\phi}_1) + \sigma_2^2\text{tr}(\bar{\boldsymbol{R}}\boldsymbol{\phi}_2\bar{\boldsymbol{R}}\boldsymbol{\phi}_2) = \boldsymbol{y}'\bar{\boldsymbol{R}}\boldsymbol{\phi}_2\bar{\boldsymbol{R}}\boldsymbol{y}, \qquad (3.8.41)$$

to see that they coincide with formula (4.4) of Rao (1972) on which his MINQUE method is based (see also Rao, 1970, 1971a, 1971b and 1979). This method in its original form is noniterative and relies on *a priori* values of the variance components σ_1^2 and σ_2^2 (stratum variances here), or their ratio, inserted into the matrix $\bar{\boldsymbol{R}}$ appearing in the equations (3.8.40) and (3.8.41). However, it has been suggested by Rao (1972, 1979) that the method can also be applied iteratively, by using the solutions of (3.8.40) and (3.8.41) as the new *a priori* values of σ_1^2 and σ_2^2, and so on until covergence. Then, the MML (REML) estimators, the iterated MINQUE estimators and the Nelder estimators of σ_1^2 and σ_2^2 will all be the same. The equivalence of the three methods just noticed is not restricted to the particular case considered here, but is general, as indicated by several authors. In particular, Patterson and Thompson (1975) not only pointed out the equivalence of the MINQUE procedure to one iteration of the MML scoring method, but also emphasized the appropriateness of the principle underlying

Nelder's (1968) method for any design (not necessarily generally balanced), as being equivalent to the MML method (shown here in Section 3.7). Also, Rao (1979) has pointed out the equivalence between the iterated MINQUE procedure and the MML (REML) method in general.

Finally, it is to be noted that for a proper block design, the equations (3.8.37) and (3.8.38) are equivalent to the equations (3.7.25) and (3.7.26), simply on account of the definitions and properties of the matrices $\boldsymbol{\phi}_\alpha$, $\alpha = 1,2,3$ (as given in Section 3.2), by which (3.8.37) and (3.8.38) can equivalently be written in form of the equations

$$\boldsymbol{y}'\bar{\boldsymbol{R}}\bar{\boldsymbol{R}}\boldsymbol{y} = \text{tr}(\bar{\boldsymbol{R}}) \quad \text{and} \quad \boldsymbol{y}'\bar{\boldsymbol{R}}\boldsymbol{D}'\boldsymbol{D}\bar{\boldsymbol{R}}\boldsymbol{y} = \text{tr}(\bar{\boldsymbol{R}}\boldsymbol{D}'\boldsymbol{D}),$$

and then because of the relation $\boldsymbol{R} = \sigma_1^2\bar{\boldsymbol{R}}$, where $\boldsymbol{R}$ is as defined in (3.7.20) and $\bar{\boldsymbol{R}}$ as in (3.8.39). That this relation holds for any proper block design can easily be seen from the expression (3.8.15), which is also true for (3.7.13) if the design is proper, i.e., from the representation

$$\boldsymbol{P}_{\Delta'(V^{-1})} = \sum_{\beta=0}^{m-1} w_{1\beta}\varepsilon_{1\beta}^{-1}\boldsymbol{P}_{\Delta' S_\beta}\boldsymbol{\phi}_1 + \sum_{\beta=1}^{m} w_{2\beta}\varepsilon_{2\beta}^{-1}\boldsymbol{P}_{\Delta' S_\beta}\boldsymbol{\phi}_2 + \boldsymbol{\phi}_3 = \boldsymbol{P}_{\Delta'(T^{-1})}.$$

Thus, for a proper block design, the stratum variances σ_1^2 and σ_2^2 can, equivalently, be estimated either as described in Section 3.7.2 or by using the method based on basic contrasts described in this section. Both methods are equivalent to the corresponding MML (REML) and the iterated MINQUE procedure.

When applying the methodology of Section 3.7.2, one has to note that in the case of a proper block design formula (3.7.8) reduces to

$$\boldsymbol{T}^{-1} = \boldsymbol{\phi}_1 + (\sigma_1^2/\sigma_2^2)(\boldsymbol{I}_n - \boldsymbol{\phi}_1) = \boldsymbol{\phi}_1 + \delta^{-1}(\boldsymbol{I}_n - \boldsymbol{\phi}_1),$$

where $\delta = \sigma_2^2/\sigma_1^2$, because of the relation $1 + k\gamma = \sigma_2^2/\sigma_1^2$. Furthermore, the iteration formula (3.7.30) can then be replaced by the equivalent formula

$$\hat{\delta} = \delta_0 + k\frac{(n-v)\boldsymbol{y}'\boldsymbol{R}_0\boldsymbol{D}'\boldsymbol{D}\boldsymbol{R}_0\boldsymbol{y} - \text{tr}(\boldsymbol{D}\boldsymbol{R}_0\boldsymbol{D}')\boldsymbol{y}'\boldsymbol{R}_0\boldsymbol{y}}{\text{tr}[(\boldsymbol{D}\boldsymbol{R}_0\boldsymbol{D}')^2]\boldsymbol{y}'\boldsymbol{R}_0\boldsymbol{y} - \text{tr}(\boldsymbol{D}\boldsymbol{R}_0\boldsymbol{D}')\boldsymbol{y}'\boldsymbol{R}_0\boldsymbol{D}'\boldsymbol{D}\boldsymbol{R}_0\boldsymbol{y}},$$

where

$$\boldsymbol{R}_0 = \boldsymbol{T}_0^{-1} - \boldsymbol{T}_0^{-1}\boldsymbol{\Delta}'(\boldsymbol{\Delta}\boldsymbol{T}_0^{-1}\boldsymbol{\Delta}')^{-1}\boldsymbol{\Delta}\boldsymbol{T}_0^{-1} \quad \text{with} \quad \boldsymbol{T}_0^{-1} = \boldsymbol{\phi}_1 + \delta_0^{-1}(\boldsymbol{I}_n - \boldsymbol{\phi}_1)$$

and δ_0 being a preliminary estimate of $\delta = \sigma_2^2/\sigma_1^2$.

3.8.3 *Properties of the combined estimators*

Now, one may ask, what can be said about the properties of the estimators of $\boldsymbol{\tau}$ and $\boldsymbol{c}'\boldsymbol{\tau}$ considered in Corollary 3.8.2 when the unknown stratum variances appearing in the estimation formulae (3.8.20)–(3.8.24) are replaced by solutions of the equations (3.8.29) and (3.8.30), i.e., by estimators obtainable by any of the three equivalent methods [Nelder's, MML (REML) and iterated MINQUE]. In trying to answer this question, let the following notation be used. Replacing σ_1^2 and σ_2^2 in (3.8.16) by their estimators $\hat{\sigma}_1^2$ and $\hat{\sigma}_2^2$, respectively, write

$$\hat{w}_{1\beta} = \frac{\varepsilon_{1\beta}\hat{\sigma}_2^2}{\varepsilon_{1\beta}\hat{\sigma}_2^2 + \varepsilon_{2\beta}\hat{\sigma}_1^2} = \frac{\varepsilon_{1\beta}\hat{\delta}}{\varepsilon_{1\beta}\hat{\delta} + \varepsilon_{2\beta}} \quad \text{and} \quad \hat{w}_{2\beta} = \frac{\varepsilon_{2\beta}\hat{\sigma}_1^2}{\varepsilon_{1\beta}\hat{\sigma}_2^2 + \varepsilon_{2\beta}\hat{\sigma}_1^2} = \frac{\varepsilon_{2\beta}}{\varepsilon_{1\beta}\hat{\delta} + \varepsilon_{2\beta}},$$

for $\beta = 1, 2, ..., m-1$, with $\hat{\delta} = \hat{\sigma}_2^2/\hat{\sigma}_1^2$ as an estimator of $\delta = \sigma_2^2/\sigma_1^2$. Accordingly, referring to (3.8.19), as an empirical estimator of $\boldsymbol{c}'_{\beta j}\boldsymbol{\tau}$, write

$$\widetilde{\boldsymbol{c}'_{\beta j}\boldsymbol{\tau}} = \hat{w}_{1\beta}(\widehat{\boldsymbol{c}'_{\beta j}\boldsymbol{\tau}})_1 + \hat{w}_{2\beta}(\widehat{\boldsymbol{c}'_{\beta j}\boldsymbol{\tau}})_2, \tag{3.8.42}$$

for all j related to $\beta = 1, 2, ..., m-1$ (as to the adjective "empirical," see the comment at the end of Section 3.7.2). Now note that

$$\widetilde{\boldsymbol{c}'_{\beta j}\boldsymbol{\tau}} - \widehat{\boldsymbol{c}'_{\beta j}\boldsymbol{\tau}} = \frac{(\varepsilon_{1\beta}/\varepsilon_{2\beta})(\hat{\delta} - \delta)}{[(\varepsilon_{1\beta}/\varepsilon_{2\beta})\hat{\delta} + 1][(\varepsilon_{1\beta}/\varepsilon_{2\beta})\delta + 1]}[(\widehat{\boldsymbol{c}'_{\beta j}\boldsymbol{\tau}})_1 - (\widehat{\boldsymbol{c}'_{\beta j}\boldsymbol{\tau}})_2],$$

from which (3.8.42) can be written as

$$\widetilde{\boldsymbol{c}'_{\beta j}\boldsymbol{\tau}} = \widehat{\boldsymbol{c}'_{\beta j}\boldsymbol{\tau}} + \frac{\varepsilon_{1\beta}/\varepsilon_{2\beta}}{(\varepsilon_{1\beta}/\varepsilon_{2\beta})\delta + 1}u_{\beta j}, \tag{3.8.43}$$

where

$$u_{\beta j} = \frac{\hat{\delta} - \delta}{(\varepsilon_{1\beta}/\varepsilon_{2\beta})\hat{\delta} + 1}[(\widehat{\boldsymbol{c}'_{\beta j}\boldsymbol{\tau}})_1 - (\widehat{\boldsymbol{c}'_{\beta j}\boldsymbol{\tau}})_2], \tag{3.8.44}$$

for $j = 1, 2, ..., \rho_\beta$ and $\beta = 1, 2, ..., m-1$. The representation (3.8.43) leads to the following result, originally from Roy and Shah (1962, Lemma 6.1) and in the present form adopted in Caliński (1996a, Lemma 3.1).

Lemma 3.8.1. *If the design is proper and the random variables $\{u_{\beta j}\}$, defined in* (3.8.44), *satisfy the conditions*

$$\mathrm{E}(u_{\beta j}) = 0, \quad \mathrm{E}(u_{\beta j}u_{\beta' j'}) = 0 \quad \textit{and} \quad \mathrm{Var}(u_{\beta j}) < \infty \tag{3.8.45}$$

for all β and j and all $(\beta, j) \neq (\beta', j')$, and for all values of $\delta = \sigma_2^2/\sigma_1^2$, then

$$\mathrm{E}(\widetilde{\boldsymbol{c}'_{\beta j}\boldsymbol{\tau}}) = \boldsymbol{c}'_{\beta j}\boldsymbol{\tau}, \tag{3.8.46}$$

$$\mathrm{Var}(\widetilde{\boldsymbol{c}'_{\beta j}\boldsymbol{\tau}}) = \mathrm{Var}(\widehat{\boldsymbol{c}'_{\beta j}\boldsymbol{\tau}}) + \left[\frac{\varepsilon_{1\beta}/\varepsilon_{2\beta}}{(\varepsilon_{1\beta}/\varepsilon_{2\beta})\delta + 1}\right]^2 \mathrm{Var}(u_{\beta j}), \tag{3.8.47}$$

for $j = 1, 2, ..., \rho_\beta$ and $\beta = 1, 2, ..., m-1$, and

$$\mathrm{Cov}(\widetilde{\boldsymbol{c}'_{\beta j}\boldsymbol{\tau}}, \widetilde{\boldsymbol{c}'_{\beta' j'}\boldsymbol{\tau}}) = 0 \quad if \quad (\beta, j) \neq (\beta', j'). \tag{3.8.48}$$

Proof. In view of (3.8.43), the result (3.8.46) follows from (3.8.45), whereas (3.8.47) and (3.8.48) follow from (3.8.45), (3.8.25) and the fact that, by Theorem 5a.2(i) of Rao (1973), $\widehat{\boldsymbol{c}'_{\beta j}\boldsymbol{\tau}}$ and $u_{\beta' j'}$ are uncorrelated for all relevant β, j, β' and j'. □

To see when the variables $\{u_{\beta j}\}$, as functions of the statistic $\hat{\delta} = \hat{\sigma}_2^2/\hat{\sigma}_1^2$, satisfy the conditions of Lemma 3.8.1, it will be helpful to note the following lemma.

Lemma 3.8.2. *Write*

$$\boldsymbol{y}'\boldsymbol{\psi}_1\boldsymbol{y} = \sum_{\ell=1}^{d_1} x_{(1)\ell}^2, \quad \boldsymbol{y}'\boldsymbol{\psi}_2\boldsymbol{y} = \sum_{\ell=d_1+1}^{d_1+d_2} x_{(2)\ell}^2, \tag{3.8.49}$$

where $x_{(1)\ell} = \boldsymbol{f}'_\ell\boldsymbol{y}$, *with* $\boldsymbol{f}_\ell$ *defined by* $\boldsymbol{f}_\ell = \boldsymbol{\psi}_1\boldsymbol{f}_\ell$ *and* $\boldsymbol{f}'_\ell\boldsymbol{f}_{\ell'} = \delta_{\ell\ell'}$ *for* $\ell, \ell' = 1, 2, ..., d_1 = n - b - h$, *and* $x_{(2)\ell} = \boldsymbol{g}'_\ell\boldsymbol{y}$, *with* $\boldsymbol{g}_\ell$ *defined by* $\boldsymbol{g}_\ell = \boldsymbol{\psi}_2\boldsymbol{g}_\ell$ *and* $\boldsymbol{g}'_\ell\boldsymbol{g}_{\ell'} = \delta_{\ell\ell'}$ *for* $\ell, \ell' = d_1 + 1, d_1 + 2, ..., d_1 + d_2 = n - v - h + \rho$ ($\delta_{\ell\ell'}$ *being the Kronecker delta*), *and introduce*

$$x_{(3)\ell} = (\widehat{\boldsymbol{c}'_\ell\boldsymbol{\tau}})_1 - (\widehat{\boldsymbol{c}'_\ell\boldsymbol{\tau}})_2 = (\varepsilon_{1i}^{-1}\boldsymbol{s}'_i\boldsymbol{\Delta}\boldsymbol{\phi}_1 - \varepsilon_{2i}^{-1}\boldsymbol{s}'_i\boldsymbol{\Delta}\boldsymbol{\phi}_2)\boldsymbol{y} \tag{3.8.50}$$

for $\ell = d_1 + d_2 + 1, d_1 + d_2 + 2, ..., n - v$. *Then,*

$$\begin{aligned}
\mathrm{E}(x_{(1)\ell}) &= E(x_{(2)\ell}) = E(x_{(3)\ell}) = 0 \quad \textit{for all} \;\; \ell,\\
\mathrm{Var}(x_{(1)\ell}) &= \sigma_1^2 \;\; \textit{for} \;\; \ell = 1, 2, ..., d_1,\\
\mathrm{Cov}(x_{(1)\ell}, x_{(1)\ell'}) &= 0 \;\; \textit{for all} \;\; \ell \neq \ell',\\
\mathrm{Var}(x_{(2)\ell}) &= \sigma_2^2 \;\; \textit{for} \;\; \ell = d_1 + 1, d_1 + 2, ..., d_1 + d_2,\\
\mathrm{Cov}(x_{(2)\ell}, x_{(2)\ell'}) &= 0 \;\; \textit{for all} \;\; \ell \neq \ell',\\
\mathrm{Cov}(x_{(1)\ell}, x_{(2)\ell'}) &= \mathrm{Cov}(x_{(1)\ell}, x_{(3)\ell''}) = \mathrm{Cov}(x_{(2)\ell'}, x_{(3)\ell''}) = 0
\end{aligned}$$

for all $\ell \neq \ell', \ell \neq \ell'', \ell' \neq \ell''$,

$$\mathrm{Var}(x_{(3)\ell}) = \varepsilon_{1\ell}^{-1}\sigma_1^2 + \varepsilon_{2\ell}^{-1}\sigma_2^2 \;\; \textit{for} \;\; \ell = d_1 + d_2 + 1, d_1 + d_2 + 2, ..., n - v$$

and

$$\mathrm{Cov}(x_{(3)\ell}, x_{(3)\ell'}) = 0 \;\; \textit{for all} \;\; \ell \neq \ell'.$$

Proof. These results follow from the known properties of the matrices $\boldsymbol{\psi}_1$ and $\boldsymbol{\psi}_2$ (see Sections 2.3 and 3.2.2) and the form of $\mathrm{Cov}(\boldsymbol{y})$ given for proper designs in (3.5.9). □

Lemma 3.8.2 implies that the Nelder (and at the same time, the MML or REML and the iterated MINQUE) estimators of σ_1^2 and σ_2^2, based on equations (3.8.29) and (3.8.30), are functions of a set of $n-v$ mutually uncorrelated random variables, each of zero expectation. Let them be written as a vector

$$\boldsymbol{x} = [\boldsymbol{x}'_{(1)}, \; \boldsymbol{x}'_{(2)}, \; \boldsymbol{x}'_{(3)}]', \tag{3.8.51}$$

where

$$\boldsymbol{x}_{(1)} = [x_{(1)1}, x_{(1)2}, ..., x_{(1)d_1}]', \quad \boldsymbol{x}_{(2)} = [x_{(2)d_1+1}, x_{(2)d_1+2}, ..., x_{(2)d_1+d_2}]'$$

and

$$\boldsymbol{x}_{(3)} = [x_{(3)d_1+d_2+1}, x_{(3)d_1+d_2+2}, ..., x_{(3)n-v}]',$$

so that Lemma 3.7.3 and Corollary 3.7.1 can easily be applied to these random variables.

With these results and notations, the following main theorem (Caliński, 1996a, Theorem 3.2), which is a simple generalization of Theorem 6.1 of Roy and Shah (1962), can be proved.

Theorem 3.8.2. *If the design is proper and the vector* $\boldsymbol{y}$ *with properties* (3.1.15) *and* (3.5.9) *has a distribution inducing all elements of the random vector* $\boldsymbol{x}$ *defined in* (3.8.51) *to have mutually independent symmetric distributions around zero, and if the statistic* $\hat{\delta} = \hat{\sigma}_2^2/\hat{\sigma}_1^2$ *used in* (3.8.44) *as an estimator of* $\delta = \sigma_2^2/\sigma_1^2$ *is completely expressible by even functions of the elements of* $\boldsymbol{x}$, *i.e., depends on* $\boldsymbol{y}$ *through such function only, then, for all nonnegative values of* δ *and* $\hat{\delta}$, *the empirical estimator*

$$\begin{aligned}\tilde{\boldsymbol{\tau}} &= \boldsymbol{r}^{-\delta}[\sum_{j=1}^{\rho_0} \boldsymbol{c}_{0j}(\widehat{\boldsymbol{c}_{0j}'\boldsymbol{\tau}}) + \sum_{\beta=1}^{m-1}\sum_{j=1}^{\rho_\beta} \boldsymbol{c}_{\beta j}(\widetilde{\boldsymbol{c}_{\beta j}'\boldsymbol{\tau}}) \\ &\quad + \sum_{j=1}^{\rho_m} \boldsymbol{c}_{mj}(\widehat{\boldsymbol{c}_{mj}'\boldsymbol{\tau}}) + \boldsymbol{c}_v(\widehat{\boldsymbol{c}_v'\boldsymbol{\tau}})] \qquad (3.8.52)\end{aligned}$$

is an unbiased estimator of $\boldsymbol{\tau}$, *with*

$$\mathrm{Cov}(\tilde{\boldsymbol{\tau}}) = \mathrm{Cov}(\hat{\boldsymbol{\tau}}) + \sum_{\beta=1}^{m-1}\left[\frac{\varepsilon_{1\beta}/\varepsilon_{2\beta}}{(\varepsilon_{1\beta}/\varepsilon_{2\beta})\delta+1}\right]^2 \sum_{j=1}^{\rho_\beta} \boldsymbol{r}^{-\delta}\boldsymbol{c}_{\beta j}\boldsymbol{c}_{\beta j}'\boldsymbol{r}^{-\delta}\mathrm{Var}(u_{\beta j}),$$

and hence,

$$\begin{aligned}\widetilde{\boldsymbol{c}'\boldsymbol{\tau}} = \boldsymbol{c}'\tilde{\boldsymbol{\tau}} &= \sum_{j=1}^{\rho_0} \boldsymbol{c}'\boldsymbol{r}^{-\delta}\boldsymbol{c}_{0j}(\widehat{\boldsymbol{c}_{0j}'\boldsymbol{\tau}}) + \sum_{\beta=1}^{m-1}\sum_{j=1}^{\rho_\beta} \boldsymbol{c}'\boldsymbol{r}^{-\delta}\boldsymbol{c}_{\beta j}(\widetilde{\boldsymbol{c}_{\beta j}'\boldsymbol{\tau}}) \\ &\quad + \sum_{j=1}^{\rho_m} \boldsymbol{c}'\boldsymbol{r}^{-\delta}\boldsymbol{c}_{mj}(\widehat{\boldsymbol{c}_{mj}'\boldsymbol{\tau}}) + \boldsymbol{c}'\boldsymbol{r}^{-\delta}\boldsymbol{c}_v(\widehat{\boldsymbol{c}_v'\boldsymbol{\tau}}) \qquad (3.8.53)\end{aligned}$$

is an unbiased estimator of any $\boldsymbol{c}'\boldsymbol{\tau}$, *with the finite variance*

$$\mathrm{Var}(\widetilde{\boldsymbol{c}'\boldsymbol{\tau}}) = \mathrm{Var}(\widehat{\boldsymbol{c}'\boldsymbol{\tau}}) + \sum_{\beta=1}^{m-1}\left[\frac{\varepsilon_{1\beta}/\varepsilon_{2\beta}}{(\varepsilon_{1\beta}/\varepsilon_{2\beta})\delta+1}\right]^2 \sum_{j=1}^{\rho_\beta} (\boldsymbol{c}'\boldsymbol{r}^{-\delta}\boldsymbol{c}_{\beta j})^2\mathrm{Var}(u_{\beta j}), \quad (3.8.54)$$

where $\{\widehat{\boldsymbol{c}_{\beta j}'\boldsymbol{\tau}}\}$ *are as in* (3.8.19), $\{\widetilde{\boldsymbol{c}_{\beta j}'\boldsymbol{\tau}}\}$ *are as in* (3.8.42), $\{u_{\beta j}\}$ *are as in* (3.8.44), *and where* $\mathrm{Var}(\widehat{\boldsymbol{c}'\boldsymbol{\tau}})$ *can be obtained from* (3.8.23).

Proof. From the assumed distribution of the vector $\boldsymbol{x}$, used in Lemma 3.8.2, and from Lemma 3.7.3 and Corollary 3.7.1, the properties (3.8.45) follow. That the expectations in (3.8.45) must be finite is evident from the relation

$$-\delta \leq \frac{\hat{\delta}-\delta}{(\varepsilon_{1\beta}/\varepsilon_{2\beta})\hat{\delta}+1} < \frac{\varepsilon_{2\beta}}{\varepsilon_{1\beta}}, \tag{3.8.55}$$

held by any nonnegative $\hat{\delta}$, assuming that δ is nonnegative and finite (which is well justified). The relation (3.8.55) implies also that

$$\text{Var}(u_{\beta j}) < \max\left\{\delta^2\left(\frac{\sigma_1^2}{\varepsilon_{1\beta}}+\frac{\sigma_2^2}{\varepsilon_{2\beta}}\right),\ \left(\frac{\varepsilon_{2\beta}}{\varepsilon_{1\beta}}\right)^2\left(\frac{\sigma_1^2}{\varepsilon_{1\beta}}+\frac{\sigma_2^2}{\varepsilon_{2\beta}}\right)\right\} < \infty. \tag{3.8.56}$$

Thus, by Lemma 3.8.1, the theorem can easily be established on account of Corollary 3.8.2, formulae (3.8.18) and (3.8.43) and the fact (seen from the proof of Lemma 3.8.1) that the random variables $\{\widehat{\boldsymbol{c}'_{\beta j}\boldsymbol{\tau}}\}$ and $\{u_{\beta' j'}\}$ are mutually uncorrelated for all relevant β, j, β' and j', which implies that not only (3.8.48), but also the equality

$$\text{Cov}(\widehat{\boldsymbol{c}'_{\beta j}\boldsymbol{\tau}}, \widetilde{\boldsymbol{c}'_{\beta' j'}\boldsymbol{\tau}}) = 0$$

holds for any $(\beta, j) \neq (\beta', j')$. □

Remark 3.8.4. Let the vector $\boldsymbol{y}$ be as considered in Theorem 3.8.2. Then, the following applies:

(a) If $\boldsymbol{y}$ has a multivariate normal distribution, then the vector $\boldsymbol{x}$ defined in (3.8.51) is also $(n-v)$-variate normal, which automatically implies that its elements have mutually independent symmetric distributions around zero.

(b) Because the functions of the vector $\boldsymbol{y}$ appearing in the equations (3.8.29) and (3.8.30) can, on account of (3.8.33) and (3.8.34), be completely expressed as quadratic even functions of the elements of the vector $\boldsymbol{x}$ defined in (3.8.51), the statistic $\hat{\delta}$, as the ratio of the solutions of these equations, is a nonnegative even function of each element of $\boldsymbol{x}$, as demanded in Theorem 3.8.2.

(c) Theorem 3.8.2 is general, not restricted to estimators of σ_1^2 and σ_2^2 obtainable through equations (3.8.29) and (3.8.30), and can equally well be applied with other estimators of the stratum variances satisfying the conditions of the theorem.

It follows from Theorem 3.8.2 that if the unknown stratum variances σ_1^2 and σ_2^2 determining the weights (3.8.16) appearing in formula (3.8.19) are replaced by their estimators $\hat{\sigma}_1^2$ and $\hat{\sigma}_2^2$, then, under the assumptions of the theorem, the unbiasedness of the estimators (3.8.18) and (3.8.22) of $\boldsymbol{\tau}$ and $\boldsymbol{c}'\boldsymbol{\tau}$, respectively, is not violated. As to the variance of such empirical estimator of $\boldsymbol{c}'\boldsymbol{\tau}$, formula (3.8.54) shows that, after this replacement, it will increase by an amount equal to

$$\sum_{\beta=1}^{m-1}\sum_{j=1}^{\rho_\beta}(\boldsymbol{c}'\boldsymbol{r}^{-\delta}\boldsymbol{c}_{\beta j})^2\left[\frac{\varepsilon_{1\beta}/\varepsilon_{2\beta}}{(\varepsilon_{1\beta}/\varepsilon_{2\beta})\delta+1}\right]^2 \text{Var}(u_{\beta j}),$$

where $u_{\beta j}$ is as defined in (3.8.44). Unfortunately, an explicit formula of $\text{Var}(u_{\beta j})$ is usually not available. Instead, an upper limit for it can be given according to (3.8.56). It may be interesting to use it to investigate the variance (3.8.47), on which the variance (3.8.54) basically depends.

From (3.8.54), (3.8.56) and (3.8.24), it is permissible to write

$$\text{Var}(\widetilde{c'_{\beta j}\boldsymbol{\tau}}) < \frac{\sigma_2^2}{\varepsilon_{2\beta}} = \text{Var}[(\widehat{c'_{\beta j}\boldsymbol{\tau}})_2] \quad \text{if } \delta > \frac{\varepsilon_{2\beta}}{\varepsilon_{1\beta}}, \tag{3.8.57}$$

$$\text{Var}(\widetilde{c'_{\beta j}\boldsymbol{\tau}}) < \frac{\sigma_1^2}{\varepsilon_{1\beta}} = \text{Var}[(\widehat{c'_{\beta j}\boldsymbol{\tau}})_1] \quad \text{if } \delta \leq \frac{\varepsilon_{2\beta}}{\varepsilon_{1\beta}}, \tag{3.8.58}$$

which can also be written as

$$\text{Var}(\widetilde{c'_{\beta j}\boldsymbol{\tau}}) < \max\{\text{Var}[(\widehat{c'_{\beta j}\boldsymbol{\tau}})_1],\ \text{Var}[(\widehat{c'_{\beta j}\boldsymbol{\tau}})_2]\}.$$

Thus, the variance of the combined estimator (3.8.42) never reaches the value of the larger of the two variances, that of the intra-block and that of the inter-block BLUE of $c'_{\beta j}\boldsymbol{\tau}$ ($j = 1, 2, ..., \rho_\beta$ and $\beta = 1, 2, ..., m-1$). This conclusion is interesting, but not very useful, because usually one would expect to have $\delta > \varepsilon_{2\beta}/\varepsilon_{1\beta}$ in a successfully designed experiment. In this case, the inequality (3.8.57) holds, whereas instead rather the inequality (3.8.58) would be desirable.

The situation with the upper limit for $\text{Var}(\widetilde{c'_{\beta j}\boldsymbol{\tau}})$ can be improved if an *a priori* knowledge on the lower limit for δ is available. Suppose that it is known (or at least expected) that $\delta \geq \delta^*$. Then, a truncated estimator of δ can be used, i.e., the estimator

$$\hat{\delta}^* = \begin{cases} \hat{\delta} & \text{if } \hat{\delta} \geq \delta^*, \\ \delta^* & \text{otherwise.} \end{cases} \tag{3.8.59}$$

Denoting by $u_{\beta j}(\hat{\delta})$ the function defined in (3.8.44) and by $u_{\beta j}(\hat{\delta}^*)$ that obtained from it when replacing $\hat{\delta}$ by $\hat{\delta}^*$, it can easily be noted that

$$u^2_{\beta j}(\hat{\delta}) \geq u^2_{\beta j}(\hat{\delta}^*),$$

the sharp inequality holding for $0 < \hat{\delta} < \delta^*$. Hence,

$$\text{E}[u^2_{\beta j}(\hat{\delta})] > \text{E}[u^2_{\beta j}(\hat{\delta}^*)]. \tag{3.8.60}$$

Now, denoting by $(\widetilde{c'_{\beta j}\boldsymbol{\tau}})^*$ the combined estimator of $c'_{\beta j}\boldsymbol{\tau}$ obtained according to (3.8.42) but with the statistic $\hat{\delta}$ in the weights $\{\hat{w}_{1\beta}\}$ and $\{\hat{w}_{2\beta}\}$ replaced by its truncated version $\hat{\delta}^*$, i.e., writing

$$(\widetilde{c'_{\beta j}\boldsymbol{\tau}})^* = \hat{w}^*_{1\beta}(\widehat{c'_{\beta j}\boldsymbol{\tau}})_1 + \hat{w}^*_{2\beta}(\widehat{c'_{\beta j}\boldsymbol{\tau}})_2, \tag{3.8.61}$$

where

$$\hat{w}^*_{1\beta} = \frac{\varepsilon_{1\beta}\hat{\delta}^*}{\varepsilon_{1\beta}\hat{\delta}^* + \varepsilon_{2\beta}} \quad \text{and} \quad \hat{w}^*_{2\beta} = \frac{\varepsilon_{2\beta}}{\varepsilon_{1\beta}\hat{\delta}^* + \varepsilon_{2\beta}}, \tag{3.8.62}$$

one can see, from (3.8.60) and on account of (3.8.47), that

$$\operatorname{Var}(\widetilde{c'_{\beta j}\tau}) > \operatorname{Var}[(\widetilde{c'_{\beta j}\tau})^*]. \tag{3.8.63}$$

This inequality was originally found by Shah (1971) in a different context of estimating the stratum variances. In fact, the result (3.8.63) will hold for any approach in estimating σ_1^2 and σ_2^2, or $\delta = \sigma_2^2/\sigma_1^2$, which satisfies the conditions of Theorem 3.8.2, and with any choice of $\delta^* > 0$ in (3.8.59). (Although it is assumed that $\hat{\delta}$ is nonnegative, formally, it would be sufficient that $\hat{\delta} > -\varepsilon_{2\beta}/\varepsilon_{1\beta}$, and then a possible choice could be $\delta^* = 0$, as considered by Shah, 1971, p. 817.) Certainly, it should be noticed that if an untruncated estimator, $\hat{\delta}$, satisfies conditions of Theorem 3.8.2, its truncated form, $\hat{\delta}^*$, also does it, provided that $\delta^* \geq 0$.

The use of the truncated estimator (3.8.59) of δ has the advantage of ensuring that the variance of the combined estimator (3.8.61) of any of the basic contrasts $\{c'_{\beta j}\tau\}$, which appear in the estimation formulae (3.8.52) and (3.8.53), is smaller than the variance of its intra-block component, provided that the true value of δ is within certain limits. To see this advantage, note that if the truncated estimator $\hat{\delta}^*$ is used, then the inequality (3.8.56) changes to

$$\operatorname{Var}[u_{\beta j}(\hat{\delta}^*)] < \max\left\{\left[\frac{\delta - \delta^*}{(\varepsilon_{1\beta}/\varepsilon_{2\beta})\delta^* + 1}\right]^2 \left(\frac{\sigma_1^2}{\varepsilon_{1\beta}} + \frac{\sigma_2^2}{\varepsilon_{2\beta}}\right), \left(\frac{\varepsilon_{2\beta}}{\varepsilon_{1\beta}}\right)^2 \left(\frac{\sigma_1^2}{\varepsilon_{1\beta}} + \frac{\sigma_2^2}{\varepsilon_{2\beta}}\right)\right\},$$

where

$$\left[\frac{\delta - \delta^*}{(\varepsilon_{1\beta}/\varepsilon_{2\beta})\delta^* + 1}\right]^2 \leq \left(\frac{\varepsilon_{2\beta}}{\varepsilon_{1\beta}}\right)^2$$

if and only if $\delta^* \leq \delta \leq 2\delta^* + \varepsilon_{2\beta}/\varepsilon_{1\beta}$. This result implies the inequality

$$\operatorname{Var}[(\widetilde{c'_{\beta j}\tau})^*] < \varepsilon_{1\beta}^{-1}\sigma_1^2 = \operatorname{Var}[(\widehat{c'_{\beta j}\tau})_1] \quad \text{if } \delta^* \leq \delta \leq 2\delta^* + \frac{\varepsilon_{2\beta}}{\varepsilon_{1\beta}}, \tag{3.8.64}$$

as originally found by Shah (1964b). So, within this range of the values of δ, the combined estimator of $c'_{\beta j}\tau$ (for $\beta = 1, 2, ..., m-1$) based on the truncated estimator (3.8.59) of δ will always have a variance smaller than that of the corresponding intra-block BLUE, i.e., will be better than the latter. Consequently, the estimation of any contrast $c'\tau$, for which the BLUE under the overall model (3.1.1) is not available, will also be improved in comparison with the intra-block estimation when using formula (3.8.53) with the estimators (3.8.61) [instead of (3.8.42)] of the involved basic contrasts $\{c'_{\beta j}\tau\}$, provided that the true value of δ belongs to $[\delta^*,\ 2\delta^* + \varepsilon_{2\beta}/\varepsilon_{1\beta}]$.

On the other hand, if $\delta > 2\delta^* + \varepsilon_{2\beta}/\varepsilon_{1\beta}$, then

$$\begin{aligned}\operatorname{Var}[(\widetilde{c'_{\beta j}\tau})^*] &< \frac{\varepsilon_{1\beta}(\delta^*)^2\sigma_1^2 + \varepsilon_{2\beta}\sigma_2^2}{(\varepsilon_{1\beta}\delta^* + \varepsilon_{2\beta})^2} \\ &= [w_{1\beta}(\delta^*)]^2 \operatorname{Var}[(\widehat{c'_{\beta j}\tau})_1] + [w_{2\beta}(\delta^*)]^2 \operatorname{Var}[(\widehat{c'_{\beta j}\tau})_2],\end{aligned} \tag{3.8.65}$$

where

$$w_{1\beta}(\delta^*) = \frac{\varepsilon_{1\beta}\delta^*}{\varepsilon_{1\beta}\delta^* + \varepsilon_{2\beta}} \quad \text{and} \quad w_{2\beta}(\delta^*) = \frac{\varepsilon_{2\beta}}{\varepsilon_{1\beta}\delta^* + \varepsilon_{2\beta}}, \tag{3.8.66}$$

$w_{1\beta}(\delta^*)$ increasing with the increase of δ^*, whereas $w_{2\beta}(\delta^*)$ decreasing with that. Thus, if δ exceeds the limit $2\delta^* + \varepsilon_{2\beta}/\varepsilon_{1\beta}$, then the upper limit of the variance of the estimator $(\widetilde{c'_{\beta j}\tau})^*$ is a combination of the variances of the intra-block and the inter-block BLUEs, with weights depending on the value δ^*, the assumed lower limit for δ. The higher the value, the more weight given to the variance of the intra-block BLUE. Equal weights will be given in the case of $\delta^* = \varepsilon_{2\beta}/\varepsilon_{1\beta}$, whereas the weights will be $\varepsilon_{1\beta}^2$ for intra-block and $\varepsilon_{2\beta}^2$ for inter-block if $\delta^* = 1$, which seems to be the smallest value that can safely be assumed in most cases (and is usually chosen, starting from Yates, 1939). Note that for $\delta \geq 1$, it is sufficient that $k\sigma_B^2 \geq \sigma_U^2$ (see Section 3.1.1 for definitions), which usually holds, unless the blocking of units is a complete failure. By increasing δ^*, one certainly increases the limit $2\delta^* + \varepsilon_{2\beta}/\varepsilon_{1\beta}$, which is a threshold for δ, above which the upper limit given in (3.8.65) exceeds $\sigma_1^2/\varepsilon_{1\beta} = \text{Var}[(\widehat{c'_{\beta j}\tau})_1]$, as

$$\frac{\varepsilon_{1\beta}(\delta^*)^2\sigma_1^2 + \varepsilon_{2\beta}\sigma_2^2}{(\varepsilon_{1\beta}\delta^* + \varepsilon_{2\beta})^2} \leq \frac{\sigma_1^2}{\varepsilon_{1\beta}}$$

if and only if $\delta \leq 2\delta^* + \varepsilon_{2\beta}/\varepsilon_{1\beta}$.

Thus, a rough conclusion that can be drawn from these considerations is that if the experiment is laid down in successfully formed blocks, for which it can be assumed that $\delta \geq \delta^*$, with δ^* well above 1, then one can expect the inequality (3.8.64) to hold in most cases. This conclusion means that in a successful experiment, for which δ^* can be chosen sufficiently large, the variance of the intra-block BLUE of $c'\tau$, given in (3.2.13), can be used as an upper limit for the variance of its empirical estimator (3.8.53), i.e., for the variance given in (3.8.54).

3.8.4 *Approximating the variance of the empirical estimator*

In general, however, instead of looking for an upper limit, it may be more interesting to try to approximate the variance (3.8.54), if an explicit formula for its components $\{\text{Var}(u_{\beta j})\}$ is not readily available. In fact, except for some simple, special cases, the exact formula for $\text{Var}(u_{\beta j})$ is practically intractable [as already mentioned in Section 3.7 with reference to the variance of the random variable (3.7.42)], even if theoretically obtainable. Therefore, following the same approach as that adopted (from Kackar and Harville, 1984) in Section 3.7, three stages of a suitable approximation of (3.8.54) will now be considered under the assumptions of Theorem 3.8.2.

First, note that because

$$u_{\beta j} = \frac{\hat{\delta} - \delta}{(\varepsilon_{1\beta}/\varepsilon_{2\beta})\hat{\delta} + 1} x_{(3)\beta j}, \tag{3.8.67}$$

where $x_{(3)\beta j} = (\widehat{\mathbf{c}'_{\beta j}\boldsymbol{\tau}})_1 - (\widehat{\mathbf{c}'_{\beta j}\boldsymbol{\tau}})_2$, one obtains

$$\frac{du_{\beta j}}{d\hat{\delta}} = \frac{(\varepsilon_{1\beta}/\varepsilon_{2\beta})\delta + 1}{[(\varepsilon_{1\beta}/\varepsilon_{2\beta})\hat{\delta} + 1]^2} x_{(3)\beta j},$$

from which (by Taylor's expansion) it is permissible to write

$$u_{\beta j} \equiv u_{\beta j}(\hat{\delta}) = u_{\beta j}(\delta) + (\hat{\delta} - \delta)\frac{du_{\beta j}}{d\hat{\delta}}\bigg|_{\hat{\delta}=\delta} + \cdots = 0 + \frac{\hat{\delta} - \delta}{(\varepsilon_{1\beta}/\varepsilon_{2\beta})\delta + 1} x_{(3)\beta j} + \cdots .$$

This result gives the approximation

$$u_{\beta j} \cong \frac{\hat{\delta} - \delta}{(\varepsilon_{1\beta}/\varepsilon_{2\beta})\delta + 1} x_{(3)\beta j}$$

of (3.8.67) and, hence, the approximation

$$u_{\beta j}^2 \cong \left[\frac{\hat{\delta} - \delta}{(\varepsilon_{1\beta}/\varepsilon_{2\beta})\delta + 1}\right]^2 x_{(3)\beta j}^2.$$

From this approximation, on account of Theorem 2b.3(iii) of Rao (1973), the expectation $\mathrm{E}(u_{\beta j}^2) = \mathrm{Var}(u_{\beta j})$ can be approximated as

$$\mathrm{E}(u_{\beta j}^2) \cong \mathrm{E}\left\{\left[\frac{\hat{\delta} - \delta}{(\varepsilon_{1\beta}/\varepsilon_{2\beta})\delta + 1}\right]^2 \mathrm{E}(x_{(3)\beta j}^2 | \hat{\delta})\right\}. \tag{3.8.68}$$

Unfortunately, the conditional expectation $\mathrm{E}(x_{(3)\beta j}^2 | \hat{\delta})$ is usually not readily obtainable, unless $\hat{\delta}$ does not involve $x_{(3)\beta j}$. The latter, however, is usually not the case, in particular if the statistic $\hat{\delta} = \hat{\sigma_2^2}/\hat{\sigma_1^2}$ is based on the solution of the equations (3.8.29) and (3.8.30), as can be seen from (3.8.33) and (3.8.34). However, in the case of a large number of the elements of the vector $\boldsymbol{x}$, defined in (3.8.51), from which $\hat{\delta}$ is calculated through the formulae (3.8.49) and (3.8.50), i.e., if the number $n - v$ is large, the dependence between $x_{(3)\beta j}^2$ and $\hat{\delta}$ can be ignored and instead of $\mathrm{E}(x_{(3)\beta j}^2 | \hat{\delta})$ the unconditional expectation $\mathrm{E}(x_{(3)\beta j}^2)$ can be used. This decision gives the second stage of approximation, which consists in replacing (3.8.68) by

$$\begin{aligned} \mathrm{E}(u_{\beta j}^2) &\cong \mathrm{E}\left\{\left[\frac{\hat{\delta} - \delta}{(\varepsilon_{1\beta}/\varepsilon_{2\beta})\delta + 1}\right]^2 \mathrm{E}(x_{(3)\beta j}^2)\right\} \\ &= \frac{(\varepsilon_{1\beta}^{-1}\sigma_1^2 + \varepsilon_{2\beta}^{-1}\sigma_2^2)}{[(\varepsilon_{1\beta}/\varepsilon_{2\beta})\delta + 1]^2}\mathrm{E}[(\hat{\delta} - \delta)^2] = \frac{\varepsilon_{1\beta}^{-1}\sigma_1^2}{(\varepsilon_{1\beta}/\varepsilon_{2\beta})\delta + 1}\mathrm{E}[(\hat{\delta} - \delta)^2] \end{aligned} \tag{3.8.69}$$

(see Lemma 3.8.2), where $E[(\hat{\delta}-\delta)^2]$, the MSE of $\hat{\delta}$, becomes $\text{Var}(\hat{\delta})$ if $\hat{\delta}$ is an unbiased estimator of δ. Now, with the approximation (3.8.69), the variance of the empirical estimator (3.8.53), given in (3.8.54), can be approximated as

$$\begin{aligned}\text{Var}(\widetilde{\boldsymbol{c}'\boldsymbol{\tau}}) &\cong \text{Var}(\widehat{\boldsymbol{c}'\boldsymbol{\tau}}) + \sigma_1^2 E[(\hat{\delta}-\delta)^2] \sum_{\beta=1}^{m-1}\sum_{j=1}^{\rho_\beta} (\boldsymbol{c}'\boldsymbol{r}^{-\delta}\boldsymbol{c}_{\beta j})^2 \frac{\varepsilon_{1\beta}\varepsilon_{2\beta}}{(\varepsilon_{1\beta}\delta+\varepsilon_{2\beta})^3} \\ &= \text{Var}_{\text{ap}}(\widetilde{\boldsymbol{c}'\boldsymbol{\tau}}) \quad \text{(say)}. \qquad (3.8.70)\end{aligned}$$

This result coincides, for a proper block design, with the approximation (3.7.56), equivalent to the two-stage approximation given by Kackar and Harville (1984).

The third stage of the approximation consists in approximating the MSE of $\hat{\delta}$, for which the exact formula is usually not easily available. However, if the distribution of $\boldsymbol{y}$ is assumed to be normal and $\hat{\delta}$ results from the solution of the equations (3.8.29) and (3.8.30), i.e., if the MML (REML) estimation procedure is used (equivalent in computations to Nelder's and to iterated MINQUE procedure), then $E[(\hat{\delta}-\delta)^2]$ in (3.8.70) can be approximated by the asymptotic variance of $\hat{\delta}$. It is obtainable from the inverse of the appropriate information matrix resulting from the normal distribution of the random vector $\boldsymbol{x}$ defined in (3.8.51). For this reason, note that the log likelihood of $\boldsymbol{x}$ can, on account of Lemma 3.8.2 and the assumed normality, be written as

$$L = \sum_{\ell=1}^{d_1} L_{(1)\ell} + \sum_{\ell=d_1+1}^{d_1+d_2} L_{(2)\ell} + \sum_{\ell=d_1+d_2+1}^{n-v} L_{(3)\ell}, \qquad (3.8.71)$$

where

$$L_{(1)\ell} = \text{const} - \frac{1}{2}\log\sigma_1^2 - \frac{1}{2\sigma_1^2}x_{(1)\ell}^2 \quad \text{for } \ell = 1,2,\ldots,d_1,$$

$$L_{(2)\ell} = \text{const} - \frac{1}{2}\log\sigma_1^2 - \frac{1}{2}\log\delta - \frac{1}{2\sigma_1^2\delta}x_{(2)\ell}^2 \quad \text{for } \ell = d_1+1, d_1+2, \ldots, d_1+d_2$$

and

$$L_{(3)\ell} = \text{const} - \frac{1}{2}\log\sigma_1^2 - \frac{1}{2}\log(\varepsilon_{1\ell}^{-1}+\varepsilon_{2\ell}^{-1}\delta) - \frac{1}{2\sigma_1^2(\varepsilon_{1\ell}^{-1}+\varepsilon_{2\ell}^{-1}\delta)}x_{(3)\ell}^2$$

$$\text{for } \ell = d_1+d_2+1, d_1+d_2+2, \ldots, n-v.$$

From (3.8.71), the information matrix for δ and σ_1^2 can easily be obtained as

$$\boldsymbol{I}(\delta,\sigma_1^2) = \frac{1}{2}\begin{bmatrix} \dfrac{d_2}{\delta^2} + \displaystyle\sum_{i=\rho+1}^{h}\left(\frac{\varepsilon_{2i}^{-1}}{\varepsilon_{1i}^{-1}+\varepsilon_{2i}^{-1}\delta}\right)^2 & \dfrac{d_2}{\sigma_1^2\delta} + \displaystyle\sum_{i=\rho+1}^{h}\frac{\varepsilon_{2i}^{-1}}{\sigma_1^2(\varepsilon_{1i}^{-1}+\varepsilon_{2i}^{-1}\delta)} \\ \dfrac{d_2}{\sigma_1^2\delta} + \displaystyle\sum_{i=\rho+1}^{h}\frac{\varepsilon_{2i}^{-1}}{\sigma_1^2(\varepsilon_{1i}^{-1}+\varepsilon_{2i}^{-1}\delta)} & \dfrac{n-v}{\sigma_1^4}\end{bmatrix}.$$

Taking the inverse of this matrix, $[\boldsymbol{I}(\delta, \sigma_1^2)]^{-1}$, one obtains the asymptotic variance of $\hat{\delta}$,

$$\mathrm{Var}_{\mathrm{as}}(\hat{\delta}) = \frac{2}{\frac{d_2}{\delta^2} + \sum_{i=\rho+1}^{h}\left(\frac{\varepsilon_{1i}}{\varepsilon_{1i}\delta + \varepsilon_{2i}}\right)^2 - \frac{1}{n-v}\left(\frac{d_2}{\delta} + \sum_{i=\rho+1}^{h}\frac{\varepsilon_{1i}}{\varepsilon_{1i}\delta + \varepsilon_{2i}}\right)^2},$$

which again coincides with formula (3.7.57) when applied to a proper block design.

Thus, finally, it can be concluded that if the ratio $\delta = \sigma_2^2/\sigma_1^2$ is estimated from the solution of the equations (3.8.29) and (3.8.30), or their equivalences (3.8.37) and (3.8.38), then, under the normality assumption, the variance of the empirical estimator (3.8.53), i.e., the variance (3.8.54), can be approximated according to the suggestions of Kackar and Harville (1984) by the formula

$$\mathrm{Var}(\widetilde{\boldsymbol{c}'\boldsymbol{\tau}}) \cong \mathrm{Var}(\widehat{\boldsymbol{c}'\boldsymbol{\tau}}) + \frac{2\sigma_1^2 \sum_{\beta=1}^{m-1} \varepsilon_{1\beta}^{-1} w_{1\beta}^2 w_{2\beta} \sum_{j=1}^{\rho_\beta} (\boldsymbol{c}'\boldsymbol{r}^{-\delta}\boldsymbol{c}_{\beta j})^2}{d_2 + \sum_{\beta=1}^{m-1} \rho_\beta w_{1\beta}^2 - (n-v)^{-1}\left(d_2 + \sum_{\beta=1}^{m-1} \rho_\beta w_{1\beta}\right)^2}. \tag{3.8.72}$$

Formula (3.8.72) can also be written as

$$\mathrm{Var}(\widetilde{\boldsymbol{c}'\boldsymbol{\tau}}) \cong \mathrm{Var}(\widehat{\boldsymbol{c}'\boldsymbol{\tau}}) + \sigma_1^2 \sum_{\beta=1}^{m-1} \varepsilon_{1\beta}^{-1} w_{1\beta}\zeta_\beta \sum_{j=1}^{\rho_\beta} (\boldsymbol{c}'\boldsymbol{r}^{-\delta}\boldsymbol{c}_{\beta j})^2,$$

where

$$\begin{aligned}\zeta_\beta &= \frac{w_{1\beta}w_{2\beta}}{\delta^2}\mathrm{Var}_{\mathrm{as}}(\hat{\delta}) = \frac{2w_{1\beta}w_{2\beta}}{d_2 + \sum_{i=\rho+1}^{h} w_{1i}^2 - (n-v)^{-1}\left(d_2 + \sum_{i=\rho+1}^{h} w_{1i}\right)^2} \\ &= \frac{2w_{1\beta}w_{2\beta}}{\sum_{\ell=1}^{n-v}(\omega_\ell - \omega_.)^2},\end{aligned} \tag{3.8.73}$$

with

$$\omega_\ell = \begin{cases} 0 & \text{for } \ell = 1, 2, ..., d_1 \quad (d_1 = n - b - h), \\ 1 & \text{for } \ell = d_1 + 1, d_1 + 2, ..., d_1 + d_2 \quad (d_2 = b - v + \rho), \\ w_{1i} & \text{for } \ell = d_1 + d_2 + 1, ..., n - v, \text{ while } i = \ell - (n - v - h), \end{cases} \tag{3.8.74}$$

$$w_{1i} = \frac{\varepsilon_{1i}\sigma_2^2}{\varepsilon_{1i}\sigma_2^2 + \varepsilon_{2i}\sigma_1^2} = \frac{\varepsilon_{1i}\delta}{\varepsilon_{1i}\delta + \varepsilon_{2i}}$$

for $i = \rho + 1, \rho + 2, ..., h$, and with $\omega_.$ defined as

$$\omega_. = \frac{1}{n-v}\sum_{\ell=1}^{n-v} \omega_\ell. \tag{3.8.75}$$

Evidently, the coefficient ζ_β depends on the values of the weights $\{w_{1i}\}$, becoming small if all the weights are close to their extreme values, 0 or 1.

Alternatively, referring to formula (3.8.23) of $\text{Var}(\widehat{\boldsymbol{c}'\boldsymbol{\tau}})$, one can write

$$\begin{aligned} \text{Var}(\widetilde{\boldsymbol{c}'\boldsymbol{\tau}}) \cong \sigma_1^2 \Bigg[&\sum_{j=1}^{\rho_0}(\boldsymbol{c}'\boldsymbol{r}^{-\delta}\boldsymbol{c}_{0j})^2 + \sum_{\beta=1}^{m-1}\frac{w_{1\beta}}{\varepsilon_{1\beta}}(1+\zeta_\beta)\sum_{j=1}^{\rho_\beta}(\boldsymbol{c}'\boldsymbol{r}^{-\delta}\boldsymbol{c}_{\beta j})^2 \\ &+ \frac{\sigma_2^2}{\sigma_1^2}\sum_{j=1}^{\rho_m}(\boldsymbol{c}'\boldsymbol{r}^{-\delta}\boldsymbol{c}_{mj})^2 + \frac{\sigma_3^2}{\sigma_1^2}(\boldsymbol{c}'\boldsymbol{r}^{-\delta}\boldsymbol{c}_v)^2\Bigg]. \end{aligned} \tag{3.8.76}$$

Also note that if $\boldsymbol{c}'\boldsymbol{\tau}$ is a basic contrast, say $\boldsymbol{c}_i'\boldsymbol{\tau}$, then formula (3.8.76) reduces to the exact formula (3.8.24) for $i = 1, 2, ..., \rho$ and for $i = h+1, ..., v-1$, or to the approximation formula

$$\text{Var}(\widetilde{\boldsymbol{c}_i'\boldsymbol{\tau}}) \cong \sigma_1^2\varepsilon_{1i}^{-1}w_{1i}(1+\zeta_i) = \text{Var}(\widehat{\boldsymbol{c}_i'\boldsymbol{\tau}})(1+\zeta_i), \tag{3.8.77}$$

where

$$\zeta_i = \frac{2w_{1i}(1-w_{1i})}{\sum_{\ell=1}^{n-v}(\omega_\ell - \omega_.)^2}, \tag{3.8.78}$$

for $i = \rho+1, \rho+2, ..., h$. In particular, if $w_{1i} = w_1$ (constant) for $i = \rho+1, \rho+2, ..., h$, then (3.8.78) gets the constant form

$$\zeta = \frac{2w_1(1-w_1)}{d_2 + (h-\rho)w_1^2 - (n-v)^{-1}[d_2 + (h-\rho)w_1]^2}$$

for each i. When applied to a BIB design, it becomes

$$\zeta = \frac{2w_1(1-w_1)}{b - v + (v-1)w_1^2 - (n-v)^{-1}[b-v+(v-1)w_1]^2}, \tag{3.8.79}$$

with

$$w_1 = \frac{v(k-1)\delta}{v(k-1)\delta + v - k} \quad [\equiv w \text{ in } (3.7.59)], \tag{3.8.80}$$

which makes (3.8.77) coinciding with the result (5.4) of Kackar and Harville (1984).

With formula (3.8.77), it is now easy to approximate the real gain resulting from the use of the empirical combined estimator

$$\widetilde{c_i'\tau} = \hat{w}_{1i}(\widehat{c_i'\tau})_1 + \hat{w}_{2i}(\widehat{c_i'\tau})_2, \tag{3.8.81}$$

with $\hat{w}_{1i}$ and $\hat{w}_{2i} = 1 - \hat{w}_{1i}$, as in (3.8.35) and (3.8.36), instead of the intra-block BLUE $(\widehat{c_i'\tau})_1$ [$\equiv (\widehat{c_i'\tau})_{\text{intra}}$ in (3.4.1)] of a basic contrast $c_i'\tau$, for $i = \rho+1, \rho+2, ..., h$. An approximate measure of this gain follows from the ratio of precisions

$$\frac{\text{Var}[(\widehat{c_i'\tau})_1]}{\text{Var}(\widetilde{c_i'\tau})} \cong \frac{\varepsilon_{1i}^{-1}}{\varepsilon_{1i}^{-1} w_{1i}(1+\zeta_i)} = 1 + \frac{1 - w_{1i}(1+\zeta_i)}{w_{1i}(1+\zeta_i)},$$

which implies that this gain is approximately equal to

$$\frac{1 - w_{1i}(1+\zeta_i)}{w_{1i}(1+\zeta_i)}, \tag{3.8.82}$$

not to $(1 - w_{1i})/w_{1i}$, as in the case of using the BLUE (3.8.19) with the known values of the weights w_{1i} and $w_{2i} = 1 - w_{1i}$, i.e., when the true value of the ratio $\delta = \sigma_2^2/\sigma_1^2$ is known.

It may be interesting to note with regard to the gain in precision given in (3.8.82) that it is positive if and only if $\zeta_i < (1 - w_{1i})/w_{1i} = w_{2i}/w_{1i}$. In particular, it can easily be proved that for a BIB design with $v > 3$, this condition, $\zeta < (1 - w_1)/w_1$, is satisfied uniformly for all values of w_1 if and only if

$$\frac{(v-1)^2}{v(v-3)} < \frac{r}{k}(k-1) \quad \text{or, equivalently,} \quad \frac{k(v-1)}{v(v-3)} < \lambda, \tag{3.8.83}$$

where $\lambda = r(k-1)/(v-1)$ (see Definition 2.4.2). It should be mentioned here that (as will be shown in Theorem 6.0.3, Volume II) the condition (3.8.83) is satisfied by all BIB designs with $v > 3$, except two cases. In fact, among the BIB designs listed by Raghavarao (1971, Table 5.10.1), only for two cases, Series 1 and 2, the inequality (3.8.83) does not hold.

In general, for a practical application, it would be advisable, before deciding whether to use the combined estimator (3.8.81) or rather the corresponding intra-block BLUE $(\widehat{c_i'\tau})_1$, to check the condition

$$\hat{\zeta}_i < \frac{\hat{w}_{2i}}{\hat{w}_{1i}} \quad (\text{with } \hat{w}_{2i} = 1 - \hat{w}_{1i}), \tag{3.8.84}$$

where $\hat{\zeta}_i = 2\hat{w}_{1i}\hat{w}_{2i}/\sum_{\ell=1}^{n-v}(\hat{\omega}_\ell - \hat{\omega}_.)^2$, with $\hat{\omega}_\ell$ and $\hat{\omega}_.$ defined as in (3.8.74) and (3.8.75), respectively, but with w_{1i} replaced by $\hat{w}_{1i}$. If the condition (3.8.84) is satisfied, i.e., $\hat{w}_{1i}(1+\hat{\zeta}_i) < 1$, then, with the same estimate of σ_1^2, the estimated approximate variance (3.8.77) becomes smaller than the estimated variance of the intra-block BLUE, i.e., a real positive gain is achieved from using (3.8.81) instead of the intra-block estimator. On the contrary, if $\hat{\zeta}_i \geq \hat{w}_{2i}/\hat{w}_{1i}$, then it is better to replace $\hat{w}_{1i}$ by 1 and, hence, $\hat{w}_{2i}$ by 0 in (3.8.81).

As already mentioned at the end of Section 3.7, not much is known at present about the closeness of the approximation (3.8.72) of the variance (3.8.54). In fact, only some comparative results concerning certain BIB designs have been published until now. Those obtained by Khatri and Shah (1974, 1975) and Kackar and Harville (1984) are shown in Table 3.2. Column 3 gives the variances of the combined estimators of elementary contrasts (i.e., paired treatment contrasts) obtainable with weights based on known variance ratio δ. These variances, as given in (3.8.23), are not inflated by the estimation of δ and, therefore, are called (after Khatri and Shah, 1975) optimum. Column 4 gives exact variances of the empirical combined estimators obtainable with weights based on δ estimated by the MML (REML) procedure, i.e., variances given in (3.8.54). Because, as already indicated by Nelder (1968, p. 309), the MML procedure coincides for any symmetric BIB design with the classical procedure introduced by Yates (1940b) (its description in terms of basic contrasts will be seen in Section 3.9.1), the generally intractable formula (3.8.54) can be replaced by the relevant formula for the Yates procedure, if available. The exact variances given in this column for the considered symmetric BIB designs are from Khatri and Shah (1974). For the asymmetric BIB designs considered here, the exact variances are from Kackar and Harville (1984), who have determined them by simulation (thus, they may be slightly different from the unavailable true exact variances). In column 5, the corresponding approximate variances are given, as found from formula (3.8.72). Finally, column 6 gives the variances of the empirical combined estimators based on δ estimated by the Yates procedure. Certainly, the entries in columns 4 and 6 coincide for the symmetric BIB designs. For the last asymmetric design (with $b = 12$ and $v = 9$), the variances for the Yates procedure are presented in two versions, one as obtained by Kackar and Harville (1984) from simulation, another as obtained by Khatri and Shah (1975) by using an exact formula. It may be added that for the purpose of this table, i.e., for estimating elementary contrasts in BIB designs, formula (3.8.23) reduces to

$$\mathrm{Var}(\widehat{\mathbf{c}'\boldsymbol{\tau}}) = \frac{2\sigma_1^2}{r\varepsilon_1} w_1,$$

and formula (3.8.72) reduces to

$$\mathrm{Var}(\widetilde{\mathbf{c}'\boldsymbol{\tau}}) \cong \frac{2\sigma_1^2}{r\varepsilon_1} w_1 (1 + \zeta), \tag{3.8.85}$$

with ζ given in (3.8.79) and w_1 in (3.8.80). On the other hand, the intra-block variance (3.3.9) is here reduced to

$$\mathrm{Var}[(\widehat{\mathbf{c}'\boldsymbol{\tau}})_{\text{intra}}] = \frac{2\sigma_1^2}{r\varepsilon_1},$$

its values for the considered designs being given in column 1.

Table 3.2
Exact and approximate variances of estimators of elementary contrasts in some BIB designs for the MML (REML) estimation procedure

Design parameters	$\delta = \sigma_2^2/\sigma_1^2$	Variance (at $\sigma_1^2 = 1$) obtained as Optimum	Exact	Approxim.	Yates	
$b = v = 5$	1	0.5000	0.5066[a]	0.5227	0.5066[a]	
$r = k = 4$		*(1.000)*	*(1.013)*	*(1.045)*	*(1.013)*	
$\lambda = 3$	3	0.5217	0.5261[a]	0.5296	0.5261[a]	
$\varepsilon_1 = 15/16 = 0.9375$		*(1.000)*	*(1.008)*	*(1.015)*	*(1.008)*	
$2/(r\varepsilon_1) = 0.5333$	5	0.5263	0.5316[a]	0.5311	0.5316[a]	
		(1.000)	*(1.010)*	*(1.009)*	*(1.010)*	
$b = v = 7$	1	0.5000	0.5098[b]	0.5333	0.5098[b]	
$r = k = 4$		*(1.000)*	*(1.020)*	*(1.067)*	*(1.020)*	
$\lambda = 2$	3	0.5455	0.5554[b]	0.5576	0.5554[b]	
$\varepsilon_1 = 7/8 = 0.8750$		*(1.000)*	*(1.018)*	*(1.022)*	*(1.018)*	
$2/(r\varepsilon_1) = 0.5714$	5	0.5556	0.5639[b]	0.5630	0.5639[b]	
		(1.000)	*(1.015)*	*(1.013)*	*(1.015)*	
$b = 15, v = 6$	1	0.4000	0.4174[a]	0.4400	0.4198[a]	
$r = 5, k = 2$		*(1.000)*	*(1.044)*	*(1.100)*	*(1.050)*	
$\lambda = 1$	3	0.5455	0.5732[a]	0.5766	0.5698[a]	
$\varepsilon_1 = 3/5 = 0.6000$		*(1.000)*	*(1.051)*	*(1.057)*	*(1.045)*	
$2/(r\varepsilon_1) = 0.6667$	5	0.5882	0.6134[a]	0.6109	0.6108[a]	
		(1.000)	*(1.043)*	*(1.038)*	*(1.038)*	
$b = 12, v = 9$	1	0.5000	0.5180[a]	0.5417	0.5184[a]	[0.5155[c]]
$r = 4, k = 3$		*(1.000)*	*(1.036)*	*(1.083)*	*(1.037)*	*(1.031)*
$\lambda = 1$	3	0.6000	0.6170[a]	0.6192	0.6164[a]	[0.6197[c]]
$\varepsilon_1 = 3/4 = 0.7500$		*(1.000)*	*(1.028)*	*(1.032)*	*(1.027)*	*(1.033)*
$2/(r\varepsilon_1) = 0.6667$	5	0.6250	0.6374[a]	0.6373	0.6372[a]	[0.6403[c]]
		(1.000)	*(1.020)*	*(1.020)*	*(1.020)*	*(1.025)*
$b = v = 15$	1	0.2857	0.2872[b]	0.2900	0.2872[b]	
$r = k = 7$		*(1.000)*	*(1.005)*	*(1.015)*	*(1.005)*	
$\lambda = 3$	3	0.3022	0.3038[b]	0.3037	0.3038[b]	
$\varepsilon_1 = 45/49 = 0.9184$		*(1.000)*	*(1.005)*	*(1.005)*	*(1.005)*	
$2/(r\varepsilon_1) = 0.3111$	5	0.3057	0.3067[b]	0.3066	0.3067[b]	
		(1.000)	*(1.003)*	*(1.003)*	*(1.003)*	

Source: Calculated from results obtained by:
[a] Kackar and Harville (1984), from simulation studies,
[b] Khatri and Shah (1974), from exact numerical computations,
[c] Khatri and Shah (1975), from exact numerical computations.

The results presented in Table 3.2 reveal that the approximated variance (3.8.85) exceeds the optimum variance by at most 10%, wheras the exact variance does it by about half of this or less. The largest discrepancies between these two percentages are at $\delta = 1$, ranging from 1% to 5.6%. They depend on two factors, the size of the experiment, more precisely, the difference $n - v$, and the efficiency factor ε_1. Discrepancies that are not negligible appear for the first four designs, where $n - v \leq 27$, the largest being at the smallest efficiency, $\varepsilon_1 = 0.6$. For the last design, for which $n - v = 90$ and $\varepsilon_1 = 0.9184$, the discrepancy at $\delta = 1$ is 1%. Notably, at $\delta \geq 3$, the discrepancies are negligible.

Thus, it can be concluded that the approximation used here performs well for sufficiently large experiments, with large $n-v$, particularly if the chosen design is of high intra-block efficiency and a successful grouping of units is accomplished, securing a possibly high value of $\delta = \sigma_2^2/\sigma_1^2$. How much this conclusion drawn from results obtained for a few BIB designs can be generalized for other block designs is difficult to say. Anyway, one can expect that high intra-block efficiency and successful blocking will always make the empirical estimator (3.8.53) have variance close to optimum.

3.8.5 Testing hypotheses in proper block designs

In Section 3.7.4, some theory of testing hypotheses applicable to a general block design has been presented. It may be interesting to reconsider it from the point of view of its application to proper block designs.

The problem of testing hypotheses will be discussed here under the usual assumption that the observed vector $\boldsymbol{y}$ has a multivariate normal distribution, with the parameters as in (3.1.15) and (3.1.16), the latter receiving the form (3.5.9) for a proper block design.

As it is known from Section 3.2.1, in the intra-block analysis, the null hypothesis H_{01} : $\mathrm{E}(\boldsymbol{y}_1) = \mathbf{0}$, i.e., the hypothesis H_{01} : $\boldsymbol{\Delta}'\boldsymbol{\tau} = \boldsymbol{D}'\boldsymbol{k}^{-\delta}\boldsymbol{D}\boldsymbol{\Delta}'\boldsymbol{\tau}$, can be tested using the statistic

$$F_1 = h^{-1}\boldsymbol{Q}_1'\boldsymbol{C}_1^-\boldsymbol{Q}_1/s_1^2, \tag{3.8.86}$$

where $s_1^2 = \boldsymbol{y}'\boldsymbol{\psi}_1\boldsymbol{y}/d_1$ and $d_1 = n-b-h$. The statistic (3.8.86) has a noncentral F distribution with h and d_1 d.f. and the noncentrality parameter $\delta_1 = \boldsymbol{\tau}'\boldsymbol{C}_1\boldsymbol{\tau}/\sigma_1^2$. The distribution becomes central when H_{01} is true. Note that the hypothesis H_{01} means that the treatment parameters are constant within each block, though they may differ from block to block, depending on treatments received by the units within the block. On the other hand, as it follows from Section 3.2.2, in the inter-block analysis of a proper block design, one can test the null hypothesis H_{02} : $\mathrm{E}(\boldsymbol{y}_2) = \mathbf{0}$, i.e., the hypothesis H_{02} : $\boldsymbol{D}'\boldsymbol{k}^{-\delta}\boldsymbol{D}\boldsymbol{\Delta}'\boldsymbol{\tau} = n^{-1}\boldsymbol{r}'\boldsymbol{\tau}\mathbf{1}_n$, with the use of the statistic

$$F_2 = (v-1-\rho)^{-1}\boldsymbol{Q}_2'\boldsymbol{C}_2^-\boldsymbol{Q}_2/s_2^2, \tag{3.8.87}$$

where $s_2^2 = \boldsymbol{y}'\boldsymbol{\psi}_2\boldsymbol{y}/d_2$ and $d_2 = b - v - \rho$. The statistic (3.8.87) has a noncentral F distribution with $v - 1 - \rho$ and d_2 d.f. and the noncentrality parameter $\delta_2 = \boldsymbol{\tau}'\boldsymbol{C}_2\boldsymbol{\tau}/\sigma_2^2$. The distribution becomes central when H_{02} is true. It is to be noted that the hypothesis H_{02} means that the within block average of treatment parameters is the same from block to block.

Thus, if H_0 denotes the hypothesis that the treatment parameters are constant within the whole experiment, i.e., H_0 : $\boldsymbol{\Delta}'\boldsymbol{\tau} = n^{-1}\boldsymbol{r}'\boldsymbol{\tau}\mathbf{1}_n$, then it can be considered as the intersection of H_{01} and H_{02}, which can be written as

$$H_0 = H_{01} \cap H_{02}, \tag{3.8.88}$$

showing that H_0 can be accepted if and only if both H_{01} and H_{02} are accepted. On the contrary, it is to be rejected if and only if at least one of the subhypotheses H_{01} or H_{02} is rejected.

To obtain an appropriate F test for testing H_0, it is essential first to note [see Sections 3.2.1 and 3.2.2, and formula (3.5.8)] that not only $\boldsymbol{Q}'_\alpha \boldsymbol{C}^-_\alpha \boldsymbol{Q}_\alpha$ is distributed independently of $\boldsymbol{y}'\boldsymbol{\psi}_\alpha \boldsymbol{y}$, for a given α, but also all four quadratic forms, for $\alpha = 1, 2$, are mutually independent. (The index α should not be confused with the significance level α.) Hence, the test statistics (3.8.86) and (3.8.87) have independent distributions, which allows the following procedure to be used.

If the hypothesis (3.8.88) is to be tested at the significance level α, choose α_1 and α_2 such that $(1-\alpha_1)(1-\alpha_2) = 1-\alpha$. Test the intra-block hypothesis H_{01} with (3.8.86) at the significance level α_1, i.e., compare F_1 with the $100\alpha_1$ percentage point of the F distribution with h and d_1 d.f., and test the inter-block hypothesis H_{02} with (3.8.87) at the significance level α_2, i.e., compare F_2 with the $100\alpha_2$ percentage point of the F distribution with $v-1-\rho$ and d_2 d.f. Reject H_0 at level α if and only if H_{01} is rejected at level α_1 or H_{02} is rejected at α_2, or both these events happen. Usually, it will be convenient and justified to take $\alpha_1 = \alpha_2 = \alpha_*$ (say), i.e., to choose $\alpha_* = 1-(1-\alpha)^{1/2}$ for each subtest. However, if from an *a priori* knowledge it can be expected that one of the two subhypotheses in (3.8.88) is more likely to be false than is the other, it may be wise to assign a lower significance level to the former and a higher to the latter. For BIB designs, an exact method of combining the two tests has been given by Cohen and Sackrowitz (1989).

Now, if H_0 is rejected, then the researcher may wish to find which contrasts in $\boldsymbol{\tau}$ are responsible for that. It is instructive in relation to this question to note that if $\{\boldsymbol{c}'_i\boldsymbol{\tau}, i = 1, 2, ..., v-1\}$ is a set of basic contrasts of the design, then the relevant treatment sums of squares can be written as

$$\boldsymbol{Q}'_1 \boldsymbol{C}^-_1 \boldsymbol{Q}_1 = \sum_{i=1}^{h} \varepsilon_{1i} (\widehat{\boldsymbol{c}'_i\boldsymbol{\tau}})^2_1 \tag{3.8.89}$$

for the F_1 statistic and

$$\boldsymbol{Q}'_2 \boldsymbol{C}^-_2 \boldsymbol{Q}_2 = \sum_{i=\rho+1}^{v-1} \varepsilon_{2i} (\widehat{\boldsymbol{c}'_i\boldsymbol{\tau}})^2_2 \tag{3.8.90}$$

for the F_2 statistic, where $(\widehat{\boldsymbol{c}'_i\boldsymbol{\tau}})_1$ and $(\widehat{\boldsymbol{c}'_i\boldsymbol{\tau}})_2$ are obtainable by (3.6.1) for $\alpha = 1$ and 2, respectively. Moreover, it can easily be checked that the components in (3.8.89) and (3.8.90), the former when divided by σ_1^2 and the latter when divided by σ_2^2, have independent noncentral χ^2 distributions with 1 d.f. each, which can be written as

$$\varepsilon_{1i}(\widehat{c_i'\tau})_1^2/\sigma_1^2 \sim \chi_1'^2(\delta_{1i}) \quad \text{with} \quad \delta_{1i} = \varepsilon_{1i}(c_i'\tau)^2/\sigma_1^2,$$

for $i = 1, 2, ..., h$, and

$$\varepsilon_{2i}(\widehat{c_i'\tau})_2^2/\sigma_2^2 \sim \chi_1'^2(\delta_{2i}) \quad \text{with} \quad \delta_{2i} = \varepsilon_{2i}(c_i'\tau)^2/\sigma_1^2,$$

for $i = \rho+1, \rho+2, ..., v-1$. These distributions allow the corresponding hypotheses on the nullity of individual basic contrasts to be tested as follows:

$$H_{01,i} : c_i'\tau = 0 \quad \text{with} \quad F_{1,i} = \varepsilon_{1i}(\widehat{c_i'\tau})_1^2/s_1^2, \tag{3.8.91}$$

which has a noncentral F distribution with 1 and d_1 d.f. and the noncentrality parameter δ_{1i}, for $i = 1, 2, ..., h$,

$$H_{02,i} : c_i'\tau = 0 \quad \text{with} \quad F_{2,i} = \varepsilon_{2i}(\widehat{c_i'\tau})_2^2/s_2^2, \tag{3.8.92}$$

which has a noncentral F distribution with 1 and d_2 d.f. and the noncentrality parameter δ_{2i}, for $i = \rho+1, \rho+2, ..., v-1$.

Note that for $i = 1, 2, ..., \rho$, and $h+1, h+2, ..., v-1$, only one hypothesis can be tested, either $H_{01,i}$, for $i \leq \rho$, or $H_{02,i}$, for $i > h$. But for $i = \rho+1, \rho+2, ..., h$, the two hypotheses $H_{01,i}$ and $H_{02,i}$ coincide. If the rejection of the overall null hypothesis H_0 is implied by the rejection of one of the two subhypotheses H_{01} or H_{02} only, the choice is simple. Having rejected H_{01}, one should test each of its component hypotheses, $H_{01,i}$, with the corresponding statistic, $F_{1,i}$, indicated in (3.8.91). If H_{02} is rejected, instead, each of its component hypotheses, $H_{02,i}$, is to be tested with the corresponding statistic, $F_{2,i}$, given in (3.8.92). The problem with the coinciding component hypotheses occurs when both H_{01} and H_{02} are rejected. In this case, it would be natural to use a single test statistic, combining the intra-block and inter-block information. (See also Mejza, 1985.)

To find such a test, suppose for a while that the ratio $\delta = \sigma_2^2/\sigma_1^2$ is known and recall, from Corollary 3.8.2 (c), that the BLUE of the basic contrast $c_i'\tau$ is then the combined estimator

$$\widehat{c_i'\tau} = w_{1i}(\widehat{c_i'\tau})_1 + w_{2i}(\widehat{c_i'\tau})_2, \tag{3.8.93}$$

with the $\text{Var}(\widehat{c_i'\tau}) = \sigma_1^2\varepsilon_{1i}^{-1}w_{1i}$, for $i = \rho+1, \rho+2, ..., h$. It may easily be found that the estimator (3.8.93) when squared and divided by its variance has a noncentral χ^2 distribution with 1 d.f., i.e.,

$$\frac{(\widehat{c_i'\tau})^2}{\sigma_1^2\varepsilon_{1i}^{-1}w_{1i}} \sim \chi_1'^2(\delta_i), \quad \text{where} \quad \delta_i = \frac{(c_i'\tau)^2}{\sigma_1^2\varepsilon_{1i}^{-1}w_{1i}}, \tag{3.8.94}$$

which provides a test of the component hypothesis $H_{0,i} : c_i'\tau = 0$ based on the statistic

$$w_{1i}^{-1}\varepsilon_{1i}(\widehat{c_i'\tau})^2/s_1^2. \tag{3.8.95}$$

From (3.8.94) it has a noncentral F distribution with 1 and d_1 d.f. and the noncentrality parameter δ_i. If the ratio of the stratum variances is not known and is to be estimated, then instead of the true weights w_{1i} and w_{2i} in (3.8.93), their estimates $\hat{w}_{1i}$ and $\hat{w}_{2i}$ [as in (3.8.35) and (3.8.36)] are to be used. This replacement leads to the empirical estimator

$$\widetilde{c_i'\tau} = \hat{w}_{1i}(\widehat{c_i'\tau})_1 + \hat{w}_{2i}(\widehat{c_i'\tau})_2 \quad \text{for} \quad i = \rho+1, \rho+2, ..., h, \tag{3.8.96}$$

considered in Sections 3.8.3 and 3.8.4 [formula (3.8.81)]. With the estimator (3.8.96), one can approximate the F statistic (3.8.95) by its empirical version

$$\tilde{F}_i = \hat{w}_{1i}^{-1}\varepsilon_{1i}(\widetilde{c_i'\tau})^2/s_1^2, \tag{3.8.97}$$

pretending that it has approximately the same distribution as (3.8.95). Properties of this approximation have been examined by Feingold (1985, 1988). On the basis of simulation studies, carried out for a wide variety of BIB and partially balanced incomplete block (PBIB) designs, for a range of values of the stratum variances and for several different variance estimators, she investigated the power of the proposed test based on the empirical statistic (3.8.97), the upper tail percentile points of this statistic and the first four moments of it. The conclusions of these investigations are that

(a) the test based on (3.8.97) is, in general, more powerful than is the usual intra-block test based on the statistic in (3.8.91);

(b) the upper quantiles of $\tilde{F}_i$, under the null hypothesis $H_{0,i} : c_i'\tau = 0$, are similar to those of a central F distribution with 1 and d_1 $(= n - b - h)$ d.f.;

(c) the moments of $\tilde{F}_i$, estimated for a wide range of alternative hypotheses, are remarkably close to those of the noncentral F distribution with 1 and d_1 d.f. and the corresponding noncentrality parameter.

It has also appeared that this close distributional correspondence of $\tilde{F}_i$ to an exact F statistic is independent of the type of design, whether balanced or partially balanced, and is also not depending on d_2 $(= b - v + \rho)$. All of these findings support the idea of using the approximated statistic (3.8.97) for testing the component hypothesis $H_{0,i}$ for $i = \rho+1, \rho+2, ..., h$ and refer it to the F distribution with 1 and d_1 d.f.

It should, however, be noticed that further improvement of the above approach is possible by taking into account the results of Section 3.8.4 and using them in accordance with the theory presented in Section 3.7.4. In particular, one has to remember that the variance of the empirical estimator (3.8.96) is larger than that of the BLUE (3.8.93), because of the estimation of the unknown variance ratio $\delta = \sigma_2^2/\sigma_1^2$. This inflation is shown in formula (3.8.77). Therefore, instead of the statistic (3.8.97), it is more justified to use the statistic

$$F_i = [\hat{w}_{1i}(1+\hat{\zeta}_i)]^{-1}\varepsilon_{1i}(\widetilde{c_i'\tau})^2/s_1^2, \tag{3.8.98}$$

where $\hat{\zeta}_i$ is the estimate of ζ_i defined in (3.8.78), obtained by replacing w_{1i} by $\hat{w}_{1i}$.

As it follows from Section 3.7.4, the unknown distribution of the statistic (3.8.98) can, under $H_{0,i}$, be approximated by the central F distribution with $d_{(i)}$ and d_1 d.f., where

$$d_{(i)} = \frac{2}{2 + 3\mathrm{Var}_{\mathrm{as}}(\hat{\delta})a_i^2}, \tag{3.8.99}$$

with

$$a_i = \frac{\sigma_1^2}{\mathrm{Var}_{\mathrm{ap}}(\widetilde{\boldsymbol{c}_i'\boldsymbol{\tau}})} \frac{(\varepsilon_{1i}\delta + \varepsilon_{2i})^2\varepsilon_{2i} - 3\varepsilon_{1i}^2\varepsilon_{2i}\mathrm{E}[(\hat{\delta} - \delta)^2]}{(\varepsilon_{1i}\delta + \varepsilon_{2i})^4}. \tag{3.8.100}$$

In fact, this result is obtained when considering the random variable

$$\sigma_1^{-2}\left\{\frac{\hat{\delta}}{\varepsilon_{1i}\hat{\delta} + \varepsilon_{2i}} + \mathrm{E}[(\hat{\delta} - \delta)^2]\frac{\varepsilon_{1i}\varepsilon_{2i}}{(\varepsilon_{1i}\hat{\delta} + \varepsilon_{2i})^3}\right\}^{-1}(\widetilde{\boldsymbol{c}_i'\boldsymbol{\tau}})^2 = Y_i \quad \text{(say)}$$

and approximating its expectation and variance, similarly as for (3.7.65). Under $H_{0,i}$, these approximations are

$$\mathrm{E}(Y_i) \cong 1 = \mathrm{E}_{\mathrm{ap}}(Y_i) \quad \text{(say)}$$

and

$$\mathrm{Var}(Y_i) \cong 2 + 3\mathrm{Var}_{\mathrm{as}}(\hat{\delta})\left\{\frac{\sigma_1^2}{\mathrm{Var}_{\mathrm{ap}}(\widetilde{\boldsymbol{c}_i'\boldsymbol{\tau}})} \frac{(\varepsilon_{1i}\delta + \varepsilon_{2i})^2\varepsilon_{2i} - 3\varepsilon_{1i}^2\varepsilon_{2i}\mathrm{E}[(\hat{\delta} - \delta)^2]}{(\varepsilon_{1i}\delta + \varepsilon_{2i})^4}\right\}$$

$$= \mathrm{Var}_{\mathrm{ap}}(Y_i) \quad \text{(say)},$$

giving, similarly to (3.7.74),

$$d_{(i)} = \frac{2[\mathrm{E}_{\mathrm{ap}}(Y_i)]^2}{\mathrm{Var}_{\mathrm{ap}}(Y_i)},$$

i.e., (3.8.99) with a_i as in (3.8.100). The latter simplifies to

$$a_i = \frac{w_{2i} - 3w_{1i}\zeta_i}{\delta(1 + \zeta_i)}$$

when using instead of the unknown MSE of $\hat{\delta}$, $\mathrm{E}[(\hat{\delta} - \delta)^2]$, its approximation by $\mathrm{Var}_{\mathrm{as}}(\hat{\delta})$, which according to the results and notation in Section 3.8.4 can be written as

$$\mathrm{Var}_{\mathrm{as}}(\hat{\delta}) = \frac{\delta^2\zeta_i}{w_{1i}w_{2i}},$$

with ζ_i defined in (3.8.78). Using it also in (3.8.99), the $d_{(i)}$ can be written as

$$d_{(i)} = \frac{2}{2 + \dfrac{3\zeta_i(w_{2i} - 3w_{1i}\zeta_i)^2}{(1 + \zeta_i)^2 w_{1i}w_{2i}}}. \tag{3.8.101}$$

If a general hypothesis is to be tested, i.e., $H_0 : \boldsymbol{U}'\boldsymbol{\tau} = \boldsymbol{q}$, then the same procedure as in Section 3.7.4 can be applied. Note, however, that the matrix $\boldsymbol{A}$ defined in (3.7.72) can be written for a proper block design with constant block size k in the form

$$\boldsymbol{A} = k\boldsymbol{W}^{-1}\boldsymbol{U}'\boldsymbol{r}^{-\delta}\left\{\sum_{\beta=1}^{m-1}\frac{(\varepsilon_{1\beta}\delta+\varepsilon_{2\beta})^2\varepsilon_{2\beta} - 3\varepsilon_{1\beta}^2\varepsilon_{2\beta}\mathrm{E}[(\hat{\delta}-\delta)^2]}{(\varepsilon_{1\beta}\delta+\varepsilon_{2\beta})^4}\sum_{j=1}^{\rho_\beta}\boldsymbol{c}_{\beta j}\boldsymbol{c}_{\beta j}' + \sum_{j=1}^{\rho_m}\boldsymbol{c}_{mj}\boldsymbol{c}_{mj}'\right\}\boldsymbol{r}^{-\delta}\boldsymbol{U}, \tag{3.8.102}$$

where $\boldsymbol{W} = \boldsymbol{U}'\boldsymbol{C}_{\mathrm{ap}}^{-1}\boldsymbol{U}$ can be written, correspondingly to (3.8.70), as

$$\boldsymbol{W} = \sigma_1^{-2}\boldsymbol{U}'\mathrm{Cov}(\hat{\boldsymbol{\tau}})\boldsymbol{U} + \mathrm{E}[(\hat{\delta}-\delta)^2]\boldsymbol{U}'\boldsymbol{r}^{-\delta}\sum_{\beta=1}^{m-1}\frac{\varepsilon_{1\beta}\varepsilon_{2\beta}}{(\varepsilon_{1\beta}\delta+\varepsilon_{2\beta})^3}\sum_{j=1}^{\rho_\beta}\boldsymbol{c}_{\beta j}\boldsymbol{c}_{\beta j}'\boldsymbol{r}^{-\delta}\boldsymbol{U}, \tag{3.8.103}$$

because of the relation $k^{1/2}\varepsilon_{2\beta}^{1/2}\boldsymbol{c}_{\beta j} = \boldsymbol{N}\boldsymbol{u}_{\beta j}$ holding for all eigenvectors $\{\boldsymbol{u}_{\beta j}\}$ in (3.7.35) that correspond to the eigenvalue $\lambda_\beta = k\varepsilon_{1\beta}$, of multiplicity ρ_β, $\beta = 1, 2, ..., m-1$. To see this relation, refer to Remark 2.3.1. Moreover, note that for a proper block design, the equality

$$\mathrm{Var}_{\mathrm{as}}(\hat{\gamma}) = k^{-2}\mathrm{Var}_{\mathrm{as}}(\hat{\delta})$$

holds, on account of the relation $1 + k\gamma = \delta$. Thus, when using the statistic (3.7.75) for a proper block design, the d.f. d of the approximating F distribution, given in (3.7.74), can be calculated using the equivalent formula

$$d \doteq \frac{2\ell^2}{2\ell + \mathrm{Var}_{\mathrm{as}}(\hat{\delta})\{2\mathrm{tr}(\bar{\boldsymbol{A}}\bar{\boldsymbol{A}}) + [\mathrm{tr}(\bar{\boldsymbol{A}})]^2\}},$$

where $\bar{\boldsymbol{A}} = k^{-1}\boldsymbol{A}$.

Finally, replacing in (3.8.102) and (3.8.103) the unknown $\mathrm{E}[(\hat{\delta}-\delta)^2]$ by $\mathrm{Var}_{\mathrm{as}}(\hat{\delta})$, the matrix $\bar{\boldsymbol{A}}$ will get the form

$$\bar{\boldsymbol{A}} = \boldsymbol{W}^{-1}\boldsymbol{U}'\boldsymbol{r}^{-\delta}\left[\sum_{\beta=1}^{m-1}\varepsilon_{2\beta}^{-1}w_{2\beta}(w_{2\beta} - 3w_{1\beta}\zeta_\beta)\sum_{j=1}^{\rho_\beta}\boldsymbol{c}_{\beta j}\boldsymbol{c}_{\beta j}' + \sum_{j=1}^{\rho_m}\boldsymbol{c}_{mj}\boldsymbol{c}_{mj}'\right]\boldsymbol{r}^{-\delta}\boldsymbol{U},$$

with

$$\boldsymbol{W} = \boldsymbol{U}'\boldsymbol{C}_{\mathrm{ap}}^{-1}\boldsymbol{U} = \boldsymbol{U}'\boldsymbol{r}^{-\delta}\left[\sum_{j=1}^{\rho_0}\boldsymbol{c}_{0j}\boldsymbol{c}_{0j}' + \sum_{\beta=1}^{m-1}\varepsilon_{1\beta}^{-1}w_{1\beta}(1+\zeta_\beta)\sum_{j=1}^{\rho_\beta}\boldsymbol{c}_{\beta j}\boldsymbol{c}_{\beta j}' + \delta\sum_{j=1}^{\rho_m}\boldsymbol{c}_{mj}\boldsymbol{c}_{mj}'\right]\boldsymbol{r}^{-\delta}\boldsymbol{U},$$

which corresponds to formula (3.8.76). If the design is connected, then the term $\delta\sum_{j=1}^{\rho_m}\boldsymbol{c}_{mj}\boldsymbol{c}_{mj}'$ disappears. Also note that the term $(\sigma_3^2/\sigma_1^2)\boldsymbol{c}_v\boldsymbol{c}_v'$ does not appear at all in the formulae above, which follows from the assumption that $\boldsymbol{U}'\boldsymbol{1}_v = \boldsymbol{0}$.

Example 3.8.1. Ceranka, Mejza and Wiśniewski (1979) have analyzed data from a plant-breeding field experiment with 12 sunflower strains compared in a block design based on the incidence matrix

$$
N = \begin{bmatrix}
1 & 0 & 0 & 0 & 0 & 0 & 0 & 0 & 0 & 1 & 0 & 0 & 0 & 0 & 0 & 0 & 0 & 0 \\
0 & 1 & 0 & 0 & 0 & 0 & 0 & 0 & 0 & 0 & 1 & 0 & 0 & 0 & 0 & 0 & 0 & 0 \\
0 & 0 & 1 & 0 & 0 & 0 & 0 & 0 & 0 & 0 & 0 & 1 & 0 & 0 & 0 & 0 & 0 & 0 \\
0 & 0 & 0 & 1 & 0 & 0 & 0 & 0 & 0 & 0 & 0 & 0 & 1 & 0 & 0 & 0 & 0 & 0 \\
0 & 0 & 0 & 0 & 1 & 0 & 0 & 0 & 0 & 0 & 0 & 0 & 0 & 1 & 0 & 0 & 0 & 0 \\
0 & 0 & 0 & 0 & 0 & 1 & 0 & 0 & 0 & 0 & 0 & 0 & 0 & 0 & 1 & 0 & 0 & 0 \\
0 & 0 & 0 & 0 & 0 & 0 & 1 & 0 & 0 & 0 & 0 & 0 & 0 & 0 & 0 & 1 & 0 & 0 \\
0 & 0 & 0 & 0 & 0 & 0 & 0 & 1 & 0 & 0 & 0 & 0 & 0 & 0 & 0 & 0 & 1 & 0 \\
0 & 0 & 0 & 0 & 0 & 0 & 0 & 0 & 1 & 0 & 0 & 0 & 0 & 0 & 0 & 0 & 0 & 1 \\
2 & 2 & 2 & 0 & 0 & 0 & 0 & 0 & 0 & 2 & 2 & 2 & 0 & 0 & 0 & 0 & 0 & 0 \\
0 & 0 & 0 & 2 & 2 & 2 & 0 & 0 & 0 & 0 & 0 & 0 & 2 & 2 & 2 & 0 & 0 & 0 \\
0 & 0 & 0 & 0 & 0 & 0 & 2 & 2 & 2 & 0 & 0 & 0 & 0 & 0 & 0 & 2 & 2 & 2
\end{bmatrix}.
$$

Evidently, this matrix represents a nonbinary, proper, nonequireplicate and disconnected block design, with $v_1 = 9$ treatments (here new strains) replicated $r_1 = 2$ times and $v_2 = 3$ treatments (standard strains) replicated $r_2 = 12$ times, each of the $b = 18$ blocks being of size $k = 3$. It is assumed that the randomizations of blocks and of plots within the blocks have been implemented according to the procedure described in Section 3.1.1. The plant trait observed on the experimental units has been again (as in Example 3.2.1) the average head diameter in centimeters. The individual plot observations are as follows:

Block	Strain	Observ.	Block	Strain	Observ.	Block	Strain	Observ.
1	1	12.3	7	7	18.4	13	4	18.5
1	10	14.5	7	12	15.5	13	11	18.8
1	10	15.0	7	12	15.8	13	11	18.7
2	2	14.5	8	8	12.5	14	5	13.0
2	10	15.0	8	12	16.0	14	11	19.2
2	10	14.8	8	12	16.1	14	11	19.0
3	3	16.4	9	9	19.5	15	6	19.0
3	10	15.1	9	12	15.8	15	11	18.0
3	10	15.2	9	12	16.0	15	11	18.3
4	4	18.4	10	1	12.0	16	7	19.5
4	11	18.8	10	10	15.1	16	12	14.9
4	11	18.5	10	10	15.3	16	12	15.2
5	5	14.5	11	2	13.5	17	8	12.0
5	11	18.8	11	10	14.8	17	12	15.0
5	11	18.7	11	10	15.0	17	12	15.5
6	6	19.6	12	3	17.4	18	9	19.0
6	11	19.0	12	10	14.5	18	12	15.6
6	11	18.9	12	10	14.9	18	12	15.9

To perform the intra-block and the inter-block analysis, first note that because the design is proper, its matrix C_1 can simply be obtained [see (3.5.15)] as

$$C_1 = \begin{bmatrix} C_{1,11} & C_{1,12} \\ C_{1,21} & C_{1,22} \end{bmatrix},$$

with

$$C_{1,11} = \frac{4}{3} I_9, \quad C_{1,22} = 4I_3,$$

$$C_{1,12} = -\frac{4}{3}\begin{bmatrix} 1_3 & 0 & 0 \\ 0 & 1_3 & 0 \\ 0 & 0 & 1_3 \end{bmatrix} \quad \text{and} \quad C_{1,21} = -\frac{4}{3}\begin{bmatrix} 1_3' & 0' & 0' \\ 0' & 1_3' & 0' \\ 0' & 0' & 1_3' \end{bmatrix}.$$

A suitable choice of a g-inverse of C_1 is the 12×12 matrix

$$\frac{3}{4}\begin{bmatrix} I_9 & O \\ O & O \end{bmatrix},$$

which has the same rank as the matrix C_1, $h = 9$. Furthermore, note that here $Q_1 = \Delta y - k^{-1} N D y$ is obtained as

$$Q_1 = [-3.76667, \ -1.20000, \ 2.63333, \ -0.33333, \ -6.90000, \ 1.00000, \\ 4.80000, \ -4.53333, \ 4.56667, \ 2.33333, \ 6.23333, \ -4.83333]'.$$

From these results, the intra-block analysis of variance (Section 3.2.1) follows as

Source	Degrees of freedom	Sum of squares	Mean square	F
Treatments	$h = 9$	$Q_1' C_1^- Q_1 = 101.7967$	11.3107	74.67
Residuals	$n - b - h = 27$	$y'\psi_1 y = 4.0900$	$s_1^2 = 0.1515$	
Total	$n - b = 36$	$y'\phi_1 y = 105.8867$		

For this value of F, the corresponding P value is lower than 10^{-6}.

As to the inter-block analysis, one needs the matrix C_2 and the vector Q_2. The first is obtained [see (3.5.16)] as

$$C_2 = \begin{bmatrix} C_{2,11} & C_{2,12} \\ C_{2,21} & C_{2,22} \end{bmatrix},$$

with

$$C_{2,11} = \frac{2}{3}\left(I_9 - \frac{1}{9} 1_9 1_9'\right), \qquad C_{2,12} = \frac{4}{3}\left(I_3 - \frac{1}{3} 1_3 1_3'\right) \otimes 1_3,$$

$$C_{2,21} = \frac{4}{3}\left(I_3 - \frac{1}{3} 1_3 1_3'\right) \otimes 1_3' \quad \text{and} \quad C_{2,22} = 8\left(I_3 - \frac{1}{3} 1_3 1_3'\right).$$

It appears that a possible simple choice of a g-inverse of $\boldsymbol{C}_2$ is the 12×12 matrix

$$\frac{3}{2}\begin{bmatrix} \boldsymbol{I}_9 & \mathbf{O} \\ \mathbf{O} & \mathbf{O} \end{bmatrix}.$$

Also note that the rank of $\boldsymbol{C}_2$ is 8 and that here $\boldsymbol{Q}_2 = k^{-1}\boldsymbol{N}\boldsymbol{D}\boldsymbol{y} - n^{-1}\boldsymbol{r}\mathbf{1}_n'\boldsymbol{y}$ is obtained as

$$\boldsymbol{Q}_2 = [-4.57037,\ -3.43704,\ -1.47037,\quad 4.59630,\ 1.76296,\ 4.96296,\ 0.46296,\ -3.60370,\ 1.29630,\ -18.95556,\ 22.64444,\ -3.68889]'.$$

These results yield the inter-block analysis of variance (Section 3.2.2), which can be presented as

Source	Degrees of freedom	Sum of squares	Mean square	F
Treatments	$v - \rho - 1 = 8$	$\boldsymbol{Q}_2'\boldsymbol{C}_2^-\boldsymbol{Q}_2 = 147.9148$	18.4894	83.20
Residuals	$b - v + \rho = 9$	$\boldsymbol{y}'\boldsymbol{\psi}_2\boldsymbol{y} = 2.0000$	$s_2^2 = 0.2222$	
Total	$b - 1 = 17$	$\boldsymbol{y}'\boldsymbol{\phi}_2\boldsymbol{y} = 149.9148$		

For the resulting value of F, the corresponding P value is here also below 10^{-6}, as in the intra-block analysis.

The results of the F test followed from the intra-block and the inter-block analysis show that both hypotheses H_{01} and H_{02} considered in connection with (3.8.86) and (3.8.87), respectively, are to be rejected for the present example. It follows then that the intersection hypothesis H_0 defined in (3.8.88) is also to be rejected, at a low significance level [smaller than $2(10^{-6})$]. Thus, it can be concluded that the treatment (strain) parameters are almost surely not constant within the whole experiment.

Now, after rejecting H_0, it may be interesting to ask which contrasts in the treatment parameters $\boldsymbol{\tau}$ are responsible for the rejection. To answer this question, it will be useful to choose a set of basic contrasts, estimate them as described in Sections 3.5 and 3.6 and then proceed as shown above (in Section 3.8.5).

Suppose that in view of the fact that the set of the first nine treatments (new strains) is divided into three subsets, of three members each, and that the remaining three treatments (standard strains) are included for comparisons, each of them to be compared with a different subset, the researcher chooses basic contrasts represented by the following vectors:

$$\begin{aligned} \boldsymbol{s}_1 &= [-2,-2,-2,\ 0,\ 0,\ 0,\ 0,\ 0,\ 0,\ 1,\ 0,\ 0]'/\sqrt{36}, \\ \boldsymbol{s}_2 &= [\ 0,\ 0,\ 0,-2,-2,-2,\ 0,\ 0,\ 0,\ 0,\ 1,\ 0]'/\sqrt{36}, \\ \boldsymbol{s}_3 &= [\ 0,\ 0,\ 0,\ 0,\ 0,\ 0,-2,-2,-2,\ 0,\ 0,\ 1]'/\sqrt{36}, \end{aligned}$$

$$\begin{aligned}
s_4 &= [-1,-1,\ 2,\ 0,\ 0,\ 0,\ 0,\ 0,\ 0,\ 0,\ 0,\ 0]'/\sqrt{12},\\
s_5 &= [-1,\ 1,\ 0,\ 0,\ 0,\ 0,\ 0,\ 0,\ 0,\ 0,\ 0,\ 0]'/\sqrt{4},\\
s_6 &= [\ 0,\ 0,\ 0,-1,-1,\ 2,\ 0,\ 0,\ 0,\ 0,\ 0,\ 0]'/\sqrt{12},\\
s_7 &= [\ 0,\ 0,\ 0,-1,\ 1,\ 0,\ 0,\ 0,\ 0,\ 0,\ 0,\ 0]'/\sqrt{4},\\
s_8 &= [\ 0,\ 0,\ 0,\ 0,\ 0,\ 0,-1,-1,\ 2,\ 0,\ 0,\ 0]'/\sqrt{12},\\
s_9 &= [\ 0,\ 0,\ 0,\ 0,\ 0,\ 0,-1,\ 1,\ 0,\ 0,\ 0,\ 0]'/\sqrt{4},\\
s_{10} &= [-1,-1,-1,-1,-1,-1,\ 2,\ 2,\ 2,-1,-1,\ 2]'/\sqrt{108},\\
s_{11} &= [-1,-1,-1,\ 1,\ 1,\ 1,\ 0,\ 0,\ 0,-1,\ 1,\ 0]'/\sqrt{36}.
\end{aligned}$$

It can easily be checked that

$$\begin{aligned}
C_1 s_i &= r^\delta s_i && \text{for } i = 1,\ 2,\ 3,\\
C_1 s_i &= \tfrac{2}{3} r^\delta s_i && \text{for } i = 4,\ 5,\ 6,\ 7,\ 8,\ 9,\\
C_1 s_i &= \mathbf{0} && \text{for } i = 10,\ 11,
\end{aligned}$$

and that $s_i' r^\delta s_{i'} = \delta_{ii'}$ (the Kronecker delta) for $i, i' = 1, 2, ..., 12$, where $s_{12} = (54)^{-1/2}\mathbf{1}_v$, which confirms that the parametric functions $\{c_i\tau = s_i r^\delta \tau,\ i = 1, 2, ..., 11\}$ form a complete set of basic contrasts (see Definition 3.4.1).

Referring to Theorem 3.6.1, it is now possible to obtain for these contrasts the within stratum BLUEs and their variances. More precisely, the intra-block BLUEs and their variances are obtainable by the formulae

$$(\widehat{c_i'\tau})_1 = \varepsilon_{1i}^{-1} s_i' Q_1 \quad \text{and} \quad \mathrm{Var}[(\widehat{c_i'\tau})_1] = \varepsilon_{1i}^{-1}\sigma_1^2 \ \text{ for } i = 1, 2, ..., 9,$$

whereas the inter-block BLUEs and their variances can be obtained by the formulae

$$(\widehat{c_i'\tau})_2 = \varepsilon_{2i}^{-1} s' Q_2 \quad \text{and} \quad \mathrm{Var}[(\widehat{c_i'\tau})_2] = \varepsilon_{2i}^{-1}\sigma_2^2 \ \text{ for } i = 4, 5, ..., 11,$$

with $\varepsilon_{1i} = \varepsilon_i$ and $\varepsilon_{2i} = 1 - \varepsilon_i$. Here, $\varepsilon_i = 1$ for $i = 1, 2, 3$, $\varepsilon_i = 2/3$ for $i = 4, 5, ..., 9$ and $\varepsilon_i = 0$ for $i = 10, 11$. Unbiased estimates of the unknown values of σ_1^2 and σ_2^2 are provided by the intra-block residual mean square, s_1^2, and the inter-block residual mean square, s_2^2, respectively. Notably, the coefficients $\{\varepsilon_{1i},\ i = 1, 2, ..., 9\}$ are the efficiency factors of the analyzed design for the corresponding basic contrasts when estimated in the intra-block analysis and $\{\varepsilon_{2i},\ i = 4, 5, ..., 11\}$ are those for the corresponding basic contrasts when estimated in the inter-block analysis. In particular, they indicate that the unique BLUEs, the intra-block BLUEs for the first three basic contrasts and the inter-block BLUEs for the last two, are simultaneously the BLUEs of those contrasts under the overall model (3.1.1), as it follows from Theorem 3.5.2.

To test the hypotheses on the nullity of the individual basic contrasts considered here, the F test statistics given in (3.8.91) in relation to the intra-block analysis and those given in (3.8.92) in relation to the inter-block analysis are to be used. Evidently, for $i = 1, 2, 3$ and $i = 10, 11$, those F statistics are unique, obtainable in the intra-block analysis for $i = 1, 2, 3$ and in the inter-block analysis for $i = 10, 11$. Note that in any case, the F statistic is equal to the square of the relevant estimate of the contrast divided by its estimated variance. Results of the intra-block and the inter-block estimation and testing obtained for the considered basic contrasts are presented in Table 3.3. (The large number of decimal places is retained to allow the subsequent calculations to be checked.)

Table 3.3
Intra-block and inter-block estimates of the basic contrasts, together with their estimated variances and the relevant F statistics, obtained in Example 3.8.1

Basic contrast	Intra-block estimate	Estimated variance	Intra-block F	Inter-block estimate	Estimated variance	Inter-block F
$c_1'\tau$	1.166667	0.151481	8.99			
$c_2'\tau$	3.116667	0.151481	64.12			
$c_3'\tau$	−2.416667	0.151481	38.55			
$c_4'\tau$	4.431163	0.227222	86.41	4.387862	0.666667	28.89
$c_5'\tau$	1.925000	0.227222	16.31	1.700000	0.666667	4.33
$c_6'\tau$	3.998151	0.227222	70.35	3.088824	0.666667	14.31
$c_7'\tau$	−4.925000	0.227222	106.75	−4.250000	0.666667	27.09
$c_8'\tau$	3.839379	0.227222	64.87	4.965212	0.666667	36.98
$c_9'\tau$	−7.000000	0.227222	215.65	−6.100000	0.666667	55.81
$c_{10}'\tau$				−1.597336	0.222222	11.48
$c_{11}'\tau$				10.400000	0.222222	486.72
Critical values		$F_{0.05;1,27}$ $F_{0.01;1,27}$	= 4.21 = 7.68		$F_{0.05;1,9}$ $F_{0.01;1,9}$	= 5.12 = 10.56

Examining these results, first note that the equalities (3.8.89) and (3.8.90) are really satisfied by the estimates of the considered basic contrasts. It may also be checked that the average value of the intra-block individual F statistics is equal to the value of F_1 defined in (3.8.86), and the average value of the inter-block individual F statistics is equal to the value of F_2 in (3.8.87). An essential conclusion from Table 3.3 is that almost every value of the individual F statistics, whether from the intra-block or the inter-block analysis, exceeds the corresponding critical value for $\alpha = 0.01$. The only exception is the inter-block F test for the hypothesis $c_5'\tau = 0$, i.e., the hypothesis of no parametric difference between the strains 1 and 2.

Because for six of the considered basic contrasts two kinds of estimates and tests have been obtained, using the intra-block and the inter-block information separately, it might now be interesting to perform for them the combined estimation and testing, based on the information from both of these strata. As to the combined estimators, the formula (3.8.96) is applicable. To test the hypotheses $H_{0,i} : c_i'\tau = 0$ for $i = 4, 5, ..., 9$, the formula (3.8.98) can be applied.

These formulae, however, involve estimates of the weights w_{1i} and w_{2i} for each i (= 4, 5, ..., 9). To obtain them, it is necessary to estimate the variance ratio $\delta = \sigma_2^2/\sigma_1^2$, as described in Sections 3.8.2 and 3.8.3. Computationally, this task can be done by employing again the GENSTAT 5 (1996) software. Its REML estimation procedure has provided for the present example the following results:

Iteration cycle	1	2	3	4	5
$\hat{\gamma}$	0.074577	0.105173	0.102279	0.102537	the same
$\hat{\sigma_1^2}$	0.151093	0.148490	0.148809	0.148781	the same

From the final estimate of γ, an estimate of δ can be obtained from the relation $\delta = 1 + k\gamma$. Hence,

$$\hat{\delta} = 1 + 3\hat{\gamma} = 1.307611.$$

With this estimate, the estimated weights have been obtained as

$$\hat{w}_{1i} = \frac{\varepsilon_{1i}\hat{\delta}}{\varepsilon_{1i}\hat{\delta} + \varepsilon_{2i}} = \frac{(2/3)1.307611}{(2/3)1.307611 + 1/3} = 0.723392$$

and

$$\hat{w}_{2i} = 1 - \hat{w}_{1i} = 0.276608 \quad \text{for } i = 4, 5, ..., 9.$$

Now, with $\hat{\zeta}_i = 0.0506411$ (equal for $i = 4, 5, ..., 9$) obtained according to (3.8.78), it can be seen that the condition (3.8.84) is satisfied, as $\hat{w}_{2i}/\hat{w}_{1i} = 0.38238$. Hence, the estimated approximate variance (3.8.77) of the empirical combined estimator (3.8.81) [or (3.8.96)] can be expected to be smaller than is the estimated variance of the intra-block BLUE for any of the contrasts $\boldsymbol{c}_i'\boldsymbol{\tau}, i = 4, 5, ..., 9$. In fact, from (3.8.77), with $\hat{\sigma_1^2} = 0.148781$, it follows that

$$\widehat{\mathrm{Var}}(\widetilde{\boldsymbol{c}'\boldsymbol{\tau}}) \cong \varepsilon_{1i}^{-1}\hat{w}_{1i}(1 + \hat{\zeta}_i)\hat{\sigma_1^2} = 1.14004\hat{\sigma_1^2},$$

which is smaller than

$$\widehat{\mathrm{Var}}[(\widehat{\boldsymbol{c}'\boldsymbol{\tau}})_1] = \varepsilon_{1i}^{-1}\hat{\sigma_1^2} = 1.5\hat{\sigma_1^2},$$

obtained from (3.4.2) for the intra-block BLUE. Furthermore, applying the formulae (3.8.96), (3.8.98) and (3.8.101), with $\hat{\sigma_1^2} = s_1^2 = 0.151481$, the following results have been obtained:

$$\begin{aligned}
\widetilde{\boldsymbol{c}_4'\boldsymbol{\tau}} &= 4.419186, & F_4 &= 113.09, & P &= 0.3735(10^{-10}),\\
\widetilde{\boldsymbol{c}_5'\boldsymbol{\tau}} &= 1.862763, & F_5 &= 20.09, & P &= 0.0001223,\\
\widetilde{\boldsymbol{c}_6'\boldsymbol{\tau}} &= 3.746624, & F_6 &= 81.28, & P &= 0.1249(10^{-8}),\\
\widetilde{\boldsymbol{c}_7'\boldsymbol{\tau}} &= -4.738290, & F_7 &= 130.01, & P &= 0.7919(10^{-11}),\\
\widetilde{\boldsymbol{c}_8'\boldsymbol{\tau}} &= 4.150793, & F_8 &= 99.77, & P &= 0.1456(10^{-9}),\\
\widetilde{\boldsymbol{c}_9'\boldsymbol{\tau}} &= -6.751053, & F_9 &= 263.92, & P &= 0.1776(10^{-14}).
\end{aligned}$$

As stated in connection with (3.8.98), under the hypothesis $H_{0,i} : \boldsymbol{c}_i'\boldsymbol{\tau} = 0$, the distribution of the suggested statistic F_i can be approximated by the central F distribution with the d.f. $d_{(i)}$ given by (3.8.101) and $d_1 = n - b - h$. For this example $\hat{d}_{(i)} = 0.990533$, obtained with w_{1i}, w_{2i} and ζ_i replaced by their above estimates, whereas $d_1 = 27$, as already used. With these d.f., the relevant P values have been obtained as given above. They all are much below 0.001, confirming that all hypotheses $H_{0,i}$, $i = 4, 5, ..., 9$, can surely be rejected.

Finally, it may be noted that as $\hat{\zeta}_i$ $(i = 4, 5, ..., 9)$ is here small, not much difference would appear in this example when applying the approximate F test with the use either of the statistic (3.8.98) or of the statistic (3.8.97).

3.9 Some other methods of estimating stratum variances in proper designs

A disadvantage of estimating the stratum variances by the method of Nelder (1968) [equivalent to MML (REML) and the iterated MINQUE procedure] described in Section 3.8.2 is that it proceeds through iterations. It is usually difficult to predict how many iterations will be needed for a given case, and even the convergence of the iterative process has not been proved in general (see also Rao, 1979, p. 149). Therefore, some users may prefer a method more straightforward, not involving iterative computations, but still suitable for application under the assumptions of Theorem 3.8.2. Several such methods have been proposed in the literature, and it may be useful to consider some of them now. The most classical is that introduced by Yates (1940b) for BIB designs with parameters v, b, r, k, λ (as in Definition 2.4.2).

3.9.1 Yates–Rao estimators

In the present notation, the estimators of the stratum variances, σ_1^2 and σ_2^2, obtainable by the classic Yates method can be written as

$$\hat{\sigma}^2_{1(Y)} = d_1^{-1} \boldsymbol{y}'\boldsymbol{\psi}_1\boldsymbol{y}, \quad d_1 = rv - v - b + 1, \tag{3.9.1}$$

and

$$\hat{\sigma}^2_{2(Y)} = \frac{k}{v(r-1)}\{\boldsymbol{y}'\boldsymbol{\psi}_2\boldsymbol{y} + \varepsilon_1\varepsilon_2\sum_{i=1}^{v-1}[(\widehat{\boldsymbol{c}_i'\boldsymbol{\tau}})_2 - (\widehat{\boldsymbol{c}_i'\boldsymbol{\tau}})_1]^2 - \frac{v-k}{kd_1}\boldsymbol{y}'\boldsymbol{\psi}_1\boldsymbol{y}\}, \tag{3.9.2}$$

where $\varepsilon_1 = \varepsilon$, the common efficiency factor of the analyzed BIB design, whereas $\varepsilon_2 = 1 - \varepsilon_1$ (see Remark 3.3.3). For comparison, note that the estimators following from formulae (3.8.35) and (3.8.36) obtain for a BIB design the forms

$$\hat{\sigma}^2_{1(N)} = (d_1')^{-1}\{\boldsymbol{y}'\boldsymbol{\psi}_1\boldsymbol{y} + \hat{w}_2^2\varepsilon_1\sum_{i=1}^{v-1}[(\widehat{\boldsymbol{c}_i'\boldsymbol{\tau}})_1 - (\widehat{\boldsymbol{c}_i'\boldsymbol{\tau}})_2]^2\} \tag{3.9.3}$$

and

$$\hat{\sigma}^2_{2(N)} = (d_2')^{-1}\{\boldsymbol{y}'\boldsymbol{\psi}_2\boldsymbol{y} + \hat{w}_1^2\varepsilon_2\sum_{i=1}^{v-1}[(\widehat{\boldsymbol{c}_i'\boldsymbol{\tau}})_2 - (\widehat{\boldsymbol{c}_i'\boldsymbol{\tau}})_1]^2\}, \tag{3.9.4}$$

where $d_1' = d_1 + \hat{w}_2(v-1)$ and $d_2' = d_2 + \hat{w}_1(v-1)$, $d_1 = n - b - h$ and $d_2 = b - v + \rho$ being the residual d.f. in the intra-block and the inter-block analysis, respectively. Evidently, in the case of a symmetric BIB design, for which $d_2 = 0$ and $\boldsymbol{y}'\boldsymbol{\psi}_2\boldsymbol{y} = 0$, the Yates estimator (3.9.2) becomes

$$\begin{aligned}\hat{\sigma^2_{2(Y)}} &= \frac{k}{v(r-1)}\{\varepsilon_1\varepsilon_2\sum_{i=1}^{v-1}[(\widehat{\boldsymbol{c}_i'\boldsymbol{\tau}})_2 - (\widehat{\boldsymbol{c}_i'\boldsymbol{\tau}})_1]^2 - \frac{v-k}{kd_1}\boldsymbol{y}'\boldsymbol{\psi}_1\boldsymbol{y}\} \\ &= \frac{1}{v-1}\{\varepsilon_2\sum_{i=1}^{v-1}[(\widehat{\boldsymbol{c}_i'\boldsymbol{\tau}})_2 - (\widehat{\boldsymbol{c}_i'\boldsymbol{\tau}})_1]^2 - \frac{\varepsilon_2}{\varepsilon_1}\frac{v-1}{d_1}\boldsymbol{y}'\boldsymbol{\psi}_1\boldsymbol{y}\}, \qquad (3.9.5)\end{aligned}$$

because for a BIB design $\varepsilon_1 = v(k-1)/[k(v-1)]$ and $\varepsilon_2 = 1-\varepsilon_1 = (v-k)/[k(v-1)]$. On the other hand, if the weights in (3.9.3) and (3.9.4) are chosen as

$$\hat{w}_1 = 1 - \frac{v-1}{d_1}\frac{\boldsymbol{y}'\boldsymbol{\psi}_1\boldsymbol{y}}{\varepsilon_1\sum_{i=1}^{v-1}[(\widehat{\boldsymbol{c}_i'\boldsymbol{\tau}})_2 - (\widehat{\boldsymbol{c}_i'\boldsymbol{\tau}})_1]^2}, \qquad (3.9.6)$$

$$\hat{w}_2 = 1 - \hat{w}_1 = \frac{v-1}{d_1}\frac{\boldsymbol{y}'\boldsymbol{\psi}_1\boldsymbol{y}}{\varepsilon_1\sum_{i=1}^{v-1}[(\widehat{\boldsymbol{c}_i'\boldsymbol{\tau}})_2 - (\widehat{\boldsymbol{c}_i'\boldsymbol{\tau}})_1]^2}, \qquad (3.9.7)$$

then for a symmetric BIB design, the Nelder estimators are reduced to

$$\hat{\sigma^2_{1(N)}} = d_1^{-1}\boldsymbol{y}'\boldsymbol{\psi}_1\boldsymbol{y}$$

and

$$\hat{\sigma^2_{2(N)}} = \frac{1}{v-1}\{\varepsilon_2\sum_{i=1}^{v-1}[(\widehat{\boldsymbol{c}_i'\boldsymbol{\tau}})_2 - (\widehat{\boldsymbol{c}_i'\boldsymbol{\tau}})_1]^2 - \frac{\varepsilon_2}{\varepsilon_1}\frac{v-1}{d_1}\boldsymbol{y}'\boldsymbol{\psi}_1\boldsymbol{y}\}.$$

Furthermore, with the weights (3.9.6) and (3.9.7) taken as the initial ones, the suggested iterative process converges in the first iteration cycle for any such design. Thus, the two types of estimators coincide in this instance (as already noticed by Nelder, 1968, p. 309).

A natural question that may be asked in connection with this result is whether an analytical solution for the weights can be found for a BIB design in general. As mentioned by Nelder (1968, Section 4.1), the answer is positive. It can easily be shown, by substituting the variances σ_1^2 and σ_2^2 appearing in (3.8.8) by their estimators (3.9.3) and (3.9.4), respectively, that the equation to be solved for the weight w_1 (or alternatively for $w_2 = 1 - w_1$) is

$$\begin{aligned}(1-w_1)w_1[(1-w_1)d_2 - w_1d_1]\varepsilon_1\varepsilon_2\sum_{i=1}^{v-1}[(\widehat{\boldsymbol{c}_i'\boldsymbol{\tau}})_2 - (\widehat{\boldsymbol{c}_i'\boldsymbol{\tau}})_1]^2 \\ + w_1[d_2 + w_1(v-1)]\varepsilon_2\boldsymbol{y}'\boldsymbol{\psi}_1\boldsymbol{y} - (1-w_1)[d_1 + (1-w_1)(v-1)]\varepsilon_1\boldsymbol{y}'\boldsymbol{\psi}_2\boldsymbol{y} = 0,\end{aligned}$$

a cubic equation in w_1, equivalent to that given by El-Shaarawi et al. (1975, p. 95). Clearly, if d_2 and $\boldsymbol{y}'\boldsymbol{\psi}_2\boldsymbol{y}$ are both equal to zero, the equation is reduced to

$$(1-w_1)w_1^2 d_1\varepsilon_1 \sum_{i=1}^{v-1}[(\widehat{\boldsymbol{c}_i'\boldsymbol{\tau}})_2-(\widehat{\boldsymbol{c}_i'\boldsymbol{\tau}})_1]^2 - w_1^2(v-1)\boldsymbol{y}'\boldsymbol{\psi}_1\boldsymbol{y}=0,$$

yielding (3.9.6) or (3.9.7) as its unique solution.

It should be mentioned at this point that the results of Yates (1940b), given here in (3.9.1) and (3.9.2), were obtained by equating the intra-block residual sum of squares and the adjusted (eliminating treatment contributions) block sum of squares to their respective expectations. This approach was also used by Yates in some other papers (Yates, 1939, 1940a) and was adopted by Nair (1944) and later by Rao (1947, 1956, 1959) to more general proper block designs.

To see this approach more closely, refer to (3.1.20), and then note (recalling from Section 3.2) that

$$\mathrm{E}(\boldsymbol{y}'\boldsymbol{\psi}_1\boldsymbol{y}) = d_1(\sigma_U^2+\sigma_e^2) = d_1\sigma_1^2 \quad (\text{where } \boldsymbol{\psi}_1 \equiv \boldsymbol{\psi}) \tag{3.9.8}$$

and that for a general proper block design

$$\begin{aligned}\mathrm{E}(\boldsymbol{y}'\boldsymbol{P}_{\boldsymbol{\phi}_*\boldsymbol{D}'}\boldsymbol{y}) &= [k^{-1}\mathrm{tr}(\boldsymbol{r}^{-\delta}\boldsymbol{N}\boldsymbol{N}')-v+h]\sigma_1^2 \\ &\quad + [b-k^{-1}\mathrm{tr}(\boldsymbol{r}^{-\delta}\boldsymbol{N}\boldsymbol{N}')]\sigma_2^2,\end{aligned} \tag{3.9.9}$$

where $\boldsymbol{P}_{\boldsymbol{\phi}_*\boldsymbol{D}'} = \boldsymbol{\phi}_*\boldsymbol{D}'(\boldsymbol{D}\boldsymbol{\phi}_*\boldsymbol{D}')^-\boldsymbol{D}\boldsymbol{\phi}_*$, with $\boldsymbol{\phi}_*$ defined as in (2.2.5). By equating $\boldsymbol{y}'\boldsymbol{\psi}_1\boldsymbol{y}$ and $\boldsymbol{y}'\boldsymbol{P}_{\boldsymbol{\phi}_*\boldsymbol{D}'}\boldsymbol{y}$ to the right-hand sides of (3.9.8) and (3.9.9), respectively, the following estimators of σ_1^2 and σ_2^2 are obtainable:

$$\frac{1}{d_1}\boldsymbol{y}'\boldsymbol{\psi}_1\boldsymbol{y} \quad \text{for} \quad \sigma_1^2$$

and

$$\frac{\boldsymbol{y}'\boldsymbol{P}_{\boldsymbol{\phi}_*\boldsymbol{D}'}\boldsymbol{y} - [k^{-1}\mathrm{tr}(\boldsymbol{r}^{-\delta}\boldsymbol{N}\boldsymbol{N}')-v+h]d_1^{-1}\boldsymbol{y}'\boldsymbol{\psi}_1\boldsymbol{y}}{b-k^{-1}\mathrm{tr}(\boldsymbol{r}^{-\delta}\boldsymbol{N}\boldsymbol{N}')} \quad \text{for} \quad \sigma_2^2.$$

Because, as can be shown, in the notation adopted here

$$\boldsymbol{y}'\boldsymbol{P}_{\boldsymbol{\phi}_*\boldsymbol{D}'}\boldsymbol{y} = \boldsymbol{y}'\boldsymbol{\psi}_2\boldsymbol{y} + \sum_{i=\rho+1}^{h}\varepsilon_{1i}\varepsilon_{2i}[(\widehat{\boldsymbol{c}_i'\boldsymbol{\tau}})_2-(\widehat{\boldsymbol{c}_i'\boldsymbol{\tau}})_1]^2, \tag{3.9.10}$$

the estimators of the stratum variances obtainable in the Yates–Rao approach (as it is usually called) can be written as

$$\hat{\sigma}_{1(YR)}^2 = \frac{1}{d_1}\boldsymbol{y}'\boldsymbol{\psi}_1\boldsymbol{y} \tag{3.9.11}$$

and

$$\begin{aligned}\hat{\sigma}_{2(YR)}^2 &= \frac{1}{b-k^{-1}\mathrm{tr}(\boldsymbol{r}^{-\delta}\boldsymbol{N}\boldsymbol{N}')}\{\boldsymbol{y}'\boldsymbol{\psi}_2\boldsymbol{y} + \sum_{i=\rho+1}^{h}\varepsilon_{1i}\varepsilon_{2i}[(\widehat{\boldsymbol{c}_i'\boldsymbol{\tau}})_2-(\widehat{\boldsymbol{c}_i'\boldsymbol{\tau}})_1]^2 \\ &\quad - [k^{-1}\mathrm{tr}(\boldsymbol{r}^{-\delta}\boldsymbol{N}\boldsymbol{N}')-v+h]d_1^{-1}\boldsymbol{y}'\boldsymbol{\psi}_1\boldsymbol{y}\},\end{aligned} \tag{3.9.12}$$

for any proper block design.

The estimator (3.9.12) reduces to

$$\hat{\sigma}^2_{2(YR)} = \frac{k}{n-v}\{y'\psi_2 y + \sum_{i=\rho+1}^{h} \varepsilon_{1i}\varepsilon_{2i}[(\widehat{c_i'\tau})_2 - (\widehat{c_i'\tau})_1]^2 - \frac{v-k(v-h)}{kd_1} y'\psi_1 y\} \tag{3.9.13}$$

for any binary proper block design. In fact, (3.9.13) is equivalent to the result given by Rao (1959, p. 330) if, in addition, the design is equireplicate.

To avoid confusion, however, one has to be aware that the meaning of the variances σ_1^2 and σ_2^2 in the quoted papers is slightly different than it is here. In the series of papers by Yates (1939, 1940a, 1940b), Nair (1944) and Rao (1947, 1956), the stratum variances σ_1^2 and σ_2^2 are understood as σ_U^2 and $k\sigma_B^2+\sigma_U^2$ (i.e., with $K_H \to \infty$), respectively, whereas in Rao's (1959) paper, the interpretation of σ_1^2 and σ_2^2 is as in the present monograph, assuming that $b = N_B$ and $k = K_H$. In all these papers, the technical error, with variance σ_e^2, is ignored, although Rao (1959, p. 336) makes a provision for it. It should be mentioned, that in most publications the assumptions such as those in the above series of papers are adopted [see, e.g., the paper by Herzberg and Wynn (1982) and the book by John, 1987, Chapter 8].

3.9.2 Other noniterative estimators

Further estimators of σ_1^2 and σ_2^2 that may be considered as belonging to the same class as the Yates–Rao estimators, are those given by Graybill and Deal (1959), Seshadri (1963), Shah (1964a), Stein (1966), Brown and Cohen (1974) and Khatri and Shah (1974). They all are some modifications of the estimators (3.9.1) and (3.9.2), if applicable to BIB designs, or of (3.9.11) and (3.9.12), if devised for more general block designs. The modifications have been introduced to secure that the obtained combined estimator, defined as in (3.8.7) but with estimated weights, like in (3.8.42), is uniformly better in the sense of preserving the property (3.8.11) for all values of the unknown ratio $\delta = \sigma_2^2/\sigma_1^2$, i.e., having uniformly smaller variance than that of its intra-block component (3.3.8). This property is usually called the uniformly smaller variance (USV) property. It appears that the Yates–Rao estimators have not been shown to provide this desirable property generally.

Let the estimators of the stratum variances obtained by the authors mentioned above be written now in the present notation, so that comparisons with (3.8.35) and (3.8.36), as well as with (3.9.1) and (3.9.2) and other formulae given subsequently here, can easily be made. In connection with these estimators, conditions under which the resulting combined estimators have the USV property are of interest. They have been established under the multivariate normality assumption imposed on the model, and are presented after Shah (1975).

For application to BIB designs, the following estimators of the stratum variances have been obtained by various authors under different approaches.

Estimators from Graybill and Deal (1959):

$$\hat{\sigma}^2_{1(GD)} = \boldsymbol{y}'\boldsymbol{\psi}_1\boldsymbol{y}/d_1, \quad \hat{\sigma}^2_{2(GD)} = \boldsymbol{y}'\boldsymbol{\psi}_2\boldsymbol{y}/d_2,$$

which secure the USV property if $d_2 \geq 10$ or $d_2 = 9$ with $d_1 \geq 18$.

Estimators from Seshadri (1963):

$$\hat{\sigma}^2_{1(Se)} = \boldsymbol{y}'\boldsymbol{\psi}_1\boldsymbol{y}/d_1,$$

$$\hat{\sigma}^2_{2(Se)} = \varepsilon_2 \sum_{i=1}^{v-1}[(\widehat{\boldsymbol{c}_i'\boldsymbol{\tau}})_2 - (\widehat{\boldsymbol{c}_i'\boldsymbol{\tau}})_1]^2/(v-3) - (\varepsilon_2/\varepsilon_1)\boldsymbol{y}'\boldsymbol{\psi}_1\boldsymbol{y}/d_1,$$

which secure the USV property if $v \geq 4$.

Estimators from Shah (1964a):

$$\hat{\sigma}^2_{1(Sh)} = \boldsymbol{y}'\boldsymbol{\psi}_1\boldsymbol{y}/d_1,$$

$$\hat{\sigma}^2_{2(Sh)} = \varepsilon_2 \sum_{i=1}^{v-1}[(\widehat{\boldsymbol{c}_i'\boldsymbol{\tau}})_2 - (\widehat{\boldsymbol{c}_i'\boldsymbol{\tau}})_1]^2/(v-1) - (\varepsilon_2/\varepsilon_1)\boldsymbol{y}'\boldsymbol{\psi}_1\boldsymbol{y}/d_1,$$

which secure the USV property if $v \geq 6$.

Estimators from Stein (1966):

$$\hat{\sigma}^2_{1(St)} = \boldsymbol{y}'\boldsymbol{\psi}_1\boldsymbol{y}/(d_1+2),$$

$$\hat{\sigma}^2_{2(St)} = \varepsilon_2 \sum_{i=1}^{v-1}[(\widehat{\boldsymbol{c}_i'\boldsymbol{\tau}})_2 - (\widehat{\boldsymbol{c}_i'\boldsymbol{\tau}})_1]^2/(v-3) - (\varepsilon_2/\varepsilon_1)\boldsymbol{y}'\boldsymbol{\psi}_1\boldsymbol{y}/(d_1+2),$$

which secure the USV property if $v \geq 4$.

Estimators from Brown and Cohen (1974):

$$\hat{\sigma}^2_{1(BC)} = \boldsymbol{y}'\boldsymbol{\psi}_1\boldsymbol{y}/(d_1+2),$$

$$\begin{aligned}\hat{\sigma}^2_{2(BC)} &= \{\boldsymbol{y}'\boldsymbol{\psi}_2\boldsymbol{y} + \varepsilon_2 \sum_{i=1}^{v-1}[(\widehat{\boldsymbol{c}_i'\boldsymbol{\tau}})_2 - (\widehat{\boldsymbol{c}_i'\boldsymbol{\tau}})_1]^2 \\ &\quad +2(\varepsilon_2/\varepsilon_1)[(2d_1+b+1)/d_1]\boldsymbol{y}'\boldsymbol{\psi}_1\boldsymbol{y}/(d_1+2)\}/(b-3),\end{aligned}$$

which secure the USV property if $b \geq 4$.

Comparing the estimators given by Shah (1964a) with those of Yates, (3.9.1) and (3.9.2), it can be seen that they coincide for symmetric BIB designs, as then $\hat{\sigma}^2_{2(Y)}$ given in (3.9.2) becomes as in (3.9.5), which is exactly equal to $\hat{\sigma}^2_{2(Sh)}$. This coincidence implies that the Yates estimators secure the USV property in all symmetric BIB designs for which $v \geq 6$.

It may be mentioned at this point that, as shown by Bhattacharya (1978), a sufficient condition for the Yates estimators to secure the USV property in asymmetric BIB designs is, under $\delta = \sigma_2^2/\sigma_1^2 \geq 1$, the inequality

$$\frac{1}{2} \leq \frac{\varepsilon_1 k d_1}{v(r-1)} \frac{b-3}{d_1+2},$$

equivalent to

$$\frac{1}{2} \leq \frac{\lambda d_1}{r(r-1)} \frac{b-3}{d_1+2}.$$

Bhattacharya (1978) has found that this condition is satisfied by all asymmetric BIB designs listed in the tables of Fisher and Yates (1963), with two exceptions. These exceptions and all others that appear among the existing BIB designs are listed in Theorem 6.0.4 (of Volume II of the present book).

As to the more general block designs, first it should be noted that the result of Shah (1964a) is essentially applicable to all proper block designs (even not necessarily connected and equireplicate) for which either all nonzero intra-block efficiency factors are equal (as in a BIB design) or all those among them smaller than 1 are equal, i.e., for which

$$1 = \varepsilon_{11} = \varepsilon_{12} = \cdots = \varepsilon_{1\rho} > \varepsilon_{1,\rho+1} = \varepsilon_{1,\rho+2} = \cdots = \varepsilon_{1,h} = \varepsilon_1 > 0.$$

In such a case, his estimator of σ_2^2 changes to

$$\hat{\sigma_2^2}_{(Sh)} = \varepsilon_2 \sum_{i=\rho+1}^{h} [(\widehat{\boldsymbol{c}_i'\boldsymbol{\tau}})_2 - (\widehat{\boldsymbol{c}_i'\boldsymbol{\tau}})_1]^2/(h-\rho) - (\varepsilon_2/\varepsilon_1)\boldsymbol{y}'\boldsymbol{\psi}_1\boldsymbol{y}/d_1, \quad (3.9.14)$$

securing the USV property if

$$(h-\rho-4)(d_1-2) \geq 8. \quad (3.9.15)$$

A similar extension of Seshadri's estimator has been given by Mejza (1978).

With reference to formula (3.9.14), it may be noticed that it coincides with (3.9.12), if $d_2 = 0$ and $\varepsilon_{1i} = \varepsilon_1$ for $i = \rho+1, \rho+2, ..., h$. To see this coincidence, it will be useful to rewrite (3.9.12) in the form [obtainable from the equality $k^{-1}\text{tr}(\boldsymbol{r}^{-\delta}\boldsymbol{N}\boldsymbol{N}') = v - \rho - \sum_{i=\rho+1}^{h} \varepsilon_{1i}$, following from (2.3.3)]

$$\begin{aligned}\hat{\sigma_2^2}_{(YR)} &= (d_2 + \sum_{i=\rho+1}^{h} \varepsilon_{1i})^{-1}\{\boldsymbol{y}'\boldsymbol{\psi}_2\boldsymbol{y} + \sum_{i=\rho+1}^{h} \varepsilon_{1i}\varepsilon_{2i}[(\widehat{\boldsymbol{c}_i'\boldsymbol{\tau}})_2 - (\widehat{\boldsymbol{c}_i'\boldsymbol{\tau}})_1]^2 \\ &\qquad - (h-\rho-\sum_{i=\rho+1}^{h} \varepsilon_{1i})d_1^{-1}\boldsymbol{y}'\boldsymbol{\psi}_1\boldsymbol{y}\},\end{aligned}$$

which reduces to (3.9.14) under the above equality, $\varepsilon_{1i} = \varepsilon_1$, when $d_2 = 0$, implying $\boldsymbol{y}'\boldsymbol{\psi}_2\boldsymbol{y} = 0$. This result shows that the Yates estimators secure the USV property not only in all symmetric BIB designs with $v \geq 6$, but also in a broader class of proper block designs, namely, that defined above by the equality of the efficiency factors $\{\varepsilon_{1i}, i = \rho+1, \rho+2, ..., h\}$ and the zero value of d_2, provided that the condition (3.9.15) holds. As indicated by Shah (1964a), all linked block designs belong to this class, and for them, (3.9.15) is satisfied whenever $b \geq 6$.

It should be added that the Yates–Rao estimators may also secure the USV property for many other block designs in which $\varepsilon_{1i} = \varepsilon_1, i = \rho+1, \rho+2, ..., h$, but $d_2 > 0$. Bhattacharya (1978) has found a sufficient condition for securing the USV property by the Yates–Rao estimators in those designs, for which these estimators are equal to (3.9.11) and

$$\begin{aligned}\hat{\sigma}^2_{2(YR)} &= [d_2 + (h-\rho)\varepsilon_1]^{-1}\{\boldsymbol{y}'\boldsymbol{\psi}_2\boldsymbol{y} + \varepsilon_1\varepsilon_2 \sum_{i=\rho+1}^{h} [(\widehat{\boldsymbol{c}_i'\boldsymbol{\tau}})_2 - (\widehat{\boldsymbol{c}_i'\boldsymbol{\tau}})_1]^2 \\ &\quad - (h-\rho)\varepsilon_2 d_1^{-1}\boldsymbol{y}'\boldsymbol{\psi}_1\boldsymbol{y}\}.\end{aligned}$$

The sufficient condition for these estimators to secure the desirable USV property in the described class of block designs is, under $\delta \geq 1$, the inequality

$$\frac{1}{2} \leq \frac{\varepsilon_1 d_1(b-v+h-2)}{[d_2+(h-\rho)\varepsilon_1](d_1+2)},$$

which for binary designs reduces to

$$\frac{1}{2} \leq \frac{\varepsilon_1 k d_1(b-v+h-2)}{(n-v)(d_1+2)} = \frac{[v(k-1)-k\rho]d_1(b-v+h-2)}{(h-\rho)(n-v)(d_1+2)},$$

with $b-v+h-2 = b-3$ for all connected designs (see Bhattacharya, 1978, Theorem 3.1). It should be mentioned that the class of designs considered here (with $\varepsilon_{1i} = \varepsilon_1$, for $i = \rho+1, \rho+2, ..., h$, and with $d_2 = 0$ or not) is broad and includes many useful designs. This class will be discussed in detail in Volume II (Chapter 7, in particular).

Results applicable to a more general class of designs have been obtained by Khatri and Shah (1974). Their estimators of the stratum variances, devised originally for all connected binary equireplicate proper block designs, can now be given (following Shah, 1992) for any proper block design in the form

$$\hat{\sigma}^2_{1(KS)} = \boldsymbol{y}'\boldsymbol{\psi}_1\boldsymbol{y}/(d_1+2), \tag{3.9.16}$$

and

$$\hat{\sigma}^2_{2(KS)} = \{\boldsymbol{y}'\boldsymbol{\psi}_2\boldsymbol{y} + \sum_{i=\rho+1}^{h} \varepsilon_{2i}[(\widehat{\boldsymbol{c}_i'\boldsymbol{\tau}})_2 - (\widehat{\boldsymbol{c}_i'\boldsymbol{\tau}})_1]^2\}/(b-v+h-2). \tag{3.9.17}$$

These estimators secure the USV property in all proper designs for which $b - v + h \geq 3$.

Finally, note that the Yates–Rao estimators as well as all other estimators of σ_1^2 and σ_2^2 considered here (in Section 3.9.2) depend on the observed vector $\boldsymbol{y}$ through even functions of the elements of the vector $\boldsymbol{x}$ defined in (3.8.51). Thus, on account of Remark 3.8.4(c), if the estimators $\hat{\sigma}_1^2$ and $\hat{\sigma}_2^2$ are obtained according to any of the discussed procedures, given by (a) Yates (1940b) and Rao (1947), (b) Graybill and Deal (1959), (c) Seshadri (1963), (d) Shah (1964a), (e) Stein (1966), (f) Brown and Cohen (1974) or (g) Khatri and Shah (1974), then the statistic $\hat{\delta} = \hat{\sigma}_2^2/\hat{\sigma}_1^2$ satisfies the relevant condition of Theorem 3.8.2 (similarly as the procedure given by Nelder, 1968). To secure that $\hat{\delta}$ is nonnegative, it is advisable to truncate the statistic according to (3.8.59), with a nonnegative δ^* (the usual suggestion, however, is to take $\delta^* = 1$). The fact that, under the normality of $\boldsymbol{y}$, the empirical estimator of a contrast, as in (3.8.53), based on the Yates–Rao estimation procedure, is unbiased (Theorem 3.8.2) has already been known since the works of Graybill and Weeks (1959), Graybill and Seshadri (1960) and Shah (1964a). An extension of this result to a larger class of estimators of δ, including all discussed here, has been obtained by Roy and Shah (1962).

3.9.3 Comparing the estimation procedures

It is important to notice that all estimators of stratum variances, those obtainable by the original Yates–Rao approach as well as those modified by Graybill and Deal (1959), Seshadri (1963), Shah (1964a), Stein (1966), Brown and Cohen (1974) and Khatri and Shah (1974), belong to the same class. To represent them by a unified formula, it is convenient to think in terms of their use in estimating the weights $w_{1\beta}$ and $w_{2\beta}$ for the combined estimator (3.8.42) of a basic contrast. As $w_{1\beta} = 1 - w_{2\beta}$, it is sufficient to consider the estimation of $w_{2\beta}$ only.

The unified formula of $\hat{w}_{2\beta}$ for the considered class of estimation procedures can be written, similarly to a formula used by Khatri and Shah (1975, p. 404), as

$$\hat{w}_{2\beta} = \frac{\boldsymbol{y}'\boldsymbol{\psi}_1\boldsymbol{y}}{a_{1\beta}\boldsymbol{y}'\boldsymbol{\psi}_1\boldsymbol{y} + a_{2\beta}\boldsymbol{y}'\boldsymbol{\psi}_2\boldsymbol{y} + \sum_{i=\rho+1}^{h} a_{i+2,\beta}[(\widehat{\boldsymbol{c}_i'\boldsymbol{\tau}})_2 - (\widehat{\boldsymbol{c}_i'\boldsymbol{\tau}})_1]^2}, \tag{3.9.18}$$

where the coefficients $a_{1\beta}, a_{2\beta}$ and $a_{i+2,\beta}$, for $i = \rho+1, \rho+2, ...h$, depend on the particular procedure of estimating the stratum variances σ_1^2 and σ_2^2. The considered estimation procedures provide these coefficients as follows:

(a) The Yates (1940b)–Rao (1947) procedure gives

$$a_{1\beta} = 1 - \frac{\varepsilon_{1\beta}}{\varepsilon_{2\beta}} \frac{k^{-1}\text{tr}(\boldsymbol{r}^{-\delta}\boldsymbol{N}\boldsymbol{N}') - v + h}{b - k^{-1}\text{tr}(\boldsymbol{r}^{-\delta}\boldsymbol{N}\boldsymbol{N}')}, \quad a_{2\beta} = \frac{\varepsilon_{1\beta}}{\varepsilon_{2\beta}} \frac{d_1}{b - k^{-1}\text{tr}(\boldsymbol{r}^{-\delta}\boldsymbol{N}\boldsymbol{N}')},$$

$$a_{i+2,\beta} = \varepsilon_{1i}\varepsilon_{2i}a_{2\beta}.$$

(b) The Graybill–Deal (1959) procedure gives (for BIB designs only)

$$a_{1\beta} = 1, \quad a_{2\beta} = \frac{\varepsilon_{1\beta}}{\varepsilon_{2\beta}} \frac{d_1}{d_2}, \quad a_{i+2,\beta} = 0 \quad \text{for all } i.$$

(c) The Seshadri (1963) procedure gives (for BIB designs only)

$$a_{1\beta} = a_{2\beta} = 0, \quad a_{i+2,\beta} = \varepsilon_{1\beta} \frac{d_1}{v-3} \quad \text{for all } i.$$

(d) The Shah (1964a) procedure gives (for designs with constant $\varepsilon_{1i}, i = \rho+1, \rho+2, ...h$, only)

$$a_{1\beta} = a_{2\beta} = 0, \quad a_{i+2,\beta} = \varepsilon_{1\beta} \frac{d_1}{h-\rho} \quad \text{for all } i.$$

(e) The Stein (1966) procedure gives (for BIB designs only)

$$a_{1\beta} = a_{2\beta} = 0, \quad a_{i+2,\beta} = \varepsilon_{1\beta} \frac{d_1+2}{v-3} \quad \text{for all } i.$$

(f) The Brown–Cohen (1974) procedure gives (for BIB designs only)

$$a_{1\beta} = 1 + 2\,\frac{2d_1+b+1}{d_1(b-3)}, \quad a_{2\beta} = \frac{\varepsilon_{1\beta}}{\varepsilon_{2\beta}} \frac{d_1+2}{b-3}, \quad a_{i+2,\beta} = \varepsilon_{2\beta} a_{2\beta} \quad \text{for all } i.$$

(g) The Khatri–Shah (1974) procedure gives

$$a_{1\beta} = 1, \quad a_{2\beta} = \frac{\varepsilon_{1\beta}}{\varepsilon_{2\beta}} \frac{d_1+2}{b-v+h-2}, \quad a_{i+2,\beta} = \varepsilon_{2i} a_{2\beta} \quad \text{for all } i.$$

With this presentation of the various estimating procedures, it is easy to see the similarities and differences between them. Evidently, procedures (c), (d) and (e) are similar, the most general among them being that of Shah (1964a). Their deficiency is the lack of using the information contained in $\boldsymbol{y}'\boldsymbol{\psi}_2\boldsymbol{y}$ the apparent residual sum of squares in the inter-block stratum, usually available, unless $d_2 = 0$. On the other hand, the procedure (b), of Graybill and Deal (1959), does not use the information in differences between the intra-block and the inter-block estimates of the basic contrasts $\boldsymbol{c}_i'\boldsymbol{\tau}, i = \rho+1, \rho+2, ..., h$. Only the procedures (a), (f) and (g) use all available information. Thus, only these three can be considered as competitive alternatives to the Nelder (1968) iterative procedure.

To make comparisons between the three procedures (a), (f) and (g), one has first to note that the a-coefficients obtainable by them for BIB designs can be rewritten as follows:

(a) The Yates–Rao procedure gives

$$a_1 = 1 - \frac{k-1}{r-1}, \quad a_2 = \frac{d_1 k(k-1)}{(v-k)(r-1)}, \quad a_{i+2} = \frac{d_1 v(k-1)^2}{k(v-1)(r-1)} \quad \text{for all } i.$$

(f) The Brown–Cohen procedure gives

$$a_1 = 1 + 2\,\frac{2d_1 + b + 1}{2d_1(b-3)}, \quad a_2 = \frac{v(k-1)(d_1+2)}{(v-k)(b-3)},$$

$$a_{i+2} = \frac{v(k-1)(d_1+2)}{k(v-1)(b-3)} \quad \text{for all } i.$$

(g) The Khatri–Shah procedure gives

$$a_1 = 1, \quad a_2 = \frac{v(k-1)(d_1+2)}{(v-k)(b-3)}, \quad a_{i+2} = \frac{v(k-1)(d_1+2)}{k(v-1)(b-3)} \quad \text{for all } i.$$

This outline shows that the procedure (f) differs from (g) with regard to a_1 only, this coefficient being larger in (f) than in (g). As a result, $\hat{w}_2$ will always be slightly smaller for the former than it is for the latter. As to the comparison between procedures (a) and (g), it may be noted that in (a) $a_{i+2} = \varepsilon_1\varepsilon_2 a_2$, whereas in (g) $a_{i+2} = \varepsilon_2 a_2$. Hence, all coefficients are smaller in (a) than they are in (g), if

$$(v - 3k)d_1 < 2(n - v), \tag{3.9.19}$$

resulting in a somewhat higher value of $\hat{w}_2$ for procedure (a) than for (g). In fact, (3.9.19) is a sufficient condition for such a relation between these two procedures in case of any connected binary proper design. A general sufficient condition for this relation, applicable to any proper block design, is

$$[\mathrm{tr}(\boldsymbol{r}^{-\delta}\boldsymbol{N}\boldsymbol{N}') - k(v - h + 2)]d_1 < 2[n - \mathrm{tr}(\boldsymbol{r}^{-\delta}\boldsymbol{N}\boldsymbol{N}')]. \tag{3.9.20}$$

Thus, in the case of a BIB design, the Khatri–Shah procedure will always give slightly higher weight to the inter-block component of the combined estimator (3.8.42) than it is obtainable from the Brown–Cohen procedure (as remarked by Shah, 1975, p. 6), but even higher weight to this component will be given in many designs, not necessarily BIB, by the Yates–Rao procedure (as indicated by Khatri and Shah, 1975, p. 405).

Note that because the Yates–Rao procedure coincides with that of Nelder (1968) when applied to a symmetric BIB design, the above observation is true for both of these procedures in that case. In general, however, such comparison with the Nelder iterative procedure is not possible, as formula (3.9.18) does not apply to it.

However, if in the Nelder (1968) iterative procedure the process starts with the initial values $w_{1\beta}^{(0)} = 1$ and $w_{2\beta}^{(0)} = 0$, then the estimators (3.8.35) and (3.8.36) obtainable in the first iteration cycle are

$$\hat{\sigma}^{2\,(1)}_{1(N)} = d_1^{-1}\boldsymbol{y}'\boldsymbol{\psi}_1\boldsymbol{y} \tag{3.9.21}$$

and

$$\hat{\sigma}^{2\,(1)}_{2(N)} = (b-1)^{-1}\{\boldsymbol{y}'\boldsymbol{\psi}_2\boldsymbol{y} + \sum_{i=\rho+1}^{h} \varepsilon_{2i}[(\widehat{\boldsymbol{c}_i'\boldsymbol{\tau}})_2 - (\widehat{\boldsymbol{c}_i'\boldsymbol{\tau}})_1]^2\}. \tag{3.9.22}$$

With such results, formula (3.9.18) is applicable giving

$$\hat{w}_{2\beta} = \frac{y'\psi_1 y}{y'\psi_1 y + \frac{\varepsilon_{1\beta}}{\varepsilon_{2\beta}}\frac{d_1}{b-1}\{y'\psi_2 y + \sum_{i=\rho+1}^{h} \varepsilon_{2i}[(\widehat{c_i'\tau})_2 - (\widehat{c_i'\tau})_1]^2\}}, \qquad (3.9.23)$$

where the a-coefficients are

$$a_{1\beta} = 1, \quad a_{2\beta} = \frac{\varepsilon_{1\beta}}{\varepsilon_{2\beta}}\frac{d_1}{b-1} \quad \text{and} \quad a_{i+2,\beta} = \varepsilon_{2i}a_{2\beta}.$$

Evidently, the weight (3.9.23) is similar to that obtainable by the procedure of Khatri and Shah (1974), the coefficients $a_{2\beta}$ and $a_{i+2,\beta}$, $i = \rho+1, \rho+2, ...h$, being slightly smaller here than they are in their procedure (for all cases with $b > v - h + 2$ for which the Khatri–Shah procedure is applicable). Also note that taking as the initial weights in the Nelder (1968) procedure $w_{1\beta}^{(0)} = 1$ and $w_{2\beta}^{(0)} = 0$ (for $\beta = 1, 2, ..., m-1$) is equivalent to taking as initial values of the stratum variances 0 for $\sigma_1^{2(0)}$ and any value for $\sigma_2^{2(0)}$. This equivalence means that using the first-cycle estimators (3.9.21) and (3.9.22) is like acting under *a priori* knowledge that σ_1^2 is practically equal to zero when compared with σ_2^2. This assumption, of course, could be justified in the case of an excellently conducted experiment (an ideal one). But, as known from Section 3.8.1, the gain acquired from the recovery of inter-block information under known stratum variances is measured by $w_{2\beta}/w_{1\beta}$. Hence, under such *a priori* knowledge as that stated above, one could not expect any real gain either from that or another recovery method. This reasoning indicates that the Khatri–Shah (1974) procedure, which gives even slightly smaller weight to the inter-block component of the combined estimator (3.8.42) than that following from (3.9.23), will perform well in cases of large values of $\delta = \sigma_2^2/\sigma_1^2$, but then not much can be gained from the recovery of inter-block information. The same remark, even more so, applies to the Brown–Cohen (1974) procedure.

On the other hand, because, as indicated above, the Yates–Rao procedure gives a higher weight to the inter-block component than does the Khatri–Shah procedure, whenever the condition (3.9.20) holds, it can be expected that the former will perform (in the sense of precision) better than the latter under smaller values of δ. This supposition can only be verified by computing the exact variances of the estimator (3.8.42) for both of the compared procedures of estimating the weights $w_{1\beta}$ and $w_{2\beta}$, under various values of δ. Before considering the question further, it should be mentioned first that, following the earlier suggestion of using a truncated rather than an untruncated estimator of $\delta = \sigma_2^2/\sigma_1^2$ (see Section 3.8.3) and choosing $\delta^* = 1$ for it (as is usually suggested), the formula for estimating $w_{2\beta}$ given in (3.9.18) is to be replaced by the formula

$$\hat{w}_{2\beta}^* = \begin{cases} \hat{w}_{2\beta} & \text{if } \hat{w}_{2\beta} \le \varepsilon_{2\beta}, \\ \varepsilon_{2\beta} & \text{otherwise.} \end{cases} \qquad (3.9.24)$$

where $\hat{w}_{2\beta}$ is the untruncated estimate of $w_{2\beta}$. In fact, this formula is used by Khatri and Shah (1975).

Example 3.9.1. To illustrate the application of formula (3.9.18), it may be interesting to use the results of Example 3.8.1. Taking from that example

$$v = 12,\ b = 18,\ k = 3,\ d_1 = 27,\ h = 9,\ \rho = 3,$$

$$\varepsilon_{1i} = \varepsilon_1 = 2/3,\ \ \varepsilon_{2i} = \varepsilon_2 = 1/3 \quad \text{for } i = 4, 5, ..., 9$$

and

$$\text{tr}(\boldsymbol{r}^{-\delta}\boldsymbol{N}\boldsymbol{N}') = 15,$$

the coefficients $a_{1\beta} = a_1, a_{2\beta} = a_2$ and $a_{i+2,\beta} = a_3$ (constant for $i = 4, 5, ..., 9$), appearing in (3.9.18), can easily be obtained for the three estimation procedures (a), (d) and (g) applicable to the design there. Using the relevant formulae given above, one obtains

(a) for the Yates–Rao procedure

$a_1 = 0,692308,\ \ a_2 = 4.153846$ and $a_3 = 0.923077,$

(d) for the Shah procedure

$a_1 = a_2 = 0$ and $a_3 = 3,$

(g) for the Khatri–Shah procedure

$a_1 = 1,\ \ a_2 = 4.461538$ and $a_3 = 1.487179.$

With them, and with the results of the intra-block and the inter-block analyses in Example 3.8.1, the three procedures provide, by (3.9.18), the estimates of $w_{2\beta} = w_2$ respectively as

(a) $\hat{w}_2 = 0.286230,$

(d) $\hat{w}_2 = 0.399512,$

(g) $\hat{w}_2 = 0.226116.$

Because here the condition (3.9.20) is clearly satisfied, it is not surprising that the estimated weight $\hat{w}_2$ obtained by the Yates–Rao procedure (a) is higher than that resulting from the Khatri–Shah procedure (g). As to the result of the Shah procedure (d), note that because it exceeds the value of ε_2, it is to be replaced in (3.8.42) by the latter, as shown in (3.9.24). The application of the obtained weights to formula (3.8.42) is left to the reader as an exercise.

It should also be mentioned that Khatri and Shah (1975) have found a general method of obtaining (in terms of incomplete beta functions) the exact variance of combined intra-block and inter-block estimators in connected equireplicate proper block designs, under the multivariate normality of the observed random vector $\boldsymbol{y}$. From the nature of their derivations, it follows that the method can be applied to all proper block designs, also disconnected and nonequireplicate.

The combined estimator considered by them [Khatri and Shah, 1975, formula (2.1)] can in the present notation be written as

$$\widetilde{c'_{\beta j}\tau} = (1 - \hat{w}_{2\beta})[(\widehat{c'_{\beta j}\tau})_1 - (\widehat{c'_{\beta j}\tau})_2] + (\widehat{c'_{\beta j}\tau})_2, \qquad (3.9.25)$$

which is equivalent to (3.8.42), or to (3.8.61) when $\hat{w}_{2\beta}$ is replaced by $\hat{w}^*_{2\beta}$ according to (3.9.24). However, the method is restricted to those estimating procedures which belong to the class characterized by formula (3.9.18), i.e., to such like the procedures (a)–(g) discussed above.

The method of Khatri and Shah (1975) devised for obtaining the exact variance of (3.9.25), and hence, by the representation (3.8.53), the exact variance of any contrast $c'\tau$, is not straightforward and, therefore, not suitable for use in most practical applications. But it can be used to study the properties of such variances under various circumstances. This possibility has been explored by the authors to make extensive numerical studies of variances of the estimators of elementary contrasts (paired treatment comparisons) obtainable through the Yates–Rao and Khatri–Shah procedures at four different values of $\delta = \sigma_2^2/\sigma_1^2$, 1, 2, 3 and 5, in four BIB designs and four PBIB designs taken from the tables given by Bose, Clatworthy and Shrikhande (1954) [later extended by Clatworthy (1973)]. The results of these studies have been presented in ratios of the variances of the combined estimators of elementary contrasts to the corresponding variances of the intra-block estimators of such contrasts, as obtained by three compared procedures: Yates–Rao, Khatri–Shah and the optimum one, the last using known stratum variances, as in (3.8.7). In the case of a BIB design, it was sufficient to give one ratio for each procedure at a given value of δ. For a PBIB design, three relevant entries had been given at each of these values: based on the variance for the first associate pair of treatments, on the variance for the second associate pair, and on the average variance for a pair of treatments.

From the ratios given by Khatri and Shah (1975, Tables 1 and 2), it is easy to calculate efficiencies of the estimators of elementary contrasts obtainable by the three procedures of estimation that can be applied in practice: the intra-block estimation, the Yates–Rao combined estimation procedure and the Khatri–Shah combined estimation procedure. The efficiency of an estimator is understood here as the ratio of the precision of that estimator to the precision of the optimum estimator, i.e., as

$$\frac{\mathrm{Var}(\widehat{c'\tau})}{\mathrm{Var}[(\widehat{c'\tau})_1]} \quad \text{for the intra-block estimator}$$

and as

$$\frac{\mathrm{Var}(\widehat{c'\tau})}{\mathrm{Var}[(\widetilde{c'\tau})^*]} \quad \text{for the combined estimator}$$

with estimated weights by the Yates–Rao or Khatri–Shah procedure, * meaning that truncation as in (3.9.24) is applied. The so-defined efficiencies have been calculated for the four BIB designs and the four PBIB designs considered by

Khatri and Shah (1975), for the latter, however, taking only efficiencies based on the average variances into account.

Results so obtained (not shown here, reported only) reveal that in all cases, the efficiency of a combined estimator is higher than that of the intra-block estimator, the efficiencies for both the Yates–Rao and Shah–Khatri procedures being high everywhere. The differences in efficiencies between the two procedures are rather small, except the cases in which $\delta = 1$, the efficiency for the Yates–Rao being then slightly higher than that for the Khatri–Shah procedure. An opposite relation may be observed at $\delta > 2$, but then the difference between the efficiencies for these two procedures is negligible. Another interesting characteristic is that a substantial increase in the efficiency of a combined estimator over that of the intra-block estimator can be observed only at cases in which the theoretical maximum gain, defined by the formula

$$\text{Maximum gain} = \frac{\text{Var}[(\widehat{c'\tau})_1]}{\text{Var}(\widehat{c'\tau})} - 1,$$

is larger than 0.05 (5%). Notably, in all of these cases, the Yates–Rao procedure performs (in the sense of efficiency) similarly to the Khatri–Shah procedure or better.

The method of obtaining the exact variances of combined estimators given by Khatri and Shah (1975) is not applicable to the Nelder (1968) procedure. Therefore, no results concerning this procedure have been given by these authors. Until now, it seems, no results on exact variances for the Nelder (i.e., also for the MML and the iterated MINQUE) procedure are available in the literature. However, some results have been obtained by simulation studies, as those published by Kackar and Harville (1984) and used in constructing Table 3.2. They are based on a large number of simulated data sets, but limited to only two asymmetric BIB designs, i.e. those really of interest, because for symmetric BIB designs, the Nelder procedure coincides with the Yates–Rao procedure. More extensive results have been published by El-Shaarawi et al. (1975). But these results are based only on 300 data sets simulated for each of $\delta = 1, 2, 4, 8$, and for each of four chosen asymmetric BIB designs. From them, the authors have obtained sampling means and variances for the estimators of δ, as calculated by the Yates–Rao and the MML (equivalent to the Nelder, 1968) estimation procedures. These results have indicated close agreement between the two procedures, with slight superiority of the former. This observation confirms the results obtained for two asymmetric BIB designs by Kackar and Harville (1984). Furthermore, they have obtained average sample variances for estimators of elementary contrasts, following from four compared estimation procedures, that of Yates (1940b), Stein (1966), Khatri and Shah (1974) and the MML procedure. Using their results, it has been possible to present efficiencies similar to those discussed above, but with the MML procedure included for comparison (see Table 3.4). Results concerning the Stein procedure have not been included, because of its rather poor performance.

Table 3.4
Efficiencies of estimators of elementary contrasts in some BIB designs for the intra-block, Yates–Rao, Khatri–Shah and MML (REML) estimation procedures

		Efficiency for				
Design parameters	$\delta = \sigma_2^2/\sigma_1^2$	Intra -block	Yates -Rao	Khatri -Shah	MML (REML)	Theoret. maximum gain
$b = 10, v = 6$	1	0.8000	0.9681	0.9571	0.9313	0.2500
$k = 3, r = 5$	2	0.8889	0.9895	0.9926	0.9721	0.1250
$\lambda = 2$	4	0.9412	0.9755	0.9849	0.9727	0.0625
	8	0.9697	0.9803	0.9823	0.9803	0.0312
$b = 18, v = 10$	1	0.8889	0.9804	0.9839	0.9769	0.1250
$k = 5, r = 9$	2	0.9412	0.9921	0.9930	0.9930	0.0625
$\lambda = 4$	4	0.9697	0.9932	0.9949	0.9940	0.0312
	8	0.9846	0.9992	0.9992	0.9992	0.0156
$b = 30, v = 10$	1	0.7407	0.9795	0.9274	0.9795	0.3500
$k = 3, r = 9$	2	0.8511	0.9809	0.9262	0.9801	0.1750
$\lambda = 2$	4	0.9195	0.9941	0.9569	0.9941	0.0875
	8	0.9581	0.9944	0.9613	0.9944	0.0437
$b = 57, v = 19$	1	0.7037	0.9868	0.9491	0.9850	0.4211
$k = 3, r = 9$	2	0.8261	0.9916	0.9565	0.9916	0.2105
$\lambda = 1$	4	0.9048	0.9915	0.9477	0.9915	0.1053
	8	0.9500	0.9980	0.9693	0.9980	0.0526

Source: Calculated from results given by El-Shaarawi et al. (1975, Table 2).

The results presented in Table 3.4 indicate that in all investigated cases, a gain in efficiency over the intra-block estimation has been acquired by each of the three compared estimation procedures: Yates–Rao, Khatri–Shah and MML. In all cases, the efficiencies for these procedures are very high, with an advantage of the Yates–Rao procedure over at least one of the other two when $\delta = 1$. However, the differences between the three methods with regard to the efficiencies are small, except for a few cases of larger designs ($b \geq 30$), in which the Khatri–Shah procedure performs slightly worse at all values of δ. A general impression is that the Yates–Rao procedure is at least as good as the MML procedure, which is in agreement with the results obtained by Kackar and Harville (1984), as can be seen in Table 3.2. Similarly as in the case of the results from Khatri and Shah (1975), a substantial gain in efficiency over the intra-block estimation is obtained in cases in which the theoretical maximum gain is larger than 0.05 (5%). It should also be mentioned that the slightly poorer performance of the Khatri–Shah procedure in some cases of Table 3.4 can partially be explained by the fact that El-Shaarawi et al. (1975) have not used truncation for the Khatri–Shah procedure.

To summarize, it can be concluded from the discussion above that each of the three procedures, the Yates–Rao, the Khatri–Shah and the MML, gives in all investigated cases an increase in efficiency over the intra-block estimation, even at high values of δ, though it is not so substantial when the theoretical maximum gain is less than 5% (which usually corresponds to $\delta > 4$). The Yates–Rao procedure is in most cases at least as good as the Khatri–Shah procedure, being evidently better at small δ (see also the comment made by Bhattacharya, 1978, p. 408). The Yates–Rao procedure can well replace the MML procedure, the evidence being that the former procedure is never not negligibly worse than the latter, and sometimes it is even better (as can also be seen in Table 3.2). Certainly these conclusions are based on limited studies. Nevertheless, some of them evidently support the theoretical observations made earlier in this section. Further studies on this subject would be welcome.

4
Balance and Efficiency: Classification of Notions

4.1 Statistical implications of the two notions of balance

In Chapter 2, two notions of balance have been considered, one based on equality of weighted concurrences (Section 2.4.1) and the other based on the proportionality of them (Section 2.4.2). These notions have been formulated and discussed in terms of constructional features of the design, leaving their statistical implications to the time when a model for the variables observed in the experiment is adopted and its properties are fully established. Because this task has been accomplished in Chapter 3, it is now possible to consider the statistical meaning of the various notions of balance, those considered in Section 2.4 in particular.

As most of the notions of balance, those related to the concept of concurrences have been developed in connection with the intra-block analysis of a block design. That is, the statistical implications should be considered mainly with regard to that analysis. From Section 3.2.1, it is known that under the intra-block submodel, obtained from the randomization model (3.1.1) by the projection $\boldsymbol{y}_1 = \boldsymbol{\phi}_1\boldsymbol{y}$, i.e., the model

$$\boldsymbol{y}_1 = \boldsymbol{\phi}_1\boldsymbol{\Delta}'\boldsymbol{\tau} + \boldsymbol{\phi}_1(\boldsymbol{\eta} + \boldsymbol{e}), \tag{4.1.1}$$

with the properties

$$\mathrm{E}(\boldsymbol{y}_1) = \boldsymbol{\phi}_1\boldsymbol{\Delta}'\boldsymbol{\tau}, \qquad \mathrm{Cov}(\boldsymbol{y}_1) = \boldsymbol{\phi}_1(\sigma_U^2 + \sigma_e^2),$$

the BLUE of any contrast $\boldsymbol{c}'\boldsymbol{\tau}$ $(\boldsymbol{c}'\mathbf{1}_v = 0)$ is of the form

$$(\widehat{\boldsymbol{c}'\boldsymbol{\tau}})_{\text{intra}} = \boldsymbol{s}'\boldsymbol{\Delta}\boldsymbol{y}_1 = \boldsymbol{c}'(\boldsymbol{\Delta}\boldsymbol{\phi}_1\boldsymbol{\Delta}')^-\boldsymbol{\Delta}\boldsymbol{\phi}_1\boldsymbol{y}, \tag{4.1.2}$$

provided that $\boldsymbol{c} = \boldsymbol{\Delta}\boldsymbol{\phi}_1\boldsymbol{\Delta}'\boldsymbol{s}$, and the variance of this BLUE is

$$\mathrm{Var}[(\widehat{\boldsymbol{c}'\boldsymbol{\tau}})_{\text{intra}}] = \boldsymbol{s}'\boldsymbol{\Delta}\boldsymbol{\phi}_1\boldsymbol{\Delta}'\boldsymbol{s}\sigma_1^2 = \boldsymbol{c}'(\boldsymbol{\Delta}\boldsymbol{\phi}_1\boldsymbol{\Delta}')^-\boldsymbol{c}\sigma_1^2, \tag{4.1.3}$$

where $\sigma_1^2 = \sigma_U^2 + \sigma_e^2$ and $\boldsymbol{\Delta}\boldsymbol{\phi}_1\boldsymbol{\Delta}' = \boldsymbol{C}_1 \equiv \boldsymbol{C}$ is known as the C-matrix (see Section 2.2).

Now, if the design is balanced in the sense of Definition 2.4.4, then on account of Theorem 2.4.2, the C-matrix can be represented as

$$\boldsymbol{C}_1 = \boldsymbol{\Delta}\boldsymbol{\phi}_1\boldsymbol{\Delta}' = \theta\left(\sum_{i=1}^{v-g} \boldsymbol{c}_i\boldsymbol{c}_i'\right),$$

where the vectors $\{\boldsymbol{c}_i\}$ are the orthonormal eigenvectors of $\boldsymbol{C}_1$ corresponding to the unique nonzero eigenvalue θ, with multiplicity $v-g$. This decomposition also implies that $\boldsymbol{c}_i'\mathbf{1}_v = 0$ for all $i \leq v-g$, and so they represent contrasts, called by Pearce (1983, p. 73) the "natural contrasts" of the design. (Some other authors have called them the "canonical contrasts," e.g., Shah, 1960.) Hence, for any contrast $\boldsymbol{c}'\boldsymbol{\tau}$ such that $\boldsymbol{c} = \theta(\sum_{i=1}^{v-g} \ell_i\boldsymbol{c}_i)$, where $\ell_i = \boldsymbol{c}_i'\boldsymbol{s}$ for some $\boldsymbol{s}$, the BLUE in the intra-block analysis is, from (4.1.2) and (4.1.3), of the form

$$(\widehat{\boldsymbol{c}'\boldsymbol{\tau}})_{\text{intra}} = \boldsymbol{s}'\boldsymbol{\Delta}\boldsymbol{\phi}_1\boldsymbol{y} = \sum_{i=1}^{v-g} \ell_i\boldsymbol{c}_i'\boldsymbol{\Delta}\boldsymbol{\phi}_1\boldsymbol{y} = \theta^{-1}\boldsymbol{c}'\boldsymbol{\Delta}\boldsymbol{\phi}_1\boldsymbol{y},$$

with

$$\text{Var}[(\widehat{\boldsymbol{c}'\boldsymbol{\tau}})_{\text{intra}}] = \theta\sum_{i=1}^{v-g} \ell_i^2\sigma_1^2 = \theta^{-1}\boldsymbol{c}'\boldsymbol{c}\sigma_1^2,$$

which can easily be proved by taking $\theta^{-1}\sum_{i=1}^{v-g} \boldsymbol{c}_i\boldsymbol{c}_i'$ as a g-inverse of $\boldsymbol{C}_1$ (see also Caliński, 1977; Pearce, 1983, Section 5.2).

Thus, a statistical consequence of the balance based on equal weighted concurrences is that any contrast $\boldsymbol{c}'\boldsymbol{\tau}$ for which the intra-block BLUE [i.e., the BLUE under the intra-block submodel (4.1.1)] exists is estimated with precision $\theta/\boldsymbol{c}'\boldsymbol{c}$, this having the same value for all contrasts that have the same norm, $\|\boldsymbol{c}\| = (\boldsymbol{c}'\boldsymbol{c})^{1/2}$. For this reason, this concept of balance is called "variance balance" (Raghavarao, 1971, p. 54; Hedayat and Federer, 1974, p. 333). The above statistical consequences of this type of balance are indeed so attractive that some authors use the term balance exclusively in the sense of variance balance (e.g., Pearce, 1983, p. 121: "to speak of a design as 'balanced' implies 'variance balance', $\cdots$").

Also note (see Theorem 2.4.1 and Remark 2.4.1) that for any equireplicate, connected orthogonal block design, the unique nonzero eigenvalue of its C-matrix, with multiplicity $v-1$, is $\theta = r$. Hence, the precision of estimating $\boldsymbol{c}'\boldsymbol{\tau}$ in such a design is $r/\boldsymbol{c}'\boldsymbol{c}$. Therefore, the efficiency factor, defined as the ratio of the precision of the intra-block BLUE obtainable in the design under consideration to that obtainable in an equireplicate, connected orthogonal block design of the same n, i.e., the efficiency factor used by Kempthorne (1956), Kshirsagar (1958), Atiqullah (1961) and Raghavarao (1971, p. 58), is for the so-called variance-balanced design equal to θ/r, where $r = n/v$, for any contrast for which the intra-block BLUE exists. Thus, from this point of view, a variance-balanced design is simultaneously balanced with regard to the efficiency.

On the other hand, if the design is balanced in the sense of Jones (Definition 2.4.6), then, by Theorem 2.4.4, its C-matrix can be represented as

$$\boldsymbol{C}_1 = \varepsilon \boldsymbol{r}^{\delta}\left(\sum_{i=1}^{v-g} \boldsymbol{s}_i \boldsymbol{s}_i'\right)\boldsymbol{r}^{\delta},$$

where the vectors $\{\boldsymbol{s}_i\}$ are defined as in Lemma 2.3.3 and correspond to the unique nonzero eigenvalue ε of $\boldsymbol{C}_1$ with respect to $\boldsymbol{r}^{\delta}$, with multiplicity $v-g$. It follows that $\boldsymbol{s}_i'\boldsymbol{r} = 0$ for all $i \leq v-g$, and so the vectors $\mathbf{c}_i = \boldsymbol{r}^{\delta}\boldsymbol{s}_i, i = 1, 2, ..., v-g$, represent contrasts, called basic contrasts by Pearce et al. (1974) (see also Definition 3.4.1). Hence, for any contrast $\boldsymbol{c}'\boldsymbol{\tau}$ such that

$$\mathbf{c} = \varepsilon \boldsymbol{r}^{\delta}\left(\sum_{i=1}^{v-g} \ell_i \boldsymbol{s}_i\right),$$

where $\ell_i = \boldsymbol{s}_i'\boldsymbol{r}^{\delta}\boldsymbol{s}$ for some $\boldsymbol{s}$, the intra-block BLUE is, from (4.1.2) and (4.1.3),

$$(\widehat{\boldsymbol{c}'\boldsymbol{\tau}})_{\text{intra}} = \boldsymbol{s}'\boldsymbol{\Delta}\boldsymbol{\phi}_1\boldsymbol{y} = \sum_{i=1}^{v-g} \ell_i \boldsymbol{s}_i'\boldsymbol{\Delta}\boldsymbol{\phi}_1\boldsymbol{y} = \varepsilon^{-1}\boldsymbol{c}'\boldsymbol{r}^{-\delta}\boldsymbol{\Delta}\boldsymbol{\phi}_1\boldsymbol{y},$$

with

$$\text{Var}[(\widehat{\boldsymbol{c}'\boldsymbol{\tau}})_{\text{intra}}] = \varepsilon\sum_{i=1}^{v-g} \ell_i^2\sigma_1^2 = \varepsilon^{-1}\boldsymbol{c}'\boldsymbol{r}^{-\delta}\mathbf{c}\sigma_1^2,$$

easily obtained by taking $\varepsilon^{-1}\sum_{i=1}^{v-g} \boldsymbol{s}_i\boldsymbol{s}_i'$ as a g-inverse of $\boldsymbol{C}_1$.

Thus, a statistical consequence of the balance based on the proportionality of weighted concurrences to the products of relevant treatment replications is that any contrast $\boldsymbol{c}'\boldsymbol{\tau}$ for which the intra-block BLUE exists is estimated with precision $\varepsilon/\boldsymbol{c}'\boldsymbol{r}^{-\delta}\mathbf{c}$, which has the same value for all contrasts of the same $\boldsymbol{r}^{-\delta}$-norm, $\|\mathbf{c}\|_{\boldsymbol{r}^{-\delta}} = (\boldsymbol{c}'\boldsymbol{r}^{-\delta}\mathbf{c})^{1/2}$. Therefore, this concept of balance could be called "$\boldsymbol{r}^{-\delta}$-balance" (see Caliński, 1977, Definition 3). But (see Remark 2.4.2), for any connected orthogonal block design with a given replication vector $\boldsymbol{r}$, the unique nonzero eigenvalue of its C-matrix with respect to $\boldsymbol{r}^{\delta}$ is $\varepsilon = 1$. Hence, the precision of estimating $\boldsymbol{c}'\boldsymbol{\tau}$ in such a design is $1/\boldsymbol{c}'\boldsymbol{r}^{-\delta}\mathbf{c}$. Now, if the efficiency factor is defined as the ratio of the precision of the intra-block BLUE obtainable in the design under consideration to that obtainable in a connected orthogonal block design with the same treatment replications, i.e., according to the definition used by Pearce (1970), James and Wilkinson (1971), Pearce et al. (1974), Gnot (1976), Jarrett (1977) and Ceranka and Mejza (1979), then ε is exactly the efficiency factor of a design balanced in the sense of Jones for any contrast $\boldsymbol{c}'\boldsymbol{\tau}$ for which the intra-block BLUE exists. (See also Remark 3.3.3.) The constancy of ε explains why this concept of balance has been called, by Williams (1975) and Puri and Nigam (1975a, 1975b), the "efficiency balance." (See also Nigam et al., 1988, Definition 3.4.)

Because of the above arguments, the two commonly used notions of balance, one referring to equal variances and the other to equal efficiencies, can now be formalized as follows, assuming that the estimation is confined to the intra-block submodel (Section 3.2.1), i.e., taking as the BLUE of a contrast its intra-block BLUE.

Definition 4.1.1. A block design is said to be variance-balanced (VB) if it provides for every contrast $\boldsymbol{c}'\boldsymbol{\tau}$ normalized with regard to the norm $\|\mathbf{c}\| = (\boldsymbol{c}'\mathbf{c})^{1/2}$, for which the BLUE in the design exists, the same variance of the BLUE.

Definition 4.1.2. A block design is said to be efficiency-balanced (EB) if it provides for every contrast $\boldsymbol{c}'\boldsymbol{\tau}$, for which the BLUE in the design exists, the same efficiency factor, defined as the ratio of the precision of the BLUE obtainable in the design under consideration to that obtainable in a connected orthogonal block design with the same replication vector $\boldsymbol{r}$.

From the preceded discussion, and Theorems 2.4.1–2.4.4, it is evident that if the design is connected, then Definition 4.1.1 is equivalent to Definition 2.4.3, although in general it is equivalent to Definition 2.4.4. Also, it is evident that Definition 4.1.2 is equivalent to Definition 2.4.5 in the connected case and equivalent to Definition 2.4.6 in general. However, some ambiguity may appear with regard to Definitions 4.1.1 and 4.1.2, as they can equivalently be expressed in the following way.

Definition 4.1.1*. A block design is said to be VB if it provides for every contrast $\boldsymbol{c}'\boldsymbol{\tau}$, for which the BLUE in the design exists, the same efficiency factor, defined as the ratio of the precision of the BLUE obtainable in the design under consideration to that obtainable in an equireplicate, connected orthogonal block design with the common replication number $r = n/v$.

Definition 4.1.2*. A block design is said to be EB if it provides for every contrast $\boldsymbol{c}'\boldsymbol{\tau}$ normalized with regard to the norm $\|\mathbf{c}\|_{\boldsymbol{r}^{-\delta}} = (\boldsymbol{c}'\boldsymbol{r}^{-\delta}\mathbf{c})^{1/2}$, for which the BLUE in the design exists, the same variance of the BLUE.

It follows, then, that the concept of balance expressed in Definition 4.1.1 and that expressed in Definition 4.1.2 can imply equal precisions (i.e., equal variances of the intra-block BLUEs) and *at the same time* equal efficiency factors for estimated contrasts, this depending on the choice of (a) the normalization of the contrasts and (b) the orthogonal design with which the comparisons of precision are to be made.

In practice, this dependence will mean, for example, that a researcher wanting to use a design with unequal replications may wish to have all of the elementary contrasts, $\tau_i - \tau_{i'}, i, i' = 1, 2, ..., v$ $(i \neq i')$, estimated with the variances either equal to $2\sigma_1^2/\theta$, as in an RBD with $r = n/v$, for which $\theta = r$, or equal to $\varepsilon^{-1}(r_i^{-1} + r_{i'}^{-1})\sigma_1^2$, as in a completely randomized design with the same vector of treatment replications $\boldsymbol{r}$, for which $\varepsilon = 1$. In the first case, a totally balanced design (in the sense of Definition 2.4.3), i.e., a VB design (Definition 4.1.1), should be recommended; in the second, the use of a totally balanced design in the sense of Jones (Definition 2.4.5), i.e., an EB design (Definition 4.1.2), will be advisable. If the different replication numbers have been chosen purposely, e.g., to reflect unequal interest in the treatments, the second concept of balance might be more appropriate.

The considerations above show that the introduction of the terms VB and EB has been to some extent arbitrary. Nevertheless, because these terms have been so widely adopted in the literature, it may be justified also to use them here, following the recommendation given by Preece (1982, p. 151). But, it may also be useful to indicate that both of these concepts of balance are in fact particular cases of a more general notion of balance, of the following form.

Definition 4.1.3. Given a diagonal $v \times v$ matrix $\boldsymbol{X}$ with the diagonal elements all positive, a block design is said to be generalized efficiency-balanced (GEB), or $\boldsymbol{X}^{-1}$-balanced, if it provides for every contrast $\boldsymbol{c}'\boldsymbol{\tau}$ normalized with regard to the norm $\|\boldsymbol{c}\|_{X^{-1}} = (\boldsymbol{c}'\boldsymbol{X}^{-1}\boldsymbol{c})^{1/2}$, for which the BLUE in the design exists, the same variance of the BLUE.

The term $\boldsymbol{X}^{-1}$-balance was introduced in Caliński (1977), whereas Das and Ghosh (1985) used the term GEB. The latter can be justified by the fact that an $\boldsymbol{X}^{-1}$-balanced design provides for every contrast $\boldsymbol{c}'\boldsymbol{\tau}$, for which the BLUE in the design exists, the same efficiency factor, defined as the ratio of the precision of the BLUE obtainable in the design under consideration to that obtainable in a connected orthogonal block design with the replication vector $\boldsymbol{r} = n(\mathbf{1}_v'\boldsymbol{X}\mathbf{1}_v)^{-1}\boldsymbol{X}\mathbf{1}_v$ (see Caliński et al., 1980).

In view of the discussion above, it may be concluded that the commonly used distinction between VB and EB is not fully convincing as long as the concept of efficiency is not defined in a unique and unambiguous way.

4.2 Within-stratum efficiency factors

A well-established notion of efficiency is that related to the resolution of the model underlying the analysis of a block design into submodels corresponding to the stratification of the experimental units. For the randomization model adopted here, the resolution is as shown in (3.2.1), where the component $\boldsymbol{y}_1$ corresponds to the intra-block stratum, $\boldsymbol{y}_2$ to the inter-block stratum, and $\boldsymbol{y}_3$ to that of the total area. As stated in Theorem 3.1.1, for a function $\boldsymbol{c}'\boldsymbol{\tau} = \boldsymbol{s}'\boldsymbol{r}^{\delta}\boldsymbol{\tau}$, the BLUE exists under the overall randomization model (3.1.1) if and only if the condition (3.1.17) is satisfied. Then, on account of Corollary 3.2.5, the BLUE, of the form $\widehat{\boldsymbol{c}'\boldsymbol{\tau}} = \boldsymbol{s}'\Delta\boldsymbol{y}$, has the variance

$$\mathrm{Var}(\widehat{\boldsymbol{c}'\boldsymbol{\tau}}) = \boldsymbol{s}'\boldsymbol{r}^{\delta}\boldsymbol{s}\sigma_1^2 = \boldsymbol{c}'\boldsymbol{r}^{-\delta}\boldsymbol{c}\sigma_1^2, \tag{4.2.1}$$

where $\sigma_1^2 = \sigma_U^2 + \sigma_e^2$, if $\boldsymbol{N}'\boldsymbol{s} = \mathbf{0}$ (implying $\boldsymbol{r}'\boldsymbol{s} = 0$),

$$\begin{aligned} \mathrm{Var}(\widehat{\boldsymbol{c}'\boldsymbol{\tau}}) &= \boldsymbol{s}'\boldsymbol{N}\boldsymbol{N}'\boldsymbol{s}(\sigma_B^2 - K_H^{-1}\sigma_U^2) + \boldsymbol{s}'\boldsymbol{r}^{\delta}\boldsymbol{s}\sigma_1^2 \\ &= \boldsymbol{s}'\boldsymbol{r}^{\delta}\boldsymbol{s}[\boldsymbol{t}'\boldsymbol{k}^{2\delta}\boldsymbol{t}(\sigma_B^2 - K_H^{-1}\sigma_U^2) + \sigma_1^2], \end{aligned} \tag{4.2.2}$$

where $t = (s'r^{\delta}s)^{-1/2}[s_1 1'_{b_1}, s_2 1'_{b_2}, ..., s_g 1'_{b_g}]'$ (for the definition of the coefficients $s_\ell, \ell = 1, 2, ..., g$, see Remark 3.2.8), if $N's \neq 0$, but $\phi_1 \Delta' s = 0$ and $r's = 0$, or

$$\mathrm{Var}(\widehat{c'\tau}) = n^{-1}[n^{-1}k'k(\sigma_B^2 - K_H^{-1}\sigma_U^2) - N_B^{-1}n\sigma_B^2 + \sigma_1^2], \tag{4.2.3}$$

if $c'\tau = n^{-1}r'\tau$ (the overall mean), i.e., if $c = n^{-1}r$ and $s = n^{-1}1_v$.

On the other hand, if the condition (3.1.17) is not satisfied, then, on account of Theorem 3.3.1, for any contrast $c'\tau = s'r^{\delta}\tau$ $(1'_v c = r's = 0)$ for which s satisfies the eigenvector condition (3.3.4) with $0 < \varepsilon < 1$, the BLUE obtainable in the intra-block analysis is

$$(\widehat{c'\tau})_{\text{intra}} = \varepsilon^{-1}s'\Delta\phi_1 y,$$

with

$$\mathrm{Var}[(\widehat{c'\tau})_{\text{intra}}] = \varepsilon^{-1}s'r^{\delta}s\sigma_1^2, \tag{4.2.4}$$

and that obtainable in the inter-block analysis is, under the conditions of Corollary 3.2.1(c),

$$(\widehat{c'\tau})_{\text{inter}} = (1-\varepsilon)^{-1}s'\Delta\phi_2 y,$$

with

$$\mathrm{Var}[(\widehat{c'\tau})_{\text{inter}}] = (1-\varepsilon)^{-1}s'r^{\delta}s[t'k^{2\delta}t(\sigma_B^2 - K_H^{-1}\sigma_U^2) + \sigma_1^2], \tag{4.2.5}$$

where t is related to s according to (3.3.7). Certainly, $s'r^{\delta}s = c'r^{-\delta}c$ here.

Thus, if s is an eigenvector of the C-matrix of the design with respect to r^{δ}, then for the same contrast $c'\tau = s'r^{\delta}\tau$, three BLUEs may exist,

$$s'\Delta y, \quad \varepsilon^{-1}s'\Delta\phi_1 y \quad \text{and} \quad (1-\varepsilon)^{-1}s'\Delta\phi_2 y,$$

the first under the overall model (3.1.1), provided that the condition (3.1.17) holds, the second in the intra-block analysis and the third in the inter-block analysis, both if the condition (3.1.17) is not satisfied, but for the inter-block analysis provided that $K_0 N's \in \mathcal{C}(N'_0)$ [for the notation, see Theorem 3.2.2 and Corollary 3.2.1(c)]. If the variance components σ_B^2 and σ_U^2 are such that $\sigma_B^2 - K_H^{-1}\sigma_U^2 = 0$, then the corresponding variances of the BLUEs are

$$s'r^{\delta}s\sigma_1^2, \quad \varepsilon^{-1}s'r^{\delta}s\sigma_1^2 \quad \text{and} \quad (1-\varepsilon)^{-1}s'r^{\delta}s\sigma_1^2,$$

i.e., are in the proportions

$$1 : \ \varepsilon^{-1} : \ (1-\varepsilon)^{-1}.$$

These relations enabled Jones (1959, p. 176) to call ε the efficiency factor of the design for the contrast represented by s and, at the same time, to call $1-\varepsilon$ the relative loss of information caused by partially confounding the contrast with blocks (see also Remark 3.3.3). Moreover, for any proper block design, the variances of the intra-block and inter-block BLUEs, i.e., those in (4.2.4) and (4.2.5), are

$$\varepsilon^{-1}\boldsymbol{s}'\boldsymbol{r}^{\delta}\boldsymbol{s}\sigma_1^2 \quad \text{and} \quad (1-\varepsilon)^{-1}\boldsymbol{s}'\boldsymbol{r}^{\delta}\boldsymbol{s}\sigma_2^2,$$

respectively, where

$$\sigma_2^2 = k\sigma_B^2 + (1 - K_H^{-1}k)\sigma_U^2 + \sigma_e^2$$

(see also Theorem 3.5.2). They attain their minimum values when $\varepsilon = 1$ or $\varepsilon = 0$, respectively, i.e., when BLUE is obtained under the overall model (3.1.1), for which the design is to be such that $\boldsymbol{N}'\boldsymbol{s} = \boldsymbol{0}$, to obtain $\varepsilon = 1$, or such that $\boldsymbol{N}\boldsymbol{k}^{-\delta}\boldsymbol{N}'\boldsymbol{s} = \boldsymbol{r}^{\delta}\boldsymbol{s}$, to obtain $\varepsilon = 0$. [To see this result, recall Corollary 3.2.5 and Remark 3.6.1, and note that when the block sizes are all equal then the variance (4.2.2) reduces to (3.2.34), i.e., to $\text{Var}(\widehat{\boldsymbol{c}'\boldsymbol{\tau}}) = \boldsymbol{s}'\boldsymbol{r}^{\delta}\boldsymbol{s}\sigma_2^2$.] Therefore, for a proper block design, in the same way as σ_1^2 and σ_2^2 in (3.5.7) are called the stratum variances, $\varepsilon_1 = \varepsilon$ and $\varepsilon_2 = 1 - \varepsilon$ can be called the "within-stratum efficiency factors." More precisely, ε_1 can be termed the efficiency factor for the contrast $\boldsymbol{c}'\boldsymbol{\tau} = \boldsymbol{s}'\boldsymbol{r}^{\delta}\boldsymbol{\tau}$ in stratum 1, i.e., intra-block, and ε_2 can be termed the efficiency factor for this contrast in stratum 2, i.e., inter-block, if the vector $\boldsymbol{s}$ representing the contrast is an eigenvector of $\boldsymbol{C}_1 = \boldsymbol{\Delta}\boldsymbol{\phi}_1\boldsymbol{\Delta}'$ with respect to $\boldsymbol{r}^{\delta}$, ε being the corresponding eigenvalue (see Remark 3.5.3 and Houtman and Speed, 1983, p. 1075).

For completeness, it may also be noted that the BLUE of the function $\boldsymbol{c}'\boldsymbol{\tau} = n^{-1}\boldsymbol{r}'\boldsymbol{\tau}$ under the total-area submodel $\boldsymbol{y}_3 = \boldsymbol{\phi}_3\boldsymbol{y}$ (Section 3.2.3), equal to $n^{-1}\mathbf{1}_v'\boldsymbol{\Delta}\boldsymbol{y}_3 = n^{-1}\mathbf{1}_v'\boldsymbol{\Delta}\boldsymbol{\phi}_3\boldsymbol{y} = n^{-1}\mathbf{1}_n'\boldsymbol{y}$, is for a proper block design also the BLUE under the overall model (3.1.1), with the variance (4.2.3) reduced to

$$\text{Var}(\widehat{\boldsymbol{c}'\boldsymbol{\tau}}) = n^{-1}[k(\sigma_B^2 - K_H^{-1}\sigma_U^2) - N_B^{-1}n\sigma_B^2 + \sigma_1^2] = \boldsymbol{s}'\boldsymbol{r}^{\delta}\boldsymbol{s}\sigma_3^2,$$

where $\boldsymbol{s} = n^{-1}\mathbf{1}_v$ and

$$\sigma_3^2 = (1 - N_B^{-1}b)k\sigma_B^2 + (1 - K_H^{-1}k)\sigma_U^2 + \sigma_e^2,$$

as in (3.5.7). Because of this fact, $\varepsilon_3 = 1$ (see also Theorem 3.5.2).

The above interpretation of the eigenvalue ε for a proper block design is, however, not so simple for nonproper block designs. The difficulty concerns the inter-block analysis, in which $1 - \varepsilon$ can be considered in general only as a notionally acceptable efficiency factor (see Remark 3.3.4 and the discussion after it). Nevertheless, the eigenvalue ε remains a well-established efficiency factor in the intra-block analysis, for proper as well as nonproper block designs. As mentioned in Remark 3.3.3, it is also called the canonical efficiency factor. It has been broadly used in the literature in describing properties of block designs. To avoid confusion, if the abbreviated term "efficiency factor" (or "canonical efficiency factor") is used, it has to be understood as the above "efficiency factor in the intra-block analysis" (i.e., as in Section 3.4).

4.3 More on Jones's concept of balance

In connection with the efficiency factors discussed in Section 4.2, it should be remembered that according to Jones (1959, p. 176), a block design is called balanced for a contrast $\boldsymbol{c}'\boldsymbol{\tau} = \boldsymbol{s}'\boldsymbol{r}^{\delta}\boldsymbol{\tau}$ when $\boldsymbol{s}$ satisfies the eigenvector condition (3.3.4), i.e., satisfies the equality

$$\boldsymbol{C}_1\boldsymbol{s} = \varepsilon\boldsymbol{r}^{\delta}\boldsymbol{s} \tag{4.3.1}$$

with some ε. The eigenvalue ε is then called the efficiency factor of the design for the contrast represented by $\boldsymbol{s}$, which means that, in Jones's terminology, a design may be called balanced for one contrast, or for a set (or rather subspace) of contrasts when all members of the set satisfy (4.3.1) with the same ε, i.e., when the design provides for them the same efficiency factor. Thus, in Jones's approach, the notions of balance and efficiency are closely interrelated. Two contrasts represented by $\boldsymbol{s}_1$ and $\boldsymbol{s}_2$ may both satisfy (4.3.1). Then, one can say that the design is balanced for each of them. But the balance may for each of them be the same or different, depending on whether the corresponding efficiency factors, ε_1 and ε_2, are equal or unequal.

Now, it may be useful to recall that contrasts represented by $\boldsymbol{r}^{\delta}$-orthonormal eigenvectors of the matrix $\boldsymbol{C}_1$ with respect to $\boldsymbol{r}^{\delta}$ have been called basic contrasts (Definition 3.4.1). Therefore, in the terminology of Jones (1959), a block design is balanced for any basic contrast taken separately, but it is also balanced jointly for a set of them, or a subspace spanned by them, if all of them correspond to the same efficiency factor. The number of distinct efficiency factors determines, then, the number of ways in which the design can be considered as balanced. One extreme case is when the C-matrix of the design has a unique positive eigenvalue with respect to $\boldsymbol{r}^{\delta}$, with multiplicity $v-1$. The design is then balanced in the sense of Jones for all possible contrasts and can really be called totally balanced in that sense. An opposite extreme case is that in which the $v-1$ eigenvalues corresponding to a complete set of basic contrasts are all different. Then, one can only say that the design is balanced for each of the basic contrasts separately. In this case, no two basic contrasts exist for which the design is balanced in the same way, i.e., with the same efficiency factor. Certainly, many possible situations can exist between these two extremes, as the eigenvalues of $\boldsymbol{C}_1$ with respect to $\boldsymbol{r}^{\delta}$ may appear in various multiplicities.

The various possibilities offered by the concept of balance in the sense of Jones (1959) are exploited in two relevant notions of balance, that of general balance introduced by Nelder (1965a, 1965b) and that of partial efficiency balance introduced by Puri and Nigam (1977a).

4.3.1 Relation to the notion of general balance

As already described in Section 3.6, the notion of general balance (GB) is strictly connected with the spectral decompositions

$$C_1 = \boldsymbol{\Delta}\boldsymbol{\phi}_1\boldsymbol{\Delta}' = \boldsymbol{r}^{\delta}\sum_{\beta=0}^{m-1}\varepsilon_\beta \boldsymbol{S}_\beta \boldsymbol{S}'_\beta \boldsymbol{r}^{\delta}, \tag{4.3.2}$$

$$C_2 = \boldsymbol{\Delta}\boldsymbol{\phi}_2\boldsymbol{\Delta}' = \boldsymbol{r}^{\delta}\sum_{\beta=1}^{m}(1-\varepsilon_\beta) \boldsymbol{S}_\beta \boldsymbol{S}'_\beta \boldsymbol{r}^{\delta} \tag{4.3.3}$$

and (for completeness)

$$C_3 = \boldsymbol{\Delta}\boldsymbol{\phi}_3\boldsymbol{\Delta}' = \boldsymbol{r}^{\delta}\boldsymbol{s}_v\boldsymbol{s}'_v\boldsymbol{r}^{\delta},$$

where $\boldsymbol{S}_\beta = [\boldsymbol{s}_{\beta 1} : \boldsymbol{s}_{\beta 2} : \cdots : \boldsymbol{s}_{\beta \rho_\beta}]$ represents a subset of basic contrasts, those corresponding to a common distinct eigenvalue ε_β, for $\beta = 0, 1, ..., m$, and $\boldsymbol{s}_v = n^{-1/2}\boldsymbol{1}_v$. These decompositions correspond to the decomposition of the subspace $\mathcal{C}(\boldsymbol{\Delta}')$ to which the expectation vector $\mathrm{E}(\boldsymbol{y})$, (3.1.15), belongs. It is based on the same $\boldsymbol{S}_\beta$ matrices and can be written as

$$\mathcal{C}(\boldsymbol{\Delta}') = \mathcal{C}(\boldsymbol{\Delta}'\boldsymbol{S}_0) \oplus \mathcal{C}(\boldsymbol{\Delta}'\boldsymbol{S}_1) \oplus \cdots \oplus \mathcal{C}(\boldsymbol{\Delta}'\boldsymbol{S}_m) \oplus \mathcal{C}(\boldsymbol{\Delta}'\boldsymbol{s}_v). \tag{4.3.4}$$

If the block design is proper, so that the covariance (dispersion) matrix $\mathrm{Cov}(\boldsymbol{y})$, (3.1.16), can be expressed as

$$\mathrm{Cov}(\boldsymbol{y}) = \sigma_1^2\boldsymbol{\phi}_1 + \sigma_2^2\boldsymbol{\phi}_2 + \sigma_3^2\boldsymbol{\phi}_3,$$

with σ_1^2, σ_2^2 and σ_3^2 defined in (3.5.7), then (see Definition 3.6.1) the design is said to be GB with respect to the decomposition (4.3.4). The eigenvalues $\{\varepsilon_\beta\}$ in (4.3.2) and $\{1 - \varepsilon_\beta\}$ in (4.3.3) can then be easily interpreted as the distinct efficiency factors, ε_β as the common efficiency factor of the design for $\boldsymbol{S}_\beta$ in the intra-block analysis and $1 - \varepsilon_\beta$ as that for $\boldsymbol{S}_\beta$ in the inter-block analysis (see Corollary 3.6.3 and Remark 3.6.4).

Note that in the terminology of Jones (1959) the design for which (4.3.4) holds can be said to be balanced for each of the subsets of basic contrasts separately. It is balanced for contrasts represented by $\boldsymbol{S}_\beta$ with the common efficiency factor given by ε_β, and with the relative loss of information, caused by partially confounding the contrasts with blocks, defined by $1 - \varepsilon_\beta, \beta = 0, 1, ..., m$. In view of this meaning of balance, it becomes natural to call the design GB with respect to (4.3.4).

It should, however, be recalled that Jones (1959) confined his considerations to the intra-block analysis only. Therefore, the efficiency factor is in his terminology always understood in the sense of efficiency of estimation in the intra-block analysis. Accordingly, $1 - \varepsilon_\beta$ is understood by him as simply the relative loss of information incurred in that analysis. He does not interpret $1 - \varepsilon_\beta$ as the efficiency factor in the inter-block analysis. This terminology would be fully justified in the case of a proper block design only (see the discussion after Remark 3.3.4), whereas Jones (1959) also takes into account in his approach nonproper block designs. On the other hand, because the difficulty with the interpretation of $1-\varepsilon_\beta$ in terms of efficiency does not appear in proper block designs, to which the definition of GB is confined, in this notion of balance, references can be made to both analyses, the intra-block and the inter-block.

It follows from this discussion that to characterize a block design from the point of view of balance, it is essential to specify the decomposition (4.3.4) and the corresponding distinct eigenvalues $\{\varepsilon_\beta\}$ appearing in (4.3.2) and (4.3.3). If the basic contrasts represented by $\{\boldsymbol{S}_\beta\}$ are of real interest to the researcher, and the corresponding efficiency factors defined by $\{\varepsilon_\beta\}$ reflect that interest properly, then the GB property has a practical meaning.

To illustrate the practical sense of the concept of GB, it may be interesting to return to one of the examples of Jones (1959, Section 8), also discussed in Caliński (1971, p. 292).

Example 4.3.1. Consider a 3 × 2 factorial experiment described by the following incidence matrix, whose rows correspond to the indicated treatment combinations:

Level of A	Level of B
0	0
1	0
2	0
0	1
1	1
2	1

$$\boldsymbol{N} = \begin{bmatrix} 1 & 0 & 1 & 0 & 1 & 1 \\ 1 & 1 & 0 & 1 & 1 & 0 \\ 0 & 1 & 1 & 1 & 0 & 1 \\ 0 & 1 & 0 & 1 & 1 & 1 \\ 1 & 1 & 1 & 0 & 0 & 1 \\ 1 & 0 & 1 & 1 & 1 & 0 \end{bmatrix}.$$

It is a proper, equireplicate, binary connected block design with the C-matrix

$$\boldsymbol{C}_1 = 4\boldsymbol{I}_6 - \frac{1}{4}\boldsymbol{N}\boldsymbol{N}',$$

where

$$\boldsymbol{N}\boldsymbol{N}' = \begin{bmatrix} 4 & 2 & 2 & 2 & 3 & 3 \\ 2 & 4 & 2 & 3 & 2 & 3 \\ 2 & 2 & 4 & 3 & 3 & 2 \\ 2 & 3 & 3 & 4 & 2 & 2 \\ 3 & 2 & 3 & 2 & 4 & 2 \\ 3 & 3 & 2 & 2 & 2 & 4 \end{bmatrix}.$$

The six basic contrasts found by Jones can be represented by the vectors

$$\begin{aligned}
\boldsymbol{s}_1 &= [1,1,1,-1,-1,-1]'/\sqrt{24} && \text{corresponding to } \varepsilon_1 = 1, \\
\boldsymbol{s}_2 &= [1,-1,0,1,-1,0]'/\sqrt{16} && \text{corresponding to } \varepsilon_2 = 15/16, \\
\boldsymbol{s}_3 &= [1,1,-2,1,1,-2]'/\sqrt{48} && \text{corresponding to } \varepsilon_3 = 15/16, \\
\boldsymbol{s}_4 &= [1,-1,0,-1,1,0]'/\sqrt{16} && \text{corresponding to } \varepsilon_4 = 13/16, \\
\boldsymbol{s}_5 &= [1,1,-2,-1,-1,2]'/\sqrt{48} && \text{corresponding to } \varepsilon_5 = 13/16.
\end{aligned}$$

Noting that $r^{\delta} = 4I_6$, it can easily be checked that the vectors $\{s_i\}$ are r^{δ}-orthonormal, i.e., satisfy the conditions $s_i' r^{\delta} s_i = 1$ and $s_i' r^{\delta} s_{i'} = 0$, if $i \neq i'$, for $i, i' = 1, 2, 3, 4, 5$, and, furthermore, that they are eigenvectors of the matrix C_1 with respect to r^{δ}, i.e., satisfy the condition $C_1 s_i = \varepsilon_i r^{\delta} s_i$ for all i, which is equivalent to the condition $M s_i = \alpha_i s_i$ of Jones (1959, p. 175), with $M = r^{-\delta} N k^{-\delta} N'$ and $\alpha_i = 1 - \varepsilon_i$, as well as to the condition $M_0 s_i = \mu_i s_i$ of Caliński (1971, p. 281), with $M_0 = M - n^{-1} 1_v r'$ and $\mu_i = \alpha_i$, as $r' s_i = 0$ for all contrasts. Moreover, it should be noted that s_1 represents the B factor contrast, s_2 and s_3 represent the A factor contrasts, and s_4 and s_5 represent the interaction contrasts. Thus, the design is orthogonal for the B factor contrast, estimated in the intra-block analysis with full efficiency, is balanced for the A factor contrasts, estimated with efficiency 15/16 in the intra-block analysis and with efficiency 1/16 in the inter-block analysis, and is balanced for the interaction contrasts, estimated with efficiency 13/16 and 3/16 intra-block and inter-block, respectively. Referring now to Remark 3.6.4, these statements can be summarized by saying that the design is GB with respect to the decomposition (4.3.4), where $S_0 = s_1$, $S_1 = [s_2, s_3]$ and $S_2 = [s_4, s_5]$. But, because of the meaning of the contrasts represented by the vectors $\{s_i\}$, in terms of the treatments actually applied in the experiment, it can also be said that the design is GB with respect to the 3×2 factorial structure of the experimental treatments.

It is essential, when using the GB terminology, always to refer to the decomposition (3.6.19) [here (4.3.4)] with respect to which the balance of the design holds. If the decomposition is coherent with the treatment structure specified by the experimental problem, then the GB property makes sense. However, if the decomposition is meaningless from the point of view of the treatment structure of the experiment, then it may be difficult to make any practical use of that concept. This difficulty will become apparent in the next example.

Example 4.3.2. Consider a 3×2 factorial experiment in which the same incidence matrix as in Example 4.3.1 is used, but now with different application to the experimental treatments, as follows:

Level of A	Level of B
0	0
0	1
1	0
1	1
2	0
2	1

$$N = \begin{bmatrix} 1 & 0 & 1 & 0 & 1 & 1 \\ 1 & 1 & 0 & 1 & 1 & 0 \\ 0 & 1 & 1 & 1 & 0 & 1 \\ 0 & 1 & 0 & 1 & 1 & 1 \\ 1 & 1 & 1 & 0 & 0 & 1 \\ 1 & 0 & 1 & 1 & 1 & 0 \end{bmatrix}$$

The C-matrix of the design remains the same, and it has the same eigenvectors $\{s_i\}$ with respect to $r^{\delta} = 4I_6$. However, now the meaning of the basic contrasts represented by these eigenvectors is different. The vector s_1 represents a contrast between the first three treatments and the remaining ones. Such contrast is of no interest from the point of view of the present factorial structure of the treatments.

Similarly, no other of the contrasts represented by $\{s_i\}$ is interesting under that structure. To obtain contrasts of interest in this experiment, one has to transform the vectors $\{s_i\}$ by the matrix

$$\begin{bmatrix} 1 & 0 & 0 & 0 & 0 & 0 \\ 0 & 0 & 0 & 1 & 0 & 0 \\ 0 & 1 & 0 & 0 & 0 & 0 \\ 0 & 0 & 0 & 0 & 1 & 0 \\ 0 & 0 & 1 & 0 & 0 & 0 \\ 0 & 0 & 0 & 0 & 0 & 1 \end{bmatrix} = \boldsymbol{A} \quad \text{(say)}.$$

The vectors $\{s_i^*\}$ so obtained are

$$\begin{aligned} s_1^* &= As_1 = [1,-1,1,-1,1,-1]'/\sqrt{24}, & s_2^* &= As_2 = [1,1,-1,-1,0,0]'/\sqrt{16}, \\ s_3^* &= As_3 = [1,1,1,1,-2,-2]'/\sqrt{48}, & s_4^* &= As_4 = [1,-1,-1,1,0,0]'/\sqrt{16}, \\ s_5^* &= As_5 = [1,-1,1,-1,-2,2]'/\sqrt{48}. \end{aligned}$$

They are still r^δ-orthonormal but, unlike the vectors $\{s_i\}$, they do not remain eigenvectors of C_1 with respect to $r^\delta = 4I_6$ anymore. Thus, the design considered in the present example is not GB with respect to the 3×2 factorial structure of the experimental treatments, as is the case in Example 4.3.1.

The two examples above show that in designing an experiment, it is not sufficient to choose a design suitable for the block structure. The design is then to be properly adopted to the experimental problem; i.e., one has to assign the design treatments to the experimental treatments in such a way that the design becomes GB with respect to contrasts that are essential from the point of view of the experimental questions under study. For further discussion, see Pearce (1983, Section 4.8), and for more illustrative examples, see Ceranka (1983, Section 7).

4.3.2 Relation to the notion of partial efficiency balance

The notion of partial efficiency balance is also strictly connected with the spectral decomposition of the C-matrix of the design, i.e., the decomposition (4.3.2). It originates from such decomposition of the matrix $M = r^{-\delta}Nk^{-\delta}N'$ used by Jones (1959, p. 175) or its modification $M_0 = M - n^{-1}1_v r'$ adopted in Caliński (1971, p. 278, 1972, p. 8). In fact, corresponding to (4.3.2), one can write

$$M = \sum_{\beta=0}^{m} \mu_\beta S_\beta S_\beta' r^\delta + n^{-1}1_v r' \qquad (4.3.5)$$

or

$$M_0 = \sum_{\beta=0}^{m} \mu_\beta S_\beta S_\beta' r^\delta, \qquad (4.3.6)$$

where the eigenvalues $\mu_\beta = 1 - \varepsilon_\beta$ are such that $0 \leq \mu_\beta \leq 1$, for $\beta = 0, 1, ..., m$, with $\mu_0 = 0$ and $\mu_m = 1$. Introducing (after Caliński, 1971, p. 289, and Puri and Nigam, 1977a, Section 1) the idempotent matrices [as in (3.4.10) and (3.6.11)]

$$\boldsymbol{L}_\beta = \boldsymbol{S}_\beta \boldsymbol{S}'_\beta \boldsymbol{r}^\delta, \quad \beta = 0, 1, ..., m, \quad \text{and} \quad \boldsymbol{L}_{m+1} = n^{-1}\boldsymbol{1}_v \boldsymbol{r}', \tag{4.3.7}$$

the last corresponding to $\mu_{m+1} = 1$, the decompositions (4.3.5) and (4.3.6) can alternatively be written as

$$\boldsymbol{M} = \sum_{\beta=0}^{m+1} \mu_\beta \boldsymbol{L}_\beta \quad \text{and} \quad \boldsymbol{M}_0 = \sum_{\beta=0}^{m} \mu_\beta \boldsymbol{L}_\beta, \tag{4.3.8}$$

respectively. Note that

$$\boldsymbol{L}_\beta^2 = \boldsymbol{L}_\beta \text{ and } \boldsymbol{L}_\beta \boldsymbol{L}_{\beta'} = \mathbf{O} \text{ if } \beta \neq \beta', \text{ for } \beta, \beta' = 0, 1, ..., m+1,$$

and that

$$\sum_{\beta=0}^{m} \boldsymbol{L}_\beta = \boldsymbol{I}_v - n^{-1}\boldsymbol{1}_v \boldsymbol{r}'.$$

It should, however, be mentioned that this notation (introduced in Section 3.4) is slightly different from that used originally (by Puri and Nigam, 1977a, in particular), because in the earlier papers the order of indexing the eigenvalues $\{\mu_\beta\}$ was different than it is here (see also Kageyama and Puri, 1985a). In those papers, the distinct eigenvalues have been ordered (at least implicitly) as

$$1 = \mu_0 > \mu_1 > \cdots > \mu_{m-1} > \mu_m \geq 0,$$

assuming that $\mu_0 = 1$ has multiplicity 1, because only connected designs have been considered there. Here, in accordance with Lemma 2.3.3, the order of the eigenvalues is opposite, namely,

$$0 = \mu_0 < \mu_1 < \cdots < \mu_{m-1} < \mu_m = \mu_{m+1} = 1, \tag{4.3.9}$$

with multiplicities ρ_β, for $\beta = 0, 1, ..., m, m+1$, where $\rho_0 = 0$ if no basic contrast is estimated with full efficiency in the intra-block analysis, i.e., the design is not orthogonal for any contrasts estimated in this analysis [see Remark 3.4.1(a), where $\rho \equiv \rho_0$], and $\rho_m = 0$ if no basic contrast is totally confounded with blocks, i.e., the design is connected, and $\rho_{m+1} = 1$ always. Thus, $m - 1$ is the number of distinct, less than 1, positive eigenvalues in (2.3.1). In correspondence to (4.3.9), the efficiency factors $\varepsilon_\beta = 1 - \mu_\beta$ are ordered as

$$1 = \varepsilon_0 > \varepsilon_1 > \cdots > \varepsilon_{m-1} > \varepsilon_m = \varepsilon_{m+1} = 0,$$

with multiplicities ρ_β for $\beta = 0, 1, ..., m, m+1$ [see again formula (2.3.2), where $\rho \equiv \rho_0$].

With this notation, the definition of Puri and Nigam (1977a, p. 755) can be expressed for any connected block design as follows (see also Caliński and Kageyama, 1996b, Section 3.2.1).

Definition 4.3.1. An arrangement of v treatments into b blocks of sizes k_1, k_2, ..., k_b is said to be a (connected) partially efficiency-balanced (PEB) design with m efficiency classes if

(i) the ith treatment is replicated r_i times, $i = 1, 2, ..., v$;

(ii) the efficiency factor associated with every contrast of the βth class is $1 - \mu_\beta$, where μ_β, $0 \leq \mu_\beta < 1$, $\beta = 0, 1, ..., m-1$, are the distinct eigenvalues of the matrix $\boldsymbol{M}_0$, with multiplicities ρ_β (> 0), such that $\sum_{\beta=0}^{m-1} \rho_\beta = v - 1$ (i.e., except the zero eigenvalue corresponding to $\boldsymbol{M}_0 \boldsymbol{1}_v = \boldsymbol{0}$);

(iii) the matrix $\boldsymbol{M}_0$ has the spectral decomposition

$$\boldsymbol{M}_0 = \sum_{\beta=0}^{m-1} \mu_\beta \boldsymbol{L}_\beta, \tag{4.3.10}$$

where $\boldsymbol{L}_\beta, \beta = 0, 1, ..., m-1$, are defined as in (4.3.7), and $\mu_0 = 0$, as in (4.3.9). However, if $\rho_0 = 0$, i.e., if $\mu_0 = 0$ does not exist, then the considered arrangement is said to have $m - 1$ efficiency classes.

In this notation, corresponding to that in Section 3.4, a contrast belonging to the βth class is understood as $\boldsymbol{c}'_{\beta j}\boldsymbol{\tau} = \boldsymbol{s}'_{\beta j}\boldsymbol{r}^\delta\boldsymbol{\tau}$, where $\boldsymbol{s}_{\beta j}$ ($\boldsymbol{r}'\boldsymbol{s}_{\beta_j} = 0$) is an eigenvector of $\boldsymbol{M}_0$ corresponding to the eigenvalue μ_β, and $\boldsymbol{\tau}$, as usual, denotes the vector of treatment parameters. Such contrasts, usually being $\boldsymbol{r}^\delta$-orthonormalized, have been called basic contrasts (Definition 3.4.1) and are simply denoted by $\boldsymbol{s}_{\beta j}$, if convenient.

In other words, for a connected PEB design with m efficiency classes, there exists a set of $v - 1$ linearly independent basic contrasts $\{\boldsymbol{s}_{\beta j}\}$ which can be partitioned into m ($< v$) disjoint classes such that all ρ_β members of the βth class, and their linear combinations, are estimated in the intra-block analysis with the same relative loss of information $\mu_\beta < 1$, i.e., are estimated under the intra-block submodel with the efficiency factor $\varepsilon_\beta = 1 - \mu_\beta$, respectively. If $\boldsymbol{s}_{\beta j}$ is $\boldsymbol{r}^\delta$-normalized (as in Section 3.4), then the BLUE in the intra-block analysis has its variance as in (3.4.2).

A special class of PEB designs is considered by Kageyama and Puri (1985b).

4.4 Classification of block designs based on efficiency balance

Because the spectral decompositions (4.3.8) hold for any block design, whether connected or disconnected, any block design is PEB in the sense that if it is connected, i.e., $\rho_m = 0$, then all ρ_β basic contrasts of the βth class ($\beta = 0, 1, ..., m-1$) are estimated intra-block with the efficiency $\varepsilon_\beta = 1 - \mu_\beta (> 0)$, but if the design is disconnected, i.e., $\rho_m > 0$, then $\varepsilon_\beta = 0$ for one β ($= m$). Therefore, for a classification of block designs in general, the specification of the multiplicities $\{\rho_\beta\}$ is important in the statistical sense. To give merely the number of efficiency classes is not enough to indicate the practical suitability of a PEB design. One wants to know how large the multiplicities are and to which efficiency factors they correspond, e.g., whether the design allows a number of contrasts to be estimated intra-block with full efficiency. In this section, more informative terms are suggested for a statistical characterization of block designs. Also, the original definition of a PEB design is extended to cover disconnected block designs as well. The statistical sense of such characterization is then discussed.

4.4.1 Generalization and a new terminology

Now, let the ordering (4.3.9) of the eigenvalues $\{\mu_\beta\}$ be used to cover connected as well as disconnected block designs in an attempt to propose a comprehensive classification of block designs.

Note first that, with the $\boldsymbol{r}^\delta$-orthonormalized vectors $\{\boldsymbol{s}_{\beta j}\}$ (as in Lemma 2.3.3), the decompositions (4.3.8) can be written for any block design in the form

$$\boldsymbol{M} = \boldsymbol{r}^{-\delta}\boldsymbol{N}\boldsymbol{k}^{-\delta}\boldsymbol{N}' = \sum_{\beta=0}^{m+1}\sum_{j=1}^{\rho_\beta}\mu_\beta \boldsymbol{s}_{\beta j}\boldsymbol{s}'_{\beta j}\boldsymbol{r}^\delta = \sum_{\beta=0}^{m+1}\mu_\beta \boldsymbol{L}_\beta \tag{4.4.1}$$

and

$$\boldsymbol{M}_0 = \boldsymbol{M} - n^{-1}\boldsymbol{1}_v\boldsymbol{r}' = \sum_{\beta=0}^{m}\sum_{j=1}^{\rho_\beta}\mu_\beta \boldsymbol{s}_{\beta j}\boldsymbol{s}'_{\beta j}\boldsymbol{r}^\delta = \sum_{\beta=0}^{m}\mu_\beta \boldsymbol{L}_\beta, \tag{4.4.2}$$

where, according to (4.3.7),

$$\boldsymbol{L}_\beta = \boldsymbol{S}_\beta \boldsymbol{S}'_\beta \boldsymbol{r}^\delta = \sum_{j=1}^{\rho_\beta}\boldsymbol{s}_{\beta j}\boldsymbol{s}'_{\beta j}\boldsymbol{r}^\delta, \quad \text{rank}(\boldsymbol{L}_\beta) = \rho_\beta, \quad \beta = 0, 1, ..., m, \tag{4.4.3}$$

with $\boldsymbol{L}_0$ redundant in the decompositions, as $\mu_0 = 0$, $\boldsymbol{L}_m$ appearing only in a disconnected block design, and $\boldsymbol{L}_{m+1} = n^{-1}\boldsymbol{1}_v\boldsymbol{r}'$ corresponding to $\mu_{m+1} = 1$, all satisfying jointly the equality $\sum_{\beta=0}^{m+1}\boldsymbol{L}_\beta = \boldsymbol{I}_v$. (The definition of $\boldsymbol{M}_0$ used here is for the disconnected case different from that used by Pearce et al., 1974, p. 455.) Note that $\boldsymbol{L}_\beta = \boldsymbol{H}_\beta \boldsymbol{r}^\delta$ for any β, with $\boldsymbol{H}_\beta$ defined in (3.4.10). Thus, $\{\varepsilon_\beta = 1 - \mu_\beta\}$ are the eigenvalues and $\{\boldsymbol{s}_{\beta j}\}$ are the corresponding $\boldsymbol{r}^\delta$-orthonormal eigenvectors of $\boldsymbol{C}_1$ with respect to $\boldsymbol{r}^\delta$, representing the basic contrasts of the design (see Section 3.4). The eigenvalues $\{\varepsilon_\beta\}$ are then the efficiency factors of the design for estimating the corresponding basic contrasts in the intra-block analysis. With the matrices (4.4.3), the spectral decomposition (4.3.2) of $\boldsymbol{C}_1$ can, alternatively, be written as in (3.4.10), i.e., in the form

$$\boldsymbol{C}_1 = \sum_{\beta=0}^{m-1}\varepsilon_\beta \boldsymbol{r}^\delta \boldsymbol{L}_\beta. \tag{4.4.4}$$

Evidently, the term for $\beta = 0$ could be deleted from (4.4.1) and (4.4.2), as $\mu_0 = 0$, but not from (4.4.4), where $\varepsilon_0 = 1$. The form (4.4.4) of the matrix $\boldsymbol{C}_1$ implies that as a possible g-inverse of $\boldsymbol{C}_1$ the matrix

$$\sum_{\beta=0}^{m-1} \varepsilon_\beta^{-1} \boldsymbol{L}_\beta \boldsymbol{r}^{-\delta}$$

can be taken (see again Section 3.4). If a g-inverse of maximum rank is required, a suitable choice is the matrix

$$\left(\sum_{\beta=0}^{m-1} \frac{1-\varepsilon_\beta}{\varepsilon_\beta} \boldsymbol{L}_\beta + \boldsymbol{I}_v \right) \boldsymbol{r}^{-\delta} = \boldsymbol{\Omega} \quad \text{(say)},$$

in this form introduced in Caliński (1971, p. 289) and used by Ceranka (1973, 1984), Puri and Nigam (1977a), and others. In the case of a connected block design, it acts as the ordinary inverse of the matrix $\boldsymbol{\Omega}^{-1} = \boldsymbol{C}_1 + n^{-1}\boldsymbol{r}\boldsymbol{r}'$ used by Tocher (1952). In general, this matrix can be written as

$$\boldsymbol{\Omega}^{-1} = \boldsymbol{C}_1 + \boldsymbol{r}^\delta \boldsymbol{L}_m + \boldsymbol{r}^\delta \boldsymbol{L}_{m+1}, \quad \text{where} \quad \boldsymbol{r}^\delta \boldsymbol{L}_{m+1} = n^{-1}\boldsymbol{r}\boldsymbol{r}'.$$

If the experimental problem has its reflection in distinguishing certain subsets of contrasts, ordered according to their importance or interest, the experiment should be designed in such a way that all members of a specified subset of basic contrasts receive a common efficiency factor, its value reflecting the importance of the contrasts in the subset. If possible, the design should allow to estimate in the intra-block analysis the most important subset of contrasts with full efficiency, i.e., with $\varepsilon_0 = 1$. In this context, one is interested in knowing how many basic contrasts are estimated with the same efficiency. So the information about the multiplicity ρ_β is essential. It can form a basis for a classification of block designs, as already suggested in Caliński and Kageyama (1996b). Thus, the characterization of a block design by the triples $\{\mu_\beta, \rho_\beta, \boldsymbol{L}_\beta\}$ for $\beta = 0, 1, ..., m$, i.e., by the idempotent matrices $\{\boldsymbol{L}_\beta\}$ defined in (4.4.3), their ranks $\{\rho_\beta\}$ and the corresponding eigenvalues $\{\mu_\beta\}$, is informative.

With the present notation, the following modification of the original Definition 4.3.1 can be given (modifying also the proposal of Ceranka and Mejza, 1980).

Definition 4.4.1. A connected block design is said to be $(\rho_0; \rho_1, ..., \rho_{m-1})$-EB if a complete set of its $v - 1$ basic contrasts can be partitioned into at most m disjoint and nonempty subsets such that all the ρ_β basic contrasts of the βth subset correspond to a common efficiency factor $\varepsilon_\beta = 1 - \mu_\beta$, different for different $\beta = 0, 1, ..., m-1$, i.e., so that the matrix $\boldsymbol{M}_0$ $(= \boldsymbol{r}^{-\delta}\boldsymbol{N}\boldsymbol{k}^{-\delta}\boldsymbol{N}' - n^{-1}\boldsymbol{1}_v\boldsymbol{r}')$ has the spectral decomposition (4.3.10), where the distinct eigenvalues $0 = \mu_0 < \mu_1 < \cdots < \mu_{m-1} < 1$ have the multiplicities $\rho_0 \geq 0$, $\rho_1 \geq 1$,..., $\rho_{m-1} \geq 1$, respectively, and the matrices $\{\boldsymbol{L}_\beta\}$ are defined as in (4.4.3).

The parameters of a $(\rho_0; \rho_1, ..., \rho_{m-1})$-EB design can be written as v, b, $\boldsymbol{r}$, $\boldsymbol{k}$, $\varepsilon_\beta = 1 - \mu_\beta$, ρ_β, $\boldsymbol{L}_\beta$ for $\beta = 0, 1, ..., m-1$. Note that $\rho_0 = 0$ if no basic contrast is estimated in the intra-block analysis with full efficiency.

Definition 4.4.1 can be extended to cover disconnected block designs. In connection with this, recall Definition 2.2.6a and Corollary 2.3.2(b).

Definition 4.4.2. A block design with disconnectedness of degree $g-1$ (connected when $g=1$) is said to be $(\rho_0; \rho_1, ..., \rho_{m-1}; \rho_m)$-EB if a complete set of its $v-1$ basic contrasts can be partitioned into at most $m+1$ disjoint and nonempty subsets such that all ρ_β basic contrasts of the βth subset correspond to a common efficiency factor $\varepsilon_\beta = 1 - \mu_\beta$, different for different $\beta = 0, 1, ..., m-1, m$, i.e., so that the matrix $\boldsymbol{M}_0$ has the spectral decomposition as in (4.4.2), where the distinct eigenvalues $0 = \mu_0 < \mu_1 < \cdots < \mu_{m-1} < \mu_m = 1$ have the multiplicities $\rho_0 \geq 0$, $\rho_1 \geq 1$, ..., $\rho_{m-1} \geq 1$, $\rho_m = g - 1 \geq 0$, respectively, and the matrices $\{\boldsymbol{L}_\beta\}$ are defined as in (4.4.3).

The parameters of a $(\rho_0; \rho_1, ..., \rho_{m-1}; \rho_m)$-EB design can be written as v, b, $\boldsymbol{r}$, $\boldsymbol{k}$, $\varepsilon_\beta = 1 - \mu_\beta$, ρ_β, $\boldsymbol{L}_\beta$ for $\beta = 0, 1, ..., m-1, m$. Hence (i) ρ_0 gives the number of basic contrasts estimated with full efficiency, i.e. not confounded with blocks, (ii) $\{\rho_1, ..., \rho_{m-1}\}$ give the numbers of basic contrasts estimated in the intra-block analysis with efficiencies $\{\varepsilon_\beta,\ \beta = 1, ..., m-1\}$ less than 1, i.e. partially confounded with blocks, and (iii) ρ_m gives the number of basic contrasts with zero efficiency, i.e. totally confounded with blocks. In a connected design $\rho_m = 0$, i.e., no basic contrast is totally confounded with blocks.

It should be noted that since the decomposition (4.4.2) holds for any block design, whether connected, i.e., with $\rho_m = 0$, or disconnected, i.e., with $\rho_m \geq 1$, any block design satisfies the condition of Definition 4.4.2. Therefore, for a classification of block designs the specification of the multiplicities $\{\rho_\beta\}$ appearing in the definition is essential. To see it better, it may be interesting to indicate some special classes of block designs specified from the general Definition 4.4.2 point of view.

4.4.2 Some special classes of block designs

A block design belongs to the class of $(0; v-1; 0)$-EB designs if it is connected and totally balanced in the sense of Jones (Definition 2.4.5), or EB (Definition 4.1.2) in the terminology of Williams (1975), and Puri and Nigam (1975a, 1975b), i.e., satisfies the conditions

$$\boldsymbol{M}_0 = \mu_1 \boldsymbol{L}_1, \quad \boldsymbol{L}_1 = \boldsymbol{I}_v - n^{-1}\boldsymbol{1}_v\boldsymbol{r}' \quad \text{and} \quad \boldsymbol{C}_1 = \varepsilon_1\left(\boldsymbol{r}^\delta - \frac{1}{n}\boldsymbol{r}\boldsymbol{r}'\right).$$

In particular, any BIB design belongs to this class.

A block design belongs to the class of $(0; v-g; g-1)$-EB designs if it is disconnected of degree $g-1$ and is balanced in the sense of Jones (Definition 2.4.6), or EB (in the general sense of Definition 4.1.2), i.e., satisfies the conditions

$$\boldsymbol{M}_0 = \mu_1 \boldsymbol{L}_1 + \boldsymbol{L}_2 \quad \text{and} \quad \boldsymbol{C}_1 = \varepsilon_1 \boldsymbol{r}^\delta \boldsymbol{S}_1 \boldsymbol{S}_1' \boldsymbol{r}^\delta = \varepsilon_1 \boldsymbol{r}^\delta \boldsymbol{L}_1.$$

A block design belongs to the class of $(v-1;0;0)$-EB designs if it is connected and orthogonal, i.e., satisfies the conditions

$$\boldsymbol{M}_0 = \mathbf{O} \quad \text{and} \quad \boldsymbol{C}_1 = \boldsymbol{r}^\delta - (1/n)\boldsymbol{r}\boldsymbol{r}'$$

(see Remark 2.3.2).

A block design belongs to the class of $(v-g;0;g-1)$-EB designs if it is disconnected of degree $g-1$ and orthogonal, i.e., satisfies the conditions

$$\boldsymbol{M}_0 = \boldsymbol{L}_1 \quad \text{and} \quad \boldsymbol{C}_1 = \boldsymbol{r}^\delta \boldsymbol{S}_0 \boldsymbol{S}_0' \boldsymbol{r}^\delta = \boldsymbol{r}^\delta \boldsymbol{L}_0$$

(see Corollary 2.3.3 and Remark 2.4.2).

A block design belongs to the class of $(\rho_0; v-1-\rho_0; 0)$-EB designs if it is a simple PEB, or PEB(s), design in the terminology of Puri and Nigam (1977a, p. 756), or belongs to the class $\mathcal{D}_1$ of designs considered by Shah (1964a), or to the class of C-designs according to Saha (1976), i.e., satisfies the conditions

$$\boldsymbol{M}_0 = \mu_1 \boldsymbol{L}_1 \quad \text{and} \quad \boldsymbol{C}_1 = \boldsymbol{r}^\delta(\varepsilon_0 \boldsymbol{S}_0 \boldsymbol{S}_0' + \varepsilon_1 \boldsymbol{S}_1 \boldsymbol{S}_1')\boldsymbol{r}^\delta = \boldsymbol{r}^\delta(\varepsilon_0 \boldsymbol{L}_0 + \varepsilon_1 \boldsymbol{L}_1), \quad (4.4.5)$$

with at least one of the eigenvalues ε_0, ε_1 present. This class can be illustrated by the following example taken from Ceranka (1983, pp. 51-53).

Example 4.4.1. Consider the case of planning a factorial experiment with two factors, A and B, where A represents three doses (concentrations), zero (0), single (1) and double (2), and the factor B represents four different chemicals, e.g., pesticides, labeled 1, 2, 3 and 4. Thus, altogether there are $3 \times 4 = 12$ treatment combinations, but four of them are only seemingly different, namely, those representing the zero doses of the four chemicals. Thus, only nine distinct treatments are to be considered, one of them being the control treatment, of no application of the chemicals. It might be reasonable to demand that the comparisons between the control and the other treatments are made with the highest possible precision. Furthermore, suppose that the researcher wants the comparison between the main effects of the two positive levels of A (the single and the double dose) as well as the interaction contrasts to be estimated with full efficiency, and agrees that all contrasts among the main effects of the levels of B (the chemicals) will be estimated with equal but lower efficiency. Then, an appropriate block design can be defined by the following incidence matrix, whose rows correspond to the indicated treatment combinations:

Level of A	Level of B
0	any
1	1
1	2
1	3
1	4
2	1
2	2
2	3
2	4

$$\boldsymbol{N} = \begin{bmatrix} 1 & 1 & 1 & 1 & 1 & 1 \\ 1 & 0 & 1 & 0 & 1 & 0 \\ 1 & 0 & 0 & 1 & 0 & 1 \\ 0 & 1 & 1 & 0 & 0 & 1 \\ 0 & 1 & 0 & 1 & 1 & 0 \\ 1 & 0 & 1 & 0 & 1 & 0 \\ 1 & 0 & 0 & 1 & 0 & 1 \\ 0 & 1 & 1 & 0 & 0 & 1 \\ 0 & 1 & 0 & 1 & 1 & 0 \end{bmatrix}.$$

It is a proper, nonequireplicate binary connected block design with

$$
\boldsymbol{NN}' = \begin{bmatrix} 6 & 3 & 3 & 3 & 3 & 3 & 3 & 3 & 3 \\ 3 & 3 & 1 & 1 & 1 & 3 & 1 & 1 & 1 \\ 3 & 1 & 3 & 1 & 1 & 1 & 3 & 1 & 1 \\ 3 & 1 & 1 & 3 & 1 & 1 & 1 & 3 & 1 \\ 3 & 1 & 1 & 1 & 3 & 1 & 1 & 1 & 3 \\ 3 & 3 & 1 & 1 & 1 & 3 & 1 & 1 & 1 \\ 3 & 1 & 3 & 1 & 1 & 1 & 3 & 1 & 1 \\ 3 & 1 & 1 & 3 & 1 & 1 & 1 & 3 & 1 \\ 3 & 1 & 1 & 1 & 3 & 1 & 1 & 1 & 3 \end{bmatrix} \quad \text{and} \quad \boldsymbol{r} = \begin{bmatrix} 6 \\ 3 \\ 3 \\ 3 \\ 3 \\ 3 \\ 3 \\ 3 \\ 3 \end{bmatrix},
$$

from which

$$
\boldsymbol{M}_0 = \boldsymbol{r}^{-\delta}\left(\frac{1}{k}\boldsymbol{NN}' - \frac{1}{n}\boldsymbol{rr}'\right)
$$

$$
= \frac{1}{30}\begin{bmatrix} 0 & 0 & 0 & 0 & 0 & 0 & 0 & 0 & 0 \\ 0 & 3 & -1 & -1 & -1 & 3 & -1 & -1 & -1 \\ 0 & -1 & 3 & -1 & -1 & -1 & 3 & -1 & -1 \\ 0 & -1 & -1 & 3 & -1 & -1 & -1 & 3 & -1 \\ 0 & -1 & -1 & -1 & 3 & -1 & -1 & -1 & 3 \\ 0 & 3 & -1 & -1 & -1 & 3 & -1 & -1 & -1 \\ 0 & -1 & 3 & -1 & -1 & -1 & 3 & -1 & -1 \\ 0 & -1 & -1 & 3 & -1 & -1 & -1 & 3 & -1 \\ 0 & -1 & -1 & -1 & 3 & -1 & -1 & -1 & 3 \end{bmatrix}
$$

and the C-matrix is

$$
\boldsymbol{C}_1 = \boldsymbol{r}^{\delta} - \frac{1}{k}\boldsymbol{NN}' = \frac{1}{5}\begin{bmatrix} 24 & -3 & -3 & -3 & -3 & -3 & -3 & -3 & -3 \\ -3 & 12 & -1 & -1 & -1 & -3 & -1 & -1 & -1 \\ -3 & -1 & 12 & -1 & -1 & -1 & -3 & -1 & -1 \\ -3 & -1 & -1 & 12 & -1 & -1 & -1 & -3 & -1 \\ -3 & -1 & -1 & -1 & 12 & -1 & -1 & -1 & -3 \\ -3 & -3 & -1 & -1 & -1 & 12 & -1 & -1 & -1 \\ -3 & -1 & -3 & -1 & -1 & -1 & 12 & -1 & -1 \\ -3 & -1 & -1 & -3 & -1 & -1 & -1 & 12 & -1 \\ -3 & -1 & -1 & -1 & -3 & -1 & -1 & -1 & 12 \end{bmatrix}.
$$

The eight $\boldsymbol{r}^{\delta}$-orthonormal eigenvectors of $\boldsymbol{M}_0$, or $\boldsymbol{r}^{\delta}$-orthonormal eigenvectors of $\boldsymbol{C}_1$ with respect to $\boldsymbol{r}^{\delta}$, are

$$
\begin{aligned}
\boldsymbol{s}_1 &= [0,\ 1,\ 1,\ 1,\ 1,\ -1,\ -1,\ -1,\ -1]'/\sqrt{24}, \\
\boldsymbol{s}_2 &= [0,\ -1,\ 1,\ 0,\ 0,\ 1,\ -1,\ 0,\ 0]'/\sqrt{12}, \\
\boldsymbol{s}_3 &= [0,\ -1,\ -1,\ 2,\ 0,\ 1,\ 1,\ -2,\ 0]'/\sqrt{36}, \\
\boldsymbol{s}_4 &= [0,\ -1,\ -1,\ -1,\ 3,\ 1,\ 1,\ 1,\ -3]'/\sqrt{72},
\end{aligned}
$$

$$s_5 = [4,\ -1,\ -1,\ -1,\ -1,\ -1,\ -1,\ -1,\ -1]'/\sqrt{120},$$

corresponding to $\mu_1 = \mu_2 = \mu_3 = \mu_4 = \mu_5 = 0$, and

$$s_6 = [0,\ -1,\ -1,\ -1,\ 3,\ -1,\ -1,\ -1,\ 3]'/\sqrt{72},$$
$$s_7 = [0,\ -1,\ -1,\ 2,\ 0,\ -1,\ -1,\ 2,\ 0]'/\sqrt{36},$$
$$s_8 = [0,\ -1,\ 1,\ 0,\ 0,\ -1,\ 1,\ 0,\ 0]'/\sqrt{12},$$

corresponding to $\mu_6 = \mu_7 = \mu_8 = 4/15$. Renumbering these eigenvectors as s_{01}, s_{02}, s_{03}, s_{04}, s_{05} and s_{11}, s_{12}, s_{13}, respectively, and denoting the first five zero eigenvalues by μ_0 and the remaining three by μ_1, one can also write the matrices $\boldsymbol{M}_0$ and $\boldsymbol{C}_1$ in their decomposed forms given in (4.4.2) and (4.4.4), respectively, which for this example reduce to those in (4.4.5), with the matrices $\boldsymbol{L}_0$ and $\boldsymbol{L}_1$ defined in (4.4.3). These matrices are

$$\boldsymbol{L}_0 = \boldsymbol{S}_0\boldsymbol{S}_0'\boldsymbol{r}^{\delta} = \sum_{j=1}^{5} \boldsymbol{s}_{0j}\boldsymbol{s}_{0j}'\boldsymbol{r}^{\delta}$$

$$= \frac{1}{40}\begin{bmatrix} 32 & -4 & -4 & -4 & -4 & -4 & -4 & -4 & -4 \\ -8 & 21 & 1 & 1 & 1 & -19 & 1 & 1 & 1 \\ -8 & 1 & 21 & 1 & 1 & 1 & -19 & 1 & 1 \\ -8 & 1 & 1 & 21 & 1 & 1 & 1 & -19 & 1 \\ -8 & 1 & 1 & 1 & 21 & 1 & 1 & 1 & -19 \\ -8 & -19 & 1 & 1 & 1 & 21 & 1 & 1 & 1 \\ -8 & 1 & -19 & 1 & 1 & 1 & 21 & 1 & 1 \\ -8 & 1 & 1 & -19 & 1 & 1 & 1 & 21 & 1 \\ -8 & 1 & 1 & 1 & -19 & 1 & 1 & 1 & 21 \end{bmatrix}$$

and

$$\boldsymbol{L}_1 = \boldsymbol{S}_1\boldsymbol{S}_1'\boldsymbol{r}^{\delta} = \sum_{j=1}^{3} \boldsymbol{s}_{1j}\boldsymbol{s}_{1j}'\boldsymbol{r}^{\delta}$$

$$= \frac{1}{8}\begin{bmatrix} 0 & 0 & 0 & 0 & 0 & 0 & 0 & 0 & 0 \\ 0 & 3 & -1 & -1 & -1 & 3 & -1 & -1 & -1 \\ 0 & -1 & 3 & -1 & -1 & -1 & 3 & -1 & -1 \\ 0 & -1 & -1 & 3 & -1 & -1 & -1 & 3 & -1 \\ 0 & -1 & -1 & -1 & 3 & -1 & -1 & -1 & 3 \\ 0 & 3 & -1 & -1 & -1 & 3 & -1 & -1 & -1 \\ 0 & -1 & 3 & -1 & -1 & -1 & 3 & -1 & -1 \\ 0 & -1 & -1 & 3 & -1 & -1 & -1 & 3 & -1 \\ 0 & -1 & -1 & -1 & 3 & -1 & -1 & -1 & 3 \end{bmatrix}.$$

Because $\mu_0 = 0$ and $\mu_1 = 4/15$ here, from which $\varepsilon_0 = 1$ and $\varepsilon_1 = 11/15$, the application of formulae (4.4.5) gives the following results:

$$
\boldsymbol{M}_0 = \mu_1 \boldsymbol{L}_1 = \left(\frac{4}{15}\right)\frac{1}{8}\begin{bmatrix}
0 & 0 & 0 & 0 & 0 & 0 & 0 & 0 & 0 \\
0 & 3 & -1 & -1 & -1 & 3 & -1 & -1 & -1 \\
0 & -1 & 3 & -1 & -1 & -1 & 3 & -1 & -1 \\
0 & -1 & -1 & 3 & -1 & -1 & -1 & 3 & -1 \\
0 & -1 & -1 & -1 & 3 & -1 & -1 & -1 & 3 \\
0 & 3 & -1 & -1 & -1 & 3 & -1 & -1 & -1 \\
0 & -1 & 3 & -1 & -1 & -1 & 3 & -1 & -1 \\
0 & -1 & -1 & 3 & -1 & -1 & -1 & 3 & -1 \\
0 & -1 & -1 & -1 & 3 & -1 & -1 & -1 & 3
\end{bmatrix}
$$

and

$$
\boldsymbol{C}_1 = \boldsymbol{r}^{\delta}\boldsymbol{L}_0 + \varepsilon_1 \boldsymbol{r}^{\delta}\boldsymbol{L}_1
$$

$$
= \frac{3}{40}\begin{bmatrix}
64 & -8 & -8 & -8 & -8 & -8 & -8 & -8 & -8 \\
-8 & 21 & 1 & 1 & 1 & -19 & 1 & 1 & 1 \\
-8 & 1 & 21 & 1 & 1 & 1 & -19 & 1 & 1 \\
-8 & 1 & 1 & 21 & 1 & 1 & 1 & -19 & 1 \\
-8 & 1 & 1 & 1 & 21 & 1 & 1 & 1 & -19 \\
-8 & -19 & 1 & 1 & 1 & 21 & 1 & 1 & 1 \\
-8 & 1 & -19 & 1 & 1 & 1 & 21 & 1 & 1 \\
-8 & 1 & 1 & -19 & 1 & 1 & 1 & 21 & 1 \\
-8 & 1 & 1 & 1 & -19 & 1 & 1 & 1 & 21
\end{bmatrix}
$$

$$
+ \left(\frac{11}{15}\right)\frac{3}{8}\begin{bmatrix}
0 & 0 & 0 & 0 & 0 & 0 & 0 & 0 & 0 \\
0 & 3 & -1 & -1 & -1 & 3 & -1 & -1 & -1 \\
0 & -1 & 3 & -1 & -1 & -1 & 3 & -1 & -1 \\
0 & -1 & -1 & 3 & -1 & -1 & -1 & 3 & -1 \\
0 & -1 & -1 & -1 & 3 & -1 & -1 & -1 & 3 \\
0 & 3 & -1 & -1 & -1 & 3 & -1 & -1 & -1 \\
0 & -1 & 3 & -1 & -1 & -1 & 3 & -1 & -1 \\
0 & -1 & -1 & 3 & -1 & -1 & -1 & 3 & -1 \\
0 & -1 & -1 & -1 & 3 & -1 & -1 & -1 & 3
\end{bmatrix},
$$

which coincide with those for $\boldsymbol{M}_0$ and $\boldsymbol{C}_1$ given earlier above, as can easily be checked.

Thus, according to the classification based on Definition 4.4.2, the design considered here belongs to the class of $(\rho_0; v - 1 - \rho_0; 0)$-EB designs. More precisely, it is a (5;3;0)-EB design with parameters $v = 9, b = 6$, $\boldsymbol{r} = [6, 3\mathbf{1}_8']'$, $\boldsymbol{k} = 5\mathbf{1}_6$, $\mu_0 = 0, \mu_1 = 4/15$ (i.e., $\varepsilon_0 = 1, \varepsilon_1 = 11/15$) and $\boldsymbol{L}_0, \boldsymbol{L}_1$ as given above.

More examples will be found in Volume II, where, as stated before (Section 1.5), different classes and subclasses of block designs, and their constructional aspects, are presented and discussed. In particular, it will be seen that any connected two-associate PBIB design with its concurrence matrix $\boldsymbol{NN}'$ having a zero eigenvalue (of multiplicity ρ_0) belongs to the class of $(\rho_0; v - 1 - \rho_0; 0)$-EB designs. Many such designs can be found in the class of group divisible, triangular or Latin-square type designs (for definitions of these terms, see Raghavarao, 1971, Section 8.4). For example, any singular group divisible and any semi-regular group divisible design is such. (See also Saha, 1976; Nigam and Puri, 1982; Ceranka, 1983; Ceranka and Kozłowska, 1983, 1984.)

Furthermore, it will be noted that any connected two-associate PBIB design with $\boldsymbol{NN}'$ having only positive eigenvalues belongs to the class of $(0;\rho_1,\rho_2;0)$-EB designs. For example, any regular group divisible design is such.

Considering designs with supplemented balance (see Pearce, 1960; Caliński, 1971; Caliński and Ceranka, 1974; Ceranka and Chudzik, 1984), i.e., designs obtained from a block design by adding to each block one or more supplementary treatments, it will be interesting to note that an orthogonally supplemented BIB design (in the sense of Caliński, 1971) belongs to the class of $(\rho_0;\rho_1;0)$-EB designs. On the other hand, an orthogonally supplemented connected two-associate PBIB design with an incidence matrix $\boldsymbol{N}$ belongs either to the class of $(\rho_0;\rho_1;\rho_2;0)$-EB or $(\rho_0;\rho_1;0)$-EB designs, depending on whether both of the eigenvalues of $\boldsymbol{NN}'$ are positive or one of them is zero.

4.4.3 Statistical sense of the classification

The characterization of a block design by the triples $\{\mu_\beta,\rho_\beta,\boldsymbol{L}_\beta\}$ for $\beta = 0,1,...,m$, i.e., by the idempotent matrices $\{\boldsymbol{L}_\beta\}$ defined in (4.4.3), their ranks $\{\rho_\beta\}$, and the corresponding eigenvalues $\{\mu_\beta\}$, in the sense of the equality

$$\boldsymbol{M}_0\boldsymbol{L}_\beta = \mu_\beta\boldsymbol{L}_\beta, \quad \beta = 0,1,2,...,m,$$

resulting from (4.4.2), is useful from the statistical point of view. First of all, it allows (4.4.4), the spectral decomposition of the C-matrix ($\boldsymbol{C} \equiv \boldsymbol{C}_1$) on which the intra-block analysis is based, to be obtained easily. Furthermore, from Corollary 3.4.2, any contrast given by $\boldsymbol{s}'_\beta\boldsymbol{L}'_\beta\boldsymbol{r}^\delta\boldsymbol{\tau}$ for some $\boldsymbol{s}_\beta$, such that $\boldsymbol{L}_\beta\boldsymbol{s}_\beta \neq \boldsymbol{0}, \beta = 0,1,...,m-1$, obtains the intra-block BLUE of the form

$$(\widehat{\boldsymbol{s}'_\beta\boldsymbol{L}'_\beta\boldsymbol{r}^\delta\boldsymbol{\tau}})_{\text{intra}} = \varepsilon_\beta^{-1}\boldsymbol{s}'_\beta\boldsymbol{L}'_\beta\boldsymbol{Q}_1 \tag{4.4.6}$$

$(\boldsymbol{Q}_1 = \boldsymbol{\Delta\phi}_1\boldsymbol{y})$, with the variance

$$\text{Var}[(\widehat{\boldsymbol{s}'_\beta\boldsymbol{L}'_\beta\boldsymbol{r}^\delta\boldsymbol{\tau}})_{\text{intra}}] = \varepsilon_\beta^{-1}\boldsymbol{s}'_\beta\boldsymbol{r}^\delta\boldsymbol{L}_\beta\boldsymbol{s}_\beta\sigma_1^2, \tag{4.4.7}$$

ε_β being the common efficiency factor of the design for all contrasts of this type. This variance is reduced to $\varepsilon_\beta^{-1}\sigma_1^2$ if $\boldsymbol{s}_\beta$ is exactly one of the columns of $\boldsymbol{S}_\beta = [\boldsymbol{s}_{\beta 1} : \boldsymbol{s}_{\beta 2} : \cdots : \boldsymbol{s}_{\beta\rho_\beta}]$, i.e., represents one of the basic contrasts corresponding to the common efficiency factor ε_β. In particular, for $\beta = 0$, a contrast $\boldsymbol{s}'_0\boldsymbol{L}'_0\boldsymbol{r}^\delta\boldsymbol{\tau}$ obtains, by Corollary 3.4.1, the BLUE of the form

$$\boldsymbol{s}'_0\widehat{\boldsymbol{L}'_0\boldsymbol{r}^\delta\boldsymbol{\tau}} = \boldsymbol{s}'_0\boldsymbol{L}'_0\boldsymbol{\Delta y} \tag{4.4.8}$$

under the overall model (3.1.1), with the variance

$$\operatorname{Var}(\widehat{s_0' L_0' r^\delta \tau}) = s_0' r^\delta L_0 s_0 \sigma_1^2, \tag{4.4.9}$$

which reduces to σ_1^2 if s_0 represents one of the basic contrasts corresponding to the efficiency factor $\varepsilon_0 = 1$.

Thus, the present terminology, using the multiplicities $\{\rho_\beta\}$, may be much more suitable for indicating the statistical advantage and utility of a block design, than is the original reference to a PEB design with m efficiency classes. Note, particularly, the difference between a $(0; \rho_1, \rho_2; 0)$-EB design and a $(\rho_0; \rho_1; 0)$-EB design, both being PEB designs with two efficiency classes according to the definition of Puri and Nigam (1977a). Also note that in the terminology proposed here, the use of the term "partial efficiency balance" can be avoided, thus meeting the demand of Preece (1982, p. 151) that the term "partial balance" should be used only in the sense of Nair and Rao (1942).

Example 4.4.2. Returning to Example 4.4.1, note that the matrix $r^\delta L_0$ given in the decomposition of C_1 spans the subspace, of dimension $\rho_0 = 5$, of all contrasts estimated in the intra-block analysis with full efficiency, i.e., those for which the BLUEs exist under the overall randomization model (3.1.1). Taking, e.g.,

$s_0 = [1, 0, 0, 0, 0, 0, 0, 0, 0]'$,

one obtains the contrast

$s_0' L_0' r^\delta \tau = \frac{3}{5}(8\tau_1 - \tau_2 - \tau_3 - \tau_4 - \tau_5 - \tau_6 - \tau_7 - \tau_8 - \tau_9)$,

which is a comparison between the control treatment and all other treatments, considered jointly. But taking

$s_0 = [3, 0, 0, 0, 0, 1, 1, 1, 1]'$,

one obtains the contrast

$s_0' L_0' r^\delta \tau = 3(4\tau_1 - \tau_2 - \tau_3 - \tau_4 - \tau_5)$,

which is a comparison between the control treatment and all four chemicals applied at the single level of concentration. Similarly, taking

$s_0 = [3, 1, 1, 1, 1, 0, 0, 0, 0]'$,

one obtains the contrast

$s_0' L_0' r^\delta \tau = 3(4\tau_1 - \tau_6 - \tau_7 - \tau_8 - \tau_9)$,

which allows the control treatment to be compared with all four chemicals applied at the double level. Furthermore, if one wants to estimate the interaction between the chemicals first and fourth and the two levels of their application, i.e., the contrast $(3/2)(\tau_2 - \tau_5 - \tau_6 + \tau_9)$, then it is sufficient to take

$s_0 = [0, 1, 0, 0, -1, 0, 0, 0, 0]'$.

All of these contrasts, as well as all others defined by $s_0' L_0' r^\delta \tau$, for any appropriate vector s_0 (i.e., such that $L_0 s_0 \neq \mathbf{0}$), will be estimated by the BLUEs of the form (4.4.8) with the variance (4.4.9), i.e., with full efficiency. On the other hand, all contrasts based on the matrix $r^\delta L_1$, i.e., contrasts of the form $s_1' L_1' r^\delta \tau$, for a suitable vector s_1, will in this example be estimated in the intra-block analysis with the efficiency $\varepsilon_1 = 11/15$. Taking, e.g.,

$s_1 = [0, 1, 0, 0, -1, 0, 0, 0, 0]'$,

one obtains the contrast

$$s_1' L_1' r^{\delta} \tau = \tfrac{3}{2}(\tau_2 - \tau_5 + \tau_6 - \tau_9),$$

which is a comparison between the first and the fourth chemical from both levels of concentration. Its intra-block BLUE is of the form (4.4.6) with the variance of the form (4.4.7), where $\beta = 1$ and $\varepsilon_1 = 11/15$.

In a similar way, the information provided by $\{\mu_\beta, \rho_\beta, L_\beta\}$, $\beta = 0, 1, ..., m$, can be used in the intra-block analysis of any other block design. Thus, if this information is available in advance for the design chosen by the researcher when planning an experiment, then it can easily be used in examining the statistical properties of the design and later in performing the analysis of the experimental data. Otherwise, the researcher has to evaluate the quantities $\{\varepsilon_\beta = 1 - \mu_\beta\}$, $\{\rho_\beta\}$ and $\{L_\beta\}$ related to the design, before deciding on its use. In these computations, the results given in Sections 2.3 (Lemma 2.3.3 and Remark 2.3.1 in particular) and 4.4.1–4.4.2 (including Example 4.4.1) will be helpful.

Finally, it should be mentioned that some people may prefer to use instead of (4.3.2) the decomposition of the matrix

$$F = r^{-\delta/2} C_1 r^{-\delta/2},$$

where $r^{\delta/2} = \mathrm{diag}[r_1^{1/2}, r_2^{1/2}, ..., r_v^{1/2}]$ and $r^{-\delta/2} = (r^{\delta/2})^{-1}$, as originally used by Pearce et al. (1974, p. 451). This decomposition can be written as

$$F = r^{\delta/2} \sum_{\beta=0}^{m-1} \varepsilon_\beta S_\beta S_\beta' r^{\delta/2} = \sum_{\beta=0}^{m-1} \varepsilon_\beta P_\beta P_\beta',$$

where $P_\beta = [p_{\beta 1} : p_{\beta 2} : \cdots : p_{\beta \rho_\beta}] = r^{\delta/2} S_\beta$, with the properties

$$P_\beta' P_\beta = S_\beta' r^{\delta} S_\beta = I_{\rho_\beta} \quad \text{and} \quad P_\beta' P_{\beta'} = \mathbf{O} \quad \text{if} \quad \beta \neq \beta',$$

for β, $\beta' = 0, 1, ..., m, m+1$, and P_{m+1} denotes the vector $p_v = r^{\delta/2} s_v = n^{-1/2} r^{\delta/2} 1_v$. Evidently, because

$$L_\beta = r^{-\delta/2} P_\beta P_\beta' r^{\delta/2} \quad \text{for} \quad \beta = 0, 1, ..., m, m+1,$$

it is also permissible to write

$$F = \sum_{\beta=0}^{m-1} \varepsilon_\beta r^{\delta/2} L_\beta r^{-\delta/2}, \tag{4.4.10}$$

with the property

$$\sum_{\beta=0}^{m+1} r^{\delta/2} L_\beta r^{-\delta/2} = \sum_{\beta=0}^{m+1} P_\beta P_\beta' = I_v.$$

To return from (4.4.10) back to the decomposition (4.4.4), which seems to be generally more convenient in view of its direct application, as shown by Example

4.4.1, one has to perform the premultiplication and postmultiplication of (4.4.10) by $\boldsymbol{r}^{\delta/2}$, i.e., to make the transformation

$$\boldsymbol{r}^{\delta/2}\boldsymbol{F}\boldsymbol{r}^{\delta/2} = \sum_{\beta=0}^{m-1} \varepsilon_\beta \boldsymbol{r}^\delta \boldsymbol{L}_\beta = \boldsymbol{C}_1.$$

However, in case of an equireplicate block design ($r_1 = r_2 = \cdots = r_v = r$), the use of the decomposition (4.4.10), getting then the form

$$\boldsymbol{F} = r^{-1}\boldsymbol{C}_1 = \sum_{\beta=0}^{m-1} \varepsilon_\beta \boldsymbol{L}_\beta, \quad \text{with} \quad \boldsymbol{L}_\beta = \boldsymbol{P}_\beta \boldsymbol{P}'_\beta, \tag{4.4.11}$$

will be more attractive than might be the application of (4.4.4). In this case, any contrast $\boldsymbol{p}'_\beta \boldsymbol{L}'_\beta \boldsymbol{\tau}$ for some $\boldsymbol{p}_\beta$, such that $\boldsymbol{L}_\beta \boldsymbol{p}_\beta \neq \boldsymbol{0}$, $\beta = 0, 1, ..., m-1$, obtains the intra-block BLUE of the form

$$(\widehat{\boldsymbol{p}'_\beta \boldsymbol{L}'_\beta \boldsymbol{\tau}})_{\text{intra}} = (\varepsilon_\beta r)^{-1} \boldsymbol{p}'_\beta \boldsymbol{L}'_\beta \boldsymbol{Q}_1 = (\varepsilon_\beta r)^{-1} \boldsymbol{p}'_\beta \boldsymbol{L}_\beta \boldsymbol{Q}_1, \tag{4.4.12}$$

with the variance

$$\text{Var}[(\widehat{\boldsymbol{p}'_\beta \boldsymbol{L}'_\beta \boldsymbol{\tau}})_{\text{intra}}] = (\varepsilon_\beta r)^{-1} \boldsymbol{p}'_\beta \boldsymbol{L}_\beta \boldsymbol{p}_\beta \sigma_1^2, \tag{4.4.13}$$

as now $\boldsymbol{L}'_\beta = \boldsymbol{L}_\beta$, in addition to the properties given in connection with definition (4.3.7). Certainly, for $\beta = 0$, where $\varepsilon_0 = 1$, formulae (4.4.12) and (4.4.13) can be written as

$$\widehat{\boldsymbol{p}'_0 \boldsymbol{L}'_0 \boldsymbol{\tau}} = r^{-1} \boldsymbol{p}'_0 \boldsymbol{L}_0 \boldsymbol{\Delta y} \tag{4.4.14}$$

and

$$\text{Var}(\widehat{\boldsymbol{p}'_0 \boldsymbol{L}'_0 \boldsymbol{\tau}}) = r^{-1} \boldsymbol{p}'_0 \boldsymbol{L}_0 \boldsymbol{p}_0 \sigma_1^2,$$

respectively, (4.4.14) being the BLUE under the overall model (3.1.1).

4.5 Some concluding remarks

The discussions in Section 4.3 concerning the concept of balance in the sense of Jones (1959) and its relations to the notion of GB, introduced by Nelder (1965a, 1965b), and the notion of PEB, introduced by Puri and Nigam (1977a), have shown that all these three concepts of balance are closely related. The relations among these concepts are particularly evident in view of the role of the basic contrasts played in defining them.

Because, as stated in Remark 3.6.4, any proper block design is GB (see also Houtman and Speed, 1983, Section 5.4), the notion of GB is interesting only from the point of view of the decomposition (4.3.4) with respect to which the balance holds. Similarly, any connected block design, not necessarily proper, is PEB by Definition 4.3.1 or the corresponding Definition 4.4.1 (see also Kageyama and

Puri, 1985a, p. 28). But, also, if disconnected, proper or not, it is PEB by the generalized Definition 4.4.2. Thus, the notion of PEB is interesting only from the point of view of the decomposition (4.4.4), related to that in (4.4.2), but also to that in (4.3.4), on account of the definition (4.3.7) of the matrices $\{\boldsymbol{L}_\beta\}$. This relation can be seen by noting that for any subset of basic contrasts, $\boldsymbol{S}'_\beta \boldsymbol{r}^\delta \boldsymbol{\tau}$, corresponding to a common ε_β, $\beta = 0, 1, ..., m-1, m$, the subspace equality $\mathcal{C}(\boldsymbol{r}^\delta \boldsymbol{S}_\beta) = \mathcal{C}(\boldsymbol{r}^\delta \boldsymbol{L}_\beta)$ holds. It follows that any block design offered for use in an experiment should be characterized with regard to these decompositions. The researcher should be informed on the subspaces $\{\mathcal{C}(\boldsymbol{\Delta}'\boldsymbol{S}_\beta)\}$ appearing in (4.3.4) or the matrices $\{\boldsymbol{r}^\delta \boldsymbol{L}_\beta\}$ appearing in (4.4.4), and on the efficiency factors $\{\varepsilon_\beta\}$ corresponding to them. This issue has already been pointed out by Houtman and Speed (1983, p. 1075), who write that these subspaces have to be discovered for each new design or class of designs. Referring directly to block designs and to PBIB designs in particular, they write (p. 1082) that although it is generally not difficult to obtain these subspaces (more precisely, orthogonal projections on them), "most writers in statistics have not taken this view point," Corsten (1976) being an notable exception. In fact, his canonical (sub)spaces are equivalent to those of the basic contrasts considered here. How these subspaces are obtainable has been shown in Section 4.4 in connection with Example 4.4.1.

The knowledge of the basic contrasts or their subspaces for which the design is GB or PEB, and of the efficiency factors assigned to them, allows the researcher to use the design for an experiment in such a way that best corresponds to the experimental problem. In particular, it helps the researcher to implement the design so that the contrasts considered as the most important can be estimated with the highest efficiency in the stratum of the smallest variance, which usually the intra-block stratum is, provided that the grouping of units into blocks is successful in the sense of eliminating, as far as possible, the intra-block heterogeneity of units. Assuming that such minimization of the intra-block stratum variance is achieved, a most desirable design, in most cases, would be such that allows the most important contrasts to be estimated in the intra-block analysis with full efficiency.

Acknowledgement. Certain parts of this chapter have been reprinted, with some changes, from Sections 2 and 3 of the article by Caliński and Kageyama (1996b), a contribution to the handbook edited by Ghosh and Rao (1996), under kind permission from Elsevier Science – NL, Sara Burgerhartstraat 25, 1055 KV Amsterdam, The Netherlands.

5
Nested Block Designs and the Concept of Resolvability

5.1 Nested block designs and their general form

In Chapter 3, a randomization model for experiments in block designs with one stratum of blocks of experimental units has been presented and discussed from the point of view of its statistical properties and their consequences for the analysis. The aim of the present chapter is to extend the model to the experimental situation in which the blocks are further grouped into some sets called "superblocks," forming in that way two strata of blocks. Such situations appear often in practice, particularly in agricultural and industrial experimentation, when several sources of local variation are present, more than can be controlled by ordinary blocking. Common examples of relevant designs are the lattice designs (introduced by Yates, 1940a) or, more generally, the so-called resolvable incomplete block designs (introduced by Bose, 1942, and generalized by Shrikhande and Raghavarao, 1964, and further by Kageyama, 1976a; see also John, 1987, Section 3.4). Whereas in an ordinary block design (as considered in Section 3.2) the stratification of the experimental units leads to three strata, of units within blocks, of blocks within the total experimental area, and of the total area, in the extended situation to be considered here, four strata can be distinguished. These strata are of units within blocks, of blocks within superblocks, of superblocks within the total experimental area, and of the total area. Consequently, the extended randomization model has to take into account three instead of two stages of randomization, i.e., of units within blocks, of blocks within superblocks, and of the latter within the total area.

A similar situation to that just described occurs when in an ordinary block design the plots (units) are divided into two or more subplots. Here, again, four strata appear, of subplots within the main plots, of the main plots within blocks, of blocks within the total experimental area, and of that total area (see Pearce, 1983, Section 6.1). If the randomization is performed independently within each of the first three strata, then virtually the same randomization model as that mentioned above will apply. Thus, although the randomization model to be considered in the present chapter is derived in the context of plots grouped into blocks and the latter grouped into superblocks, it can perfectly well be thought of as suitable for subplots grouped into main plots and those into blocks (see, e.g., Mejza, 1987; I. Mejza, 1996).

In both of these situations, the relations between the strata are of nesting type. One can speak in this context of two systems of blocks, one nested in the other, either of blocks nested in the superblocks or main plots nested in the blocks. As to the arrangement of treatments in blocks and superblocks, usually a kind of balance is desirable. In particular, Preece (1967) introduced a class of nested balanced incomplete block (NBIB) designs for experimental situations in which two sources of the variability of experimental units exist and one source is nested within the other. In accordance with this type of variability, an NBIB design has two systems of blocks of units, the second nested within the first (so that each block from the first system, a superblock, contains some blocks, called "subblocks," from the second), in which the arrangement of treatments is such that ignoring either system leaves a BIB design whose blocks are those of the other system. If, however, ignoring the first system, of superblocks, leaves a BIB design but ignoring the second, of subblocks, leaves, instead of a BIB, a connected proper binary orthogonal block design (an RBD, see Section 2.1), then the considered nested design is a resolvable BIB design in the sense of Bose (1942). For more general concepts of resolvability, see Kageyama (1972, 1973, 1976a), and in Volume II, see Chapters 6 (Section 6.0.3) and 9.

The statistical analysis for NBIB designs is based on an extension of that developed by Yates (1940b) for BIB designs. Interesting applications of NBIB designs for laboratory studies have been given by Kleczkowski (1960) and Kassanis and Kleczkowski (1965), and for plant breeding experiments by Gupta and Kageyama (1994), who have shown that some NBIB designs provide universally optimal complete diallel crosses. Systematic methods of constructing NBIB designs have been considered by Agrawal and Prasad (1983), Jimbo and Kuriki (1983), Dey, Das and Banerjee (1986), and Kageyama and Miao (1997, 1998b). The idea of NBIB designs has been generalized to nested partially balanced incomplete block designs by Homel and Robinson (1975) and discussed by Banerjee and Kageyama (1990, 1993). It should also be remarked that another type of nested designs has been discussed by Longyear (1981), Colbourn and Colbourn (1983), Gupta (1984), and others. For further references concerning NBIB designs and certain extensions of them, see Morgan (1996), and for those concerning some methods of constructing resolvable block designs, see Kageyama and Miao (1998a).

Following the papers by Caliński (1994, 1997) and Mejza and Kageyama (1995, 1998), in which the concept of NBIB designs has been treated in a generalized form, the term "nested block design" (NB design, for short) will be adopted for designs considered in the present chapter. As in those papers, the NB designs will also be considered here in a most general framework. The usual NBIB design can then be seen as a special case of the general NB design.

As in Chapter 3, the discussion of statistical consequences of the derived randomization model will concern mainly the best linear unbiased estimation of treatment parametric functions under the overall model and under the submodels related to the different strata. In the traditional block designs with two strata of blocks, in which the superblock stratum is usually composed of single replicates (see, e.g., John, 1987, Table 8.3), only two analyses are of interest, the intra-block and the inter-block analysis, the superblock stratum contributing nothing to the estimation of parametric functions. In the present chapter, however, a general situation is to be considered, in which the inter-superblock analysis may also be of interest, at least in some cases.

The basic notation and terminology of the present chapter follow that of Section 2.2. As usual, a block design for v treatments in b blocks is described by its $v \times b$ incidence matrix $\boldsymbol{N} = \boldsymbol{\Delta}\boldsymbol{D}'$, where $\boldsymbol{\Delta}'$ is the $n \times v$ design matrix for treatments and $\boldsymbol{D}'$ is the $n \times b$ design matrix for blocks. Let such design be denoted here by $\mathcal{D}^*$. Suppose that the blocks of $\mathcal{D}^*$, i.e., the so-called subblocks, are grouped into a superblocks, say. This can be reflected by the partitions $\boldsymbol{\Delta} = [\boldsymbol{\Delta}_1 : \boldsymbol{\Delta}_2 : \cdots : \boldsymbol{\Delta}_a]$, $\boldsymbol{D} = \text{diag}[\boldsymbol{D}_1 : \boldsymbol{D}_2 : \cdots : \boldsymbol{D}_a]$, and, consequently,

$$\boldsymbol{N} = [\boldsymbol{N}_1 : \boldsymbol{N}_2 : \cdots : \boldsymbol{N}_a],$$

where $\boldsymbol{\Delta}_h$, $\boldsymbol{D}_h$ and $\boldsymbol{N}_h = \boldsymbol{\Delta}_h\boldsymbol{D}_h'$ describe a component design, denoted by $\mathcal{D}_h$, confined to superblock h $(= 1, 2, ..., a)$. Let the resulting block design by which the v treatments are arranged into the a superblocks be then denoted by $\mathcal{D}$. Its $v \times a$ incidence matrix can be written as (say) $\boldsymbol{\Re} = \boldsymbol{\Delta}\boldsymbol{G}'$, where $\boldsymbol{G}'$ is the $n \times a$ design matrix for superblocks of the form

$$\boldsymbol{G}' = \boldsymbol{D}'\text{diag}[\boldsymbol{1}_{b_1} : \boldsymbol{1}_{b_2} : \cdots : \boldsymbol{1}_{b_a}] = \text{diag}[\boldsymbol{1}_{n_1} : \boldsymbol{1}_{n_2} : \cdots : \boldsymbol{1}_{n_a}],$$

with b_h denoting the number of blocks in superblock h (i.e., in the component design $\mathcal{D}_h$), and n_h denoting the number of units (plots) in that superblock, i.e., its size. Note that

$$n_h = \boldsymbol{1}_{b_h}'\boldsymbol{k}_h, \quad \text{where} \quad \boldsymbol{k}_h = [k_{1(h)}, k_{2(h)}, ..., k_{b_h(h)}]' = \boldsymbol{N}_h'\boldsymbol{1}_v$$

denotes the vector of block sizes in $\mathcal{D}_h$. The matrix $\boldsymbol{\Re}$ can also be written as

$$\boldsymbol{\Re} = [\boldsymbol{r}_1 : \boldsymbol{r}_2 : \cdots : \boldsymbol{r}_a], \quad \text{where} \quad \boldsymbol{r}_h = [r_{1(h)}, r_{2(h)}, ..., r_{v(h)}]' = \boldsymbol{N}_h\boldsymbol{1}_{b_h} \quad (5.1.1)$$

denotes the vector of treatment replications in $\mathcal{D}_h$, for $h = 1, 2, ..., a$. [Note that the symbol $\boldsymbol{\Re}$ used here to denote the matrix (5.1.1) should be distinguished from $\boldsymbol{R}$ introduced in Section 3.7 to represent the matrix (3.7.20).]

Evidently, $\boldsymbol{N}\boldsymbol{1}_b = \boldsymbol{\Re}\boldsymbol{1}_a = \boldsymbol{r} = [r_1, r_2, ..., r_v]'$ is the vector of treatment replications in the whole NB design, as well as in $\mathcal{D}^*$ and $\mathcal{D}$, $\boldsymbol{N}'\boldsymbol{1}_v = \boldsymbol{k} = [\boldsymbol{k}_1', \boldsymbol{k}_2', ..., \boldsymbol{k}_a']'$ is the vector of block sizes in $\mathcal{D}^*$, and $\boldsymbol{\Re}'\boldsymbol{1}_v = \boldsymbol{n} = [n_1, n_2, ..., n_a]'$

is the vector of superblock sizes in $\mathcal{D}$. As $\boldsymbol{k}^{\delta} = \boldsymbol{D}\boldsymbol{D}'$ is the diagonal matrix of block sizes in $\mathcal{D}^*$, $\boldsymbol{n}^{\delta} = \boldsymbol{G}\boldsymbol{G}'$ is a diagonal matrix with the numbers n_h on the diagonal, representing the superblock sizes in $\mathcal{D}$. The total number of units (plots) used in the experiment is $n = \mathbf{1}_v'\boldsymbol{r} = \mathbf{1}_b'\boldsymbol{k} = \mathbf{1}_a'\boldsymbol{n}$.

As the intention is to consider a general case, disconnected designs are also to be taken into account. However, because of the partitioned structure of $\boldsymbol{N}$, with its submatrices $\boldsymbol{N}_h$, $h = 1, 2, ..., a$, corresponding naturally to the superblocks, as shown above, the assumption made in Section 2.2, that $\boldsymbol{N}$ is to be considered as having a quasi-diagonal structure reflecting the disconnectedness of the design, cannot be adopted here. Nevertheless, the incidence matrix of a disconnected design $\mathcal{D}^*$ or $\mathcal{D}$ can always be visualized in that form after proper reordering of its rows and columns.

In a similar way as in Section 3.1, distinction is to be made between the potential (or available) number of superblocks, N_A, from which a choice can be made, and the number, a, of those actually chosen for the experiment. The usual situation is that $a = N_A$, but in general $a \leq N_A$. Also, it is convenient to make a distinction between the potential (available) number of blocks within a superblock, denoting it by B with appropriate subscript, and the number of blocks actually chosen among them to the experiment, denoting it by b with a relevant subscript. Finally, a distinction is to be made between the potential number of units (plots) within a block, denoting it by K with a subscript, and the number of the units of the block actually used in the experiment, denoting it by k with a subscript.

5.2 A randomization model for a nested block design

The same approach as that used in Section 3.1 will be adopted here to derive and investigate the model of the variables observed on the n units actually used in the experiment. The extension consists in applying a threefold instead of a twofold randomization of superblocks within a total area of them, of blocks within the superblocks, and of units within the blocks.

5.2.1 Derivation of the model

Suppose that N_A superblocks are available, originally labeled $\nu = 1, 2, ..., N_A$, and that superblock ν contains B_ν blocks, which are originally labeled $\xi(\nu) = 1, 2, ..., B_\nu$. Further suppose that block $\xi(\nu)$ contains $K_{\xi(\nu)}$ units (plots), which are originally labeled $\pi[\xi(\nu)] = 1, 2, ..., K_{\xi(\nu)}$. The randomization of superblocks can then be understood as choosing at random a permutation of numbers $1, 2, ...$, N_A, and then renumbering the superblocks with $h = 1, 2, ..., N_A$ according to the positions of their original labels taken in the random permutation. Similarly, the randomization of blocks within superblock ν consists in selecting at random a permutation of numbers $1, 2, ..., B_\nu$, and then renumbering the blocks with $j = 1, 2, ..., B_\nu$ accordingly. Finally, the randomization of units within block $\xi(\nu)$ can be seen as selecting at random a permutation of numbers $1, 2, ..., K_{\xi(\nu)}$, and then renumbering the units of the block with $\ell = 1, 2, ..., K_{\xi(\nu)}$ accordingly (see Section 3.1.1). The usual assumption will be made that any permutation of superblock labels can be selected with equal probability, that any permutation of block labels within a superblock can be selected with equal probability, as well as that any permutation of unit labels within a block can be selected in that way. Finally, it will be assumed that the randomizations of units within the blocks are among the blocks independent, that the randomizations of blocks within the superblocks are among the superblocks independent and also independent of the randomizations of units, and that all of these randomizations are independent of the randomization of superblocks.

Using the concept of a "null" experiment (see Nelder, 1965a), let the response of the unit labeled $\pi[\xi(\nu)]$ be denoted by $\mu_{\pi[\xi(\nu)]}$, and let it be denoted by $m_{\ell[j(h)]}$ if in result of the randomizations the superblock originally labeled ν receives label h, the block originally labeled $\xi(\nu)$, i.e., the block ξ in the superblock ν, receives label j and the unit originally labeled $\pi[\xi(\nu)]$, i.e., the unit π in the block $\xi(\nu)$, receives label ℓ. Now, introducing the linear identity

$$\mu_{\pi[\xi(\nu)]} = \mu_{\cdot[\cdot(\cdot)]} + (\mu_{\cdot[\cdot(\nu)]} - \mu_{\cdot[\cdot(\cdot)]}) + (\mu_{\cdot[\xi(\nu)]} - \mu_{\cdot[\cdot(\nu)]}) + (\mu_{\pi[\xi(\nu)]} - \mu_{\cdot[\xi(\nu)]}),$$

where (according to the usual dot notation)

$$\mu_{\cdot[\xi(\nu)]} = K_{\xi(\nu)}^{-1} \sum_{\pi[\xi(\nu)]=1}^{K_{\xi(\nu)}} \mu_{\pi[\xi(\nu)]}, \quad \mu_{\cdot[\cdot(\nu)]} = B_\nu^{-1} \sum_{\xi(\nu)=1}^{B_\nu} \mu_{\cdot[\xi(\nu)]}$$

and

$$\mu_{\cdot[\cdot(\cdot)]} = N_A^{-1} \sum_{\nu=1}^{N_A} \mu_{\cdot[\cdot(\nu)]},$$

and defining the variance components (following Nelder, 1977)

$$\sigma_A^2 = (N_A - 1)^{-1} \sum_{\nu=1}^{N_A} (\mu_{\cdot[\cdot(\nu)]} - \mu_{\cdot[\cdot(\cdot)]})^2, \quad \sigma_B^2 = N_A^{-1} \sum_{\nu=1}^{N_A} \sigma_{B,\nu}^2,$$

where

$$\sigma_{B,\nu}^2 = (B_\nu - 1)^{-1} \sum_{\xi(\nu)=1}^{B_\nu} (\mu_{\cdot[\xi(\nu)]} - \mu_{\cdot[\cdot(\nu)]})^2,$$

and

$$\sigma_U^2 = N_A^{-1} \sum_{\nu=1}^{N_A} B_\nu^{-1} \sum_{\xi(\nu)=1}^{B_\nu} \sigma_{U,\xi(\nu)}^2,$$

where

$$\sigma_{U,\xi(\nu)}^2 = (K_{\xi(\nu)} - 1)^{-1} \sum_{\pi[\xi(\nu)]=1}^{K_{\xi(\nu)}} (\mu_{\pi[\xi(\nu)]} - \mu_{\cdot[\xi(\nu)]})^2,$$

and introducing the weighted harmonic averages B_H and K_H defined as

$$B_H^{-1} = N_A^{-1} \sum_{\nu=1}^{N_A} B_\nu^{-1} \sigma^2_{B,\nu} / \sigma^2_B$$

and

$$K_H^{-1} = N_A^{-1} \sum_{\nu=1}^{N_A} B_\nu^{-1} \sum_{\xi(\nu)=1}^{B_\nu} K_{\xi(\nu)}^{-1} \sigma^2_{U,\xi(\nu)} / \sigma^2_U,$$

accordingly, one can write the linear model

$$m_{\ell[j(h)]} = \mu + \alpha_h + \beta_{j(h)} + \eta_{\ell[j(h)]}, \tag{5.2.1}$$

for any indices h, j and ℓ resulting from the randomizations, where $\mu = \mu_{\cdot[\cdot(\cdot)]}$ is a constant parameter, and α_h, $\beta_{j(h)}$ and $\eta_{\ell[j(h)]}$ are random variables, the first representing a superblock random effect, the second representing a block random effect and the third representing a unit error. The following moments of these random variables are easily obtainable:

$$\mathrm{E}(\alpha_h) = \mathrm{E}(\beta_{j(h)}) = \mathrm{E}(\eta_{\ell[j(h)]}) = 0,$$

$$\mathrm{Cov}(\alpha_h, \beta_{j(h')}) = \mathrm{Cov}(\alpha_h, \eta_{\ell[j(h')]}) = 0, \quad \text{whether} \quad h = h' \text{ or } h \neq h',$$

$$\mathrm{Cov}(\beta_{j(h)}, \eta_{\ell[j'(h')]}) = 0, \quad \text{whether} \quad j(h) = j'(h') \text{ or } j(h) \neq j'(h'),$$

$$\mathrm{Cov}(\alpha_h, \alpha_{h'}) = \begin{cases} N_A^{-1}(N_A - 1)\sigma^2_A & \text{if } h = h', \\ -N_A^{-1}\sigma^2_A & \text{if } h \neq h', \end{cases}$$

$$\mathrm{Cov}(\beta_{j(h)}, \beta_{j'(h')}) = \begin{cases} B_H^{-1}(B_H - 1)\sigma^2_B & \text{if } h = h' \text{ and } j = j', \\ -B_H^{-1}\sigma^2_B & \text{if } h = h' \text{ and } j \neq j', \\ 0 & \text{if } h \neq h', \end{cases}$$

and

$$\mathrm{Cov}(\eta_{\ell[j(h)]}, \eta_{\ell'[j'(h')]}) = \begin{cases} K_H^{-1}(K_H - 1)\sigma^2_U & \text{if } h = h' \;\; j = j' \text{ and } \ell = \ell', \\ -K_H^{-1}\sigma^2_U & \text{if } h = h' \;\; j = j' \text{ and } \ell \neq \ell', \\ 0 & \text{if } j(h) \neq j'(h'). \end{cases}$$

The derivations are straightforward, similar to those used in Section 3.1.1.

Thus, the responses $\{m_{\ell[j(h)]}\}$ have in the conceptual null experiment the model (5.2.1) with the properties

$$\mathrm{E}(m_{\ell[j(h)]}) = \mu$$

and

$$\begin{aligned}\mathrm{Cov}(m_{\ell[j(h)]}, m_{\ell'[j'(h')]}) &= (\delta_{hh'} - N_A^{-1})\sigma_A^2 + \delta_{hh'}(\delta_{jj'} - B_H^{-1})\sigma_B^2 \\ &\quad + \delta_{hh'}\delta_{jj'}(\delta_{\ell\ell'} - K_H^{-1})\sigma_U^2,\end{aligned}$$

where the δ's are the usual Kronecker deltas, denoting 1 if the indices coincide, and 0 otherwise.

Furthermore, taking into account [as in (3.1.9)] the technical error and denoting it by $e_{\ell[j(h)]}$, the model of the variable observed on the unit $\ell[j(h)]$ of the null experiment can be written as

$$y_{\ell[j(h)]} = m_{\ell[j(h)]} + e_{\ell[j(h)]} = \mu + \alpha_h + \beta_{j(h)} + \eta_{\ell[j(h)]} + e_{\ell[j(h)]}, \qquad (5.2.2)$$

for any h, j and ℓ. Starting from similar assumptions as those made in connection with formula (3.1.9), it can be shown that the technical errors $\{e_{\ell[j(h)]}\}$ are uncorrelated, with zero expectation and constant variance, σ_e^2, and that they are also not correlated with the remaining random variables in the model. Hence, the first and second moments of the random variables $\{y_{\ell[j(h)]}\}$ defined in (5.2.2) have the forms

$$\mathrm{E}(y_{\ell[j(h)]}) = \mu$$

and

$$\begin{aligned}\mathrm{Cov}(y_{\ell[j(h)]}, y_{\ell'[j'(h')]}) &= (\delta_{hh'} - N_A^{-1})\sigma_A^2 + \delta_{hh'}(\delta_{jj'} - B_H^{-1})\sigma_B^2 \\ &\quad + \delta_{hh'}\delta_{jj'}(\delta_{\ell\ell'} - K_H^{-1})\sigma_U^2 + \delta_{hh'}\delta_{jj'}\delta_{\ell\ell'}\sigma_e^2 \qquad (5.2.3)\end{aligned}$$

for all $\ell[j(h)]$ and all $\ell'[j'(h')]$.

As in the model considered in Section 3.1.1, it follows from these moments that the superblocks, the blocks within superblocks and the units within the blocks, can be regarded as "homogeneous" in the sense that the observed responses of the units may, under the same treatment, be considered as observations on random variables $\{y_{\ell[j(h)]}\}$ exchangeable, without affecting their moments, individually within a block, in sets among the blocks within a superblock, as far as the sizes of the blocks allow for this, and in sets of such sets among the superblocks, as far as the block sizes and the numbers of blocks within the superblocks allow for this.

With regard to this type of homogeneity of units, blocks and superblocks, the randomization principle can be obeyed in designing an experiment according to a chosen incidence matrix $\boldsymbol{N} = [\boldsymbol{N}_1 : \boldsymbol{N}_2 : \cdots : \boldsymbol{N}_a]$ by adopting the following rule. The a submatrices $\boldsymbol{N}_h, h = 1, 2, ..., a$, are assigned to a out of the N_A available superblocks by assigning the hth submatrix to that superblock which after the randomization has label h, and the b_h columns of $\boldsymbol{N}_h$ are assigned to b_h out of the B_h blocks available in the hth superblock by assigning the $j(h)$th column of $\boldsymbol{N}_h$ to that block which after the randomization has label $j(h)$. This assignment is to be accomplished for $h = 1, 2, ..., a$. Then, the treatments indicated by the $j(h)$th column of $\boldsymbol{N}$ are assigned to the experimental units of the block labeled $j(h)$, in numbers defined by the corresponding elements of $\boldsymbol{N}_h = [n_{ij(h)}]$, i.e., the ith treatment to $n_{ij(h)}$ units, in the order determined by the labels the units of the block have received after the randomization. This rule implies not only that $a \leq N_A$, but also that $b_h \leq B_h$ for $h = 1, 2, ..., a$, and that the units in the available blocks are in sufficient numbers with regard to the vector of block sizes $\boldsymbol{k}$, i.e., that $\max_{j(h)} k_{j(h)} \leq \min_{\xi(\nu)} K_{\xi(\nu)}$. This requirement means that either the choice of $\boldsymbol{N}$ is to be conditioned by the above constraints, or an adjustment of $\boldsymbol{N}$ is to be made after accomplishing the randomization of blocks and superblocks (as suggested by White, 1975, p. 558).

Now, adopting the assumption of complete additivity, as postulated in Sections 1.3 and 3.1, equivalent here to the assumption that the variances and covariances of the variables $\{\alpha_h\}$, $\{\beta_j\}$, $\{\eta_{\ell[j(h)]}\}$ and $\{e_{\ell[j(h)]}\}$ do not depend on the treatment applied, the adjustment of the model (5.2.2) to a real experimental situation of comparing several treatments, v, in the same experiment can be made by changing the constant term only. Thus, the model gets the form

$$y_{\ell[j(h)]}(i) = \mu(i) + \alpha_h + \beta_{j(h)} + \eta_{\ell[j(h)]} + e_{\ell[j(h)]} \tag{5.2.4}$$

$(i = 1, 2, ..., v; h = 1, 2, ..., a; j = 1, 2, ..., b_h; \ell = 1, 2, ..., k_{j(h)})$, with

$$\mathrm{E}[y_{\ell[(j(h)]}(i)] = \mu(i) = N_A^{-1} \sum_{\nu=1}^{N_A} B_\nu^{-1} \sum_{\xi(\nu)=1}^{B_\nu} K_{\xi(\nu)}^{-1} \sum_{\pi[\xi(\nu)]=1}^{K_{\xi(\nu)}} \mu_{\pi[\xi(\nu)]}(i), \tag{5.2.5}$$

where $\mu_{\pi[\xi(\nu)]}(i)$ is the true response of unit π in block ξ within superblock ν to treatment i, and with

$$\mathrm{Cov}[y_{\ell[j(h)]}(i), y_{\ell'[j'(h')]}(i')] = \mathrm{Cov}(y_{\ell[j(h)]}, y_{\ell'[j'(h')]}), \tag{5.2.6}$$

as given explicitly in (5.2.3).

Finally, writing the observed variables $\{y_{\ell[j(h)]}(i)\}$ as an $n \times 1$ vector $\boldsymbol{y} = [\boldsymbol{y}_1', \boldsymbol{y}_2', ..., \boldsymbol{y}_a']'$, where the subvector $\boldsymbol{y}_h$ represents the variables observed on the $n_h = \sum_{j(h)=1}^{b_h} k_{j(h)}$ units of the superblock h, and the corresponding unit error and technical error variables in form of $n \times 1$ vectors $\boldsymbol{\eta}$ and $\boldsymbol{e}$, respectively, and writing the treatment parameters as $\boldsymbol{\tau} = [\tau_1, \tau_2, ..., \tau_v]'$, where $\tau_i = \mu(i)$, the superblock variables as $\boldsymbol{\alpha} = [\alpha_1, \alpha_2, ..., \alpha_a]'$ and the block variables as $\boldsymbol{\beta} = [\boldsymbol{\beta}_1', \boldsymbol{\beta}_2', ..., \boldsymbol{\beta}_a']'$, where $\boldsymbol{\beta}_h = [\beta_{1(h)}, \beta_{2(h)}, ..., \beta_{b_h(h)}]'$, one can express the model (5.2.4) in matrix notation as

$$\boldsymbol{y} = \boldsymbol{\Delta}'\boldsymbol{\tau} + \boldsymbol{G}'\boldsymbol{\alpha} + \boldsymbol{D}'\boldsymbol{\beta} + \boldsymbol{\eta} + \boldsymbol{e}, \tag{5.2.7}$$

and the corresponding moments (5.2.5) and (5.2.6) in the form of the expectation vector

$$\mathrm{E}(\boldsymbol{y}) = \boldsymbol{\Delta}'\boldsymbol{\tau} \tag{5.2.8}$$

and the covariance (dispersion) matrix

$$\begin{aligned}\text{Cov}(\boldsymbol{y}) &= (\boldsymbol{G}'\boldsymbol{G} - N_A^{-1}\mathbf{1}_n\mathbf{1}_n')\sigma_A^2 + (\boldsymbol{D}'\boldsymbol{D} - B_H^{-1}\boldsymbol{G}'\boldsymbol{G})\sigma_B^2 \\ &\quad + (\boldsymbol{I}_n - K_H^{-1}\boldsymbol{D}'\boldsymbol{D})\sigma_U^2 + \boldsymbol{I}_n\sigma_e^2, \end{aligned} \tag{5.2.9}$$

respectively (see Section 5.1 for definitions of the matrices involved).

Note that the model (5.2.7), with properties (5.2.8) and (5.2.9), can be seen as a generalization of the model used by Patterson and Thompson (1971), when writing the covariance matrix (5.2.9) in the form

$$\text{Cov}(\boldsymbol{y}) = \sigma_1^2(\boldsymbol{G}'\boldsymbol{\Gamma}_1\boldsymbol{G} + \boldsymbol{D}'\boldsymbol{\Gamma}_2\boldsymbol{D} + \boldsymbol{I}_n), \tag{5.2.10}$$

where

$$\boldsymbol{\Gamma}_1 = \boldsymbol{I}_a\gamma_1 - N_A^{-1}\mathbf{1}_a\mathbf{1}_a'\sigma_A^2/\sigma_1^2, \quad \gamma_1 = (\sigma_A^2 - B_H^{-1}\sigma_B^2)/\sigma_1^2 \tag{5.2.11}$$

and

$$\boldsymbol{\Gamma}_2 = \boldsymbol{I}_b\gamma_2, \quad \gamma_2 = (\sigma_B^2 - K_H^{-1}\sigma_U^2)/\sigma_1^2, \quad \sigma_1^2 = \sigma_U^2 + \sigma_e^2. \tag{5.2.12}$$

Furthermore, note that if $B_1 = B_2 = \cdots = B_{N_A} = B$ (say) and $K_{1(\nu)} = K_{2(\nu)} = \cdots = K_{B_\nu(\nu)} = K$ (say) for each ν $(= 1, 2, ..., N_A)$, then the present model is comparable with that considered by Mejza (1992, p. 269).

5.2.2 Main estimation results

Under the present model, the following main results concerning the linear estimation of treatment parametric functions are obtainable. These results are relevant generalizations of those given in Section 3.1.2.

Theorem 5.2.1. *Under the model* (5.2.7), *with properties* (5.2.8) *and* (5.2.9), *a function $\boldsymbol{w}'\boldsymbol{y}$ is uniformly the BLUE of $\boldsymbol{c}'\boldsymbol{\tau}$ if and only if $\boldsymbol{w} = \boldsymbol{\Delta}'\boldsymbol{s}$, where $\boldsymbol{s} = \boldsymbol{r}^{-\delta}\boldsymbol{c}$ satisfies the conditions*

$$(\boldsymbol{k}^\delta - \boldsymbol{N}'\boldsymbol{r}^{-\delta}\boldsymbol{N})\boldsymbol{N}'\boldsymbol{s} = \mathbf{0} \tag{5.2.13}$$

and

$$(\boldsymbol{n}^\delta - \boldsymbol{\mathfrak{R}}'\boldsymbol{r}^{-\delta}\boldsymbol{\mathfrak{R}})\boldsymbol{\mathfrak{R}}'\boldsymbol{s} = \mathbf{0}. \tag{5.2.14}$$

Proof. On account of Theorem 1.2.1, a function $\boldsymbol{w}'\boldsymbol{y}$ is, under the considered model, the BLUE of its expectation, i.e., of $\boldsymbol{w}'\boldsymbol{\Delta}'\boldsymbol{\tau}$, if and only if

$$\begin{aligned}(\boldsymbol{I}_n - \boldsymbol{P}_{\Delta'})[(\boldsymbol{G}'\boldsymbol{G} - N_A^{-1}\mathbf{1}_n\mathbf{1}_n')\sigma_A^2 + (\boldsymbol{D}'\boldsymbol{D} - B_H^{-1}\boldsymbol{G}'\boldsymbol{G})\sigma_B^2 \\ + (\boldsymbol{I}_n - K_H^{-1}\boldsymbol{D}'\boldsymbol{D})\sigma_U^2 + \boldsymbol{I}_n\sigma_e^2]\boldsymbol{w} = \mathbf{0},\end{aligned}$$

where $\boldsymbol{P}_{\Delta'} = \boldsymbol{\Delta}'\boldsymbol{r}^{-\delta}\boldsymbol{\Delta}$ denotes the orthogonal projector on $\mathcal{C}(\boldsymbol{\Delta}')$, the column space of $\boldsymbol{\Delta}'$. If this equation is to hold uniformly for any set of the variance components $\sigma_A^2, \sigma_B^2, \sigma_U^2$ and σ_e^2, it is necessary and sufficient that

$$(\boldsymbol{I}_n - \boldsymbol{P}_{\Delta'})\boldsymbol{w} = \mathbf{0}, \quad (\boldsymbol{I}_n - \boldsymbol{P}_{\Delta'})\boldsymbol{D}'\boldsymbol{D}\boldsymbol{w} = \mathbf{0} \quad \text{and} \quad (\boldsymbol{I}_n - \boldsymbol{P}_{\Delta'})\boldsymbol{G}'\boldsymbol{G}\boldsymbol{w} = \mathbf{0}.$$

For these equations to hold simultaneously, it is necessary and sufficient that $w = \boldsymbol{\Delta}'s$, $(\boldsymbol{I}_n - \boldsymbol{P}_{\Delta'})\boldsymbol{D}'\boldsymbol{D}\,\boldsymbol{\Delta}'s = \mathbf{0}$ and $(\boldsymbol{I}_n - \boldsymbol{P}_{\Delta'})\boldsymbol{G}'\boldsymbol{G}\boldsymbol{\Delta}'s = \mathbf{0}$ for some vector s. However, the latter two equations are equivalent to (5.2.13) and (5.2.14), respectively. □

In connection with this proof, note that the component $N_A^{-1}\mathbf{1}_n\mathbf{1}_n'\sigma_A^2$ in (5.2.9) does not play any role in establishing Theorem 5.2.1, as $(\boldsymbol{I}_n - \boldsymbol{P}_{\Delta'})\mathbf{1}_n = \mathbf{0}$. For the same reason, the simplification of $\boldsymbol{\Gamma}_1$ to the form $\boldsymbol{I}_a\gamma_1$, as suggested by Patterson and Thompson (1971), does not affect the BLUE of $\boldsymbol{c}'\boldsymbol{\tau}$.

Corollary 5.2.1. *For the estimation of $\boldsymbol{c}'\boldsymbol{\tau} = \boldsymbol{s}'\boldsymbol{r}^{\delta}\boldsymbol{\tau}$, under the model as in Theorem* 5.2.1, *the following applies:*

(a) *If $\boldsymbol{N}'\boldsymbol{s} = \mathbf{0}$, which implies $\boldsymbol{\Re}'\boldsymbol{s} = \mathbf{0}$, the conditions* (5.2.13) *and* (5.2.14) *are satisfied, and the estimated function is a contrast, i.e., $\boldsymbol{c}'\mathbf{1}_v = \boldsymbol{s}'\boldsymbol{r} = 0$.*

(b) *If $\boldsymbol{N}'\boldsymbol{s} \neq \mathbf{0}$ and $\boldsymbol{\Re}'\boldsymbol{s} = \mathbf{0}$, then* (5.2.14) *is satisfied, but to satisfy* (5.2.13) *it is necessary and sufficient that the elements of $\boldsymbol{N}'\boldsymbol{s}$ obtained from the same connected subdesign of $\mathcal{D}^*$ are all equal.*

(c) *If $\boldsymbol{\Re}'\boldsymbol{s} \neq \mathbf{0}$, which implies that $\boldsymbol{N}'\boldsymbol{s} \neq \mathbf{0}$, then to satisfy* (5.2.14) *in addition to* (5.2.13), *it is necessary and sufficient that not only the elements of $\boldsymbol{N}'\boldsymbol{s}$ from the same connected subdesign of $\mathcal{D}^*$ are all equal, but also the elements of $\boldsymbol{\Re}'\boldsymbol{s}$ obtained from the same connected subdesign of $\mathcal{D}$ are all equal.*

Proof. The results can be proved following the proof of Corollary 3.1.1. □

Now a question may be raised, under which design conditions any function $\boldsymbol{s}'\boldsymbol{\Delta y}$ is the BLUE of its expectation. An answer to this question can be given as follows.

Theorem 5.2.2. *Under the model as in Theorem* 5.2.1, *any function $\boldsymbol{w}'\boldsymbol{y} = \boldsymbol{s}'\boldsymbol{\Delta y}$, i.e., with any $\boldsymbol{s}$, is uniformly the BLUE of* $\mathrm{E}(\boldsymbol{w}'\boldsymbol{y}) = \boldsymbol{s}'\boldsymbol{r}^{\delta}\boldsymbol{\tau}$ *if and only if*

(i) *both of the designs, $\mathcal{D}^*$ and $\mathcal{D}$, are orthogonal (in the sense of Definition* 2.2.7 *or the equivalent Definition* 2.2.8) *and*

(ii) *the block sizes are constant within any connected subdesign of $\mathcal{D}^*$ and the sizes of superblocks are constant within any connected subdesign of $\mathcal{D}$.*

Proof. From the proof of Theorem 5.2.1, it is evident that $\boldsymbol{s}'\boldsymbol{\Delta y}$ is the BLUE of its expectation for any $\boldsymbol{s}$ if and only if

$$(\boldsymbol{I}_n - \boldsymbol{P}_{\Delta'})\boldsymbol{D}'\boldsymbol{D}\boldsymbol{\Delta}' = \mathbf{O} \tag{5.2.15}$$

and

$$(\boldsymbol{I}_n - \boldsymbol{P}_{\Delta'})\boldsymbol{G}'\boldsymbol{G}\boldsymbol{\Delta}' = \mathbf{O}. \tag{5.2.16}$$

Adopting the same reasoning as that used in the proof of Theorem 3.1.2, the present results can be proved. □

Note that Remark 3.1.2 applies here as well, with an obvious extension to the design $\mathcal{D}$. In particular, if the latter is connected, which usually is the case, then the orthogonality condition for it can be written as $\mathfrak{R} = n^{-1}\boldsymbol{r}\boldsymbol{n}'$, which under the additional condition (ii) reduces to $\mathfrak{R} = (1/a)\boldsymbol{r}\mathbf{1}'_a$.

Remark 5.2.1. If the conditions (i) and (ii) stated in Theorem 5.2.2 are satisfied, i.e., if (5.2.15) and (5.2.16) hold, then

$$\begin{aligned}\mathrm{Cov}(\boldsymbol{y})\boldsymbol{\Delta}' &= \boldsymbol{\Delta}'\boldsymbol{r}^{-\delta}[(\mathfrak{R}\mathfrak{R}' - N_A^{-1}\boldsymbol{r}\boldsymbol{r}')\sigma_A^2 + (\boldsymbol{N}\boldsymbol{N}' - B_H^{-1}\mathfrak{R}\mathfrak{R}')\sigma_B^2 \\ &\quad + (\boldsymbol{r}^{\delta} - K_H^{-1}\boldsymbol{N}\boldsymbol{N}')\sigma_U^2 + \boldsymbol{r}^{\delta}\sigma_e^2],\end{aligned}$$

which implies that both $\boldsymbol{\Delta}'\boldsymbol{s}$ and $\mathrm{Cov}(\boldsymbol{y})\boldsymbol{\Delta}'\boldsymbol{s}$ belong to $\mathcal{C}(\boldsymbol{\Delta}')$ for any $\boldsymbol{s}$, and thus, by Theorem 1.2.2, the BLUEs obtainable under the model (5.2.7), with the moments (5.2.8) and (5.2.9), can equivalently be obtained under a simple alternative model in which the covariance matrix (5.2.9) is reduced to $\boldsymbol{I}_n$, multiplied by a positive scalar (see also Rao and Mitra, 1971, Section 8.2). Moreover, it can be shown (applying, e.g., Theorem 2.3.2 of Rao and Mitra, 1971) that the equalities (5.2.15) and (5.2.16) are not only sufficient, but also necessary conditions for the BLUEs obtainable under the two alternative models to be the same. Thus, (5.2.15) and (5.2.16) are necessary and sufficient for $\boldsymbol{s}'\boldsymbol{\Delta}\boldsymbol{y}$ to be both the SLSE and the BLUE of its expectation, $\boldsymbol{s}'\boldsymbol{r}^{\delta}\boldsymbol{\tau}$, whichever vector $\boldsymbol{s}$ is used.

5.3 Resolving into stratum submodels

The results of Section 5.2 are discouraging (as in Section 3.1) bacause, according to them, in many NB designs, the BLUEs will exist under the model (5.2.7) for only exceptional parametric functions of interest, or for none of them.

This difficulty with the model (5.2.7) can be evaded by resolving it into four submodels (one more than in the case of an ordinary block design; see Section 3.2), in accordance with the stratification of the experimental units. In fact, the units of a NB design can be seen as being grouped according to a nested classification with four strata. The strata may be defined as follows:

1st stratum—of units within blocks, called "intra-block,"

2nd stratum—of blocks within superblocks, called "inter-block-intra-superblock,"

3rd stratum—of superblocks within the experimental area, called "inter-superblock,"

4th stratum—of the total experimental area.

Using Nelder's (1965a) notation, this "block-structure" can be represented by the relation

Units (plots) → Blocks → Superblocks → Total area.

In accordance to this stratification, the observed vector $\boldsymbol{y}$ can be decomposed as

$$\boldsymbol{y} = \boldsymbol{y}_1 + \boldsymbol{y}_2 + \boldsymbol{y}_3 + \boldsymbol{y}_4, \tag{5.3.1}$$

where each of the four components is related to one of the strata. The component vectors $\boldsymbol{y}_\alpha$, $\alpha = 1, 2, 3, 4$, are thus obtainable by projecting $\boldsymbol{y}$ orthogonally onto relevant subspaces, mutually orthogonal. The first component in (5.3.1) can be written as

$$\boldsymbol{y}_1 = \tilde{\boldsymbol{\phi}}_1 \boldsymbol{y}, \tag{5.3.2}$$

where $\tilde{\boldsymbol{\phi}}_1 = \boldsymbol{I}_n - \boldsymbol{D}'\boldsymbol{k}^{-\delta}\boldsymbol{D} = \boldsymbol{I}_n - \boldsymbol{P}_{D'}$ is exactly as $\boldsymbol{\phi}_1$ defined in (3.2.3); i.e., $\boldsymbol{y}_1$ is the orthogonal projection of $\boldsymbol{y}$ on $\mathcal{C}^\perp(\boldsymbol{D}')$, the orthogonal complement of $\mathcal{C}(\boldsymbol{D}')$. The second component is

$$\boldsymbol{y}_2 = \tilde{\boldsymbol{\phi}}_2 \boldsymbol{y}, \tag{5.3.3}$$

where now $\tilde{\boldsymbol{\phi}}_2 = \boldsymbol{D}'\boldsymbol{k}^{-\delta}\boldsymbol{D} - \boldsymbol{G}'\boldsymbol{n}^{-\delta}\boldsymbol{G} = \boldsymbol{P}_{D'} - \boldsymbol{P}_{G'}$, with $\boldsymbol{n}^{-\delta} = (\boldsymbol{G}\boldsymbol{G}')^{-1}$; i.e., $\boldsymbol{y}_2$ is the orthogonal projection of $\boldsymbol{y}$ on $\mathcal{C}^\perp(\boldsymbol{G}') \cap \mathcal{C}(\boldsymbol{D}')$, the orthogonal complement of $\mathcal{C}(\boldsymbol{G}')$ in $\mathcal{C}(\boldsymbol{D}')$. The third component is

$$\boldsymbol{y}_3 = \tilde{\boldsymbol{\phi}}_3 \boldsymbol{y}, \tag{5.3.4}$$

where $\tilde{\boldsymbol{\phi}}_3 = \boldsymbol{G}'\boldsymbol{n}^{-\delta}\boldsymbol{G} - n^{-1}\mathbf{1}_n\mathbf{1}_n' = \boldsymbol{P}_{G'} - \boldsymbol{P}_{\mathbf{1}_n}$; i.e., $\boldsymbol{y}_3$ is the orthogonal projection of $\boldsymbol{y}$ on $\mathcal{C}^\perp(\mathbf{1}_n) \cap \mathcal{C}(\boldsymbol{G}')$, the orthogonal complement of $\mathcal{C}(\mathbf{1}_n)$ in $\mathcal{C}(\boldsymbol{G}')$. Finally, the fourth component is

$$\boldsymbol{y}_4 = \tilde{\boldsymbol{\phi}}_4 \boldsymbol{y}, \tag{5.3.5}$$

where $\tilde{\boldsymbol{\phi}}_4 = n^{-1}\mathbf{1}_n\mathbf{1}_n' = \boldsymbol{P}_{\mathbf{1}_n}$; i.e., $\boldsymbol{y}_4$ is the orthogonal projection of $\boldsymbol{y}$ on $\mathcal{C}(\mathbf{1}_n)$.

Evidently, the four matrices $\tilde{\boldsymbol{\phi}}_1$, $\tilde{\boldsymbol{\phi}}_2$, $\tilde{\boldsymbol{\phi}}_3$ and $\tilde{\boldsymbol{\phi}}_4$ (comparable with those of Pearce, 1983, p. 153) satisfy the conditions

$$\tilde{\boldsymbol{\phi}}_\alpha = \tilde{\boldsymbol{\phi}}_\alpha', \quad \tilde{\boldsymbol{\phi}}_\alpha\tilde{\boldsymbol{\phi}}_\alpha = \tilde{\boldsymbol{\phi}}_\alpha, \quad \tilde{\boldsymbol{\phi}}_\alpha\tilde{\boldsymbol{\phi}}_{\alpha'} = \mathbf{O} \text{ for } \alpha \neq \alpha', \tag{5.3.6}$$

where $\alpha, \alpha' = 1, 2, 3, 4$, and the condition $\tilde{\boldsymbol{\phi}}_1 + \tilde{\boldsymbol{\phi}}_2 + \tilde{\boldsymbol{\phi}}_3 + \tilde{\boldsymbol{\phi}}_4 = \boldsymbol{I}_n$, the third equality in (5.3.6) implying, in particular, that

$$\tilde{\boldsymbol{\phi}}_1\boldsymbol{D}' = \mathbf{O},\ \tilde{\boldsymbol{\phi}}_\alpha\boldsymbol{G}' = \mathbf{O} \text{ for } \alpha = 1, 2, \text{ and } \tilde{\boldsymbol{\phi}}_\alpha\mathbf{1}_n = \mathbf{0} \text{ for } \alpha = 1, 2, 3, \tag{5.3.7}$$

whereas the first two equalities in (5.3.6) imply that

$$\text{rank}(\tilde{\boldsymbol{\phi}}_1) = n - b,\ \text{rank}(\tilde{\boldsymbol{\phi}}_2) = b - a,\ \text{rank}(\tilde{\boldsymbol{\phi}}_3) = a - 1 \text{ and } \text{rank}(\tilde{\boldsymbol{\phi}}_4) = 1.$$

The projections (5.3.2), (5.3.3), (5.3.4) and (5.3.5) can be considered as submodels of the original overall model (5.2.7). They are of particular interest when the conditions (5.2.13) and (5.2.14) are not satisfied. Similarly as shown in Section 3.2, the submodel (5.3.2) leads to the intra-block analysis, the submodel (5.3.3) leads to the inter-block-intra-superblock analysis, the submodel (5.3.4) to the inter-superblock analysis. The submodel (5.3.5) underlies the total-area analysis, suitable mainly for estimating the general parametric mean (hence, the last stratum is sometimes called the "mean stratum"; see John, 1987, p. 184).

It will be now interesting to examine properties of the four submodels and their implications for the resulting analyses. Because close similarities to the results presented in Section 3.2 exist, in most cases it will be possible to give the new results without detailed proofs.

5.3.1 Intra-block submodel

The submodel (5.3.2) has the properties

$$\mathrm{E}(\boldsymbol{y}_1) = \tilde{\boldsymbol{\phi}}_1 \boldsymbol{\Delta}' \boldsymbol{\tau} \quad \text{and} \quad \mathrm{Cov}(\boldsymbol{y}_1) = \tilde{\boldsymbol{\phi}}_1(\sigma_U^2 + \sigma_e^2), \quad \text{with } \tilde{\boldsymbol{\phi}}_1 \equiv \boldsymbol{\phi}_1,$$

which are exactly the same as in (3.2.11) and (3.2.12). Hence, the results concerning estimation of parameters and testing of relevant hypotheses under this submodel are exactly as presented in Section 3.2.1. In particular, on account of Theorem 3.2.1 and formula (3.2.13), any contrast $\boldsymbol{c}'\boldsymbol{\tau}$ such that $\boldsymbol{c} = \tilde{\boldsymbol{C}}_1\boldsymbol{s}$ for some $\boldsymbol{s}$, where $\tilde{\boldsymbol{C}}_1 = \boldsymbol{\Delta}\tilde{\boldsymbol{\phi}}_1\boldsymbol{\Delta}' = \boldsymbol{r}^{\delta} - \boldsymbol{N}\boldsymbol{k}^{-\delta}\boldsymbol{N}' \equiv \boldsymbol{C}_1$ (Section 3.2.1), receives the BLUE under this submodel in the form $\widehat{\boldsymbol{c}'\boldsymbol{\tau}} = \boldsymbol{s}'\boldsymbol{\Delta}\boldsymbol{y}_1$. Its variance has the form

$$\mathrm{Var}(\widehat{\boldsymbol{c}'\boldsymbol{\tau}}) = \boldsymbol{s}'\tilde{\boldsymbol{C}}_1\boldsymbol{s}(\sigma_U^2 + \sigma_e^2) = \boldsymbol{c}'\tilde{\boldsymbol{C}}_1^{-}\boldsymbol{c}(\sigma_U^2 + \sigma_e^2), \tag{5.3.8}$$

where $\tilde{\boldsymbol{C}}_1^{-}$ is any generalized inverse (g-inverse) of $\tilde{\boldsymbol{C}}_1$.

The intra-block analysis of variance can be presented as

$$\boldsymbol{y}'\tilde{\boldsymbol{\phi}}_1\boldsymbol{y} = \tilde{\boldsymbol{Q}}_1'\tilde{\boldsymbol{C}}_1^{-}\tilde{\boldsymbol{Q}}_1 + \boldsymbol{y}'\tilde{\boldsymbol{\psi}}_1\boldsymbol{y},$$

where $\tilde{\boldsymbol{Q}}_1'\tilde{\boldsymbol{C}}_1^{-}\tilde{\boldsymbol{Q}}_1$, with $\tilde{\boldsymbol{Q}}_1 = \boldsymbol{\Delta}\tilde{\boldsymbol{\phi}}_1\boldsymbol{y}$, is the intra-block treatment sum of squares, and $\boldsymbol{y}'\tilde{\boldsymbol{\psi}}_1\boldsymbol{y}$, with

$$\tilde{\boldsymbol{\psi}}_1 = \tilde{\boldsymbol{\phi}}_1 - \tilde{\boldsymbol{\phi}}_1\boldsymbol{\Delta}'\tilde{\boldsymbol{C}}_1^{-}\boldsymbol{\Delta}\tilde{\boldsymbol{\phi}}_1 = \tilde{\boldsymbol{\phi}}_1(\boldsymbol{I}_n - \boldsymbol{\Delta}'\tilde{\boldsymbol{C}}_1^{-}\boldsymbol{\Delta})\tilde{\boldsymbol{\phi}}_1 \equiv \boldsymbol{\psi}_1$$

(see Section 3.2.1), is the intra-block residual sum of squares, the first on $\tilde{h}_1 = \mathrm{rank}(\tilde{\boldsymbol{C}}_1)$ d.f., the second on $n - b - \tilde{h}_1 = \mathrm{rank}(\tilde{\boldsymbol{\psi}}_1)$ d.f. (where $\tilde{h}_1 \equiv h$ of Section 3.2.1). The resulting intra-block residual mean square $s_1^2 = \boldsymbol{y}'\tilde{\boldsymbol{\psi}}_1\boldsymbol{y}/(n - b - \tilde{h}_1)$ is the MINQUE of

$$\sigma_1^2 = \sigma_U^2 + \sigma_e^2,$$

giving an unbiased estimator of (5.3.8), in the form

$$\widehat{\mathrm{Var}(\widehat{\boldsymbol{c}'\boldsymbol{\tau}})} = \boldsymbol{s}'\tilde{\boldsymbol{C}}_1\boldsymbol{s}s_1^2 = \boldsymbol{c}'\tilde{\boldsymbol{C}}_1^{-}\boldsymbol{c}s_1^2. \tag{5.3.9}$$

Furthermore (as in Section 3.2.1), under the multivariate normal distribution of $\boldsymbol{y}$, the hypothesis $\boldsymbol{\tau}'\tilde{\boldsymbol{C}}_1\boldsymbol{\tau} = 0$, equivalent to $\mathrm{E}(\boldsymbol{y}_1) = \boldsymbol{0}$ or $\mathrm{E}(\boldsymbol{y}) \in \mathcal{C}(\boldsymbol{D}')$, can be tested by the variance ratio criterion $\tilde{h}_1^{-1}\tilde{\boldsymbol{Q}}_1'\tilde{\boldsymbol{C}}_1^{-}\tilde{\boldsymbol{Q}}_1/s_1^2$, which has then the F distribution with $\tilde{h}_1$ and $n - b - \tilde{h}_1$ d.f., central when the hypothesis is true.

Also, it should be noted that the results presented in Sections 3.4 and 4.1 apply here as well.

5.3.2 Inter-block-intra-superblock submodel

The submodel (5.3.3) has the properties

$$\mathrm{E}(\boldsymbol{y}_2) = \tilde{\boldsymbol{\phi}}_2\boldsymbol{\Delta}'\boldsymbol{\tau} = \boldsymbol{D}'\boldsymbol{k}^{-\delta}\boldsymbol{D}\boldsymbol{\Delta}'\boldsymbol{\tau} - \boldsymbol{G}'\boldsymbol{n}^{-\delta}\boldsymbol{G}\boldsymbol{\Delta}'\boldsymbol{\tau}$$

and

$$\mathrm{Cov}(\boldsymbol{y}_2) = \tilde{\boldsymbol{\phi}}_2\boldsymbol{D}'\boldsymbol{D}\tilde{\boldsymbol{\phi}}_2(\sigma_B^2 - K_H^{-1}\sigma_U^2) + \tilde{\boldsymbol{\phi}}_2(\sigma_U^2 + \sigma_e^2),$$

similar to those of the inter-block submodel considered in Section 3.2.2, but now with $\tilde{\boldsymbol{\phi}}_2$ different from $\boldsymbol{\phi}_2$. In fact, the submodel (3.2.4) can be seen as a special case of (5.3.3) for $a = 1$, as then $\boldsymbol{G}' = \mathbf{1}_n$ and $\boldsymbol{n} = n$, giving $\tilde{\boldsymbol{\phi}}_2 = \boldsymbol{\phi}_2 = \boldsymbol{D}'\boldsymbol{k}^{-\delta}\boldsymbol{D} - n^{-1}\mathbf{1}_n\mathbf{1}_n'$, as in (3.2.5).

The main estimation result under (5.3.3) can be expressed as follows.

Theorem 5.3.2. *Under* (5.3.3), *a function* $\boldsymbol{w}'\boldsymbol{y}_2 = \boldsymbol{w}'\tilde{\boldsymbol{\phi}}_2\boldsymbol{y}$ *is uniformly the BLUE of* $\boldsymbol{c}'\boldsymbol{\tau}$ *if and only if* $\tilde{\boldsymbol{\phi}}_2\boldsymbol{w} = \tilde{\boldsymbol{\phi}}_2\boldsymbol{\Delta}'\boldsymbol{s}$, *where the vectors* $\mathbf{c}$ *and* $\boldsymbol{s}$ *are in the relation* $\mathbf{c} = \tilde{\boldsymbol{C}}_2\boldsymbol{s}$, *with* $\tilde{\boldsymbol{C}}_2 = \boldsymbol{\Delta}\tilde{\boldsymbol{\phi}}_2\boldsymbol{\Delta}' = \boldsymbol{N}\boldsymbol{k}^{-\delta}\boldsymbol{N}' - \boldsymbol{\mathfrak{R}}\boldsymbol{n}^{-\delta}\boldsymbol{\mathfrak{R}}'$, *and* $\boldsymbol{s}$ *satisfies the condition*

$$[\tilde{\boldsymbol{K}}_0 - \tilde{\boldsymbol{N}}_0'(\tilde{\boldsymbol{N}}_0\boldsymbol{k}^{-\delta}\tilde{\boldsymbol{N}}_0')^-\tilde{\boldsymbol{N}}_0]\tilde{\boldsymbol{N}}_0'\boldsymbol{s} = \mathbf{0}, \tag{5.3.10}$$

or its equivalence

$$[\tilde{\boldsymbol{K}}_0 - \tilde{\boldsymbol{N}}_0'\boldsymbol{r}^{-\delta}\tilde{\boldsymbol{N}}_0(\tilde{\boldsymbol{N}}_0'\boldsymbol{r}^{-\delta}\tilde{\boldsymbol{N}}_0)^-\tilde{\boldsymbol{K}}_0]\tilde{\boldsymbol{N}}_0'\boldsymbol{s} = \mathbf{0}, \tag{5.3.11}$$

where

$$\tilde{\boldsymbol{K}}_0 = \mathrm{diag}[\boldsymbol{K}_{01} : \boldsymbol{K}_{02} : \cdots : \boldsymbol{K}_{0a}], \quad \boldsymbol{K}_{0h} = \boldsymbol{k}_h^{\delta} - n_h^{-1}\boldsymbol{k}_h\boldsymbol{k}_h',$$

and

$$\tilde{\boldsymbol{N}}_0 = [\boldsymbol{N}_{01} : \boldsymbol{N}_{02} : \cdots : \boldsymbol{N}_{0a}], \quad \boldsymbol{N}_{0h} = \boldsymbol{N}_h - n_h^{-1}\boldsymbol{r}_h\boldsymbol{k}_h'.$$

Proof. A proof similar to that of Theorem 3.2.2 can be adopted. Here too, the following relations hold: $\boldsymbol{D}'\boldsymbol{k}^{-\delta}\boldsymbol{D}\tilde{\boldsymbol{\phi}}_2 = \tilde{\boldsymbol{\phi}}_2$, $\boldsymbol{D}\tilde{\boldsymbol{\phi}}_2\boldsymbol{D}' = \tilde{\boldsymbol{K}}_0$, $\boldsymbol{\Delta}\tilde{\boldsymbol{\phi}}_2\boldsymbol{D}' = \tilde{\boldsymbol{N}}_0$, $\tilde{\boldsymbol{C}}_2 = \boldsymbol{\Delta}\tilde{\boldsymbol{\phi}}_2\boldsymbol{\Delta}' = \tilde{\boldsymbol{N}}_0\boldsymbol{k}^{-\delta}\tilde{\boldsymbol{N}}_0'$ and $\tilde{\boldsymbol{N}}_0\boldsymbol{k}^{-\delta}\tilde{\boldsymbol{K}}_0' = \tilde{\boldsymbol{N}}_0$. □

Corollary 5.3.1. *For the estimation of* $\boldsymbol{c}'\boldsymbol{\tau} = \boldsymbol{s}'\tilde{\boldsymbol{N}}_0\boldsymbol{k}^{-\delta}\tilde{\boldsymbol{N}}_0'\boldsymbol{\tau}$ *under* (5.3.2), *the following applies:*

(a) *The case* $\tilde{\boldsymbol{N}}_0'\boldsymbol{s} = \mathbf{0}$ *is to be excluded.*

(b) *If* $\tilde{\boldsymbol{N}}_0'\boldsymbol{s} \neq \mathbf{0}$, *then* $\boldsymbol{c}'\boldsymbol{\tau}$ *is a contrast, and to satisfy* (5.3.10) *or* (5.3.11) *by the vector* $\boldsymbol{s}$, *it is necessary and sufficient that* $\tilde{\boldsymbol{K}}_0\tilde{\boldsymbol{N}}_0'\boldsymbol{s} \in \mathcal{C}(\tilde{\boldsymbol{N}}_0'\boldsymbol{r}^{-\delta}\tilde{\boldsymbol{N}}_0) = \mathcal{C}(\tilde{\boldsymbol{N}}_0')$.

(c) *If* $\boldsymbol{s}$ *is such that* $\boldsymbol{r}_h'\boldsymbol{s} = 0$ *for* $h = 1, 2, ..., a$, *then the conditions* (5.3.10) *and* (5.3.11) *can be replaced by*

$$\tilde{\boldsymbol{K}}_0\boldsymbol{N}'\boldsymbol{s} = \tilde{\boldsymbol{N}}_0'(\tilde{\boldsymbol{N}}_0\boldsymbol{k}^{-\delta}\tilde{\boldsymbol{N}}_0')^-\tilde{\boldsymbol{N}}_0\boldsymbol{N}'\boldsymbol{s}$$

and

$$\tilde{\boldsymbol{K}}_0\boldsymbol{N}'\boldsymbol{s} = \tilde{\boldsymbol{N}}_0'\boldsymbol{r}^{-\delta}\tilde{\boldsymbol{N}}_0(\tilde{\boldsymbol{N}}_0'\boldsymbol{r}^{-\delta}\tilde{\boldsymbol{N}}_0)^-\tilde{\boldsymbol{K}}_0\boldsymbol{N}'\boldsymbol{s},$$

respectively. To satisfy any of them it is then necessary and sufficient that $\tilde{\boldsymbol{K}}_0\boldsymbol{N}'\boldsymbol{s} \in \mathcal{C}(\tilde{\boldsymbol{N}}_0')$.

(d) *If all block sizes are equal, i.e.,* $k_1 = k_2 = \cdots = k_b = k$ *(say), then any of the conditions* (5.3.10) *and* (5.3.11) *is satisfied automatically by any* $\boldsymbol{s}$.

Proof. The results (a), (b) and (c) can be proved exactly as in Corollary 3.2.1. The result (d) is obtainable from the equality $\tilde{\boldsymbol{K}}_0\tilde{\boldsymbol{N}}' = k\tilde{\boldsymbol{N}}_0'$, where k is the constant block size, and the use of Lemma 2.2.6(c) of Rao and Mitra (1971). □

Now, it may be noted that if the conditions of Theorem 5.3.2 are satisfied, then any contrast $\boldsymbol{c}'\boldsymbol{\tau}$, such that $\boldsymbol{c} = \tilde{\boldsymbol{C}}_2\boldsymbol{s}$, for some $\boldsymbol{s}$, receives the BLUE under the submodel (5.3.3) in the form $\widehat{\boldsymbol{c}'\boldsymbol{\tau}} = \boldsymbol{s}'\boldsymbol{\Delta}\boldsymbol{y}_2 = \boldsymbol{c}'\tilde{\boldsymbol{C}}_2^-\boldsymbol{\Delta}\tilde{\boldsymbol{\phi}}_2\boldsymbol{y}$. Its variance has the form

$$\mathrm{Var}(\widehat{\boldsymbol{c}'\boldsymbol{\tau}}) = \boldsymbol{s}'\tilde{\boldsymbol{N}}_0\tilde{\boldsymbol{N}}_0'\boldsymbol{s}(\sigma_B^2 - K_H^{-1}\sigma_U^2) + \boldsymbol{s}'\tilde{\boldsymbol{N}}_0\boldsymbol{k}^{-\delta}\tilde{\boldsymbol{N}}_0'\boldsymbol{s}(\sigma_U^2 + \sigma_e^2). \quad (5.3.12)$$

Evidently, if $k_1 = k_2 = \cdots = k_b = k$, the variance (5.3.12) reduces to

$$\begin{aligned}\mathrm{Var}(\widehat{\boldsymbol{c}'\boldsymbol{\tau}}) &= k^{-1}\boldsymbol{s}'\tilde{\boldsymbol{N}}_0\tilde{\boldsymbol{N}}_0'\boldsymbol{s}[k\sigma_B^2 + (1 - K_H^{-1}k)\sigma_U^2 + \sigma_e^2] \\ &= \boldsymbol{c}'(k^{-1}\tilde{\boldsymbol{N}}_0\tilde{\boldsymbol{N}}_0')^-\boldsymbol{c}[k\sigma_B^2 + (1 - K_H^{-1}k)\sigma_U^2 + \sigma_e^2]. \quad (5.3.13)\end{aligned}$$

Note that $k^{-1}\tilde{\boldsymbol{N}}_0\tilde{\boldsymbol{N}}_0' = \tilde{\boldsymbol{C}}_2$, then.

An answer to the question what is necessary and sufficient for the condition (5.3.10), or (5.3.11), of Theorem 5.3.2 to be satisfied by any $\boldsymbol{s}$ is as follows.

Corollary 5.3.2. *The condition* (5.3.10) *holds for any* $\boldsymbol{s}$, *i.e., the equality*

$$\tilde{\boldsymbol{K}}_0\tilde{\boldsymbol{N}}_0' = \tilde{\boldsymbol{N}}_0'(\tilde{\boldsymbol{N}}_0\boldsymbol{k}^{-\delta}\tilde{\boldsymbol{N}}_0')^-\tilde{\boldsymbol{N}}_0\tilde{\boldsymbol{N}}_0' \quad (5.3.14)$$

holds, if and only if for any vector $\boldsymbol{t}$ *that satisfies the equality* $\tilde{\boldsymbol{N}}_0\boldsymbol{t} = \boldsymbol{0}$, *the equality* $\tilde{\boldsymbol{N}}_0\tilde{\boldsymbol{K}}_0\boldsymbol{t} = \boldsymbol{0}$ *is also satisfied.*

Proof. This result can be proved exactly as Corollary 3.2.2. □

Remark 5.3.1. Note that for $\tilde{\boldsymbol{N}}_0\boldsymbol{t} = \boldsymbol{0}$ to imply $\tilde{\boldsymbol{N}}_0\tilde{\boldsymbol{K}}_0\boldsymbol{t} = \boldsymbol{0}$, it is sufficient that $\boldsymbol{N}_{0h}\boldsymbol{t}_h = \boldsymbol{0}$ implies $\boldsymbol{N}_{0h}\boldsymbol{K}_{0h}\boldsymbol{t}_h = \boldsymbol{0}$ for $h = 1, 2, ..., a$, with $\boldsymbol{t} = [\boldsymbol{t}_1', \boldsymbol{t}_2', ..., \boldsymbol{t}_a]'$. Moreover, for $\boldsymbol{N}_{0h}\boldsymbol{t}_h = \boldsymbol{0}$ to imply $\boldsymbol{N}_{0h}\boldsymbol{K}_{0h}\boldsymbol{t}_h = \boldsymbol{0}$, it is sufficient that $\boldsymbol{k}_h = k_h\boldsymbol{1}_{b_h}$, i.e., that the block sizes within the superblock h are all equal, for $h = 1, 2, ..., a$.

Remark 5.3.2. (a) Because the equalities $\tilde{\boldsymbol{N}}_0\boldsymbol{1}_b = \boldsymbol{0}$ and $\tilde{\boldsymbol{N}}_0\tilde{\boldsymbol{K}}_0\boldsymbol{1}_b = \boldsymbol{0}$ hold always, the necessary and sufficient condition for the equality (5.3.14) of Corollary 5.3.2 can be replaced by the condition that $\boldsymbol{N}\boldsymbol{t}_0 = \boldsymbol{0}$ implies $\tilde{\boldsymbol{N}}_0\boldsymbol{k}^\delta\boldsymbol{t}_0 = \boldsymbol{0}$ for any vector $\boldsymbol{t}_0 = [\boldsymbol{t}_{01}', \boldsymbol{t}_{02}', ..., \boldsymbol{t}_{0a}']'$ such that $\boldsymbol{t}_{0h}$ is $\boldsymbol{k}_h^\delta$-orthogonal to $\boldsymbol{1}_{b_h}$ for any h $(= 1, 2, ..., a)$.

(b) If $\text{rank}(\boldsymbol{N}_{0h}) = b_h - 1$ for any h, i.e., the columns of each $\boldsymbol{N}_h$ are linearly independent [as $\text{rank}(\boldsymbol{N}_{0h}) = \text{rank}(\boldsymbol{N}_h) - 1$], then a vector $\boldsymbol{t}_h$ satisfying $\boldsymbol{N}_{0h}\boldsymbol{t}_h = \boldsymbol{0}$ must be equal or proportional to $\boldsymbol{1}_{b_h}$ [i.e., $\boldsymbol{t}_h \in \mathcal{C}(\boldsymbol{1}_{b_h})$], and so satisfy also the equality $\boldsymbol{N}_{0h}\boldsymbol{K}_{0h}\boldsymbol{t}_h = \boldsymbol{0}$. Thus, the condition of Corollary 5.3.2 is then satisfied automatically, independently of the block sizes.

Remark 5.3.3. If the equation (5.3.10) holds for any $\boldsymbol{s}$, i.e., if (5.3.14) holds, then

$$\text{Cov}(\boldsymbol{y}_2)\tilde{\boldsymbol{\phi}}_2\boldsymbol{\Delta}' = \tilde{\boldsymbol{\phi}}_2\boldsymbol{\Delta}'[(\tilde{\boldsymbol{N}}_0'\boldsymbol{k}^{-\delta}\tilde{\boldsymbol{N}}_0')^-\tilde{\boldsymbol{N}}_0\tilde{\boldsymbol{N}}_0'(\sigma_B^2 - K_H^{-1}\sigma_U^2) + \boldsymbol{I}_v(\sigma_U^2 + \sigma_e^2)],$$

which implies that both $\tilde{\boldsymbol{\phi}}_2\boldsymbol{\Delta}'\boldsymbol{s}$ and $\text{Cov}(\boldsymbol{y}_2)\tilde{\boldsymbol{\phi}}_2\boldsymbol{\Delta}'\boldsymbol{s}$ belong to $\mathcal{C}(\tilde{\boldsymbol{\phi}}_2\boldsymbol{\Delta}')$ for any $\boldsymbol{s}$, and thus, the conditions stated in Theorem 1.2.2, when applied to Theorem 5.3.2, are satisfied. The implications of this result are exactly as those indicated in Remark 3.2.5. In particular, it follows that (5.3.14) is necessary and sufficient for the BLUEs and SLSEs to coincide under this submodel.

Thus, if the equality (5.3.14) holds, then the inter-block-intra-superblock analysis of variance can be obtained, in the form

$$\boldsymbol{y}'\tilde{\boldsymbol{\phi}}_2\boldsymbol{y} = \tilde{\boldsymbol{Q}}_2'\tilde{\boldsymbol{C}}_2^-\tilde{\boldsymbol{Q}}_2 + \boldsymbol{y}'\tilde{\boldsymbol{\psi}}_2\boldsymbol{y},$$

where $\tilde{\boldsymbol{Q}}_2'\tilde{\boldsymbol{C}}_2^-\tilde{\boldsymbol{Q}}_2$, with $\tilde{\boldsymbol{Q}}_2 = \boldsymbol{\Delta}\tilde{\boldsymbol{\phi}}_2\boldsymbol{y}$, is the inter-block-intra-superblock treatment sum of squares, and $\boldsymbol{y}'\tilde{\boldsymbol{\psi}}_2\boldsymbol{y}$, with

$$\tilde{\boldsymbol{\psi}}_2 = \tilde{\boldsymbol{\phi}}_2 - \tilde{\boldsymbol{\phi}}_2\boldsymbol{\Delta}'\tilde{\boldsymbol{C}}_2^-\boldsymbol{\Delta}\tilde{\boldsymbol{\phi}}_2 = \tilde{\boldsymbol{\phi}}_2(\boldsymbol{I}_n - \boldsymbol{\Delta}'\tilde{\boldsymbol{C}}_2^-\boldsymbol{\Delta})\tilde{\boldsymbol{\phi}}_2,$$

is the inter-block-intra-superblock residual sum of squares, the first on $\tilde{h}_2 = \text{rank}(\tilde{\boldsymbol{C}}_2) = \text{rank}(\tilde{\boldsymbol{N}}_0)$ d.f., the second on $b - a - \tilde{h}_2 = \text{rank}(\tilde{\boldsymbol{\psi}}_2)$ d.f. The resulting inter-block-intra-superblock residual mean square $s_2^2 = \boldsymbol{y}'\tilde{\boldsymbol{\psi}}_2\boldsymbol{y}/(b - a - \tilde{h}_2)$ is an unbiased estimator of

$$\sigma_2^2 = (b - a - \tilde{h}_2)^{-1}\text{tr}(\tilde{\boldsymbol{K}}_0 - \tilde{\boldsymbol{N}}_0'\tilde{\boldsymbol{C}}_2^-\tilde{\boldsymbol{N}}_0)(\sigma_B^2 - K_H^{-1}\sigma_U^2) + \sigma_U^2 + \sigma_e^2. \quad (5.3.15)$$

In the case of all k_j equal ($k_1 = k_2 = \cdots = k_b = k$), the mean square s_2^2 is in fact the MINQUE of σ_2^2, which then reduces from (5.3.15) to

$$\sigma_2^2 = k\sigma_B^2 + (1 - K_H^{-1}k)\sigma_U^2 + \sigma_e^2, \quad (5.3.16)$$

further reducing to $\sigma_2^2 = k\sigma_B^2 + \sigma_e^2$ if $k = K_H$. It should be noted, however, that $b - a - \tilde{h}_2 = 0$ if $b - a = \tilde{h}_2$ (obviously $b - a \geq \tilde{h}_2$ always). In that case, no estimator for σ_2^2 exists in the inter-block-intra-superblock analysis.

Thus, in the case of equal k_j's and $b - a > \tilde{h}_2$, the mean square s_2^2 can be used to obtain an unbiased estimator of (5.3.13), in the form

$$\widehat{\text{Var}(\boldsymbol{c}'\boldsymbol{\tau})} = k^{-1}\boldsymbol{s}'\tilde{\boldsymbol{N}}_0\tilde{\boldsymbol{N}}_0'\boldsymbol{s}s_2^2 = k\boldsymbol{c}'(\tilde{\boldsymbol{N}}_0\tilde{\boldsymbol{N}}_0')^-\boldsymbol{c}s_2^2.$$

In general, the estimation of (5.3.12) is not so simple.

Furthermore, if $k_1 = k_2 = \cdots = k_b = k$, then $\text{Cov}(\boldsymbol{y}_2) = \tilde{\boldsymbol{\phi}}_2\sigma_2^2$, where σ_2^2 is as defined in (5.3.16), and under the multivariate normal distribution of $\boldsymbol{y}$, it is possible (following the same argument as in Section 3.2.2) to test the hypothesis $\boldsymbol{\tau}'\tilde{\boldsymbol{C}}_2\boldsymbol{\tau} = 0$, equivalent to $\text{E}(\boldsymbol{y}_2) = \boldsymbol{0}$ or $\boldsymbol{P}_{D'}\text{E}(\boldsymbol{y}) \in \mathcal{C}(\boldsymbol{G}')$, by the variance ratio criterion $\tilde{h}_2^{-1}\tilde{\boldsymbol{Q}}_2'\tilde{\boldsymbol{C}}_2^{-}\tilde{\boldsymbol{Q}}_2/s_2^2$, which has then the F distribution with $\tilde{h}_2$ and $b - a - \tilde{h}_2$ d.f., central when the hypothesis is true. This result, however, does not apply to the general case.

5.3.3 Inter-superblock submodel

The submodel (5.3.4) has the properties

$$\text{E}(\boldsymbol{y}_3) = \tilde{\boldsymbol{\phi}}_3\boldsymbol{\Delta}'\boldsymbol{\tau} = \boldsymbol{G}'\boldsymbol{n}^{-\delta}\boldsymbol{G}\boldsymbol{\Delta}'\boldsymbol{\tau} - n^{-1}\boldsymbol{1}_n\boldsymbol{r}'\boldsymbol{\tau} \tag{5.3.17}$$

and

$$\begin{aligned}\text{Cov}(\boldsymbol{y}_3) &= \tilde{\boldsymbol{\phi}}_3\boldsymbol{G}'\boldsymbol{G}\tilde{\boldsymbol{\phi}}_3(\sigma_A^2 - B_H^{-1}\sigma_B^2) \\ &\quad + \tilde{\boldsymbol{\phi}}_3\boldsymbol{D}'\boldsymbol{D}\tilde{\boldsymbol{\phi}}_3(\sigma_B^2 - K_H^{-1}\sigma_U^2) + \tilde{\boldsymbol{\phi}}_3(\sigma_U^2 + \sigma_e^2).\end{aligned} \tag{5.3.18}$$

The main estimation result under (5.3.4) can be expressed as follows.

Theorem 5.3.3. *Under* (5.3.4), *a function* $\boldsymbol{w}'\boldsymbol{y}_3 = \boldsymbol{w}'\tilde{\boldsymbol{\phi}}_3\boldsymbol{y}$ *is uniformly the BLUE of* $\boldsymbol{c}'\boldsymbol{\tau}$ *if and only if* $\tilde{\boldsymbol{\phi}}_3\boldsymbol{w} = \tilde{\boldsymbol{\phi}}_3\boldsymbol{\Delta}'\boldsymbol{s}$, *where the vectors* $\mathbf{c}$ *and* $\boldsymbol{s}$ *are in the relation* $\mathbf{c} = \tilde{\boldsymbol{C}}_3\boldsymbol{s}$, *with* $\tilde{\boldsymbol{C}}_3 = \boldsymbol{\Delta}\tilde{\boldsymbol{\phi}}_3\boldsymbol{\Delta}' = \boldsymbol{\mathfrak{R}}\boldsymbol{n}^{-\delta}\boldsymbol{\mathfrak{R}}' - n^{-1}\boldsymbol{r}\boldsymbol{r}'$, *and* $\boldsymbol{s}$ *satisfies the conditions*

$$\{\boldsymbol{K}_0 - \tilde{\boldsymbol{K}}_0 - (\boldsymbol{N}_0 - \tilde{\boldsymbol{N}}_0)'\tilde{\boldsymbol{C}}_3^{-}(\boldsymbol{N}_0 - \tilde{\boldsymbol{N}}_0)\}(\boldsymbol{N}_0 - \tilde{\boldsymbol{N}}_0)'\boldsymbol{s} = \boldsymbol{0} \tag{5.3.19}$$

and

$$(\boldsymbol{\aleph}_0 - \boldsymbol{\mathfrak{R}}_0'\tilde{\boldsymbol{C}}_3^{-}\boldsymbol{\mathfrak{R}}_0)\boldsymbol{\mathfrak{R}}_0'\boldsymbol{s} = \boldsymbol{0}, \tag{5.3.20}$$

where $\boldsymbol{K}_0 = \boldsymbol{k}^\delta - n^{-1}\boldsymbol{k}\boldsymbol{k}'$ *and* $\boldsymbol{N}_0 = \boldsymbol{N} - n^{-1}\boldsymbol{r}\boldsymbol{k}'$, *as in Section* 3.2.2, *and* $\tilde{\boldsymbol{K}}_0$ *and* $\tilde{\boldsymbol{N}}_0$ *are as in Section* 5.3.2, *and where* $\boldsymbol{\aleph}_0 = \boldsymbol{n}^\delta - n^{-1}\boldsymbol{n}\boldsymbol{n}'$ *and* $\boldsymbol{\mathfrak{R}}_0 = \boldsymbol{\mathfrak{R}} - n^{-1}\boldsymbol{r}\boldsymbol{n}'$, *with* $\boldsymbol{\mathfrak{R}}$ *as defined in Section* 5.1. (*Certainly, one has to distinguish between the symbols* $\boldsymbol{N}_0$ *and* $\boldsymbol{\aleph}_0$, *as they represent different matrices.*)

Proof. Under (5.3.4), with (5.3.17) and (5.3.18), the necessary and sufficient condition of Theorem 1.2.1 for a function $\boldsymbol{w}'\boldsymbol{y}_3 = \boldsymbol{w}'\tilde{\boldsymbol{\phi}}_3\boldsymbol{y}$ to be the BLUE of $\text{E}(\boldsymbol{w}'\tilde{\boldsymbol{\phi}}_3) = \boldsymbol{w}'\tilde{\boldsymbol{\phi}}_3\boldsymbol{\Delta}'\boldsymbol{\tau}$ is the equality

$$\begin{aligned}(\boldsymbol{I}_n - \boldsymbol{P}_{\tilde{\phi}_3\Delta'})[&\tilde{\boldsymbol{\phi}}_3\boldsymbol{G}'\boldsymbol{G}\tilde{\boldsymbol{\phi}}_3(\sigma_A^2 - B_H^{-1}\sigma_B^2) + \tilde{\boldsymbol{\phi}}_3\boldsymbol{D}'\boldsymbol{D}\tilde{\boldsymbol{\phi}}_3(\sigma_B^2 - K_H^{-1}\sigma_U^2) \\ &+\tilde{\boldsymbol{\phi}}_3(\sigma_U^2 + \sigma_e^2)]\boldsymbol{w} = \boldsymbol{0}.\end{aligned}$$

It holds uniformly if and only if the equalities

$$(\boldsymbol{I}_n - \boldsymbol{P}_{\tilde{\phi}_3\Delta'})\tilde{\boldsymbol{\phi}}_3\boldsymbol{w} = \boldsymbol{0}, \quad (\boldsymbol{I}_n - \boldsymbol{P}_{\tilde{\phi}_3\Delta'})\tilde{\boldsymbol{\phi}}_3\boldsymbol{D}'\boldsymbol{D}\tilde{\boldsymbol{\phi}}_3\boldsymbol{w} = \boldsymbol{0}$$

and

$$(\boldsymbol{I}_n - \boldsymbol{P}_{\tilde{\phi}_3\Delta'})\tilde{\boldsymbol{\phi}}_3\boldsymbol{G}'\boldsymbol{G}\tilde{\boldsymbol{\phi}}_3\boldsymbol{w} = \mathbf{0}$$

hold simultaneously. The first equality holds if and only if $\tilde{\boldsymbol{\phi}}_3\boldsymbol{w} = \tilde{\boldsymbol{\phi}}_3\boldsymbol{\Delta}'\boldsymbol{s}$ for some $\boldsymbol{s}$, which holds if and only if $\boldsymbol{D}\tilde{\boldsymbol{\phi}}_3\boldsymbol{w} = \boldsymbol{D}\tilde{\boldsymbol{\phi}}_3\boldsymbol{\Delta}'\boldsymbol{s}$ as well as if and only if $\boldsymbol{G}\tilde{\boldsymbol{\phi}}_3\boldsymbol{w} = \boldsymbol{G}\tilde{\boldsymbol{\phi}}_3\boldsymbol{\Delta}'\boldsymbol{s}$ for that $\boldsymbol{s}$. With these conditions, the remaining two equalities read

$$(\boldsymbol{I}_n - \boldsymbol{P}_{\tilde{\phi}_3\Delta'})\tilde{\boldsymbol{\phi}}_3\boldsymbol{D}'\boldsymbol{D}\tilde{\boldsymbol{\phi}}_3\boldsymbol{\Delta}'\boldsymbol{s} = \mathbf{0} \quad \text{and} \quad (\boldsymbol{I}_n - \boldsymbol{P}_{\tilde{\phi}_3\Delta'})\tilde{\boldsymbol{\phi}}_3\boldsymbol{G}'\boldsymbol{G}\tilde{\boldsymbol{\phi}}_3\boldsymbol{\Delta}'\boldsymbol{s} = \mathbf{0},$$

which are equivalent to (5.3.19) and (5.3.20), respectively, because of the relations $\boldsymbol{D}'\boldsymbol{k}^{-\delta}\boldsymbol{D}\tilde{\boldsymbol{\phi}}_3 = \tilde{\boldsymbol{\phi}}_3$, $\boldsymbol{G}'\boldsymbol{n}^{-\delta}\boldsymbol{G}\tilde{\boldsymbol{\phi}}_3 = \tilde{\boldsymbol{\phi}}_3$, $\boldsymbol{D}\tilde{\boldsymbol{\phi}}_3\boldsymbol{D}' = \boldsymbol{K}_0 - \tilde{\boldsymbol{K}}_0$, $\boldsymbol{\Delta}\tilde{\boldsymbol{\phi}}_3\boldsymbol{D}' = \boldsymbol{N}_0 - \tilde{\boldsymbol{N}}_0$, $\boldsymbol{G}\tilde{\boldsymbol{\phi}}_3\boldsymbol{G}' = \boldsymbol{\aleph}_0$, $\boldsymbol{\Delta}\tilde{\boldsymbol{\phi}}_3\boldsymbol{G}' = \boldsymbol{\mathfrak{R}}_0$ and $\tilde{\boldsymbol{C}}_3 = \boldsymbol{\Delta}\tilde{\boldsymbol{\phi}}_3\boldsymbol{\Delta}' = (\boldsymbol{N}_0 - \tilde{\boldsymbol{N}}_0)\boldsymbol{k}^{-\delta}(\boldsymbol{N}_0 - \tilde{\boldsymbol{N}}_0)'$ $= \boldsymbol{\mathfrak{R}}_0\boldsymbol{n}^{-\delta}\boldsymbol{\mathfrak{R}}_0'$. Finally, the relation between $\boldsymbol{c}$ and $\boldsymbol{s}$ follows from the fact that $\mathrm{E}(\boldsymbol{s}'\boldsymbol{\Delta}\boldsymbol{y}_3) = \boldsymbol{s}'\boldsymbol{\Delta}\tilde{\boldsymbol{\phi}}_3\boldsymbol{\Delta}'\boldsymbol{\tau} = \boldsymbol{c}'\boldsymbol{\tau}$. □

Note that if $\boldsymbol{\mathfrak{R}} = n^{-1}\boldsymbol{r}\boldsymbol{n}'$, i.e., the design $\mathcal{D}$ is connected and orthogonal (see Corollary 2.3.4), then for no function $\boldsymbol{c}'\boldsymbol{\tau}$ (with $\boldsymbol{c} \neq \mathbf{0}$), the BLUE under the submodel (5.3.4) exists. Therefore, further considerations under this submodel are relevant only to cases in which $\boldsymbol{\mathfrak{R}} \neq n^{-1}\boldsymbol{r}\boldsymbol{n}'$, i.e., $\boldsymbol{\mathfrak{R}}_0 \neq \mathbf{O}$.

Corollary 5.3.3. *For the estimation of $\boldsymbol{c}'\boldsymbol{\tau} = \boldsymbol{s}'\tilde{\boldsymbol{C}}_3\boldsymbol{\tau}$ under* (5.3.4) *the following applies:*

(a) *The case $\boldsymbol{\mathfrak{R}}_0'\boldsymbol{s} = \mathbf{0}$ is to be excluded.*

(b) *If $\boldsymbol{\mathfrak{R}}_0'\boldsymbol{s} \neq \mathbf{0}$, then $\boldsymbol{c}'\boldsymbol{\tau}$ is a contrast, and to satisfy* (5.3.19) *and* (5.3.20) *by the vector $\boldsymbol{s}$, it is necessary and sufficient that $(\boldsymbol{K}_0 - \tilde{\boldsymbol{K}}_0)(\boldsymbol{N}_0 - \tilde{\boldsymbol{N}}_0)'\boldsymbol{s} \in \mathcal{C}(\boldsymbol{N}_0' - \tilde{\boldsymbol{N}}_0')$ and, simultaneously, $\boldsymbol{\aleph}_0\boldsymbol{\mathfrak{R}}_0'\boldsymbol{s} \in \mathcal{C}(\boldsymbol{\mathfrak{R}}_0')$.*

(c) *If $\boldsymbol{s}$ is such that $\boldsymbol{r}'\boldsymbol{s} = 0$, then the conditions* (5.3.19) *and* (5.3.20) *can be replaced by*

$$(\boldsymbol{K}_0 - \tilde{\boldsymbol{K}}_0)(\boldsymbol{N} - \tilde{\boldsymbol{N}}_0)'\boldsymbol{s} = (\boldsymbol{N}_0 - \tilde{\boldsymbol{N}}_0)'\tilde{\boldsymbol{C}}_3^-(\boldsymbol{N}_0 - \tilde{\boldsymbol{N}}_0)(\boldsymbol{N} - \tilde{\boldsymbol{N}}_0)'\boldsymbol{s}$$

and

$$\boldsymbol{\aleph}_0\boldsymbol{\mathfrak{R}}'\boldsymbol{s} = \boldsymbol{\mathfrak{R}}_0'\tilde{\boldsymbol{C}}_3^-\boldsymbol{\mathfrak{R}}_0\boldsymbol{\mathfrak{R}}'\boldsymbol{s},$$

respectively. To satisfy them, it is then necessary and sufficient that both $(\boldsymbol{K}_0 - \tilde{\boldsymbol{K}}_0)(\boldsymbol{N} - \tilde{\boldsymbol{N}}_0)'\boldsymbol{s} \in \mathcal{C}(\boldsymbol{N}_0' - \tilde{\boldsymbol{N}}_0')$ and $\boldsymbol{\aleph}_0\boldsymbol{\mathfrak{R}}'\boldsymbol{s} \in \mathcal{C}(\boldsymbol{\mathfrak{R}}_0')$.

(d) *If all block sizes are equal, i.e., $k_1 = k_2 = \cdots = k_b = k$, and all superblock sizes are equal, i.e., $n_1 = n_2 = \cdots = n_a = n_0$ (say, $n_0 = n/a$), then the conditions* (5.3.19) *and* (5.3.20) *are both satisfied automatically by any $\boldsymbol{s}$.*

Proof. The result (a) is obvious, as $\boldsymbol{\mathfrak{R}}_0'\boldsymbol{s} = \mathbf{0}$ implies $\boldsymbol{c} = \mathbf{0}$. To prove (b), note that $\boldsymbol{\mathfrak{R}}_0'\mathbf{1}_v = \mathbf{0}$, and that the equations $(\boldsymbol{N}_0' - \tilde{\boldsymbol{N}}_0')\boldsymbol{x}_1 = (\boldsymbol{K}_0 - \tilde{\boldsymbol{K}}_0)(\boldsymbol{N}_0' - \tilde{\boldsymbol{N}}_0')\boldsymbol{s}$ and $\boldsymbol{\mathfrak{R}}_0'\boldsymbol{x}_2 = \boldsymbol{\aleph}_0\boldsymbol{\mathfrak{R}}_0'\boldsymbol{s}$ are consistent if and only if (5.3.19) and (5.3.20) hold, respectively, because $[(\boldsymbol{N}_0 - \tilde{\boldsymbol{N}}_0)\boldsymbol{k}^{-\delta}(\boldsymbol{N}_0 - \tilde{\boldsymbol{N}}_0)']^-(\boldsymbol{N}_0 - \tilde{\boldsymbol{N}}_0)\boldsymbol{k}^{-\delta}$ can be used as a g-inverse of $(\boldsymbol{N}_0 - \tilde{\boldsymbol{N}}_0)'$ and $(\boldsymbol{\mathfrak{R}}_0\boldsymbol{n}^{-\delta}\boldsymbol{\mathfrak{R}}_0')^-\boldsymbol{\mathfrak{R}}_0\boldsymbol{n}^{-\delta}$ as a g-inverse of $\boldsymbol{\mathfrak{R}}_0'$, and because $(\boldsymbol{N}_0 - \tilde{\boldsymbol{N}}_0)\boldsymbol{k}^{-\delta}(\boldsymbol{K}_0 - \tilde{\boldsymbol{K}}_0) = \boldsymbol{N}_0 - \tilde{\boldsymbol{N}}_0$ and $\boldsymbol{\mathfrak{R}}_0\boldsymbol{n}^{-\delta}\boldsymbol{\aleph}_0 = \boldsymbol{\mathfrak{R}}_0$. The result

(c) is obvious, as $(\boldsymbol{N}_0 - \tilde{\boldsymbol{N}}_0)'\boldsymbol{s} = (\boldsymbol{N} - \tilde{\boldsymbol{N}}_0)'\boldsymbol{s}$ and $\mathfrak{R}_0'\boldsymbol{s} = \mathfrak{R}'\boldsymbol{s}$ if $\boldsymbol{r}'\boldsymbol{s} = 0$. The result (d) can easily be checked similarly as Corollary 5.3.1(d), by noting that, here, $(\boldsymbol{K}_0 - \tilde{\boldsymbol{K}}_0)(\boldsymbol{N}_0 - \tilde{\boldsymbol{N}}_0)' = k(\boldsymbol{N}_0 - \tilde{\boldsymbol{N}}_0)'$ and $\aleph_0\mathfrak{R}_0' = n_0\mathfrak{R}_0'$. □

Now it may be noted that if the conditions of Theorem 5.3.3 are satisfied, then $\widehat{\boldsymbol{c}'\boldsymbol{\tau}} = \boldsymbol{s}'\boldsymbol{\Delta y}_3 = \boldsymbol{c}'\tilde{\boldsymbol{C}}_3^-\boldsymbol{\Delta\phi}_3\boldsymbol{y}$ is the BLUE of the contrast $\boldsymbol{c}'\boldsymbol{\tau}$ under (5.3.4), and that its variance has the form

$$\begin{aligned}\operatorname{Var}(\widehat{\boldsymbol{c}'\boldsymbol{\tau}}) &= \boldsymbol{s}'\mathfrak{R}_0\mathfrak{R}_0'\boldsymbol{s}(\sigma_A^2 - B_H^{-1}\sigma_B^2) \\ &\quad + \boldsymbol{s}'(\boldsymbol{N}_0 - \tilde{\boldsymbol{N}}_0)(\boldsymbol{N}_0 - \tilde{\boldsymbol{N}}_0)'\boldsymbol{s}(\sigma_B^2 - K_H^{-1}\sigma_U^2) \\ &\quad + \boldsymbol{s}'\mathfrak{R}_0\boldsymbol{n}^{-\delta}\mathfrak{R}_0'\boldsymbol{s}(\sigma_U^2 + \sigma_e^2). \end{aligned} \tag{5.3.21}$$

Evidently, if $k_1 = k_2 = \cdots = k_b = k$ and $n_1 = n_2 = \cdots = n_a = n_0$, then

$$\operatorname{Var}(\widehat{\boldsymbol{c}'\boldsymbol{\tau}}) = n_0^{-1}\boldsymbol{s}'\mathfrak{R}_0\mathfrak{R}_0'\boldsymbol{s}[n_0\sigma_A^2 + (k - B_H^{-1}n_0)\sigma_B^2 + (1 - K_H^{-1}k)\sigma_U^2 + \sigma_e^2], \tag{5.3.22}$$

where $n_0^{-1}\boldsymbol{s}'\mathfrak{R}_0\mathfrak{R}_0'\boldsymbol{s} = \boldsymbol{c}'(n_0^{-1}\mathfrak{R}_0\mathfrak{R}_0')^-\boldsymbol{c}$. Note that $n_0^{-1}\mathfrak{R}_0\mathfrak{R}_0' = \tilde{\boldsymbol{C}}_3$, then. The formula (5.3.22) reduces further to

$$\operatorname{Var}(\widehat{\boldsymbol{c}'\boldsymbol{\tau}}) = \boldsymbol{c}'\tilde{\boldsymbol{C}}_3^-\boldsymbol{c}(n_0\sigma_A^2 + \sigma_e^2)$$

if $k = B_H^{-1}n_0 = K_H$, i.e., if the number of available blocks in each superblock is constant, equal to n_0/k $(= b/a)$, and the size of each available block is constant, equal to k.

An answer to the question what is necessary and sufficient for the conditions of Theorem 3.3.3 to be satisfied by any $\boldsymbol{s}$ can be given as follows.

Corollary 5.3.4. *The conditions* (5.3.19) *and* (5.3.20) *hold for any* $\boldsymbol{s}$, *i.e., the equalities*

$$(\boldsymbol{K}_0 - \tilde{\boldsymbol{K}}_0)(\boldsymbol{N}_0 - \tilde{\boldsymbol{N}}_0)' = (\boldsymbol{N}_0 - \tilde{\boldsymbol{N}}_0)'\tilde{\boldsymbol{C}}_3^-(\boldsymbol{N}_0 - \tilde{\boldsymbol{N}}_0)(\boldsymbol{N}_0 - \tilde{\boldsymbol{N}}_0)' \tag{5.3.23}$$

and

$$\aleph_0\mathfrak{R}_0' = \mathfrak{R}_0'\tilde{\boldsymbol{C}}_3^-\mathfrak{R}_0\mathfrak{R}_0' \tag{5.3.24}$$

hold, if and only if for any $b \times 1$ *vector* $\boldsymbol{t}$ *satisfying the equality* $(\boldsymbol{N}_0 - \tilde{\boldsymbol{N}}_0)\boldsymbol{t} = \boldsymbol{0}$, *the equality* $(\boldsymbol{N}_0 - \tilde{\boldsymbol{N}}_0)(\boldsymbol{K}_0 - \tilde{\boldsymbol{K}}_0)\boldsymbol{t} = \boldsymbol{0}$ *holds too, and for any* $a \times 1$ *vector* $\boldsymbol{u}$ *satisfying the equality* $\mathfrak{R}_0\boldsymbol{u} = \boldsymbol{0}$, *the equality* $\mathfrak{R}_0\aleph_0\boldsymbol{u} = \boldsymbol{0}$ *also holds.*

Proof. This corollary can be proved as Corollary 3.2.2. □

Remark 5.3.4. In connection with Corollary 5.3.4, the following can be noted:

(a) Because of the relations $\boldsymbol{N}_0 - \tilde{\boldsymbol{N}}_0 = \mathfrak{R}_0\boldsymbol{n}^{-\delta}\boldsymbol{G}\boldsymbol{D}'$, $\boldsymbol{K}_0 - \tilde{\boldsymbol{K}}_0 = \boldsymbol{D}\boldsymbol{G}'(\boldsymbol{I}_a - n^{-1}\boldsymbol{1}_a\boldsymbol{n}')\boldsymbol{n}^{-\delta}\boldsymbol{G}\boldsymbol{D}'$ and $\mathfrak{R}_0 = \mathfrak{R}(\boldsymbol{I}_a - n^{-1}\boldsymbol{1}_a\boldsymbol{n}')$, the conditions given in Corollary 5.3.4 for the equalities (5.3.23) and (5.3.24) can be reduced as follows. The equalities (5.3.23) and (5.3.24) hold simultaneously if and only if for any (nonzero) vector $\boldsymbol{n}^{\delta}$-orthogonal to $\boldsymbol{1}_a$, $\boldsymbol{u}_0$, say (i.e., such that $\boldsymbol{n}'\boldsymbol{u}_0 = 0$), the equality $\mathfrak{R}\boldsymbol{u}_0 = \boldsymbol{0}$ implies the equalities $(\boldsymbol{N}_0 - \tilde{\boldsymbol{N}}_0)\boldsymbol{D}\boldsymbol{G}'\boldsymbol{u}_0 = \boldsymbol{0}$ and $\mathfrak{R}_0\boldsymbol{n}^{\delta}\boldsymbol{u}_0 = \boldsymbol{0}$.

(b) If $\text{rank}(\mathfrak{R}_0) = a - 1$, i.e., the columns of the matrix $\mathfrak{R}$ are linearly independent [as $\text{rank}(\mathfrak{R}_0) = \text{rank}(\mathfrak{R}) - 1$], then a nonzero vector $\boldsymbol{u}$ satisfying $\mathfrak{R}_0\boldsymbol{u} = \mathbf{0}$ must be equal or proportional to $\mathbf{1}_a$ [i.e., $\boldsymbol{u} \in \mathcal{C}(\mathbf{1}_a)$], and so satisfy also the equality $\mathfrak{R}_0\aleph_0\boldsymbol{u} = \mathbf{0}$. Moreover, with this rank of $\mathfrak{R}_0$, any vector $\boldsymbol{t}$ such that $\boldsymbol{n}^{-\delta}\boldsymbol{G}\boldsymbol{D}'\boldsymbol{t} \in \mathcal{C}(\mathbf{1}_a)$ satisfies simultaneously the equalities $(\boldsymbol{N}_0 - \tilde{\boldsymbol{N}}_0)\boldsymbol{t} = \mathbf{0}$ and $(\boldsymbol{N}_0 - \tilde{\boldsymbol{N}}_0)(\boldsymbol{K}_0 - \tilde{\boldsymbol{K}}_0)\boldsymbol{t} = \mathbf{0}$. Any other $\boldsymbol{t}$ for which the first of these two holds has to be such that $\boldsymbol{G}\boldsymbol{D}'\boldsymbol{t} = \mathbf{0}$, and then the second equality also holds. Thus, the conditions of Corollary 5.3.4 are then satisfied automatically, independently of the k_j's and n_h's.

Remark 5.3.5. From the proof of Theorem 5.3.3, it is evident that the equalities (5.3.19) and (5.3.20) hold simultaneously for any $\boldsymbol{s}$, i.e., (5.3.23) and (5.3.24) hold, if and only if

$$\text{Cov}(\boldsymbol{y}_3)\tilde{\boldsymbol{\phi}}_3\boldsymbol{\Delta}' = \boldsymbol{P}_{\tilde{\phi}_3\Delta'}\text{Cov}(\boldsymbol{y}_3)\tilde{\boldsymbol{\phi}}_3\boldsymbol{\Delta}',$$

which shows that not only $\tilde{\boldsymbol{\phi}}_3\boldsymbol{\Delta}'\boldsymbol{s}$, but also $\text{Cov}(\boldsymbol{y}_3)\tilde{\boldsymbol{\phi}}_3\boldsymbol{\Delta}'\boldsymbol{s}$ belongs to $\mathcal{C}(\tilde{\boldsymbol{\phi}}_3\boldsymbol{\Delta}')$ for any $\boldsymbol{s}$, and thus, the conditions stated in Theorem 1.2.2, when applied to Theorem 5.3.3, are satisfied. This result means that the BLUE of any function $\boldsymbol{c}'\boldsymbol{\tau}$, where $\boldsymbol{c} \in \mathcal{C}(\tilde{\boldsymbol{C}}_3)$, obtainable under the inter-superblock submodel (5.3.4) is simultaneously the SLSE.

Remark 5.3.5 implies in particular that if the equalities (5.3.23) and (5.3.24) hold, then the inter-superblock analysis of variance is obtainable in the form

$$\boldsymbol{y}'\tilde{\boldsymbol{\phi}}_3\boldsymbol{y} = \tilde{\boldsymbol{Q}}_3'\tilde{\boldsymbol{C}}_3^-\tilde{\boldsymbol{Q}}_3 + \boldsymbol{y}'\tilde{\boldsymbol{\psi}}_3\boldsymbol{y},$$

where $\tilde{\boldsymbol{Q}}_3'\tilde{\boldsymbol{C}}_3^-\tilde{\boldsymbol{Q}}_3$, with $\tilde{\boldsymbol{Q}}_3 = \boldsymbol{\Delta}\tilde{\boldsymbol{\phi}}_3\boldsymbol{y}$, is the inter-superblock treatment sum of squares, and $\boldsymbol{y}'\tilde{\boldsymbol{\psi}}_3\boldsymbol{y}$, with

$$\tilde{\boldsymbol{\psi}}_3 = \tilde{\boldsymbol{\phi}}_3 - \tilde{\boldsymbol{\phi}}_3\boldsymbol{\Delta}'\tilde{\boldsymbol{C}}_3^-\boldsymbol{\Delta}\tilde{\boldsymbol{\phi}}_3 = \tilde{\boldsymbol{\phi}}_3(\boldsymbol{I}_n - \boldsymbol{\Delta}'\tilde{\boldsymbol{C}}_3^-\boldsymbol{\Delta})\tilde{\boldsymbol{\phi}}_3,$$

is the inter-superblock residual sum of squares, the first on $\tilde{h}_3 = \text{rank}(\tilde{\boldsymbol{C}}_3) = \text{rank}(\mathfrak{R}_0)$ d.f., the second on $a - 1 - \tilde{h}_3 = \text{rank}(\tilde{\boldsymbol{\psi}}_3)$ d.f. The resulting inter-superblock residual mean square $s_3^2 = \boldsymbol{y}'\tilde{\boldsymbol{\psi}}_3\boldsymbol{y}/(a-1-\tilde{h}_3)$ is an unbiased estimator of

$$\begin{aligned}\sigma_3^2 &= \frac{1}{a-1-\tilde{h}_3}\{\text{tr}(\aleph_0 - \mathfrak{R}_0'\tilde{\boldsymbol{C}}_3^-\mathfrak{R}_0)(\sigma_A^2 - B_H^{-1}\sigma_B^2) + \text{tr}[(\boldsymbol{K}_0 - \tilde{\boldsymbol{K}}_0) \\ &\qquad - (\boldsymbol{N}_0 - \tilde{\boldsymbol{N}}_0)'\tilde{\boldsymbol{C}}_3^-(\boldsymbol{N}_0 - \tilde{\boldsymbol{N}}_0)](\sigma_B^2 - K_H^{-1}\sigma_U^2)\} \\ &\quad + \sigma_U^2 + \sigma_e^2. \end{aligned} \tag{5.3.25}$$

In the case of

$$k_1 = k_2 = \cdots = k_b = k \quad \text{and} \quad n_1 = n_2 = \cdots = n_a = n_0, \tag{5.3.26}$$

the mean square s_3^2 is in fact the MINQUE of the estimated variance parametric function (5.3.25), which then becomes

$$\sigma_3^2 = n_0\sigma_A^2 + (k - B_H^{-1}n_0)\sigma_B^2 + (1 - K_H^{-1}k)\sigma_U^2 + \sigma_e^2, \tag{5.3.27}$$

further reducing to $n_0\sigma_A^2 + \sigma_e^2$ if $B_H^{-1}n_0 = k = K_H$, i.e., if $B_H = n_0/k$ $(= b/a)$ and $K_H = k$. It should be noted, however, that $a - 1 - \tilde{h}_3 = 0$ if $a - 1 = \tilde{h}_3$ (that $a - 1 \geq \tilde{h}_3$ is obvious). In that case, no estimator for σ_3^2 exists in the inter-superblock analysis.

Thus, in the case of equal k_j's, equal n_h's and $a - 1 > \tilde{h}_3$, the mean square s_3^2 can be used to obtain an unbiased estimator of the variance (5.3.22), as

$$\widehat{\text{Var}(c'\tau)} = s'\tilde{C}_3 s s_3^2 = c'\tilde{C}_3^- c s_3^2.$$

In general, the estimation of (5.3.21) is not so simple.

Furthermore, if the equalities (5.3.26) hold, then $\text{Cov}(y_3) = \tilde{\phi}_3\sigma_3^2$, and, as for the intra-block analysis, it can be shown that under the multivariate normality assumption the quadratic functions $\tilde{Q}_3'\tilde{C}_3^-\tilde{Q}_3/\sigma_3^2$ and $y'\tilde{\psi}_3 y/\sigma_3^2$ have independent χ^2 distributions, the first being noncentral with $\tilde{h}_3$ d.f. and with the noncentrality parameter $\tilde{\delta}_3 = \tau'\tilde{C}_3\tau/\sigma_3^2$, and the second being central with $a - 1 - \tilde{h}_3$ d.f. Hence, the hypothesis $\tau'\tilde{C}_3\tau = 0$, equivalent to $\text{E}(y_3) = \mathbf{0}$ [or $\tilde{\phi}_3\Delta'\tau = \mathbf{0}$, or $P_{G'}\text{E}(y) \in \mathcal{C}(\mathbf{1}_n)$], can be tested by the variance ratio criterion

$$\tilde{h}_3^{-1}\tilde{Q}_3'\tilde{C}_3^-\tilde{Q}_3/s_3^2,$$

which under the assumed normality has then the F distribution with $\tilde{h}_3$ and $a - 1 - \tilde{h}_3$ d.f., central when the hypothesis is true. This result, however, does not apply to the general case, when the equalities (5.3.26) do not hold.

5.3.4 Total-area submodel

Considering the fourth submodel, given in (5.3.5), it is evident that its properties are

$$\text{E}(y_4) = \tilde{\phi}_4\Delta'\tau = n^{-1}\mathbf{1}_n r'\tau,$$

$$\text{Cov}(y_4) = \tilde{\phi}_4[(n^{-1}n'n - N_A^{-1}n)\sigma_A^2 + (n^{-1}k'k - B_H^{-1}n^{-1}n'n)\sigma_B^2 + (1 - K_H^{-1}n^{-1}k'k)\sigma_U^2 + \sigma_e^2],$$

the latter reducing in the case of equal k_j's and equal n_h's [i.e., in the case of (5.3.26)] to $\text{Cov}(y_4) = \tilde{\phi}_4\sigma_4^2$, where

$$\sigma_4^2 = (n_0 - N_A^{-1}n)\sigma_A^2 + (k - B_H^{-1}n_0)\sigma_B^2 + (1 - K_H^{-1}k)\sigma_U^2 + \sigma_e^2.$$

In the general case, the following main result concerning estimation under (5.3.5) is obtainable.

Theorem 5.3.4. *Under* (5.3.5), *a function* $w'y_4 = w'\tilde{\phi}_4 y$ *is uniformly the BLUE of* $c'\tau$ *if and only if* $\tilde{\phi}_4 w = \tilde{\phi}_4\Delta's$, *where the vectors* c *and* s *are in the relation* $c = \tilde{C}_4 s$, *with* $\tilde{C}_4 = \Delta\tilde{\phi}_4\Delta' = n^{-1}rr'$.

Proof. This theorem can be proved exactly as Theorem 3.2.1. □

Remark 5.3.6. (a) The only parametric functions for which the BLUEs under (5.3.5) exist are those defined as $\boldsymbol{c}'\boldsymbol{\tau} = (\boldsymbol{s}'\boldsymbol{r})n^{-1}\boldsymbol{r}'\boldsymbol{\tau}$, where $\boldsymbol{s}'\boldsymbol{r} \neq 0$, i.e., the general parametric mean and its multiplicities, contrasts being excluded *a fortiori* (as $\mathbf{1}_v'\boldsymbol{c} = \boldsymbol{r}'\boldsymbol{s}$).

(b) Because

$$\text{Cov}(\boldsymbol{y}_4)\tilde{\boldsymbol{\phi}}_4\boldsymbol{\Delta}' = \boldsymbol{\phi}_4\boldsymbol{\Delta}'[(n^{-1}\boldsymbol{n}'\boldsymbol{n} - N_A^{-1}n)\sigma_A^2 + (n^{-1}\boldsymbol{k}'\boldsymbol{k} - B_H^{-1}n^{-1}\boldsymbol{n}'\boldsymbol{n})\sigma_B^2 + (1 - K_H^{-1}n^{-1}\boldsymbol{k}'\boldsymbol{k})\sigma_U^2 + \sigma_e^2],$$

the BLUEs under (5.3.5) and the SLSEs are the same (on account of Theorem 1.2.2 applied to Theorem 5.3.4).

If $\boldsymbol{c}'\boldsymbol{\tau} = (\boldsymbol{s}'\boldsymbol{r})n^{-1}\boldsymbol{r}'\boldsymbol{\tau} = (\boldsymbol{c}'\mathbf{1}_v)n^{-1}\boldsymbol{r}'\boldsymbol{\tau}$, then the variance of its BLUE under (5.3.5), i.e., of $\widehat{\boldsymbol{c}'\boldsymbol{\tau}} = \boldsymbol{s}'\boldsymbol{\Delta}\boldsymbol{y}_4 = \boldsymbol{c}'(\boldsymbol{\Delta}\tilde{\boldsymbol{\phi}}_4\boldsymbol{\Delta}')^-\boldsymbol{\Delta}\tilde{\boldsymbol{\phi}}_4\boldsymbol{y} = (\boldsymbol{c}'\mathbf{1}_v)n^{-1}\mathbf{1}_n'\boldsymbol{y}$, is of the form

$$\begin{aligned}\text{Var}(\widehat{\boldsymbol{c}'\boldsymbol{\tau}}) &= n^{-1}(\boldsymbol{s}'\boldsymbol{r})^2[(n^{-1}\boldsymbol{n}'\boldsymbol{n} - N_A^{-1}n)\sigma_A^2 + (n^{-1}\boldsymbol{k}'\boldsymbol{k} - B_H^{-1}n^{-1}\boldsymbol{n}'\boldsymbol{n})\sigma_B^2 \\ &\quad + (1 - K_H^{-1}n^{-1}\boldsymbol{k}'\boldsymbol{k})\sigma_U^2 + \sigma_e^2], \qquad (5.3.28)\end{aligned}$$

where $\boldsymbol{s}'\boldsymbol{r} = \boldsymbol{c}'\mathbf{1}_v$. Evidently, if the equalities (5.3.26) hold, then the variance (5.3.28) reduces to

$$\text{Var}(\widehat{\boldsymbol{c}'\boldsymbol{\tau}}) = n^{-1}(\boldsymbol{c}'\mathbf{1}_v)^2\sigma_4^2, \qquad (5.3.29)$$

and if, in addition, $a = N_A$, $B_H = n_0/k$ $(= b/a)$ and $K_H = k$, which may be considered as the usual case, then

$$\text{Var}(\widehat{\boldsymbol{c}'\boldsymbol{\tau}}) = n^{-1}(\boldsymbol{c}'\mathbf{1}_v)^2\sigma_e^2. \qquad (5.3.30)$$

Finally, it may be noted that because $\boldsymbol{P}_{\tilde{\phi}_4\Delta'} = \tilde{\boldsymbol{\phi}}_4$ (as $n^{-1}\mathbf{1}_v\mathbf{1}_v'$ is a g-inverse of $\boldsymbol{\Delta}\tilde{\boldsymbol{\phi}}_4\boldsymbol{\Delta}' = n^{-1}\boldsymbol{r}\boldsymbol{r}'$) and, hence,

$$(\boldsymbol{I}_n - \boldsymbol{P}_{\tilde{\phi}_4\Delta'})\boldsymbol{y}_4 = \mathbf{0} \quad \text{and} \quad (\boldsymbol{I}_n - \boldsymbol{P}_{\tilde{\phi}_4\Delta'})\text{Cov}(\boldsymbol{y}_4) = \mathbf{O},$$

the vector $\boldsymbol{P}_{\tilde{\phi}_4\Delta'}\boldsymbol{y}_4 = \boldsymbol{y}_4 = n^{-1}\mathbf{1}_n\mathbf{1}_n'\boldsymbol{y}$ is the BLUE of its expectation, i.e., of $n^{-1}\mathbf{1}_n\boldsymbol{r}'\boldsymbol{\tau}$, leaving no residuals.

5.3.5 Some special cases

It follows from the considerations above that any function $\boldsymbol{s}'\boldsymbol{\Delta}\boldsymbol{y}$ can be resolved into four components in the form

$$\boldsymbol{s}'\boldsymbol{\Delta}\boldsymbol{y} = \boldsymbol{s}'\tilde{\boldsymbol{Q}}_1 + \boldsymbol{s}'\tilde{\boldsymbol{Q}}_2 + \boldsymbol{s}'\tilde{\boldsymbol{Q}}_3 + \boldsymbol{s}'\tilde{\boldsymbol{Q}}_4, \qquad (5.3.31)$$

with $\tilde{\boldsymbol{Q}}_1 = \boldsymbol{\Delta}\tilde{\boldsymbol{\phi}}_1\boldsymbol{y}$, $\tilde{\boldsymbol{Q}}_2 = \boldsymbol{\Delta}\tilde{\boldsymbol{\phi}}_2\boldsymbol{y}$, $\tilde{\boldsymbol{Q}}_3 = \boldsymbol{\Delta}\tilde{\boldsymbol{\phi}}_3\boldsymbol{y}$ and $\tilde{\boldsymbol{Q}}_4 = \boldsymbol{\Delta}\tilde{\boldsymbol{\phi}}_4\boldsymbol{y}$. Each of the components in (5.3.31) represents a contribution to $\boldsymbol{s}'\boldsymbol{\Delta}\boldsymbol{y}$ from a different stratum. As in Section 3.2.4, the components $\boldsymbol{s}'\tilde{\boldsymbol{Q}}_1$, $\boldsymbol{s}'\tilde{\boldsymbol{Q}}_2$, $\boldsymbol{s}'\tilde{\boldsymbol{Q}}_3$ and $\boldsymbol{s}'\tilde{\boldsymbol{Q}}_4$ may then be called the intra-block, the inter-block-intra-superblock, the inter-superblock and the total-area component, respectively.

In connection with formula (5.3.31), it is interesting to consider some special cases of the vector $\boldsymbol{s}$ (and, hence, of $\boldsymbol{c} = \boldsymbol{r}^{\delta}\boldsymbol{s}$), as in Section 3.2.4. As the first case, suppose that $\boldsymbol{N}'\boldsymbol{s} = \boldsymbol{0}$, i.e., $\boldsymbol{s}$ is orthogonal to the columns of $\boldsymbol{N}$ [or, equivalently, that $\boldsymbol{\Delta}'\boldsymbol{s} \in \mathcal{C}^{\perp}(\boldsymbol{D}')$], which also implies that $\boldsymbol{r}'\boldsymbol{s} = 0$ (i.e., $\boldsymbol{1}_v'\boldsymbol{c} = 0$). Then, $\boldsymbol{s}'\tilde{\boldsymbol{Q}}_2 = 0$ and $\boldsymbol{s}'\tilde{\boldsymbol{Q}}_3 = -\boldsymbol{s}'\tilde{\boldsymbol{Q}}_4$, giving the equality $\boldsymbol{s}'\boldsymbol{\Delta y} = \boldsymbol{s}'\tilde{\boldsymbol{Q}}_1$. Thus, in this case, only the intra-block stratum contributes. As the second case, suppose that $\boldsymbol{s}$ is such that $\boldsymbol{N}'\boldsymbol{s} \neq \boldsymbol{0}$, but the conditions $\tilde{\boldsymbol{\phi}}_1\boldsymbol{\Delta}'\boldsymbol{s} = \boldsymbol{0}$ and $\boldsymbol{G\Delta}'\boldsymbol{s} = \boldsymbol{0}$ are satisfied [or, equivalently, that $\boldsymbol{\Delta}'\boldsymbol{s} \in \mathcal{C}^{\perp}(\boldsymbol{G}') \cap \mathcal{C}(\boldsymbol{D}')$], which also implies that $\boldsymbol{\Re}'\boldsymbol{s} = \boldsymbol{0}$. Then, $\boldsymbol{s}'\tilde{\boldsymbol{Q}}_1 = 0$ and $\boldsymbol{s}'\tilde{\boldsymbol{Q}}_3 = -\boldsymbol{s}'\tilde{\boldsymbol{Q}}_4$, which implies that $\boldsymbol{s}'\boldsymbol{\Delta y} = \boldsymbol{s}'\tilde{\boldsymbol{Q}}_2$, showing that the contribution comes from the inter-block-intra-superblock stratum only. As the third case, suppose that $\boldsymbol{s}$ is such that $\boldsymbol{\Re}'\boldsymbol{s} \neq \boldsymbol{0}$, but it satisfies the conditions $\tilde{\boldsymbol{\phi}}_1\boldsymbol{\Delta}'\boldsymbol{s} = \boldsymbol{0}$, $\tilde{\boldsymbol{\phi}}_2\boldsymbol{\Delta}'\boldsymbol{s} = \boldsymbol{0}$ and $\boldsymbol{1}_n'\boldsymbol{\Delta}'\boldsymbol{s} = 0$ [or, equivalently, that $\boldsymbol{\Delta}'\boldsymbol{s} \in \mathcal{C}^{\perp}(\boldsymbol{1}_n) \cap \mathcal{C}(\boldsymbol{G}')$], which also implies that $\boldsymbol{r}'\boldsymbol{s} = 0$ (i.e., $\boldsymbol{1}_v'\boldsymbol{c} = 0$). Then, $\boldsymbol{s}'\tilde{\boldsymbol{Q}}_1 = \boldsymbol{s}'\tilde{\boldsymbol{Q}}_2 = \boldsymbol{s}'\tilde{\boldsymbol{Q}}_4 = 0$, giving the equality $\boldsymbol{s}'\boldsymbol{\Delta y} = \boldsymbol{s}'\tilde{\boldsymbol{Q}}_3$. Thus, in this case, the contribution comes from the inter-superblock stratum only. Finally, suppose that $\boldsymbol{s} \in \mathcal{C}(\boldsymbol{1}_v)$, i.e., that $\boldsymbol{s}$ is proportional to the vector $\boldsymbol{1}_v$ [or, equivalently, that $\boldsymbol{\Delta}'\boldsymbol{s} \in \mathcal{C}(\boldsymbol{1}_n)$]. Then, on account of (5.3.7), $\boldsymbol{s}'\tilde{\boldsymbol{Q}}_1 = \boldsymbol{s}'\tilde{\boldsymbol{Q}}_2 = \boldsymbol{s}'\tilde{\boldsymbol{Q}}_3 = 0$, giving $\boldsymbol{s}'\boldsymbol{\Delta y} = \boldsymbol{s}'\tilde{\boldsymbol{Q}}_4$. This equality means that the only contribution is then from the total-area stratum.

Applying to the above cases the results of Section 5.2.2, one can prove the following corollary.

Corollary 5.3.5. *The function $\boldsymbol{s}'\boldsymbol{\Delta y}$ is the BLUE of $\boldsymbol{c}'\boldsymbol{\tau} = \boldsymbol{s}'\boldsymbol{r}^{\delta}\boldsymbol{\tau}$ under the overall model* (5.2.7) *in the following four cases:*

(a) $\boldsymbol{N}'\boldsymbol{s} = \boldsymbol{0}$ *(implying $\boldsymbol{r}'\boldsymbol{s} = 0$); the BLUE is then equal to $\boldsymbol{s}'\tilde{\boldsymbol{Q}}_1$, and its variance is of the form*

$$\mathrm{Var}(\widehat{\boldsymbol{c}'\boldsymbol{\tau}}) = \boldsymbol{s}'\boldsymbol{r}^{\delta}\boldsymbol{s}(\sigma_U^2 + \sigma_e^2) = \boldsymbol{c}'\boldsymbol{r}^{-\delta}\boldsymbol{c}(\sigma_U^2 + \sigma_e^2). \tag{5.3.32}$$

(b) $\boldsymbol{N}'\boldsymbol{s} \neq \boldsymbol{0}$, *but $\tilde{\boldsymbol{\phi}}_1\boldsymbol{\Delta}'\boldsymbol{s} = \boldsymbol{0}$ and $\boldsymbol{\Re}'\boldsymbol{s} = \boldsymbol{0}$, provided that the first part of the condition* (ii) *of Theorem* 5.2.2 *holds with regard to those connected subdesigns of $\mathcal{D}^*$ to which the nonzero elements of $\boldsymbol{s}$ correspond; the BLUE is then equal to $\boldsymbol{s}'\tilde{\boldsymbol{Q}}_2$, and its variance is of the form*

$$\mathrm{Var}(\widehat{\boldsymbol{c}'\boldsymbol{\tau}}) = \boldsymbol{s}'\boldsymbol{N}\boldsymbol{N}'\boldsymbol{s}(\sigma_B^2 - K_H^{-1}\sigma_U^2) + \boldsymbol{s}'\boldsymbol{r}^{\delta}\boldsymbol{s}(\sigma_U^2 + \sigma_e^2). \tag{5.3.33}$$

(c) $\boldsymbol{\Re}'\boldsymbol{s} \neq \boldsymbol{0}$, *but $\tilde{\boldsymbol{\phi}}_1\boldsymbol{\Delta}'\boldsymbol{s} = \boldsymbol{0}$, $\tilde{\boldsymbol{\phi}}_2\boldsymbol{\Delta}'\boldsymbol{s} = \boldsymbol{0}$ and $\boldsymbol{r}'\boldsymbol{s} = 0$, provided that the whole condition* (ii) *of Theorem* 5.2.2 *holds with regard to those connected subdesigns of $\mathcal{D}^*$ and $\mathcal{D}$ to which the nonzero elements of $\boldsymbol{s}$ correspond; the BLUE is then equal to $\boldsymbol{s}'\tilde{\boldsymbol{Q}}_3$, and its variance is of the form*

$$\begin{aligned}\mathrm{Var}(\widehat{\boldsymbol{c}'\boldsymbol{\tau}}) = {} & \boldsymbol{s}'\boldsymbol{\Re}\boldsymbol{\Re}'\boldsymbol{s}(\sigma_A^2 - B_H^{-1}\sigma_B^2) + \boldsymbol{s}'(\boldsymbol{N} - \tilde{\boldsymbol{N}}_0)(\boldsymbol{N} - \tilde{\boldsymbol{N}}_0)'\boldsymbol{s}(\sigma_B^2 - K_H^{-1}\sigma_U^2) \\ & + \boldsymbol{s}'\boldsymbol{r}^{\delta}\boldsymbol{s}(\sigma_U^2 + \sigma_e^2).\end{aligned} \tag{5.3.34}$$

(d) $\boldsymbol{s} = n^{-1}(\boldsymbol{s}'\boldsymbol{r})\mathbf{1}_v = n^{-1}(\boldsymbol{c}'\mathbf{1}_v)\mathbf{1}_v$, *provided that the whole condition* (ii) *of Theorem* 5.2.2 *holds; the BLUE is then equal to* $\boldsymbol{s}'\tilde{\boldsymbol{Q}}_4$, *and its variance is of the form* (5.3.28).

Proof. These results can be proved in a similar way as those in Corollary 3.2.5. See also Remarks 3.2.8 and 3.2.9. □

For the case (a) of Corollary 5.3.5, see Remark 3.2.7.

Remark 5.3.7. For the case (b) of Corollary 5.3.5, it should be noted that the conditions $\tilde{\boldsymbol{\phi}}_1\boldsymbol{\Delta}'\boldsymbol{s} = \mathbf{0}$ and $\boldsymbol{\Re}'\boldsymbol{s} = \mathbf{0}$ (i.e., $\boldsymbol{r}_h'\boldsymbol{s} = 0$ for all h) imply that the design $\mathcal{D}^*$ is disconnected, and any component design $\mathcal{D}_h$ ($h = 1, 2, ..., a$) is disconnected or such that treatments to which nonzero elements of $\boldsymbol{s}$ correspond are not present in it (i.e., the corresponding rows of $\boldsymbol{N}_h$ are void).

Remark 5.3.8. For the case (c) of Corollary 5.3.5, it should be noted that the conditions $\tilde{\boldsymbol{\phi}}_1\boldsymbol{\Delta}'\boldsymbol{s} = \mathbf{0}$, $\tilde{\boldsymbol{\phi}}_2\boldsymbol{\Delta}'\boldsymbol{s} = \mathbf{0}$ and $\boldsymbol{r}'\boldsymbol{s} = 0$ imply that the design $\mathcal{D}^*$ is disconnected, and that the design $\mathcal{D}$ is disconnected.

Also note that if $k_1 = k_2 = \cdots = k_b = k$, then (5.3.33) reduces to

$$\mathrm{Var}(\widehat{\boldsymbol{c}'\boldsymbol{\tau}}) = \boldsymbol{c}'\boldsymbol{r}^{-\delta}\boldsymbol{c}[k\sigma_B^2 + (1 - K_H^{-1}k)\sigma_U^2 + \sigma_e^2].$$

If, in addition to the equality of all k_j's, the equalities $n_1 = n_2 = \cdots = n_a = n_0$ hold, then the formula (5.3.34) reduces to

$$\mathrm{Var}(\widehat{\boldsymbol{c}'\boldsymbol{\tau}}) = \boldsymbol{c}'\boldsymbol{r}^{-\delta}\boldsymbol{c}[n_0\sigma_A^2 + (k - B_H^{-1}n_0)\sigma_B^2 + (1 - K_H^{-1}k)\sigma_U^2 + \sigma_e^2].$$

Finally, if all k_j's are equal and all n_h's are equal, then the variance given in Corollary 5.3.5 for the case (d) reduces from (5.3.28) to (5.3.29).

5.3.6 Examples

The theory presented in Sections 5.3.1–5.3.5 will now be illustrated by some examples.

Example 5.3.1. Let the block design considered in Example 3.2.1 (taken from Pearce, 1983, p. 102) be redefined as $\boldsymbol{N} = [\boldsymbol{N}_1 : \boldsymbol{N}_2 : \boldsymbol{N}_3]$, where

Superblock 1 *Superblock 2* *Superblock 3*

$$\boldsymbol{N}_1 = \begin{bmatrix} 1 & 1 \\ 1 & 1 \\ 1 & 1 \\ 1 & 1 \\ 0 & 0 \\ 0 & 0 \end{bmatrix}, \quad \boldsymbol{N}_2 = \begin{bmatrix} 1 & 1 \\ 1 & 1 \\ 1 & 1 \\ 1 & 1 \\ 1 & 0 \\ 0 & 1 \end{bmatrix}, \quad \boldsymbol{N}_3 = \begin{bmatrix} 1 & 1 \\ 1 & 1 \\ 1 & 1 \\ 1 & 1 \\ 1 & 1 \\ 1 & 1 \end{bmatrix},$$

i.e., as an NB design with three superblocks ($a = 3$), each composed of two blocks ($b_1 = b_2 = b_3 = 2$). The blocks within a superblock are of equal sizes,

$k_{1(1)} = k_{2(1)} = 4$, $k_{1(2)} = k_{2(2)} = 5$ and $k_{1(3)} = k_{2(3)} = 6$. Whereas the incidence submatrices $\boldsymbol{N}_1, \boldsymbol{N}_2, \boldsymbol{N}_3$ describe the component designs within the three superblocks, $\mathcal{D}_1, \mathcal{D}_1, \mathcal{D}_3$, the design for superblocks, $\mathcal{D}$, is described by the incidence matrix

$$\boldsymbol{\Re} = \begin{bmatrix} 2 & 2 & 2 \\ 2 & 2 & 2 \\ 2 & 2 & 2 \\ 2 & 2 & 2 \\ 0 & 1 & 2 \\ 0 & 1 & 2 \end{bmatrix}.$$

Evidently, the columns of $\boldsymbol{\Re}$ are the vectors of treatment replications in the component designs. The treatment replications for the whole design are then given by the vector $\boldsymbol{r} = [6,6,6,6,3,3]'$ $(= \boldsymbol{\Re}\mathbf{1}_3)$, and the superblock sizes are given by the vector $\boldsymbol{n} = [8,10,12]'$ $(= \boldsymbol{\Re}'\mathbf{1}_6)$. Suppose that this design is applied to available experimental units grouped into blocks and those into superblocks, all of conformable sizes, to allow the threefold randomization to be performed as described in Section 5.2.

To see for which contrasts the BLUEs under the intra-block submodel exist, it may be helpful to find the matrix $\tilde{\boldsymbol{C}}_1 = \boldsymbol{r}^\delta - \boldsymbol{N}\boldsymbol{k}^{-\delta}\boldsymbol{N}'$. Here

$$\tilde{\boldsymbol{C}}_1 = \frac{1}{30}\begin{bmatrix} 143 & -37 & -37 & -37 & -16 & -16 \\ -37 & 143 & -37 & -37 & -16 & -16 \\ -37 & -37 & 143 & -37 & -16 & -16 \\ -37 & -37 & -37 & 143 & -16 & -16 \\ -16 & -16 & -16 & -16 & 74 & -10 \\ -16 & -16 & -16 & -16 & -10 & 74 \end{bmatrix}.$$

Evidently, $\text{rank}(\tilde{\boldsymbol{C}}_1) = v - 1 = 5$, which also follows directly from the fact that the design is connected (see Lemma 2.3.2). This matrix implies that the columns of $\tilde{\boldsymbol{C}}_1$ span the subspace of all contrasts (of all vectors $\boldsymbol{c}$ representing contrasts). Hence, on account of Theorem 3.2.1, for any contrast $\boldsymbol{c}'\boldsymbol{\tau}$, there exists the BLUE under the intra-block submodel. It is of the form $\widehat{\boldsymbol{c}'\boldsymbol{\tau}} = \boldsymbol{s}'\boldsymbol{\Delta}\boldsymbol{y}_1$, where $\boldsymbol{s}$ is such that $\boldsymbol{c} = \tilde{\boldsymbol{C}}_1\boldsymbol{s}$. In particular, to estimate the contrast between treatment 1 and treatment 2, one can use the vector $\boldsymbol{s} = (1/6)[1,-1,0,0,0,0]'$. It should, however, be noted that for this $\boldsymbol{s}$ the equality $\boldsymbol{N}'\boldsymbol{s} = \mathbf{0}$ holds, which on account of Corollary 5.3.5(a) implies that $\boldsymbol{s}'\boldsymbol{\Delta}\boldsymbol{y}_1 = \boldsymbol{s}'\boldsymbol{\Delta}\boldsymbol{y}$, and this function is the BLUE of $\boldsymbol{c}'\boldsymbol{\tau}$ under the overall model (5.2.7). The same is true for any contrast among the first four treatments, as can easily be seen from the form of the matrix $\boldsymbol{N}$ of the considered design.

Contrasts for which the BLUEs under the inter-block-intra-superblock submodel may possibly exist are to be searched by considering the matrix $\tilde{\boldsymbol{C}}_2 = \tilde{\boldsymbol{N}}_0\boldsymbol{k}^{-\delta}\tilde{\boldsymbol{N}}_0'$ (see Theorem 5.3.2). For this reason, note that in the present example

$$\tilde{\boldsymbol{N}}_0 = [\boldsymbol{N}_{01} : \boldsymbol{N}_{02} : \boldsymbol{N}_{03}] = [\mathbf{O} : \boldsymbol{N}_{02} : \mathbf{O}],$$

with

$$N_{02} = \frac{1}{2}\begin{bmatrix} 0 & 0 & 0 & 0 & 1 & -1 \\ 0 & 0 & 0 & 0 & -1 & 1 \end{bmatrix}',$$

from which

$$\tilde{C}_2 = \frac{1}{10}\begin{bmatrix} 0 & 0 & 0 & 0 & 0 & 0 \\ 0 & 0 & 0 & 0 & 0 & 0 \\ 0 & 0 & 0 & 0 & 0 & 0 \\ 0 & 0 & 0 & 0 & 0 & 0 \\ 0 & 0 & 0 & 0 & 1 & -1 \\ 0 & 0 & 0 & 0 & -1 & 1 \end{bmatrix}, \quad \text{of rank 1.}$$

This matrix shows that there is only one contrast for which the BLUE under the considered submodel may exist, the contrast between treatment 5 and treatment 6. It can be represented by the vector $\boldsymbol{c} = \tilde{\boldsymbol{C}}_2\boldsymbol{s}$, where $\boldsymbol{s} = 10[0, 0, 0, 0, 1, 0]'$. Because the block sizes within the second superblock are equal ($k_{1(2)} = k_{2(2)} = 5$), and this is the only superblock contributing to $\tilde{\boldsymbol{N}}_0$ and, hence, to $\tilde{\boldsymbol{C}}_2$, the BLUE under the inter-block-intra-superblock submodel really exists for this contrast, as it follows from Corollary 5.3.2 and Remark 5.3.1. It has the form $\widehat{\boldsymbol{c}'\boldsymbol{\tau}} = \boldsymbol{s}'\boldsymbol{\Delta}\boldsymbol{y}_2$.

Turning now to the estimation of contrasts under the inter-superblock submodel, it should be recalled (from Theorem 5.3.3) that the BLUEs may exist only for such contrasts that are generated by the matrix $\tilde{\boldsymbol{C}}_3 = \boldsymbol{\Delta}\tilde{\boldsymbol{\phi}}_3\boldsymbol{\Delta}' = (\boldsymbol{N}_0 - \tilde{\boldsymbol{N}}_0)\boldsymbol{k}^{-\delta}(\boldsymbol{N}_0 - \tilde{\boldsymbol{N}}_0)'$. Because here

$$\boldsymbol{N}_0 - \tilde{\boldsymbol{N}}_0 = \frac{1}{5}\begin{bmatrix} 1 & 1 & 0 & 0 & -1 & -1 \\ 1 & 1 & 0 & 0 & -1 & -1 \\ 1 & 1 & 0 & 0 & -1 & -1 \\ 1 & 1 & 0 & 0 & -1 & -1 \\ -2 & -2 & 0 & 0 & 2 & 2 \\ -2 & -2 & 0 & 0 & 2 & 2 \end{bmatrix},$$

it follows that

$$\tilde{C}_3 = \frac{1}{30}\begin{bmatrix} 1 & 1 & 1 & 1 & -2 & -2 \\ 1 & 1 & 1 & 1 & -2 & -2 \\ 1 & 1 & 1 & 1 & -2 & -2 \\ 1 & 1 & 1 & 1 & -2 & -2 \\ -2 & -2 & -2 & -2 & 4 & 4 \\ -2 & -2 & -2 & -2 & 4 & 4 \end{bmatrix}, \quad \text{of rank 1.}$$

This matrix shows that there is only one contrast for which the BLUE under this submodel could possibly exist, the contrast between the treatments 1, 2, 3, 4 and the treatments 5, 6. To see whether the BLUE for this contrast really exists, refer to Corollary 5.3.3(b). To make use of it, note that

$$(\boldsymbol{K}_0 - \tilde{\boldsymbol{K}}_0)(\boldsymbol{N}_0 - \tilde{\boldsymbol{N}}_0)' = \frac{2}{75}\begin{bmatrix} 34 & 34 & 34 & 34 & -68 & -68 \\ 34 & 34 & 34 & 34 & -68 & -68 \\ 5 & 5 & 5 & 5 & -10 & -10 \\ 5 & 5 & 5 & 5 & -10 & -10 \\ -39 & -39 & -39 & -39 & 78 & 78 \\ -39 & -39 & -39 & -39 & 78 & 78 \end{bmatrix},$$

of rank 1. Evidently, no (nonzero) linear combination of the columns of this matrix can be a linear combination of the columns of the matrix $(\boldsymbol{N}_0 - \tilde{\boldsymbol{N}}_0)'$. Also, it can be noted that

$$\mathfrak{R}_0' = \frac{2}{5}\begin{bmatrix} 1 & 1 & 1 & 1 & -2 & -2 \\ 0 & 0 & 0 & 0 & 0 & 0 \\ -1 & -1 & -1 & -1 & 2 & 2 \end{bmatrix}, \quad \text{of rank 1,}$$

from which

$$\aleph_0\mathfrak{R}_0' = \frac{8}{75}\begin{bmatrix} 34 & 34 & 34 & 34 & -68 & -68 \\ 5 & 5 & 5 & 5 & -10 & -10 \\ -39 & -39 & -39 & -39 & 78 & 78 \end{bmatrix}, \quad \text{of rank 1.}$$

Certainly, no (nonzero) linear combination of the columns of the latter matrix can be a linear combination of the columns of the former. Thus, the conditions of Corollary 5.3.3(b) are not satisfied, and so the BLUE does not exist under the inter-superblock submodel for any contrast, for that indicated above in particular.

Finally, if one is interested in estimating the general parameter mean, $\boldsymbol{c}'\boldsymbol{\tau} = n^{-1}\boldsymbol{r}'\boldsymbol{\tau}$, the BLUE of it is obtainable under the total-area submodel in the form $\widehat{\boldsymbol{c}'\boldsymbol{\tau}} = \boldsymbol{s}'\boldsymbol{\Delta}\boldsymbol{y}_4$, where $\boldsymbol{s} = n^{-1}\mathbf{1}_v$, i.e., simply in the form $n^{-1}\mathbf{1}_n'\boldsymbol{y}$. This estimator is equal to $\boldsymbol{s}'\boldsymbol{\Delta}\boldsymbol{y}$. However, it is not the BLUE under the overall model (5.2.7), as the condition of Corollary 5.3.5(d) is not satisfied.

Example 5.3.2. Let the block design considered in Example 3.2.4 (taken from Pearce, 1983, p. 225) for a 2^3 factorial structure of treatments, with the three factors denoted by X, Y and Z, be now redefined as an NB design with two superblocks ($a = 2$), according to the partitioned incidence matrix $\boldsymbol{N} = [\boldsymbol{N}_1 : \boldsymbol{N}_2]$, where

$$\begin{array}{cccccc} \textit{Treatment} & & \textit{Superblock 1} & & & \textit{Superblock 2} \\ \begin{array}{c} 1 \\ \mathrm{X} \\ \mathrm{Y} \\ \mathrm{Z} \\ \mathrm{XY} \\ \mathrm{XZ} \\ \mathrm{YZ} \\ \mathrm{XYZ} \end{array} & \boldsymbol{N}_1 = & \begin{bmatrix} 0 & 1 & 1 & 1 \\ 0 & 0 & 0 & 0 \\ 0 & 0 & 0 & 0 \\ 0 & 0 & 0 & 0 \\ 1 & 1 & 1 & 0 \\ 1 & 1 & 0 & 1 \\ 1 & 0 & 1 & 1 \\ 0 & 0 & 0 & 0 \end{bmatrix} & \text{and} & \boldsymbol{N}_2 = & \begin{bmatrix} 0 & 0 & 0 & 0 \\ 2 & 1 & 1 & 1 \\ 1 & 2 & 1 & 1 \\ 1 & 1 & 2 & 1 \\ 0 & 0 & 0 & 0 \\ 0 & 0 & 0 & 0 \\ 0 & 0 & 0 & 0 \\ 1 & 1 & 1 & 2 \end{bmatrix} \end{array}$$

(treatment 1 representing the combination of all three factors at the lower level, treatment X representing the combination of factor X at the upper level with Y and Z at the lower level, etc.). Each of the two superblocks is composed of four blocks ($b_1 = b_2 = 4$), of size 3 in the first superblock ($k_{1(1)} = k_{2(1)} = k_{3(1)} = k_{4(1)} = 3$) and of size 5 in the second ($k_{1(2)} = k_{2(2)} = k_{3(2)} = k_{4(2)} = 5$). From the incidence matrices $\boldsymbol{N}_h$, $h = 1, 2$, describing the component designs $\mathcal{D}_1$ and $\mathcal{D}_2$, the incidence matrix describing the design $\mathcal{D}$ is obtainable, as

$$\mathfrak{R} = \begin{bmatrix} 3 & 0 & 0 & 0 & 3 & 3 & 3 & 0 \\ 0 & 5 & 5 & 5 & 0 & 0 & 0 & 5 \end{bmatrix}'.$$

From this matrix, the vector of treatment replications for the whole design is $\boldsymbol{r} = [3, 5, 5, 5, 3, 3, 3, 5]'$, and the vector of superblock sizes is $\boldsymbol{n} = [12, 20]'$. As for the first example, it is assumed that this design is applied to available experimental units grouped into blocks and those joined into superblocks, all of them of conformable sizes, so that the appropriate threefold randomization can be performed.

For the estimation under the intra-block submodel, one needs to examine the matrix $\boldsymbol{C}_1$. Here, it is

$$\tilde{\boldsymbol{C}}_1 = \begin{bmatrix} 15 & 0 & 0 & 0 & -5 & -5 & -5 & 0 \\ 0 & 27 & -9 & -9 & 0 & 0 & 0 & -9 \\ 0 & -9 & 27 & -9 & 0 & 0 & 0 & -9 \\ 0 & -9 & -9 & 27 & 0 & 0 & 0 & -9 \\ -5 & 0 & 0 & 0 & 15 & -5 & -5 & 0 \\ -5 & 0 & 0 & 0 & -5 & 15 & -5 & 0 \\ -5 & 0 & 0 & 0 & -5 & -5 & 15 & 0 \\ 0 & -9 & -9 & -9 & 0 & 0 & 0 & 27 \end{bmatrix}, \quad \text{of rank 6.}$$

It shows that because of the disconnectedness, not all contrasts can be estimated under the intra-block submodel. In fact, the BLUEs under this submodel exist for contrasts giving the main effects and two factor interactions, but not for the contrast giving the three factor interaction, i.e., not for the contrast represented by the vector $\boldsymbol{c} = [-1, 1, 1, 1, -1, -1, -1, 1]'$.

As to the estimation under the inter-block-intra-superblock submodel, it involves the matrix $\tilde{\boldsymbol{N}}_0 = [\boldsymbol{N}_{01} : \boldsymbol{N}_{02}]$, where

$$\boldsymbol{N}_{01} = \frac{1}{4}\begin{bmatrix} -3 & 1 & 1 & 1 \\ 0 & 0 & 0 & 0 \\ 0 & 0 & 0 & 0 \\ 0 & 0 & 0 & 0 \\ 1 & 1 & 1 & -3 \\ 1 & 1 & -3 & 1 \\ 1 & -3 & 1 & 1 \\ 0 & 0 & 0 & 0 \end{bmatrix} \quad \text{and} \quad \boldsymbol{N}_{02} = \frac{1}{4}\begin{bmatrix} 0 & 0 & 0 & 0 \\ 3 & -1 & -1 & -1 \\ -1 & 3 & -1 & -1 \\ -1 & -1 & 3 & -1 \\ 0 & 0 & 0 & 0 \\ 0 & 0 & 0 & 0 \\ 0 & 0 & 0 & 0 \\ -1 & -1 & -1 & 3 \end{bmatrix}.$$

Hence,

$$\tilde{C}_2 = \frac{1}{60}\begin{bmatrix} 15 & 0 & 0 & 0 & -5 & -5 & -5 & 0 \\ 0 & 9 & -3 & -3 & 0 & 0 & 0 & -3 \\ 0 & -3 & 9 & -3 & 0 & 0 & 0 & -3 \\ 0 & -3 & -3 & 9 & 0 & 0 & 0 & -3 \\ -5 & 0 & 0 & 0 & 15 & -5 & -5 & 0 \\ -5 & 0 & 0 & 0 & -5 & 15 & -5 & 0 \\ -5 & 0 & 0 & 0 & -5 & -5 & 15 & 0 \\ 0 & -3 & -3 & -3 & 0 & 0 & 0 & 9 \end{bmatrix}, \quad \text{of rank 6.}$$

From the comparison of $\tilde{C}_2$ with $\tilde{C}_1$, one can see that the same contrasts for which the BLUEs exist under the intra-block submodel can be considered for estimating under the inter-block-intra-superblock submodel. On account of Remark 5.3.1, the equality of block sizes within the superblocks implies that the BLUEs under this submodel really exist for all of these contrasts. Thus, all of the main effects and all of the two-factor interactions receive BLUEs under both submodels.

Now, as to the estimation under the inter-superblock submodel, one has to take into account the matrix $\tilde{C}_3$, which can be obtained (alternatively to the formula used in Example 5.3.1) from the formula $\tilde{C}_3 = \mathfrak{R}_0 \boldsymbol{n}^{-\delta} \mathfrak{R}_0'$ (see the proof of Theorem 5.3.3). Because here

$$\mathfrak{R}_0' = \frac{15}{8}\begin{bmatrix} 1 & -1 & -1 & -1 & 1 & 1 & 1 & -1 \\ -1 & 1 & 1 & 1 & -1 & -1 & -1 & 1 \end{bmatrix},$$

it follows that

$$\tilde{C}_3 = \frac{15}{32}\begin{bmatrix} 1 & -1 & -1 & -1 & 1 & 1 & 1 & -1 \\ -1 & 1 & 1 & 1 & -1 & -1 & -1 & 1 \\ -1 & 1 & 1 & 1 & -1 & -1 & -1 & 1 \\ -1 & 1 & 1 & 1 & -1 & -1 & -1 & 1 \\ 1 & -1 & -1 & -1 & 1 & 1 & 1 & -1 \\ 1 & -1 & -1 & -1 & 1 & 1 & 1 & -1 \\ 1 & -1 & -1 & -1 & 1 & 1 & 1 & -1 \\ -1 & 1 & 1 & 1 & -1 & -1 & -1 & 1 \end{bmatrix}, \quad \text{of rank 1.}$$

Evidently, there is only one contrast that can be considered for estimation under the inter-superblock submodel. It is the contrast giving the three-factor interaction, that which cannot be estimated under the previous two submodels. Because the columns of the incidence matrix $\mathfrak{R}$ are linearly independent, the conditions of Corollary 5.3.4 are satisfied, on account of Remark 5.3.4(b). Thus, the BLUE of this contrast really exists under the inter-superblock submodel. Also, it should be noted that the corresponding vector $\boldsymbol{s} = \boldsymbol{r}^{-\delta}\boldsymbol{c} = [-1/3, 1/5, 1/5, 1/5, -1/3, -1/3, -1/3, 1/5]'$ satisfies the equalities $\tilde{C}_1\boldsymbol{s} = \boldsymbol{0}$, $\tilde{C}_2\boldsymbol{s} = \boldsymbol{0}$ and $\boldsymbol{r}'\boldsymbol{s} = 0$, whereas $\mathfrak{R}\boldsymbol{s} \neq \boldsymbol{0}$, and that the condition (ii) of Theorem 5.2.2 is satisfied completely. Thus, by Corollary 5.3.5(c), $\boldsymbol{s}'\boldsymbol{\Delta y}$ is the BLUE of the three-factor interaction, $\boldsymbol{c}'\boldsymbol{\tau}$, under the overall model (5.2.7).

Finally, because both parts of the condition (ii) of Theorem 5.2.2 are satisfied, $n^{-1}\mathbf{1}_n'\boldsymbol{y}$ is the BLUE of $n^{-1}\boldsymbol{r}'\boldsymbol{\tau}$ under the overall model (5.2.7), as it follows from Corollary 5.3.5(d).

Example 5.3.3. Consider a design with the incidence matrix

$$\boldsymbol{N} = \begin{bmatrix} 0 & 0 & 1 & 0 & 0 & 1 & 0 & 1 & 1 & 0 \\ 1 & 0 & 0 & 0 & 1 & 0 & 0 & 1 & 0 & 1 \\ 0 & 1 & 1 & 0 & 0 & 0 & 1 & 0 & 0 & 1 \\ 0 & 1 & 0 & 1 & 1 & 0 & 0 & 0 & 1 & 0 \\ 1 & 0 & 0 & 1 & 0 & 1 & 1 & 0 & 0 & 0 \end{bmatrix}.$$

Evidently, the matrix describes an arrangement of five treatments in ten blocks, each containing two treatments in such a way that every treatment occurs at most once in a block and in exactly four blocks, and that every two treatments concur in exactly one block. This arrangement means (see Definition 2.4.2) that it is a BIB design specified by parameters $v = 5, b = 10, r = 4, k = 2, \lambda = 1$. Let the blocks of the design, denoted by $\mathcal{D}^*$, be grouped into superblocks defined by the partition $\boldsymbol{N} = [\boldsymbol{N}_1 : \boldsymbol{N}_2 : \boldsymbol{N}_3 : \boldsymbol{N}_4 : \boldsymbol{N}_5]$, where

Superblock 1 *Superblock 2* *Superblock 3* *Superblock 4* *Superblock 5*

$$\boldsymbol{N}_1 = \begin{bmatrix} 0\,0 \\ 1\,0 \\ 0\,1 \\ 0\,1 \\ 1\,0 \end{bmatrix}, \boldsymbol{N}_2 = \begin{bmatrix} 1\,0 \\ 0\,0 \\ 1\,0 \\ 0\,1 \\ 0\,1 \end{bmatrix}, \boldsymbol{N}_3 = \begin{bmatrix} 0\,1 \\ 1\,0 \\ 0\,0 \\ 1\,0 \\ 0\,1 \end{bmatrix}, \boldsymbol{N}_4 = \begin{bmatrix} 0\,1 \\ 0\,1 \\ 1\,0 \\ 0\,0 \\ 1\,0 \end{bmatrix}, \boldsymbol{N}_5 = \begin{bmatrix} 1\,0 \\ 0\,1 \\ 0\,1 \\ 1\,0 \\ 0\,0 \end{bmatrix}.$$

This partition gives a NB design with five superblocks ($a = 5$), each composed of two blocks ($b_1 = b_2 = b_3 = b_4 = b_5 = 2$), of the same size of two units in all superblocks ($k_{1(h)} = k_{2(h)} = 2$ for all h). From the incidence matrices $\boldsymbol{N}_h$ describing the component designs $\mathcal{D}_h$, $h = 1, 2, 3, 4, 5$, the incidence matrix of the design $\mathcal{D}$ can be obtained as

$$\mathfrak{R} = \begin{bmatrix} 0 & 1 & 1 & 1 & 1 \\ 1 & 0 & 1 & 1 & 1 \\ 1 & 1 & 0 & 1 & 1 \\ 1 & 1 & 1 & 0 & 1 \\ 1 & 1 & 1 & 1 & 0 \end{bmatrix}.$$

It clearly shows that $\mathcal{D}$ is again a BIB design with parameters $v = 5, b = 5, r = 4, k = 4, \lambda = 3$. However, in the present notation for $\mathcal{D}$ [see formula (5.1.1) and Corollary 5.3.3(d)], one has to replace here b by a and k by n_0 $(= n/a)$. Thus, the considered NB design is an NBIB design (see Section 5.1). In fact, it is the first design on the list of NBIB designs given by Preece (1967, Table 3). It also coincides with the so-called "nearly-resolvable unreduced design for 5 varieties" considered by Bailey, Goldrei and Holt (1984, p. 259).

As usual, it is supposed that this design is applied to available experimental units grouped into blocks and those into superblocks, all of conformable sizes allowing the threefold randomization to be performed as described in Section 5.2.1.

It is easy to obtain the matrix $\tilde{C}_1$, which gets the form

$$\tilde{C}_1 = \frac{1}{2}\begin{bmatrix} 4 & -1 & -1 & -1 & -1 \\ -1 & 4 & -1 & -1 & -1 \\ -1 & -1 & 4 & -1 & -1 \\ -1 & -1 & -1 & 4 & -1 \\ -1 & -1 & -1 & -1 & 4 \end{bmatrix} = \frac{5}{2}\left(I_5 - \frac{1}{5}\mathbf{1}_5\mathbf{1}_5'\right),$$

as in (2.4.4) with $\theta = 5/2$. Its rank is evidently $v-1 = 4$ (the design is connected), which implies that the columns of $\tilde{C}_1$ span the subspace of all contrasts (c vectors such that $c'\mathbf{1}_v = 0$). This implication means that for any contrast $c'\tau$, the BLUE under the intra-block submodel exists (on account of Theorem 3.2.1). It is of the form $\widehat{c'\tau} = s'\Delta y_1$, with $c = \tilde{C}_1 s$, which can also be written as $\widehat{c'\tau} = c'\tilde{C}_1^- \Delta\phi_1 y$ (see Section 3.2.1). The variance of any $\widehat{c'\tau}$ is of the form $\text{Var}(\widehat{c'\tau}) = c'\tilde{C}_1^- c(\sigma_U^2 + \sigma_e^2)$, as given in (5.3.8). Note that as a g-inverse of $\tilde{C}_1$ the matrix $(2/5)I_5$ can be taken, giving $\widehat{c'\tau} = (2/5)c'\Delta\tilde{\phi}_1 y = (2/5)c'\tilde{Q}_1$, with $\tilde{Q}_1 = \Delta\tilde{\phi}_1 y$, and $\text{Var}(\widehat{c'\tau}) = (2/5)c'c\sigma_1^2$, with $\sigma_1^2 = \sigma_U^2 + \sigma_e^2$.

To consider the estimation of contrasts under the inter-block-intra-superblock submodel, one needs the matrix $\tilde{C}_2$. To obtain it, first note that here

$$\tilde{N}_0 = \frac{1}{2}\begin{bmatrix} 0 & 0 & 1 & -1 & -1 & 1 & -1 & 1 & 1 & -1 \\ 1 & -1 & 0 & 0 & 1 & -1 & -1 & 1 & -1 & 1 \\ -1 & 1 & 1 & -1 & 0 & 0 & 1 & -1 & -1 & 1 \\ -1 & 1 & -1 & 1 & 1 & -1 & 0 & 0 & 1 & -1 \\ 1 & -1 & -1 & 1 & -1 & 1 & 1 & -1 & 0 & 0 \end{bmatrix},$$

which, because of the equality $k_1 = k_2 = \cdots = k_{10} = k = 2$, gives the matrix

$$\tilde{C}_2 = \frac{1}{k}\tilde{N}_0\tilde{N}_0' = \frac{5}{4}\left(I_5 - \frac{1}{5}\mathbf{1}_5\mathbf{1}_5'\right) = \frac{1}{2}\tilde{C}_1, \quad \text{of rank 4.}$$

This formula shows that, as before, the columns of $\tilde{C}_2$ span the subspace of all contrasts. Moreover, because on account of Corollary 5.3.1(d), any vector s from the relation $c = \tilde{C}_2 s$ satisfies the conditions (5.3.10) and (5.3.11) of Theorem 5.3.2, the BLUE under the inter-block-intra-superblock submodel exists for any contrast $c'\tau$. It has the form $\widehat{c'\tau} = s'\Delta y_2 = c'\tilde{C}_2^-\tilde{Q}_2 = (4/5)c'\tilde{Q}_2$, with $\tilde{Q}_2 = \Delta\tilde{\phi}_2 y$, and its variance obtains from (5.3.13) the form

$$\text{Var}(\widehat{c'\tau}) = c'(k^{-1}\tilde{N}_0\tilde{N}_0')^- c\sigma_2^2 = \frac{4}{5}c'c\sigma_2^2,$$

with σ_2^2 as in (5.3.16).

Going next to the estimation of contrasts under the inter-superblock submodel, one needs the matrix $\mathfrak{R}_0 = \mathfrak{R} - n^{-1}\boldsymbol{r}\boldsymbol{n}'$. It is here of the form

$$\mathfrak{R}_0 = \frac{1}{5}\begin{bmatrix} -4 & 1 & 1 & 1 & 1 \\ 1 & -4 & 1 & 1 & 1 \\ 1 & 1 & -4 & 1 & 1 \\ 1 & 1 & 1 & -4 & 1 \\ 1 & 1 & 1 & 1 & -4 \end{bmatrix}.$$

From it and the equality $n_1 = n_2 = n_3 = n_4 = n_5 = n_0 = 4$, the relevant matrix $\tilde{\boldsymbol{C}}_3$ is obtainable as

$$\tilde{\boldsymbol{C}}_3 = \frac{1}{n_0}\mathfrak{R}_0\mathfrak{R}_0' = \frac{5}{20}\left(\boldsymbol{I}_5 - \frac{1}{5}\mathbf{1}_5\mathbf{1}_5'\right) = \frac{1}{10}\tilde{\boldsymbol{C}}_1, \quad \text{of rank 4.}$$

This formula shows that, again, the columns of $\tilde{\boldsymbol{C}}_3$ span the subspace of all contrasts. Moreover, as on account of Corollary 5.3.3(d) the conditions (5.3.19) and (5.3.20) of Theorem 5.3.3 are satisfied, the BLUE under the inter-superblock submodel exists for any contrast $\boldsymbol{c}'\boldsymbol{\tau}$, where $\boldsymbol{c} = \tilde{\boldsymbol{C}}_3\boldsymbol{s}$ for some $\boldsymbol{s}$. It has the form $\widehat{\boldsymbol{c}'\boldsymbol{\tau}} = \boldsymbol{s}'\boldsymbol{\Delta}\boldsymbol{y}_3 = \boldsymbol{c}'\tilde{\boldsymbol{C}}_3^-\tilde{\boldsymbol{Q}}_3 = 4\boldsymbol{c}'\tilde{\boldsymbol{Q}}_3$, with $\tilde{\boldsymbol{Q}}_3 = \boldsymbol{\Delta}\tilde{\boldsymbol{\phi}}_3\boldsymbol{y}$, and its variance has the form

$$\mathrm{Var}(\widehat{\boldsymbol{c}'\boldsymbol{\tau}}) = \boldsymbol{c}'\tilde{\boldsymbol{C}}_3^-\boldsymbol{c}\sigma_3^2 = 4\boldsymbol{c}'\boldsymbol{c}\sigma_3^2,$$

with σ_3^2 as in (5.3.27).

Finally, applying Corollary 5.3.5(d) note that, because both parts of the condition (ii) of Theorem 5.2.2 are satisfied in this example, $n^{-1}\mathbf{1}_n'\boldsymbol{y}$ is the BLUE of the general parameter mean, $n^{-1}\boldsymbol{r}'\boldsymbol{\tau}$, under the overall model (5.2.7). The relevant variance is of the form (5.3.29) with $\boldsymbol{c}'\mathbf{1}_v = 1$.

Example 5.3.4. Consider a design obtained from that used in Example 5.3.3 by extending the number of treatments from five to six and adding one block to each of the five component designs in such a way that the superblocks become complete. The resulting component designs have then the following incidence matrices:

Superblock 1	*Superblock 2*	*Superblock 3*	*Superblock 4*	*Superblock 5*
$\boldsymbol{N}_1 = \begin{bmatrix} 0&0&1\\1&0&0\\0&1&0\\0&1&0\\1&0&0\\0&0&1 \end{bmatrix}$,	$\boldsymbol{N}_2 = \begin{bmatrix} 1&0&0\\0&0&1\\1&0&0\\0&1&0\\0&1&0\\0&0&1 \end{bmatrix}$,	$\boldsymbol{N}_3 = \begin{bmatrix} 0&1&0\\1&0&0\\0&0&1\\1&0&0\\0&1&0\\0&0&1 \end{bmatrix}$,	$\boldsymbol{N}_4 = \begin{bmatrix} 0&1&0\\0&1&0\\1&0&0\\0&0&1\\1&0&0\\0&0&1 \end{bmatrix}$,	$\boldsymbol{N}_5 = \begin{bmatrix} 1&0&0\\0&1&0\\0&1&0\\1&0&0\\0&0&1\\0&0&1 \end{bmatrix}$.

Note that the design $\mathcal{D}^*$, i.e., the design described by the incidence matrix $\boldsymbol{N} = [\boldsymbol{N}_1 : \boldsymbol{N}_2 : \boldsymbol{N}_3 : \boldsymbol{N}_4 : \boldsymbol{N}_5]$, is a BIB design with parameters $v = 6, b =$

$15, r = 5, k = 2, \lambda = 1$. On the other hand, the design $\mathcal{D}$, given by the incidence matrix

$$\mathfrak{R} = [\boldsymbol{N}_1\boldsymbol{1}_3 : \boldsymbol{N}_2\boldsymbol{1}_3 : \boldsymbol{N}_3\boldsymbol{1}_3 : \boldsymbol{N}_4\boldsymbol{1}_3 : \boldsymbol{N}_5\boldsymbol{1}_3] = [\boldsymbol{1}_6 : \boldsymbol{1}_6 : \boldsymbol{1}_6 : \boldsymbol{1}_6 : \boldsymbol{1}_6] = \boldsymbol{1}_6\boldsymbol{1}_5',$$

is an equireplicate connected orthogonal and proper block design (an RBD; see Section 2.1 and Corollary 2.3.4). Because of these properties of the designs $\mathcal{D}^*$ and $\mathcal{D}$, the considered NB design belongs to the class of resolvable BIB designs. In fact, it is exactly the design No. 2 on the list given by Kageyama (1972, Table). It also coincides with the "resolvable unreduced design for 6 varieties" considered by Bailey et al. (1984, p. 259).

As in the previous examples, it is assumed that the design presented here is applied to available experimental units grouped into blocks and those further into superblocks, all of them of conformable size, so that the appropriate randomization can be performed, as described in Section 5.2.1.

The matrices $\tilde{\boldsymbol{C}}_1, \tilde{\boldsymbol{C}}_2$ and $\tilde{\boldsymbol{C}}_3$ for this design can easily be obtained as

$$\tilde{\boldsymbol{C}}_1 = 3\left(\boldsymbol{I}_6 - \frac{1}{6}\boldsymbol{1}_6\boldsymbol{1}_6'\right), \quad \tilde{\boldsymbol{C}}_2 = \frac{2}{3}\tilde{\boldsymbol{C}}_1 \quad \text{and} \quad \tilde{\boldsymbol{C}}_3 = \mathbf{O}.$$

They show that for any contrast, there exists the BLUE under the intra-block submodel and under the inter-block-intra-superblock submodel, whereas no contrast can be estimated under the inter-superblock submodel, as the design $\mathcal{D}$ is orthogonal and connected, giving $\mathfrak{R}_0 = \mathbf{O}$ [see Corollary 5.3.3(a)]. To find the appropriate formulae for the intra-block BLUEs and the inter-block-intra-superblock BLUEs, and their variances, one can proceed as in Example 5.3.3. This task is left to the reader as an exercise. Also note that the BLUE of the general parameter mean, $n^{-1}\boldsymbol{r}'\boldsymbol{\tau}$, is $n^{-1}\boldsymbol{1}_n'\boldsymbol{y}$, obtainable under the overall model (5.2.7), exactly as in Example 5.3.3. Why ?

In connection with Example 5.3.4, the following general remark is worth noting.

Remark 5.3.9. If the design $\mathcal{D}$ of an NB design is connected and orthogonal, i.e., $\mathfrak{R} = n^{-1}\boldsymbol{r}\boldsymbol{n}'$, then $\boldsymbol{\Delta}\tilde{\boldsymbol{\phi}}_2 = \boldsymbol{\Delta}\boldsymbol{\phi}_2$ and, hence, the inter-block-intra-superblock analysis of variance becomes similar to the inter-block analysis of variance presented in Section 3.2.2, with the treatment sum of squares identical, $\tilde{\boldsymbol{Q}}_2'\tilde{\boldsymbol{C}}_2^-\tilde{\boldsymbol{Q}}_2 = \boldsymbol{Q}_2'\boldsymbol{C}_2^-\boldsymbol{Q}_2$, as $\tilde{\boldsymbol{C}}_2 = \boldsymbol{C}_2$ and $\tilde{\boldsymbol{Q}}_2 = \boldsymbol{Q}_2$ then, and with the residual sum of squares

$$\boldsymbol{y}'\tilde{\boldsymbol{\psi}}_2\boldsymbol{y} = \boldsymbol{y}'\tilde{\boldsymbol{\phi}}_2\boldsymbol{y} - \boldsymbol{y}'\boldsymbol{\phi}_2\boldsymbol{\Delta}'\boldsymbol{C}_2^-\boldsymbol{\Delta}\boldsymbol{\phi}_2\boldsymbol{y} = \boldsymbol{y}'\boldsymbol{\psi}_2\boldsymbol{y} - \boldsymbol{y}'\tilde{\boldsymbol{\phi}}_3\boldsymbol{y},$$

as in general $\tilde{\boldsymbol{\phi}}_2 = \boldsymbol{\phi}_2 - \tilde{\boldsymbol{\phi}}_3$. Futhermore, because in this case $\tilde{\boldsymbol{C}}_3 = \mathbf{O}$, the inter-superblock sum of squares $\boldsymbol{y}'\tilde{\boldsymbol{\phi}}_3\boldsymbol{y}$ has no treatment component, and so the inter-superblock analysis of variance is then immaterial (see the comment preceding Corollary 5.3.3).

5.4 Generally balanced nested block designs

To complete the presentation of the properties of NB designs resulting from the randomization model, it may also be useful to consider briefly the notion of GB for these designs. For block designs with one stratum of blocks, this notion has been presented in Section 3.6, preceded by Section 3.5 giving the definition of the OBS. Here, let these notions be extended for NB designs.

In connection with the decomposition (5.3.1) of the observed vector $\boldsymbol{y}$, note that the expectation of $\boldsymbol{y}$ can be written, using the notation of Section 5.3, as

$$\mathrm{E}(\boldsymbol{y}) = \mathrm{E}(\boldsymbol{y}_1) + \mathrm{E}(\boldsymbol{y}_2) + \mathrm{E}(\boldsymbol{y}_3) + \mathrm{E}(\boldsymbol{y}_4) = \tilde{\boldsymbol{\phi}}_1\boldsymbol{\Delta}'\boldsymbol{\tau} + \tilde{\boldsymbol{\phi}}_2\boldsymbol{\Delta}'\boldsymbol{\tau} + \tilde{\boldsymbol{\phi}}_3\boldsymbol{\Delta}'\boldsymbol{\tau} + \tilde{\boldsymbol{\phi}}_4\boldsymbol{\Delta}'\boldsymbol{\tau}.$$

The question of interest is then, under which conditions also the covariance matrix of $\boldsymbol{y}$, of the general form

$$\mathrm{Cov}(\boldsymbol{y}) = \sum_{\alpha=1}^{4}\mathrm{Cov}(\boldsymbol{y}_\alpha) + \sum\sum_{\alpha\neq\alpha'}\mathrm{Cov}(\boldsymbol{y}_\alpha, \boldsymbol{y}_{\alpha'}),$$

can be reduced to a simple form, similar to (3.5.9), i.e., to the formula

$$\begin{aligned}\mathrm{Cov}(\boldsymbol{y}) &= \mathrm{Cov}(\boldsymbol{y}_1) + \mathrm{Cov}(\boldsymbol{y}_2) + \mathrm{Cov}(\boldsymbol{y}_3) + \mathrm{Cov}(\boldsymbol{y}_4)\\ &= \tilde{\boldsymbol{\phi}}_1\sigma_1^2 + \tilde{\boldsymbol{\phi}}_2\sigma_2^2 + \tilde{\boldsymbol{\phi}}_3\sigma_3^2 + \tilde{\boldsymbol{\phi}}_4\sigma_4^2, \end{aligned} \tag{5.4.1}$$

with the scalars $\sigma_1^2, \sigma_2^2, \sigma_3^2$ and σ_4^2 representing the relevant unknown stratum variances, in forms similar to those in (3.5.7).

According to Definition 3.5.1, an experiment in an NB design will be said to have the OBS if the covariance matrix of $\boldsymbol{y}$ receives the representation (5.4.1), where the matrices $\{\tilde{\boldsymbol{\phi}}_\alpha\}$ are symmetric, idempotent and pairwise orthogonal, summing to the identity matrix, as given in (5.3.6). Extending Lemma 3.5.1, the following result can be obtained.

Lemma 5.4.1. *An experiment in an NB design has the OBS property, corresponding to* (5.4.1), *if and only if both* $\mathcal{D}^*$ *and* $\mathcal{D}$ *of the NB design are proper, i.e., the equalities* (5.3.26), $k_j = k$ *for* $j = 1, 2, ..., b$ *and* $n_h = n_0$ *for* $h = 1, 2, ..., a$, *hold.*

Proof. This lemma can be proved similarly as Lemma 3.5.1, by considering the conditions $\tilde{\boldsymbol{\phi}}_\alpha\mathrm{Cov}(\boldsymbol{y})\tilde{\boldsymbol{\phi}}_{\alpha'} = \mathbf{O}$ for any $\alpha \neq \alpha'$, $\alpha, \alpha' = 1, 2, 3, 4$. □

Evidently, if the conditions of Lemma 5.4.1 hold, then the projectors $\{\tilde{\boldsymbol{\phi}}_\alpha\}$ and the corresponding variances $\{\sigma_\alpha^2\}$ can be written as

$$\begin{aligned}\tilde{\boldsymbol{\phi}}_1 &= \boldsymbol{I}_n - k^{-1}\boldsymbol{D}'\boldsymbol{D},\\ \tilde{\boldsymbol{\phi}}_2 &= k^{-1}\boldsymbol{D}'\boldsymbol{D} - n_0^{-1}\boldsymbol{G}'\boldsymbol{G},\\ \tilde{\boldsymbol{\phi}}_3 &= n_0^{-1}\boldsymbol{G}'\boldsymbol{G} - n^{-1}\mathbf{1}_n\mathbf{1}_n',\\ \tilde{\boldsymbol{\phi}}_4 &= n^{-1}\mathbf{1}_n\mathbf{1}_n'\end{aligned}$$

and

$$\begin{aligned}
\sigma_1^2 &= \sigma_U^2 + \sigma_e^2, && (5.4.2)\\
\sigma_2^2 &= k\sigma_B^2 + (1 - K_H^{-1}k)\sigma_U^2 + \sigma_e^2, && (5.4.3)\\
\sigma_3^2 &= n_0\sigma_A^2 + (k - B_H^{-1}n_0)\sigma_B^2 + (1 - K_H^{-1}k)\sigma_U^2 + \sigma_e^2, && (5.4.4)\\
\sigma_4^2 &= (n_0 - N_A^{-1}n)\sigma_A^2 + (k - B_H^{-1}n_0)\sigma_B^2 + (1 - K_H^{-1}k)\sigma_U^2 + \sigma_e^2, && (5.4.5)
\end{aligned}$$

respectively. Hence, the matrices $\tilde{\boldsymbol{C}}_\alpha = \boldsymbol{\Delta}\tilde{\boldsymbol{\phi}}_\alpha\boldsymbol{\Delta}', \alpha = 1,2,3,4$, obtain the forms

$$\begin{aligned}
\tilde{\boldsymbol{C}}_1 &= \boldsymbol{\Delta}\tilde{\boldsymbol{\phi}}_1\boldsymbol{\Delta}' = \boldsymbol{r}^\delta - k^{-1}\boldsymbol{N}\boldsymbol{N}', && (5.4.6)\\
\tilde{\boldsymbol{C}}_2 &= \boldsymbol{\Delta}\tilde{\boldsymbol{\phi}}_2\boldsymbol{\Delta}' = k^{-1}\boldsymbol{N}\boldsymbol{N}' - n_0^{-1}\boldsymbol{\Re}\boldsymbol{\Re}', && (5.4.7)\\
\tilde{\boldsymbol{C}}_3 &= \boldsymbol{\Delta}\tilde{\boldsymbol{\phi}}_3\boldsymbol{\Delta}' = n_0^{-1}\boldsymbol{\Re}\boldsymbol{\Re}' - n^{-1}\boldsymbol{r}\boldsymbol{r}' && (5.4.8)
\end{aligned}$$

and

$$\tilde{\boldsymbol{C}}_4 = \boldsymbol{\Delta}\tilde{\boldsymbol{\phi}}_4\boldsymbol{\Delta}' = n^{-1}\boldsymbol{r}\boldsymbol{r}' \qquad (5.4.9)$$

(compare with Section 3.5). Note that the covariance matrix (5.4.1) coincides with that presented by Morgan (1996, Section 2.3), except that the above stratum variances coincide with those defined by him only in the special case of N_A, B_H and K_H tending to infinity (this is often assumed in the literature, though seldom realistic).

Now the notion of GB defined in Definition 3.6.1 can easily be extended for NB designs, simply by taking into account the above four matrices $\tilde{\boldsymbol{C}}_\alpha, \alpha = 1,2,3,4$, instead of the three considered in Section 3.6. However, to meet the conditions of Definition 3.6.1, it is required that all four matrices have a common set of eigenvectors with respect to $\boldsymbol{r}^\delta$, as can be seen from Lemma 3.6.1. To fulfil this requirement, it is necessary and sufficient that the matrices $\{\boldsymbol{r}^{-\delta/2}\tilde{\boldsymbol{C}}_\alpha\boldsymbol{r}^{-\delta/2}\}$ commute in pairs or, equivalently, that the equalities

$$\tilde{\boldsymbol{C}}_\alpha\boldsymbol{r}^{-\delta}\tilde{\boldsymbol{C}}_{\alpha'} = (\tilde{\boldsymbol{C}}_\alpha\boldsymbol{r}^{-\delta}\tilde{\boldsymbol{C}}_{\alpha'})' \quad \text{for any} \quad \alpha \neq \alpha', \quad \alpha, \alpha' = 1,2,3,4, \qquad (5.4.10)$$

hold. It is straightforward to show the following result.

Lemma 5.4.2. *The matrices* (5.4.6)–(5.4.9) *satisfy the equalities* (5.4.10) *if and only if the equality*

$$\boldsymbol{N}\boldsymbol{N}'\boldsymbol{r}^{-\delta}\boldsymbol{\Re}\boldsymbol{\Re}' = \boldsymbol{\Re}\boldsymbol{\Re}'\boldsymbol{r}^{-\delta}\boldsymbol{N}\boldsymbol{N}' \qquad (5.4.11)$$

holds.

Proof. The reader is advised to prove this lemma as an exercise. □

Remark 5.4.1. For an NB design to satisfy the condition (5.4.11), it is sufficient that the equality $\boldsymbol{D}'\boldsymbol{D}\boldsymbol{\Delta}'\boldsymbol{r}^{-\delta}\boldsymbol{\Delta}\boldsymbol{G}'\boldsymbol{G} = \boldsymbol{G}'\boldsymbol{G}\boldsymbol{\Delta}'\boldsymbol{r}^{-\delta}\boldsymbol{\Delta}\boldsymbol{D}'\boldsymbol{D}$ holds, and this equality holds if and only if

$$\boldsymbol{N}_h'\boldsymbol{r}^{-\delta}\boldsymbol{r}_{h'}\boldsymbol{1}_{b_{h'}}' = \boldsymbol{1}_{b_h}\boldsymbol{r}_h'\boldsymbol{r}^{-\delta}\boldsymbol{N}_{h'} \quad \text{for} \quad h, h' = 1,2,...,a. \qquad (5.4.12)$$

Thus, the results presented in Section 3.6 can be extended to an NB design, provided that the conditions given in Lemma 5.4.1 hold (i.e., $\mathcal{D}^*$ and $\mathcal{D}$ of the considered NB design are proper) and, in addition, the matrices (5.4.6)–(5.4.9) have a common set of eigenvectors with respect to $\boldsymbol{r}^\delta$, for which it is necessary and sufficient that the equality (5.4.11) in Lemma 5.4.2 holds. If the necessary and sufficient conditions in both of these lemmas are satisfied and the eigenvectors are chosen so that they represent basic contrasts, defined in Section 3.4.1 and ordered as in Theorem 3.6.2, then the NB design can be said to be GB in the sense of Definition 3.6.1, properly extended from three to four strata.

Note, however, that when extending the results of Section 3.6 to an NB design having the GB property in the above sense, one has to take into account the fact that for each contrast there will not be two, as in Theorem 3.6.1, but three efficiency factors, related to the estimation of the contrast in three different strata (see Section 5.3 for the definitions of the strata). To see this, let $\{\boldsymbol{s}_i, \boldsymbol{s}_i'\boldsymbol{r} = \boldsymbol{0}, i = 1, 2, ..., v-1\}$ be a common set of $\boldsymbol{r}^\delta$-orthonormal eigenvectors of the matrices $\tilde{\boldsymbol{C}}_1$, $\tilde{\boldsymbol{C}}_2$ and $\tilde{\boldsymbol{C}}_3$ of the design with respect to $\boldsymbol{r}^\delta$. Then, writing

$$\tilde{\boldsymbol{C}}_1\boldsymbol{s}_i = \varepsilon_{1i}\boldsymbol{r}^\delta\boldsymbol{s}_i, \quad \tilde{\boldsymbol{C}}_2\boldsymbol{s}_i = \varepsilon_{2i}\boldsymbol{r}^\delta\boldsymbol{s}_i \quad \text{and} \quad \tilde{\boldsymbol{C}}_3\boldsymbol{s}_i = \varepsilon_{3i}\boldsymbol{r}^\delta\boldsymbol{s}_i \quad \text{for } i = 1, 2, ..., v-1,$$

and using arguments similar to those used in Section 3.5 (see Remark 3.5.3 in particular), one can interpret the eigenvalues ε_{1i}, ε_{2i} and ε_{3i} as the efficiency factors for the relevant basic contrasts $\boldsymbol{c}_i'\boldsymbol{\tau} = \boldsymbol{s}_i'\boldsymbol{r}^\delta\boldsymbol{\tau}$ when it is estimated in the intra-block, the inter-block-intra-superblock and the inter-superblock strata, respectively. Because the matrices $\tilde{\boldsymbol{C}}_2$ and $\tilde{\boldsymbol{C}}_3$ can be seen as resulting from the partition of the matrix $\boldsymbol{C}_2$ (from Chapter 3), of the form $\boldsymbol{C}_2 = \tilde{\boldsymbol{C}}_2 + \tilde{\boldsymbol{C}}_3$, it can easily be checked that the equality $\varepsilon_{1i} + \varepsilon_{2i} + \varepsilon_{3i} = 1$ holds for any i $(= 1, 2, ..., v-1)$. With this in mind, and using the results of Section 5.3, no difficulty should be encountered in extending the results of Sections 3.5 and 3.6 to any NB design with the OBS and GB properties. In particular, Theorem 3.6.1 can be extended as follows.

Theorem 5.4.1. *Let* $\{\boldsymbol{c}_i'\boldsymbol{\tau} = \boldsymbol{s}_i'\boldsymbol{r}^\delta\boldsymbol{\tau}, i = 1, 2, ..., v-1\}$ *be any set of basic contrasts of a NB design having the OBS and GB properties, i.e., satisfying the conditions* (5.3.26) *and* (5.4.11), *and let* $\{\varepsilon_{\alpha i}, i = 1, 2, ..., v-1\}$ *be the corresponding eigenvalues of the matrix* $\tilde{\boldsymbol{C}}_\alpha$ *with respect to* $\boldsymbol{r}^\delta$. *Then, the analysis within stratum* α $(= 1, 2, 3)$ *provides the BLUEs, of the form*

$$(\widehat{\boldsymbol{c}_i'\boldsymbol{\tau}})_\alpha = \varepsilon_{\alpha i}^{-1}\boldsymbol{s}_i'\tilde{\boldsymbol{Q}}_\alpha = \varepsilon_{\alpha i}^{-1}\boldsymbol{c}_i'\boldsymbol{r}^{-\delta}\tilde{\boldsymbol{Q}}_\alpha, \quad \textit{with } \tilde{\boldsymbol{Q}}_\alpha = \boldsymbol{\Delta}\tilde{\boldsymbol{\phi}}_\alpha\boldsymbol{y}, \tag{5.4.13}$$

and their variances

$$\mathrm{Var}[(\widehat{\boldsymbol{c}_i'\boldsymbol{\tau}})_\alpha] = \varepsilon_{\alpha i}^{-1}\sigma_\alpha^2 \tag{5.4.14}$$

and covariances

$$\mathrm{Cov}[(\widehat{\boldsymbol{c}_i'\boldsymbol{\tau}})_\alpha, (\widehat{\boldsymbol{c}_{i'}'\boldsymbol{\tau}})_\alpha] = 0, \quad i \neq i', \tag{5.4.15}$$

for those of the basic contrasts for which the efficiency factors in stratum α $\{\varepsilon_{\alpha i}\}$ *are nonzero (positive). Also, the correlations between* $(\widehat{c_i'\tau})_\alpha$ *and* $(\widehat{c_{i'}'\tau})_{\alpha'}$ *are zero for* $\alpha \neq \alpha'$, *whether* $i = i'$ *or* $i \neq i'$ $(i, i' = 1, 2, ..., v-1; \alpha, \alpha' = 1, 2, 3)$.

Proof. These results can be proved in the same way as those of Theorem 3.6.1, as (5.4.13), (5.4.14) and (5.4.15) correspond to (3.6.1), (3.6.2) and (3.6.3), respectively, being their extensions from two to three strata. □

In a similar way, other results of Sections 3.5 and 3.6 can be extended to cover NB designs with the desirable properties of OBS and GB. In making such extensions, attention should be paid to appropriate spectral decompositions of the matrices $\tilde{C}_\alpha$, $\alpha = 1, 2, 3, 4$. In view of Definition 3.6.1, and its appropriate extension, it is required that these decompositions are based on a common set of eigenvectors of all four matrices with respect to the diagonal matrix r^δ. Note that in case of an NB design, this cannot be accomplished so straightforwardly as for a block design with one stratum of blocks, considered in Chapter 3, i.e., as in the representation (3.6.16).

To be more precise, suppose that $\{s_i, i = 1, 2, ..., v-1\}$ is a set of r^δ-orthonormal eigenvectors of $\tilde{C}_1$, $(\equiv C_1)$ with respect to r^δ (as in Lemma 2.3.3) and, simultaneously, of $\tilde{C}_2$, $\tilde{C}_3$ and $\tilde{C}_4$ with respect to r^δ. Furthermore, let $s_{\beta 1}, s_{\beta 2}, ..., s_{\beta \tilde{\rho}_\beta}$ be a subset of $\{s_i\}$ such that all its members correspond to a common eigenvalue $\varepsilon_{\alpha\beta}$ (not necessarily positive) of the matrix $\tilde{C}_\alpha$ with respect to r^δ for any α, i.e., that

$$\tilde{C}_\alpha s_{\beta j} = \varepsilon_{\alpha\beta} r^\delta s_{\beta j} \quad \text{for } j = 1, 2, ..., \tilde{\rho}_\beta \quad \text{and} \quad \text{for } \alpha = 1, 2, 3, 4.$$

Writing the subset as $\tilde{S}_\beta = [s_{\beta 1} : s_{\beta 2} : \cdots : s_{\beta \tilde{\rho}_\beta}]$, this property can be expressed as

$$\tilde{C}_\alpha \tilde{S}_\beta = \varepsilon_{\alpha\beta} r^\delta \tilde{S}_\beta \quad \text{for } \alpha = 1, 2, 3, 4. \tag{5.4.16}$$

Note, however, that the matrix $\tilde{S}_\beta$ used in (5.4.16) may not coincide with the matrix S_β used in Chapter 3, particularly in Corollary 3.6.3, as there does not exist such a direct relation among the eigenvalues $\varepsilon_{1\beta}$, $\varepsilon_{2\beta}$ and $\varepsilon_{3\beta}$ considered here, as it is between $\varepsilon_{1\beta}$ and $\varepsilon_{2\beta}$ $(= 1 - \varepsilon_{1\beta})$ in Chapter 3, unless $\varepsilon_{3\beta} = 0$.

Thus, let the eigenvectors of the set $\{s_i, i = 1, 2, ..., v-1\}$ be grouped into subsets to form the matrices

$$\tilde{S}_0, \tilde{S}_1, ..., \tilde{S}_{\tilde{m}} \quad \text{and} \quad \tilde{S}_{\tilde{m}+1} = s_v = n^{-1/2} \mathbf{1}_v \tag{5.4.17}$$

satisfying the conditions (5.4.16), with the corresponding eigenvalues $\{\varepsilon_{\alpha\beta}\}$ ordered so that $\varepsilon_{10} = 1$, implying $\varepsilon_{20} = \varepsilon_{30} = 0$, and so that $\varepsilon_{1\tilde{m}} = 0$, implying $\varepsilon_{2\tilde{m}} + \varepsilon_{3\tilde{m}} = 1$. Then, the matrices $\tilde{C}_\alpha$, $\alpha = 1, 2, 3, 4$, can be presented in the following spectral decomposition forms:

$$\tilde{C}_1 = \boldsymbol{\Delta}\tilde{\phi}_1\boldsymbol{\Delta}' = \boldsymbol{r}^{\delta}\sum_{\beta=0}^{\tilde{m}-1}\varepsilon_{1\beta}\tilde{\boldsymbol{H}}_{\beta}\boldsymbol{r}^{\delta}, \tag{5.4.18}$$

$$\tilde{C}_2 = \boldsymbol{\Delta}\tilde{\phi}_2\boldsymbol{\Delta}' = \boldsymbol{r}^{\delta}\sum_{\beta=1}^{\tilde{m}}\varepsilon_{2\beta}\tilde{\boldsymbol{H}}_{\beta}\boldsymbol{r}^{\delta}, \tag{5.4.19}$$

$$\tilde{C}_3 = \boldsymbol{\Delta}\tilde{\phi}_3\boldsymbol{\Delta}' = \boldsymbol{r}^{\delta}\sum_{\beta=1}^{\tilde{m}}\varepsilon_{3\beta}\tilde{\boldsymbol{H}}_{\beta}\boldsymbol{r}^{\delta} \tag{5.4.20}$$

and

$$\tilde{C}_4 = \boldsymbol{\Delta}\tilde{\phi}_4\boldsymbol{\Delta}' = \boldsymbol{r}^{\delta}\tilde{\boldsymbol{H}}_{\tilde{m}+1}\boldsymbol{r}^{\delta} = \boldsymbol{r}^{\delta}\boldsymbol{s}_v\boldsymbol{s}_v'\boldsymbol{r}^{\delta}, \tag{5.4.21}$$

where $\tilde{\boldsymbol{H}}_{\beta} = \tilde{\boldsymbol{S}}_{\beta}\tilde{\boldsymbol{S}}_{\beta}'$ for $\beta = 0, 1, ..., \tilde{m}, \tilde{m}+1$ ($\tilde{\boldsymbol{H}}_{\tilde{m}+1} = \boldsymbol{s}_v\boldsymbol{s}_v' = n^{-1}\mathbf{1}_v\mathbf{1}_v'$), and, evidently,

$$\tilde{\boldsymbol{S}}_{\beta}'\boldsymbol{r}^{\delta}\tilde{\boldsymbol{S}}_{\beta} = \boldsymbol{I}_{\tilde{\rho}_{\beta}} \quad \text{for any } \beta \text{ and } \tilde{\boldsymbol{S}}_{\beta}'\boldsymbol{r}^{\delta}\tilde{\boldsymbol{S}}_{\beta'} = \mathbf{O} \quad \text{for } \beta \neq \beta'. \tag{5.4.22}$$

With this notation, Definition 3.6.1 becomes also applicable to any NB design inducing the OBS property, a relevant extension of the definition being obvious. In connection with this, Lemma 3.6.1 can also be extended easily, as follows.

Lemma 5.4.3. *An NB design having the OBS property is GB with respect to the decomposition*

$$\mathcal{C}(\boldsymbol{\Delta}') = \oplus_{\beta}\mathcal{C}(\boldsymbol{\Delta}'\tilde{\boldsymbol{S}}_{\beta})$$

if and only if the matrices $\{\tilde{\boldsymbol{S}}_{\beta}\}$ *satisfy the conditions* (5.4.16) *and* (5.4.22) *for all* α $(= 1, 2, 3, 4)$ *and* β $(= 0, 1, ..., \tilde{m}, \tilde{m}+1)$.

Proof. This result can be proved similarly as Lemma 3.6.1, taking into account the notation introduced above. □

Similarly, Remark 3.6.4 can also be extended, remembering however that now not only the conditions (5.3.26) but also the condition (5.4.11) is to be satisfied.

The discussion above shows that any further extension of the theory of the GB property, aimed at covering block designs with even more than two strata of blocks, could also be achieved straightforwardly.

Considering now the application of the notion of GB to various NB designs, it can easily be shown that any NBIB design is GB. Also, some known generalizations of NBIB designs are GB. In particular, any NB design constructed so that its $\mathcal{D}^*$ and $\mathcal{D}$ are proper block designs and that, additionally, the condition (5.4.12) in Remark 5.4.1 is satisfied, has the GB property.

It may also be noted that the condition (5.4.11) is satisfied automatically if $\mathcal{D}^*$ and $\mathcal{D}$ are proper block designs and, in addition, $\mathcal{D}$ is connected and orthogonal, i.e., $\mathfrak{R} = n^{-1}\boldsymbol{r}\boldsymbol{n}'$ (see Corollary 2.3.4). Therefore, it is interesting to pay special attention to NB designs of this type. The class of NB designs so defined contains all resolvable BIB designs, in the sense of Bose (1942), and the most common generalizations of them, the α-resolvable block designs of Shrikhande and Raghavarao (1964) in particular. Further examples of designs that belong to this class are the square lattice design, as noticed by Nelder (1968, Section 4.2), and the rectangular lattice design, as shown by Bailey and Speed (1986, Section 5). Thus, this class seems to be large, although some more general resolvable block designs, such as those considered by Kageyama (1976a) do not belong to this class, unless their $\mathcal{D}^*$ and $\mathcal{D}$ are proper. For example, even a resolvable BIB design generalized in the sense of Kageyama (1976a) is not GB, and so does not belong to this class, if its $\mathcal{D}^*$ and $\mathcal{D}$ are not proper.

Another feature worth noting is that if an NB design belongs to the above class, i.e., is such that its $\mathcal{D}^*$ is proper and $\mathcal{D}$ is connected proper and orthogonal, then it is GB (in the sense of Definition 3.6.1 extended to four strata) with $\tilde{C}_1$ as in (5.4.6),

$$\tilde{C}_2 = k^{-1}NN' - n^{-1}rr', \quad \tilde{C}_3 = \mathbf{O}$$

and $\tilde{C}_4$ as in (5.4.9). This simplifies the analysis of experimental data considerably. In fact, because in this case the matrices $\tilde{C}_\alpha, \alpha = 1, 2$ coincide with those in (3.5.15) and (3.5.16), respectively, the theory of the analysis presented in Sections 3.5 and 3.6 can be adopted directly for designs belonging to the class just described.

Example 5.4.1. Consider again the design discussed in Example 5.3.3. It is an NBIB design and, as such, is based on designs $\mathcal{D}^*$ and $\mathcal{D}$ that are both proper. Thus, it has the OBS property. Furthermore, because its matrices $\{\tilde{C}_\alpha\}$ are

$$\tilde{C}_1 = \frac{5}{2}\left(I_5 - \frac{1}{5}\mathbf{1}_5\mathbf{1}_5'\right), \quad \tilde{C}_2 = \frac{1}{2}\tilde{C}_1, \quad \tilde{C}_3 = \frac{1}{10}\tilde{C}_1 \quad \text{and} \quad \tilde{C}_4 = \frac{4}{5}\mathbf{1}_5\mathbf{1}_5',$$

it can easily be shown that the condition (5.4.10) holds for all pairs, which means that the design is GB. As already noted in Example 5.3.3, the columns of $\tilde{C}_1$ ($\equiv C_1$) span the subspace of all contrasts, which implies that the columns of $\tilde{C}_2$ and $\tilde{C}_3$ also span such subspace. Now, referring to Theorem 5.3.2, it can be seen that for any contrast $c'\tau = s'r^\delta\tau$, the BLUE is obtainable in the analysis within each of the first three strata ($\alpha = 1, 2, 3$), the relevant within stratum efficiency factors (see Remark 3.5.3) being as follows:

$\varepsilon_1 = 5/8$	in the intra-block analysis (1st stratum),
$\varepsilon_2 = 5/16$	in the inter-block-intra-superblock analysis (2nd stratum),
$\varepsilon_3 = 5/80$	in the inter-superblock analysis (3rd stratum).

Hence, the corresponding variances of the BLUEs can be written according to (3.5.24) in the form

$$\mathrm{Var}[(\widehat{c'\tau})_\alpha] = \varepsilon_\alpha^{-1}4s's\sigma_\alpha^2 = \varepsilon_\alpha^{-1}4^{-1}c'c\sigma_\alpha^2 \quad \text{for} \quad \alpha = 1, 2, 3,$$

where the stratum variances σ_1^2, σ_2^2 and σ_3^2 are as given in (5.4.2), (5.4.3) and (5.4.4), respectively. If basic contrasts $\{\boldsymbol{c}_i'\boldsymbol{\tau} = \boldsymbol{s}_i'\boldsymbol{r}^\delta\boldsymbol{\tau}\}$ are considered, as in Theorem 5.4.1, then the above formula of variance simplifies to that in (5.4.14), i.e., to

$$\text{Var}[(\widehat{\boldsymbol{c}_i'\boldsymbol{\tau}})_\alpha] = \varepsilon_\alpha^{-1}\sigma_\alpha^2 \quad \text{for} \quad \alpha = 1, 2, 3,$$

equally applicable for all basic contrasts.

Also note that because here all basic contrasts receive a common efficiency factor within each stratum, it is permissible to say, by referring to Definition 3.6.1 and Lemma 5.4.3, that the considered design is GB with respect to the decomposition

$$\mathcal{C}(\boldsymbol{\Delta}') = \mathcal{C}(\boldsymbol{\Delta}'\tilde{\boldsymbol{S}}_1) \oplus \mathcal{C}(\boldsymbol{\Delta}'\boldsymbol{s}_v),$$

where $\tilde{\boldsymbol{S}}_1 = [\boldsymbol{s}_1 : \boldsymbol{s}_2 : \cdots : \boldsymbol{s}_{v-1}]$ represents a complete set of basic contrasts of the design and $\boldsymbol{s}_v = n^{-1/2}\mathbf{1}_v$. Because the columns of the matrix $\tilde{\boldsymbol{S}}_1$ span the subspace of all contrasts (independently of the choice of basic contrasts), one can say that the design is GB in the same way for all contrasts of treatment parameters. This feature is characteristic for all NBIB designs.

Example 5.4.2. Consider again the design presented and discussed in Example 5.3.4. It is a resolvable BIB design based on designs $\mathcal{D}^*$ and $\mathcal{D}$ that are both proper. This condition ensures that the design has the OBS property. It is also GB, as can be seen from the structures of the matrices

$$\tilde{\boldsymbol{C}}_1 = 3\left(\boldsymbol{I}_6 - \frac{1}{6}\mathbf{1}_6\mathbf{1}_6'\right), \quad \tilde{\boldsymbol{C}}_2 = \frac{2}{3}\tilde{\boldsymbol{C}}_1, \quad \tilde{\boldsymbol{C}}_3 = \mathbf{O} \quad \text{and} \quad \tilde{\boldsymbol{C}}_4 = \frac{5}{6}\mathbf{1}_6\mathbf{1}_6',$$

which evidently satisfy the conditions (5.4.10). Because here $\tilde{\boldsymbol{C}}_3 = \mathbf{O}$, for no contrast the BLUE is obtainable in the analysis within the third stratum, i.e., in the inter-superblock analysis. On the other hand, because of the forms of the matrices $\tilde{\boldsymbol{C}}_1$ and $\tilde{\boldsymbol{C}}_2$, and hence, their column spaces, for any contrast $\boldsymbol{c}'\boldsymbol{\tau} = \boldsymbol{s}'\boldsymbol{r}^\delta\boldsymbol{\tau}$, the BLUE is obtainable in the analysis within the first and the second stratum (see Lemma 3.5.2), the relevant within stratum efficiency factors being as follows:

$\varepsilon_1 = 3/5$ in the intra-block analysis (1st stratum),

$\varepsilon_2 = 2/5$ in the inter-block-intra-superblock analysis (2nd stratum).

The corresponding variances of the BLUEs (obtainable according to Theorem 3.5.2) are then

$$\text{Var}[(\widehat{\boldsymbol{c}'\boldsymbol{\tau}})_\alpha] = \varepsilon_\alpha^{-1}5\boldsymbol{s}'\boldsymbol{s}\sigma_\alpha^2 = \varepsilon_\alpha^{-1}5^{-1}\boldsymbol{c}'\boldsymbol{c}\sigma_\alpha^2 \quad \text{for} \quad \alpha = 1, 2,$$

with the stratum variances σ_1^2 and σ_2^2 as given in (5.4.2) and (5.4.3), respectively. This variance formula reduces to

$$\text{Var}[(\widehat{\boldsymbol{c}'\boldsymbol{\tau}})_\alpha] = \varepsilon_\alpha^{-1}\sigma_\alpha^2 \quad \text{for} \quad \alpha = 1, 2,$$

if basic contrasts $\{\boldsymbol{c}_i'\boldsymbol{\tau}\}$ are considered (see Theorem 5.4.1).

Here, as in Example 5.4.1, the efficiency factors ε_1 and ε_2 are the same for all contrasts, and hence, the considered design can be seen as being GB with respect to the decomposition

$$\mathcal{C}(\boldsymbol{\Delta}') = \mathcal{C}(\boldsymbol{\Delta}'\tilde{\boldsymbol{S}}_1) \oplus \mathcal{C}(\boldsymbol{\Delta}'\boldsymbol{s}_v),$$

where $\tilde{\boldsymbol{S}}_1 = [\boldsymbol{s}_1 : \boldsymbol{s}_2 : \cdots : \boldsymbol{s}_{v-1}]$ represents a complete set of basic contrasts of the design and $\boldsymbol{s}_v = n^{-1/2}\mathbf{1}_v$. Thus, the GB property of the design applies in the same way to all contrasts, which follows from the BIB property of the design $\mathcal{D}^*$ (see Theorem 2.4.1) and the fact that $\mathcal{D}$ is a connected proper and orthogonal block design.

Example 5.4.3. Suppose that a $v_A \times v_B$ factorial experiment is arranged in an NB design in such a way that its $\mathcal{D}^*$ and $\mathcal{D}$ are both proper equireplicate block designs of the following structures. The incidence matrix of $\mathcal{D}$ is of the form

$$\mathfrak{R} = \boldsymbol{N}_A \otimes \boldsymbol{N}_B, \tag{5.4.23}$$

where $\boldsymbol{N}_A$ is an incidence matrix of a binary, equireplicate proper block design for v_A treatments, here levels of factor A, and where $\boldsymbol{N}_B$ is an incidence matrix of a binary equireplicate proper block design for v_B treatments, levels of factor B. The incidence matrix of $\mathcal{D}^*$, $\boldsymbol{N} = [\boldsymbol{N}_1 : \boldsymbol{N}_2 : \cdots : \boldsymbol{N}_a]$, is related to the matrix $\mathfrak{R}$ according to (5.1.1), as required for an NB design, and in addition, it is structured so that

$$\boldsymbol{N}\boldsymbol{N}' = \kappa \boldsymbol{I}_{v_A} \otimes \boldsymbol{N}_B\boldsymbol{N}_B',$$

where κ is a constant. It can easily be shown that this NB design satisfies the condition (5.4.11) in Lemma 5.4.2, i.e., that the design is GB.

It may be noted that the NB design considered in Example 5.4.3 can be regarded as a split-plot type design. In the terminology of this type of designs, the superblocks are called blocks, the blocks are called main plots (or whole-plots) and the plots are called subplots (or split-plots), as already mentioned in Section 5.1 (see also Hinkelmann and Kempthorne, 1994, Chapter 3). In fact, the idea of Example 5.4.3 has been taken from Brzeskwiniewicz (1994), who considers such split-plot designs with incidence matrices of BIB or PBIB designs used for $\boldsymbol{N}_A$ and $\boldsymbol{N}_B$ in (5.4.23).

5.5 Recovery of information from higher strata in a nested block design

The main estimation results obtained for NB designs in Section 5.2.2 show that unless the estimated function $\boldsymbol{c}'\boldsymbol{\tau}$ satisfies certain quite restrictive conditions, there does not exist, in general, the BLUE of it under the randomization model (5.2.7). However (see Section 5.3), in many, though not in all, cases the estimation of a contrast of treatment parameters, $\boldsymbol{c}'\boldsymbol{\tau}$, can be based on information available in two or three of the experimental strata. Unfortunately, each of them provides a separate estimate of the contrast, often of different value. Therefore, a natural question originating in this context is whether and how it is possible to use the information from various strata to obtain a single estimate in a somehow optimal way. In Section 3.7, this has been considered for a general, but ordinary, block design, where the combination of information concerns two strata only. In the present section, an attempt will be made (following Caliński, 1997) to extend the results given there, so that they become applicable to any NB design for which the general randomization model (5.2.7) is appropriate.

5.5.1 BLUEs under known variance components

First, let the problem be considered under an unrealistic assumption that the variance components appearing in the dispersion matrix (5.2.9) are known, or at least their ratios γ_1 and γ_2 defined in (5.2.11) and (5.2.12) are known. Then, the following results are essential (extending those given in Section 3.7.1).

Lemma 5.5.1. *Let the model be as in* (5.2.7), *with the expectation vector* (5.2.8) *and the covariance matrix* (5.2.9), *the latter written equivalently as in* (5.2.10). *Further, suppose that the true values of* γ_1 *and* γ_2 *are known. Then:*

(a) *any function* $\boldsymbol{w}'\boldsymbol{y}$ *that is the BLUE of its expectation* $\boldsymbol{w}'\boldsymbol{\Delta}'\boldsymbol{\tau}$,

(b) *a vector that is the BLUE of* $\mathrm{E}(\boldsymbol{y}) = \boldsymbol{\Delta}'\boldsymbol{\tau}$,

(c) *a vector that gives the residuals,*

all remain unchanged when altering the present model by deleting the term $N_A^{-1}(\sigma_A^2/\sigma_1^2)\mathbf{1}_a\mathbf{1}_a'$ *in* (5.2.11), *i.e., by reducing* (5.2.10) *to*

$$\mathrm{Cov}(\boldsymbol{y}) = \sigma_1^2(\gamma_1\boldsymbol{G}'\boldsymbol{G} + \gamma_2\boldsymbol{D}'\boldsymbol{D} + \boldsymbol{I}_n) = \sigma_1^2\boldsymbol{T}, \tag{5.5.1}$$

where $\boldsymbol{T} = \gamma_1\boldsymbol{G}'\boldsymbol{G} + \gamma_2\boldsymbol{D}'\boldsymbol{D} + \boldsymbol{I}_n$. *The matrix* $\boldsymbol{T}$ *is p.d. if* $\gamma_1 \geq 0$ *and* $\gamma_2 > -1/k_{\max}$, *where* $k_{\max} = \max_{j(h)} k_{j(h)}$.

Proof. Note that, with $\boldsymbol{P}_{\Delta'} = \boldsymbol{\Delta}'\boldsymbol{r}^{-\delta}\boldsymbol{\Delta}$, the equality

$$(\boldsymbol{G}'\boldsymbol{\Gamma}_1\boldsymbol{G} + \boldsymbol{D}'\boldsymbol{\Gamma}_2\boldsymbol{D} + \boldsymbol{I}_n)(\boldsymbol{I}_n - \boldsymbol{P}_{\Delta'}) = (\gamma_1\boldsymbol{G}'\boldsymbol{G} + \gamma_2\boldsymbol{D}'\boldsymbol{D} + \boldsymbol{I}_n)(\boldsymbol{I}_n - \boldsymbol{P}_{\Delta'})$$

holds. This implies that (a) the relevant condition for $\boldsymbol{w}$, given in Theorem 1.2.1, is satisfied under the original model if and only if it is satisfied under the alternative model with the covariance matrix (5.5.1), and that (b) the BLUE of $\boldsymbol{\Delta}'\boldsymbol{\tau}$, as given by Rao (1974, Theorem 3.2), remains unchanged when (5.2.10) is replaced by (5.5.1), so that also (c) the residual vector is unchanged then. The matrix $\boldsymbol{T}$ in (5.5.1) is p.d. if $\gamma_1 \geq 0$ and $\gamma_2 > -\boldsymbol{x}'\boldsymbol{x}/\boldsymbol{x}'\boldsymbol{D}'\boldsymbol{D}\boldsymbol{x}$, for any vector $\boldsymbol{x}$, and the latter holds if $\gamma_2 > -1/\kappa_{\max}$, where $\kappa_{\max}$ is the maximum eigenvalue of $\boldsymbol{D}'\boldsymbol{D}$ (and of $\boldsymbol{D}\boldsymbol{D}'$), this being $k_{\max}$. (These assumptions on γ_1 and γ_2 are reasonable, as $\gamma_1 \geq 0$ if and only if $\sigma_A^2 \geq B_H^{-1}\sigma_B^2$, and for $\gamma_2 > -1/k_{\max}$, it is sufficient that $K_H \geq k_{\max}$ and $K_H\sigma_B^2 + \sigma_e^2 > 0$; of course, $\gamma_2 \geq 0$ if and only if $\sigma_B^2 \geq K_H^{-1}\sigma_U^2$.) □

Theorem 5.5.1. *Under the model and assumptions as those adopted in Lemma 5.5.1, including the assumption that $\gamma_1 \geq 0$ and $\gamma_2 > -1/k_{\max}$:*

(a) *the BLUE of $\boldsymbol{\tau}$ is of the form*

$$\hat{\boldsymbol{\tau}} = (\boldsymbol{\Delta}\boldsymbol{T}^{-1}\boldsymbol{\Delta}')^{-1}\boldsymbol{\Delta}\boldsymbol{T}^{-1}\boldsymbol{y} = \boldsymbol{C}_c^{-1}\boldsymbol{Q}_c, \tag{5.5.2}$$

where $\boldsymbol{C}_c = \boldsymbol{\Delta}\boldsymbol{T}^{-1}\boldsymbol{\Delta}', \boldsymbol{Q}_c = \boldsymbol{\Delta}\boldsymbol{T}^{-1}\boldsymbol{y}$, and $\boldsymbol{T}^{-1}$ can be taken as

$$\begin{aligned} \boldsymbol{T}^{-1} &= \tilde{\boldsymbol{\phi}}_1 \\ &\quad + \boldsymbol{D}'\boldsymbol{k}^{-\delta}(\boldsymbol{k}^{-\delta} + \gamma_2\boldsymbol{I}_b + \gamma_1\text{diag}[\mathbf{1}_{b_1}\mathbf{1}'_{b_1} : \mathbf{1}_{b_2}\mathbf{1}'_{b_2} : \cdots : \mathbf{1}_{b_a}\mathbf{1}'_{b_a}])^{-1}\boldsymbol{k}^{-\delta}\boldsymbol{D}, \end{aligned} \tag{5.5.3}$$

with $\tilde{\boldsymbol{\phi}}_1 = \boldsymbol{I}_n - \boldsymbol{D}'\boldsymbol{k}^{-\delta}\boldsymbol{D}$;

(b) *the covariance matrix of $\hat{\boldsymbol{\tau}}$ is*

$$\text{Cov}(\hat{\boldsymbol{\tau}}) = \sigma_1^2(\boldsymbol{\Delta}\boldsymbol{T}^{-1}\boldsymbol{\Delta}')^{-1} - N_A^{-1}\sigma_A^2\mathbf{1}_v\mathbf{1}'_v; \tag{5.5.4}$$

(c) *the BLUE of $\boldsymbol{c}'\boldsymbol{\tau}$ for any $\boldsymbol{c}$ is $\boldsymbol{c}'\hat{\boldsymbol{\tau}}$, with the variance $\boldsymbol{c}'\text{Cov}(\hat{\boldsymbol{\tau}})\boldsymbol{c}$, which reduces to*

$$\text{Var}(\widehat{\boldsymbol{c}'\boldsymbol{\tau}}) = \sigma_1^2\boldsymbol{c}'(\boldsymbol{\Delta}\boldsymbol{T}^{-1}\boldsymbol{\Delta}')^{-1}\boldsymbol{c}, \tag{5.5.5}$$

if $\boldsymbol{c}'\boldsymbol{\tau}$ is a contrast;

(d) *the MINQUE of σ_1^2 is*

$$\hat{\sigma_1^2} = (n-v)^{-1}\|\boldsymbol{y} - \boldsymbol{\Delta}'\hat{\boldsymbol{\tau}}\|^2_{\boldsymbol{T}^{-1}} = (n-v)^{-1}(\boldsymbol{y} - \boldsymbol{\Delta}'\hat{\boldsymbol{\tau}})'\boldsymbol{T}^{-1}(\boldsymbol{y} - \boldsymbol{\Delta}'\hat{\boldsymbol{\tau}}). \tag{5.5.6}$$

Proof. From Theorem 3.2(c) of Rao (1974), and Lemma 5.5.1 above, the BLUE of $\boldsymbol{\Delta}'\boldsymbol{\tau}$ is $\widehat{\boldsymbol{\Delta}'\boldsymbol{\tau}} = \boldsymbol{P}_{\Delta'(T^{-1})}\boldsymbol{y}$, where $\boldsymbol{P}_{\Delta'(T^{-1})} = \boldsymbol{\Delta}'(\boldsymbol{\Delta}\boldsymbol{T}^{-1}\boldsymbol{\Delta}')^{-1}\boldsymbol{\Delta}\boldsymbol{T}^{-1}$. Hence, the result (5.5.2). This then implies (5.5.4), because (5.2.10) can be written as $\text{Cov}(\boldsymbol{y}) = \sigma_1^2\boldsymbol{T} - N_A^{-1}\sigma_A^2\mathbf{1}_n\mathbf{1}'_n$. Formula (5.5.5) follows from (5.5.4) directly, and formula (5.5.3) can easily be checked. Formula (5.5.6) follows from Theorem 3.4(c) of Rao (1974) by noting that the residual sum of squares is of the form

$$\|(\boldsymbol{I}_n - \boldsymbol{P}_{\Delta'(T^{-1})})\boldsymbol{y}\|^2_{\boldsymbol{T}^{-1}} = \boldsymbol{y}'(\boldsymbol{I}_n - \boldsymbol{P}_{\Delta'(T^{-1})})'\boldsymbol{T}^{-1}(\boldsymbol{I}_n - \boldsymbol{P}_{\Delta'(T^{-1})})\boldsymbol{y}, \tag{5.5.7}$$

equivalent to (3.13) of Rao (1974), which provides the MINQUE of $d\sigma_1^2$, where $d = \text{rank}(\boldsymbol{T} : \boldsymbol{\Delta}') - \text{rank}(\boldsymbol{\Delta}') = n - v$. □

Note that $\boldsymbol{\phi}_1$ in (5.5.3) is responsible for the intra-block information, whereas the term following it takes care of the inter-block information, with γ_2 responsible for that within the superblocks and with γ_1 for that between the superblocks, the maximum recovery of the former information being achieved at $\gamma_2 \leq 0$ and that of the latter at $\gamma_1 = 0$.

Also, note that if the equalities (5.3.26) hold, i.e., the NB design induces the OBS property (5.4.1), then (5.5.3) becomes equal to

$$\boldsymbol{T}^{-1} = \tilde{\boldsymbol{\phi}}_1 + (\sigma_1^2/\sigma_2^2)\tilde{\boldsymbol{\phi}}_2 + (\sigma_1^2/\sigma_3^2)(\tilde{\boldsymbol{\phi}}_3 + \tilde{\boldsymbol{\phi}}_4),$$

giving

$$C_c = \Delta[\tilde{\phi}_1 + (\sigma_1^2/\sigma_2^2)\tilde{\phi}_2 + (\sigma_1^2/\sigma_3^2)(\tilde{\phi}_3 + \tilde{\phi}_4)]\Delta'$$

and

$$Q_c = \Delta[\tilde{\phi}_1 + (\sigma_1^2/\sigma_2^2)\tilde{\phi}_2 + (\sigma_1^2/\sigma_3^2)(\tilde{\phi}_3 + \tilde{\phi}_4)]y,$$

which with

$$\delta_2 = \sigma_2^2/\sigma_1^2 \quad \text{and} \quad \delta_3 = \sigma_3^2/\sigma_1^2 \tag{5.5.8}$$

can also be written as

$$C_c = \Delta\tilde{\phi}_1\Delta' + \delta_2^{-1}\Delta\tilde{\phi}_2\Delta' + \delta_3^{-1}(\Delta\tilde{\phi}_3\Delta' + \Delta\tilde{\phi}_4\Delta') \tag{5.5.9}$$

and

$$Q_c = \Delta\tilde{\phi}_1 y + \delta_2^{-1}\Delta\tilde{\phi}_2 y + \delta_3^{-1}(\Delta\tilde{\phi}_3 y + \Delta\tilde{\phi}_4 y). \tag{5.5.10}$$

Now, if the design is GB, i.e., in addition to (5.3.26), the equality (5.4.11) holds, then, by the use of the decomposition formulae (5.4.18)–(5.4.21), it becomes possible to write the matrix (5.5.9) as

$$\begin{aligned} C_c &= r^\delta \sum_{\beta=0}^{\tilde{m}-1} \varepsilon_{1\beta}\tilde{H}_\beta r^\delta + \delta_2^{-1} r^\delta \sum_{\beta=1}^{\tilde{m}} \varepsilon_{2\beta}\tilde{H}_\beta r^\delta \\ &\quad + \delta_3^{-1} r^\delta \sum_{\beta=1}^{\tilde{m}} \varepsilon_{3\beta}\tilde{H}_\beta r^\delta + \delta_3^{-1} r^\delta \tilde{H}_{\tilde{m}+1} r^\delta \\ &= r^\delta\{\tilde{H}_0 + \sum_{\beta=1}^{\tilde{m}} (\varepsilon_{1\beta} + \delta_2^{-1}\varepsilon_{2\beta} + \delta_3^{-1}\varepsilon_{3\beta})\tilde{H}_\beta + \delta_3^{-1}\tilde{H}_{\tilde{m}+1}\} r^\delta \end{aligned}$$

and, hence, its inverse as

$$C_c^{-1} = \tilde{H}_0 + \sum_{\beta=1}^{\tilde{m}} (\varepsilon_{1\beta} + \delta_2^{-1}\varepsilon_{2\beta} + \delta_3^{-1}\varepsilon_{3\beta})^{-1}\tilde{H}_\beta + \delta_3\tilde{H}_{\tilde{m}+1}.$$

This result allows the BLUE (5.5.2) to be written as

$$\begin{aligned} \hat{\tau} &= C_c^{-1}Q_c \\ &= \tilde{H}_0 Q_c + \sum_{\beta=1}^{\tilde{m}} (\varepsilon_{1\beta} + \delta_2^{-1}\varepsilon_{2\beta} + \delta_3^{-1}\varepsilon_{3\beta})^{-1}\tilde{H}_\beta Q_c + \delta_3\tilde{H}_{\tilde{m}+1}Q_c, \end{aligned}$$

where, from (5.5.10),

$$Q_c = \tilde{Q}_1 + \delta_2^{-1}\tilde{Q}_2 + \delta_3^{-1}\tilde{Q}_3 + \delta_3^{-1}\tilde{Q}_4,$$

with $\tilde{\boldsymbol{Q}}_\alpha = \boldsymbol{\Delta}\tilde{\boldsymbol{\phi}}_\alpha \boldsymbol{y}$, $\alpha = 1,2,3,4$, as in (5.4.13). However, it can easily be shown that $\tilde{\boldsymbol{H}}_0\tilde{\boldsymbol{Q}}_\alpha = \boldsymbol{0}$ for $\alpha = 2,3,4$, $\tilde{\boldsymbol{H}}_{\tilde{m}+1}\tilde{\boldsymbol{Q}}_\alpha = \boldsymbol{0}$ for $\alpha = 1,2,3$ and $\tilde{\boldsymbol{H}}_{\tilde{m}+1}\tilde{\boldsymbol{Q}}_4 = \boldsymbol{s}_v\boldsymbol{s}_v'\tilde{\boldsymbol{Q}}_4 = n^{-1}\boldsymbol{1}_v\boldsymbol{1}_n'\boldsymbol{y}$.

Thus, for an NB design that is GB, the BLUE of $\boldsymbol{\tau}$ can finally be written in the form

$$\begin{aligned}\hat{\boldsymbol{\tau}} = \boldsymbol{C}_c^{-1}\boldsymbol{Q}_c &= \sum_{\beta=0}^{\tilde{m}-1}(\varepsilon_{1\beta} + \delta_2^{-1}\varepsilon_{2\beta} + \delta_3^{-1}\varepsilon_{3\beta})^{-1}\tilde{\boldsymbol{H}}_\beta\tilde{\boldsymbol{Q}}_1 \\ &+ \sum_{\beta=1}^{\tilde{m}}(\delta_2\varepsilon_{1\beta} + \varepsilon_{2\beta} + \delta_3^{-1}\delta_2\varepsilon_{3\beta})^{-1}\tilde{\boldsymbol{H}}_\beta\tilde{\boldsymbol{Q}}_2 \\ &+ \sum_{\beta=1}^{\tilde{m}}(\delta_3\varepsilon_{1\beta} + \delta_2^{-1}\delta_3\varepsilon_{2\beta} + \varepsilon_{3\beta})^{-1}\tilde{\boldsymbol{H}}_\beta\tilde{\boldsymbol{Q}}_3 \\ &+ \boldsymbol{s}_v\boldsymbol{s}_v'\tilde{\boldsymbol{Q}}_4, \end{aligned} \tag{5.5.11}$$

and the BLUE of any function $\boldsymbol{c}'\boldsymbol{\tau} = \boldsymbol{s}'\boldsymbol{r}^\delta\boldsymbol{\tau}$ as

$$\begin{aligned}\widehat{\boldsymbol{c}'\boldsymbol{\tau}} = \boldsymbol{c}'\hat{\boldsymbol{\tau}} &= \sum_{\beta=0}^{\tilde{m}-1}(\varepsilon_{1\beta} + \delta_2^{-1}\varepsilon_{2\beta} + \delta_3^{-1}\varepsilon_{3\beta})^{-1}\boldsymbol{s}'\boldsymbol{r}^\delta\tilde{\boldsymbol{H}}_\beta\tilde{\boldsymbol{Q}}_1 \\ &+ \sum_{\beta=1}^{\tilde{m}}(\delta_2\varepsilon_{1\beta} + \varepsilon_{2\beta} + \delta_3^{-1}\delta_2\varepsilon_{3\beta})^{-1}\boldsymbol{s}'\boldsymbol{r}^\delta\tilde{\boldsymbol{H}}_\beta\tilde{\boldsymbol{Q}}_2 \\ &+ \sum_{\beta=1}^{\tilde{m}}(\delta_3\varepsilon_{1\beta} + \delta_2^{-1}\delta_3\varepsilon_{2\beta} + \varepsilon_{3\beta})^{-1}\boldsymbol{s}'\boldsymbol{r}^\delta\tilde{\boldsymbol{H}}_\beta\tilde{\boldsymbol{Q}}_3 \\ &+ n^{-1}\boldsymbol{s}'\boldsymbol{r}\boldsymbol{1}_n'\boldsymbol{y}, \end{aligned} \tag{5.5.12}$$

remembering that, for $\beta = 0$, $\varepsilon_{10} = 1$ and $\varepsilon_{20} = \varepsilon_{30} = 0$. In particular, for any basic contrast $\boldsymbol{c}_{\beta j}'\boldsymbol{\tau} = \boldsymbol{s}_{\beta j}'\boldsymbol{r}^\delta\boldsymbol{\tau}$, (5.5.12) reduces to

$$\begin{aligned}\widehat{\boldsymbol{c}_{\beta j}'\boldsymbol{\tau}} &= (\varepsilon_{1\beta} + \delta_2^{-1}\varepsilon_{2\beta} + \delta_3^{-1}\varepsilon_{3\beta})^{-1}\boldsymbol{s}_{\beta j}'\tilde{\boldsymbol{Q}}_1 \\ &+ (\delta_2\varepsilon_{1\beta} + \varepsilon_{2\beta} + \delta_3^{-1}\delta_2\varepsilon_{3\beta})^{-1}\boldsymbol{s}_{\beta j}'\tilde{\boldsymbol{Q}}_2 \\ &+ (\delta_3\varepsilon_{1\beta} + \delta_2^{-1}\delta_3\varepsilon_{2\beta} + \varepsilon_{3\beta})^{-1}\boldsymbol{s}_{\beta j}'\tilde{\boldsymbol{Q}}_3, \end{aligned}$$

which can also be written as

$$\widehat{\boldsymbol{c}_{\beta j}'\boldsymbol{\tau}} = w_{1\beta}(\widehat{\boldsymbol{c}_{\beta j}'\boldsymbol{\tau}})_1 + w_{2\beta}(\widehat{\boldsymbol{c}_{\beta j}'\boldsymbol{\tau}})_2 + w_{3\beta}(\widehat{\boldsymbol{c}_{\beta j}'\boldsymbol{\tau}})_3, \tag{5.5.13}$$

where $(\widehat{\boldsymbol{c}_{\beta j}'\boldsymbol{\tau}})_\alpha$ denotes the BLUE of $\boldsymbol{c}_{\beta j}'\boldsymbol{\tau}$ obtained within stratum α $(= 1,2,3)$, as given in Theorem 5.4.1, and accordingly,

$$w_{\alpha\beta} = \begin{cases} \varepsilon_{1\beta}/(\varepsilon_{1\beta} + \delta_2^{-1}\varepsilon_{2\beta} + \delta_3^{-1}\varepsilon_{3\beta}) & \text{for } \alpha = 1, \\ \varepsilon_{2\beta}/(\delta_2\varepsilon_{1\beta} + \varepsilon_{2\beta} + \delta_3^{-1}\delta_2\varepsilon_{3\beta}) & \text{for } \alpha = 2, \\ \varepsilon_{3\beta}/(\delta_3\varepsilon_{1\beta} + \delta_2^{-1}\delta_3\varepsilon_{2\beta} + \varepsilon_{3\beta}) & \text{for } \alpha = 3. \end{cases} \tag{5.5.14}$$

Evidently, (5.5.13) is an extension of (3.8.19), with which it coincides when $\varepsilon_{3\beta} = 0$.

With this notation, (5.5.11) can be seen as an extension of (3.8.18), and its covariance matrix, given in (5.5.4) and now written as

$$\text{Cov}(\hat{\tau}) = \sigma_1^2 \tilde{H}_0 + \sigma_1^2 \sum_{\beta=1}^{\tilde{m}-1} \frac{w_{1\beta}}{\varepsilon_{1\beta}} \tilde{H}_\beta + \sigma_2^2 \frac{w_{2\tilde{m}}}{\varepsilon_{2\tilde{m}}} \tilde{H}_{\tilde{m}} + \sigma_4^2 \tilde{H}_{\tilde{m}+1}, \qquad (5.5.15)$$

can evidently be considered as an extension of (3.8.21), to which it reduces when $\varepsilon_{3\tilde{m}} = 0$, getting then the form

$$\text{Cov}(\hat{\tau}) = \sigma_1^2 \tilde{H}_0 + \sigma_1^2 \sum_{\beta=1}^{\tilde{m}-1} \frac{w_{1\beta}}{\varepsilon_{1\beta}} \tilde{H}_\beta + \sigma_2^2 \tilde{H}_{\tilde{m}} + \sigma_4^2 \tilde{H}_{\tilde{m}+1}.$$

From (5.5.15), the variance of the BLUE $\widehat{c'\tau}$, given in (5.5.12), can be written as

$$\begin{aligned} \text{Var}(\widehat{c'\tau}) &= c'\text{Cov}(\hat{\tau})c = s'r^\delta \text{Cov}(\hat{\tau}) r^\delta s \\ &= \sigma_1^2 s'r^\delta \tilde{H}_0 r^\delta s + \sigma_1^2 \sum_{\beta=1}^{\tilde{m}-1} \frac{w_{1\beta}}{\varepsilon_{1\beta}} s'r^\delta \tilde{H}_\beta r^\delta s \\ &\quad + \sigma_2^2 \frac{w_{2\tilde{m}}}{\varepsilon_{2\tilde{m}}} s'r^\delta \tilde{H}_{\tilde{m}} r^\delta s + \sigma_4^2 s'r^\delta \tilde{H}_{\tilde{m}+1} r^\delta s, \end{aligned} \qquad (5.5.16)$$

which for the BLUE $\widehat{c'_{\beta j}\tau}$, given in (5.5.13), becomes

$$\text{Var}(\widehat{c'_{\beta j}\tau}) = \begin{cases} \sigma_1^2 & \text{for } \beta = 0, \\ \sigma_1^2 w_{1\beta}/\varepsilon_{1\beta} & \text{for } \beta = 1, 2, ..., \tilde{m} - 1, \\ \sigma_2^2 w_{2\tilde{m}}/\varepsilon_{2\tilde{m}} = \sigma_3^2 w_{3\tilde{m}}/\varepsilon_{3\tilde{m}} & \text{for } \beta = \tilde{m}, \end{cases} \qquad (5.5.17)$$

as an obvious extension of (3.8.24). Certainly, from (5.5.15), it also follows that all of the BLUEs $\widehat{c'_{\beta j}\tau}, j = 1, 2, ..., \tilde{\rho}_\beta, \beta = 0, 1, ..., \tilde{m}$, are uncorrelated, which means that the equality (3.8.25) holds here as well.

Referring to the weights (5.5.14) used in the formulae (5.5.13), (5.5.15), (5.5.16) and (5.5.17), it should be noted that they satisfy the equality $w_{1\beta} + w_{2\beta} + w_{3\beta} = 1$ and are in agreement with Proposition 4.1 of Houtman and Speed (1983). This conclusion means that in an NB design with the GB property a contrast is simply combinable, in the sense of Martin and Zyskind (1966), if and only if it is a basic contrast or is a linear combination of basic contrasts from the same subset $\tilde{S}_\beta$ in (5.4.17). (See again Corollary 3.8.1 and Remark 3.8.1.)

5.5.2 Estimation of unknown variance ratios

Results established in Section 5.5.1 are based on the assumption that the ratios γ_1 and γ_2 are known (see Lemma 5.5.1). In practice, however, this is usually not the case. Therefore, to make the theory applicable, estimators not only of σ_1^2, but also of γ_1 and γ_2 are needed. Although various approaches may be adopted for finding these estimators, that used in Section 3.7.2 seems to be particularly suitable. Let its extension for NB designs be presented here in details.

First, let the residual sum of squares (5.5.7) be written as

$$\begin{aligned}\|(\boldsymbol{I}_n - \boldsymbol{P}_{\Delta'(T^{-1})})\boldsymbol{y}\|^2_{T^{-1}} &= \boldsymbol{y}'\boldsymbol{RTRy} \\ &= \gamma_1\boldsymbol{y}'\boldsymbol{RG}'\boldsymbol{GRy} + \gamma_2\boldsymbol{y}'\boldsymbol{RD}'\boldsymbol{DRy} \\ &\quad + \boldsymbol{y}'\boldsymbol{RRy},\end{aligned} \tag{5.5.18}$$

where

$$\boldsymbol{R} = \boldsymbol{T}^{-1}(\boldsymbol{I}_n - \boldsymbol{P}_{\Delta'(T^{-1})}) = \boldsymbol{T}^{-1} - \boldsymbol{T}^{-1}\boldsymbol{\Delta}'(\boldsymbol{\Delta T}^{-1}\boldsymbol{\Delta}')^{-1}\boldsymbol{\Delta T}^{-1}. \tag{5.5.19}$$

Equating then the partial sums of squares in (5.5.18) to their expectations, one obtains the set of equations

$$\begin{bmatrix} \mathrm{tr}(\boldsymbol{RG}'\boldsymbol{GRG}'\boldsymbol{G}) & \mathrm{tr}(\boldsymbol{RG}'\boldsymbol{GRD}'\boldsymbol{D}) & \mathrm{tr}(\boldsymbol{RG}'\boldsymbol{GR}) \\ \mathrm{tr}(\boldsymbol{RD}'\boldsymbol{DRG}'\boldsymbol{G}) & \mathrm{tr}(\boldsymbol{RD}'\boldsymbol{DRD}'\boldsymbol{D}) & \mathrm{tr}(\boldsymbol{RD}'\boldsymbol{DR}) \\ \mathrm{tr}(\boldsymbol{RG}'\boldsymbol{GR}) & \mathrm{tr}(\boldsymbol{RD}'\boldsymbol{DR}) & \mathrm{tr}(\boldsymbol{RR}) \end{bmatrix} \begin{bmatrix} \sigma_1^2\gamma_1 \\ \sigma_1^2\gamma_2 \\ \sigma_1^2 \end{bmatrix}$$

$$= \begin{bmatrix} \boldsymbol{y}'\boldsymbol{RG}'\boldsymbol{GRy} \\ \boldsymbol{y}'\boldsymbol{RD}'\boldsymbol{DRy} \\ \boldsymbol{y}'\boldsymbol{RRy} \end{bmatrix}, \tag{5.5.20}$$

from which estimators of $\sigma_1^2\gamma_1$, $\sigma_1^2\gamma_2$ and σ_1^2, and, hence, of γ_1 and γ_2, can be obtained. Exactly the same equations, as those in (5.5.20), follow from the MINQUE approach of Rao (1971a). Also note that, on account of the equalities $\mathrm{tr}(\boldsymbol{RG}'\boldsymbol{GRT}) = \mathrm{tr}(\boldsymbol{RG}'\boldsymbol{G})$, $\mathrm{tr}(\boldsymbol{RD}'\boldsymbol{DRT}) = \mathrm{tr}(\boldsymbol{RD}'\boldsymbol{D})$ and $\mathrm{tr}(\boldsymbol{RRT}) = \mathrm{tr}(\boldsymbol{R})$, the equations (5.5.20) can equivalently be written as

$$\begin{aligned}\boldsymbol{y}'\boldsymbol{RG}'\boldsymbol{GRy} &= \sigma_1^2\mathrm{tr}(\boldsymbol{RG}'\boldsymbol{G}), \\ \boldsymbol{y}'\boldsymbol{RD}'\boldsymbol{DRy} &= \sigma_1^2\mathrm{tr}(\boldsymbol{RD}'\boldsymbol{D}), \\ \boldsymbol{y}'\boldsymbol{RRy} &= \sigma_1^2\mathrm{tr}(\boldsymbol{R}),\end{aligned} \tag{5.5.21}$$

and the equations (5.5.21) can be shown to coincide with those on which the MML (REML) estimation method is based (see Section 3.7.2).

Clearly, the equations (5.5.20) have no direct analytic solution, because the matrix $\boldsymbol{R}$ itself contains the unknown parameters γ_1 and γ_2, as can be seen from (5.5.3) and (5.5.19). Therefore, to solve these equations, or any equivalence of them, an iterative procedure is to be applied (similarly as in Section 3.7.2). It starts with some preliminary estimates $\gamma_{1,0}$ and $\gamma_{2,0}$ incorporated into the equations (5.5.20) by changing the matrix $\boldsymbol{R}$ there to

$$\boldsymbol{R}_0 = \boldsymbol{T}_0^{-1} - \boldsymbol{T}_0^{-1}\boldsymbol{\Delta}'(\boldsymbol{\Delta T}_0^{-1}\boldsymbol{\Delta}')^{-1}\boldsymbol{\Delta T}_0^{-1}, \tag{5.5.22}$$

where

$$T_0^{-1} = \tilde{\phi}_1 + D'k^{-\delta}(k^{-\delta} + \gamma_{2,0} I_b + \gamma_{1,0}\text{diag}[1_{b_1}1'_{b_1} : 1_{b_2}1'_{b_2} : \cdots : 1_{b_a}1'_{b_a}])^{-1}k^{-\delta}D$$

(with $T_0^{-1} \to \tilde{\phi}_1$ if $\gamma_{2,0} \to \infty$). However, instead of the equations so obtained, it is more convenient to solve iteratively their equivalence of the form

$$\begin{bmatrix} \text{tr}(R_0G'GR_0G'G) & \text{tr}(R_0G'GR_0D'D) & \text{tr}(R_0G'G) \\ \text{tr}(R_0D'DR_0G'G) & \text{tr}(R_0D'DR_0D'D) & \text{tr}(R_0D'D) \\ \text{tr}(R_0G'G) & \text{tr}(R_0D'D) & n-v \end{bmatrix} \begin{bmatrix} \sigma_1^2(\gamma_1 - \gamma_{1,0}) \\ \sigma_1^2(\gamma_2 - \gamma_{2,0}) \\ \sigma_1^2 \end{bmatrix}$$

$$= \begin{bmatrix} y'R_0G'GR_0y \\ y'R_0D'DR_0y \\ y'R_0y \end{bmatrix}. \tag{5.5.23}$$

By solving the equations (5.5.23), one obtains the revised estimates of γ_1 and γ_2, which can be written as

$$\hat{\gamma}_1 = \gamma_{1,0} + \frac{a^{00}y'R_0G'GR_0y + a^{01}y'R_0D'DR_0y + a^{02}y'R_0y}{a^{20}y'R_0G'GR_0y + a^{21}y'R_0D'DR_0y + a^{22}y'R_0y} \tag{5.5.24}$$

and

$$\hat{\gamma}_2 = \gamma_{2,0} + \frac{a^{10}y'R_0G'GR_0y + a^{11}y'R_0D'DR_0y + a^{12}y'R_0y}{a^{20}y'R_0G'GR_0y + a^{21}y'R_0D'DR_0y + a^{22}y'R_0y}, \tag{5.5.25}$$

where

$$\begin{aligned} a^{00} &= a_{11}a_{22} - a_{12}^2, & a^{01} &= a^{10} = a_{12}a_{02} - a_{01}a_{22}, \\ a^{02} &= a^{20} = a_{01}a_{12} - a_{11}a_{02}, & a^{11} &= a_{00}a_{22} - a_{02}^2, \\ a^{12} &= a^{21} = a_{01}a_{02} - a_{00}a_{12}, & a^{22} &= a_{00}a_{11} - a_{01}^2, \end{aligned}$$

with

$$\begin{aligned} a_{00} &= \text{tr}[(GR_0G')^2], & a_{01} &= \text{tr}(DR_0G'GR_0D'), & a_{02} &= \text{tr}(GR_0G'), \\ a_{11} &= \text{tr}[(DR_0D')^2], & a_{12} &= \text{tr}(DR_0D') \text{ and} & a_{22} &= n - v. \end{aligned}$$

Thus, a single iteration of the iterative method (which extends that of Patterson and Thompson, 1971, p. 550) consists here of the following two steps:

(0) One starts with some preliminary estimates $\gamma_{1,0}$ (≥ 0) and $\gamma_{2,0}$ ($> -1/k_{\max}$) of γ_1 and γ_2, respectively, to obtain the equations (5.5.23).

(1) By solving (5.5.23), revised estimates of γ_1 and γ_2 are obtained in the form (5.5.24) and (5.5.25), respectively, and these are then used as new preliminary estimates in step (0) of the next iteration.

However, it should always be observed that the original as well as the new preliminary estimates satisfy the conditions $\gamma_{1,0} \geq 0$ and $\gamma_{2,0} > -1/k_{\max}$. Therefore, if any of the formulae (5.5.24) and (5.5.25) gives a revised estimate not satisfying these bounds, the result is to be adjusted before entering step (0), in a way similar to that given in Section 3.7.2, following the suggestion of Rao and Kleffe (1988, p. 237).

The described iteration is to be repeated until convergence, i.e., until the equalities

$$\frac{\boldsymbol{y}'\boldsymbol{R}_0\boldsymbol{G}'\boldsymbol{G}\boldsymbol{R}_0\boldsymbol{y}}{\mathrm{tr}(\boldsymbol{G}\boldsymbol{R}_0\boldsymbol{G}')} = \frac{\boldsymbol{y}'\boldsymbol{R}_0\boldsymbol{D}'\boldsymbol{D}\boldsymbol{R}_0\boldsymbol{y}}{\mathrm{tr}(\boldsymbol{D}\boldsymbol{R}_0\boldsymbol{D}')} = \frac{\boldsymbol{y}'\boldsymbol{R}_0\boldsymbol{y}}{n-v}$$

are reached. The values $\hat{\gamma}_1 = \gamma_{1,0}$ and $\hat{\gamma}_2 = \gamma_{2,0}$ satisfying them are then considered as the final estimates of γ_1 and γ_2, respectively, and the resulting ratio $\hat{\sigma}_1^2 = \boldsymbol{y}'\hat{\boldsymbol{R}}\boldsymbol{y}/(n-v)$, with $\hat{\boldsymbol{R}} = \boldsymbol{R}_0$, can be considered as the final estimate of σ_1^2. Clearly, the matrix $\hat{\boldsymbol{R}}$ is obtained according to (5.5.22), but after replacing $\gamma_{1,0}$ and $\gamma_{2,0}$ by the final values $\hat{\gamma}_1$ and $\hat{\gamma}_2$, respectively. Although the convergence of this process has not been proved, some experiences (see, e.g., Rao and Kleffe, 1988, p. 226) indicate that the procedure should work well in practice.

Now, inserting in formula (5.5.2), instead of $\boldsymbol{T}^{-1}$, the matrix $\hat{\boldsymbol{T}}^{-1}$ resulting from the replacements of γ_1 and γ_2 in (5.5.3) by their final estimates $\hat{\gamma}_1$ and $\hat{\gamma}_2$, respectively, an empirical estimator

$$\tilde{\boldsymbol{\tau}} = (\boldsymbol{\Delta}\hat{\boldsymbol{T}}^{-1}\boldsymbol{\Delta}')^{-1}\boldsymbol{\Delta}\hat{\boldsymbol{T}}^{-1}\boldsymbol{y} \tag{5.5.26}$$

is obtained. The adjective "empirical" is used here to indicate that the unknown variance ratios γ_1 and γ_2 appearing in $\boldsymbol{T}$ have been replaced by their empirical estimates (see Rao and Kleffe, 1988, Section 10.5). Of course, (5.5.26) is not the same as the BLUE in (5.5.2) obtainable with the exact values of γ_1 and γ_2.

Finally, note that for an NB design that is GB, it will be more convenient to use, instead of (5.5.26), the notation

$$\tilde{\boldsymbol{\tau}} = \hat{\boldsymbol{C}}_c^{-1}\hat{\boldsymbol{Q}}_c,$$

where $\hat{\boldsymbol{C}}_c$ and $\hat{\boldsymbol{Q}}_c$ are defined as in (5.5.9) and (5.5.10), respectively, but with δ_2 and δ_3 replaced by their estimates $\hat{\delta}_2$ and $\hat{\delta}_3$. To obtain those estimates, use is to be made of the relations

$$\delta_2 = \sigma_2^2/\sigma_1^2 = k\gamma_2 + 1 \quad \text{and} \quad \delta_3 = \sigma_3^2/\sigma_1^2 = n_0\gamma_1 + k\gamma_2 + 1,$$

holding under the conditions of Lemma 5.4.1. So, if the estimates $\hat{\gamma}_1$ and $\hat{\gamma}_2$ are obtained from (5.5.24) and (5.5.25), respectively, then the estimates of δ_2 and δ_3 are readily obtainable as

$$\hat{\delta}_2 = k\hat{\gamma}_2 + 1 \quad \text{and} \quad \hat{\delta}_3 = n_0\hat{\gamma}_1 + \hat{\delta}_2,$$

provided that $\hat{\gamma}_1 \geq 0$ and $\hat{\gamma}_2 > -1/k$ (in accordance with the assumptions of Theorem 5.5.1) or, equivalently, $\hat{\delta}_3 \geq \hat{\delta}_2 > 0$. Certainly, with these estimates, the weights (5.5.14) can in the various formulae be replaced by their estimates

$$\hat{w}_{\alpha\beta} = \begin{cases} \varepsilon_{1\beta}/(\varepsilon_{1\beta} + \hat{\delta}_2^{-1}\varepsilon_{2\beta} + \hat{\delta}_3^{-1}\varepsilon_{3\beta}) & \text{for } \alpha = 1, \\ \varepsilon_{2\beta}/(\hat{\delta}_2\varepsilon_{1\beta} + \varepsilon_{2\beta} + \hat{\delta}_3^{-1}\hat{\delta}_2\varepsilon_{3\beta}) & \text{for } \alpha = 2, \\ \varepsilon_{3\beta}/(\hat{\delta}_3\varepsilon_{1\beta} + \hat{\delta}_2^{-1}\hat{\delta}_3\varepsilon_{2\beta} + \varepsilon_{3\beta}) & \text{for } \alpha = 3, \end{cases}$$

to obtain the relevant empirical estimators, as in Section 3.8; see (3.8.42) in particular.

5.5.3 Properties of the empirical estimators

To gain an insight into the properties of the empirical estimator (5.5.26) of $\boldsymbol{\tau}$ and, hence, of such estimator of any parametric function $\boldsymbol{c}'\boldsymbol{\tau}$, the approximation suggested by Kackar and Harville (1984) can be used. Its application becomes straightforward when it is possible to represent $\tilde{\boldsymbol{\tau}}$ in terms of some simple linear functions of the observed vector $\boldsymbol{y}$, as shown in Section 3.7.3 for experiments with one stratum of blocks. For NB designs, such representation is not readily available in the general case.

More precisely, the possibility of obtaining a simple representation of the estimator (5.5.26) in terms of linear functions of $\boldsymbol{y}$ depends on the availability of a suitable spectral decomposition of the matrix $\boldsymbol{\phi}_*\boldsymbol{T}\boldsymbol{\phi}_*$, where $\boldsymbol{\phi}_* = \boldsymbol{I}_n - \boldsymbol{\Delta}'\boldsymbol{r}^{-\delta}\boldsymbol{\Delta}$ and $\boldsymbol{T}$ is as in (5.5.1). For this reason, it is required that the component matrices in

$$\boldsymbol{\phi}_*\boldsymbol{T}\boldsymbol{\phi}_* = \gamma_1\boldsymbol{\phi}_*\boldsymbol{G}'\boldsymbol{G}\boldsymbol{\phi}_* + \gamma_2\boldsymbol{\phi}_*\boldsymbol{D}'\boldsymbol{D}\boldsymbol{\phi}_* + \boldsymbol{\phi}_* \qquad (5.5.27)$$

commute in pairs. Each of the first two matrices on the right-hand side in (5.5.27) commutes with $\boldsymbol{\phi}_*$ always, but these two matrices cannot be shown to commute one with the other in general. However, it can easily be seen that for $\boldsymbol{\phi}_*\boldsymbol{G}'\boldsymbol{G}\boldsymbol{\phi}_*$ to commute with $\boldsymbol{\phi}_*\boldsymbol{D}'\boldsymbol{D}\boldsymbol{\phi}_*$, it is sufficient that the matrix $\boldsymbol{G}'\boldsymbol{G}\boldsymbol{\phi}_*\boldsymbol{D}'\boldsymbol{D}$ is symmetric. Conditions for this matrix to be such can be given as follows.

Lemma 5.5.2. *The matrix $\boldsymbol{G}'\boldsymbol{G}\boldsymbol{\phi}_*\boldsymbol{D}'\boldsymbol{D}$ is symmetric if and only if the following conditions hold:*

$$k_{1(h)} = k_{2(h)} = \cdots = k_{b_h(h)} = k_{(h)} \quad \textit{for} \quad h = 1, 2, ..., a, \qquad (5.5.28)$$

where $k_{(h)} = n_h/b_h$, and simultaneously, the equalities (5.4.12), *i.e.,*

$$\boldsymbol{N}'_h\boldsymbol{r}^{-\delta}\boldsymbol{r}_{h'}\boldsymbol{1}'_{b_{h'}} = \boldsymbol{1}_{b_h}\boldsymbol{r}'_h\boldsymbol{r}^{-\delta}\boldsymbol{N}_{h'} \quad \textit{for} \quad h, h' = 1, 2, ..., a. \qquad (5.5.29)$$

Proof. Recalling (from Section 5.1) that $\boldsymbol{G}' = \boldsymbol{D}'\text{diag}[\boldsymbol{1}_{b_1} : \boldsymbol{1}_{b_2} : \cdots : \boldsymbol{1}_{b_a}]$, note that the equality $\boldsymbol{D}'\boldsymbol{D}\boldsymbol{\phi}_*\boldsymbol{G}'\boldsymbol{G} = \boldsymbol{G}'\boldsymbol{G}\boldsymbol{\phi}_*\boldsymbol{D}'\boldsymbol{D}$ holds if and only if the matrix $\boldsymbol{D}\boldsymbol{\phi}_*\boldsymbol{D}'\text{diag}[\boldsymbol{1}_{b_1}\boldsymbol{1}'_{b_1} : \boldsymbol{1}_{b_2}\boldsymbol{1}'_{b_2} : \cdots : \boldsymbol{1}_{b_a}\boldsymbol{1}'_{b_a}]$ is symmetric, and this condition can easily be seen to hold if and only if (5.5.28) and (5.5.29) hold. □

Remark 5.5.1. The condition (5.5.29) implies that

$$b_1 = b_2 = \cdots = b_a = b_0, \quad \text{where} \quad b_0 = b/a. \qquad (5.5.30)$$

Although the conditions (5.5.28) and (5.5.29), sufficient for $\boldsymbol{\phi}_*\boldsymbol{G}'\boldsymbol{G}\boldsymbol{\phi}_*$ to commute with $\boldsymbol{\phi}_*\boldsymbol{D}'\boldsymbol{D}\boldsymbol{\phi}_*$, may seem restrictive, it can be claimed that many

NB designs used in practice satisfy them. When considering NB designs based on the concept of resolvability, the following remark is relevant.

Remark 5.5.2. The matrices $\boldsymbol{\phi}_*\boldsymbol{G}'\boldsymbol{G}\boldsymbol{\phi}_*$ and $\boldsymbol{\phi}_*\boldsymbol{D}'\boldsymbol{D}\boldsymbol{\phi}_*$ in (5.5.27) commute one with the other if, in particular, the NB design satisfies the following two conditions:

(a) The component designs, $\mathcal{D}_h$, $h = 1, 2, ..., a$, are all proper with the same number of blocks, i.e., the equalities (5.5.28) and (5.5.30) hold.

(b) The design with respect to superblocks, $\mathcal{D}$, is connected and orthogonal, i.e., $\boldsymbol{\mathfrak{R}} = n^{-1}\boldsymbol{r}\boldsymbol{n}'$.

It can easily be shown that the conditions (a) and (b) of Remark 5.5.2 are sufficient for satisfying the conditions of Lemma 5.5.2. [Note that the conditions (a) and (b) of Remark 5.5.2 are less restrictive than those given as (i) and (ii) in the earlier paper by Caliński (1997, Section 3.3).]

Example 5.5.1. One may ask, which of the designs considered in Examples 5.3.1–5.3.4 satisfy the conditions of Lemma 5.5.2. The answer is that those in Examples 5.3.1, 5.3.2 and 5.3.4 satisfy the conditions, whereas that in Example 5.3.3 does not. The reader is advised to check this finding as an exercise.

Thus, from now on, it will be assumed that the NB design under consideration satisfies the conditions (5.5.28) and (5.5.29), i.e., that its matrix $\boldsymbol{G}'\boldsymbol{G}\boldsymbol{\phi}_*\boldsymbol{D}'\boldsymbol{D}$ is symmetric. Under this assumption one can obtain not only the decomposition

$$\boldsymbol{\phi}_*\boldsymbol{D}'\boldsymbol{D}\boldsymbol{\phi}_* = \sum_{j=1}^{h_*} \lambda_j \boldsymbol{v}_j \boldsymbol{v}_j', \tag{5.5.31}$$

as in (3.7.36), but also

$$\boldsymbol{\phi}_*\boldsymbol{G}'\boldsymbol{G}\boldsymbol{\phi}_* = \sum_{j=1}^{f_*} \vartheta_j \boldsymbol{v}_j \boldsymbol{v}_j', \tag{5.5.32}$$

where, simultaneously, the vectors $\boldsymbol{v}_j$, $j = 1, 2, ..., f_*, ..., h_*, ..., n - v$, are orthonormal eigenvectors of the matrix $\boldsymbol{\phi}_*$, corresponding to the unit eigenvalues of it, and where $f_* = \text{rank}(\boldsymbol{G}\boldsymbol{\phi}_*\boldsymbol{G}')$ and $h_* = \text{rank}(\boldsymbol{D}\boldsymbol{\phi}_*\boldsymbol{D}')$. Note that $\boldsymbol{G}\boldsymbol{\phi}_*\boldsymbol{G}' = \boldsymbol{n}^{\delta} - \boldsymbol{\mathfrak{R}}'\boldsymbol{r}^{-\delta}\boldsymbol{\mathfrak{R}}$ is a dual of the matrix $\boldsymbol{r}^{\delta} - \boldsymbol{\mathfrak{R}}\boldsymbol{n}^{-\delta}\boldsymbol{\mathfrak{R}}'$ of rank f (say), and $\boldsymbol{D}\boldsymbol{\phi}_*\boldsymbol{D}' = \boldsymbol{k}^{\delta} - \boldsymbol{N}'\boldsymbol{r}^{-\delta}\boldsymbol{N}$ is a dual of the matrix $\boldsymbol{r}^{\delta} - \boldsymbol{N}\boldsymbol{k}^{-\delta}\boldsymbol{N}'$ of rank $\tilde{h}_1$ (see Section 5.3.1), the spectral decomposition of $\boldsymbol{D}\boldsymbol{\phi}_*\boldsymbol{D}'$ being as shown in (3.7.35). Hence, in general, $f_* = a - v + f$ and $h_* = b - v + \tilde{h}_1$ (see also Section 3.7.3). If the design $\mathcal{D}$ is connected (as assumed in Remark 5.5.2), then $f = v - 1$ and, hence, $f_* = a - 1$. If the design $\mathcal{D}^*$ is connected, then $\tilde{h}_1 = v - 1$ and $h_* = b - 1$. Also, it is important to note with regard to the eigenvalues in (5.5.31) and (5.5.32) that the equality $\vartheta_j = b_0\lambda_j$ holds for $j = 1, 2, ..., f_*$, as it follows from the proof of Lemma 5.5.2, Remark 5.5.1, and the relation

$$\begin{aligned}\boldsymbol{G}\boldsymbol{\phi}_*\boldsymbol{G}' &= \text{diag}[\mathbf{1}_{b_1}' : \mathbf{1}_{b_2}' : \cdots : \mathbf{1}_{b_a}']\boldsymbol{D}\boldsymbol{\phi}_*\boldsymbol{D}'\text{diag}[\mathbf{1}_{b_1} : \mathbf{1}_{b_2} : \cdots : \mathbf{1}_{b_a}] \\ &= (\boldsymbol{I}_a \otimes \mathbf{1}_{b_0}')\boldsymbol{D}\boldsymbol{\phi}_*\boldsymbol{D}'(\boldsymbol{I}_a \otimes \mathbf{1}_{b_0}).\end{aligned}$$

Now, with (5.5.31) and (5.5.32), obtainable under the conditions (5.5.28) and (5.5.29), the spectral decomposition of $\boldsymbol{\phi}_*\boldsymbol{T}\boldsymbol{\phi}_*$ gets from (5.5.27) the form

$$\begin{aligned}\boldsymbol{\phi}_*\boldsymbol{T}\boldsymbol{\phi}_* &= \sum_{j=1}^{f_*}(\gamma_1 b_0\lambda_j + \gamma_2\lambda_j + 1)\boldsymbol{v}_j\boldsymbol{v}_j' \\ &\quad + \sum_{j=f_*+1}^{h_*}(\gamma_2\lambda_j + 1)\boldsymbol{v}_j\boldsymbol{v}_j' + \sum_{j=h_*+1}^{n-v}\boldsymbol{v}_j\boldsymbol{v}_j' \qquad (5.5.33)\end{aligned}$$

and, because the Moore–Penrose inverse of $\boldsymbol{\phi}_*\boldsymbol{T}\boldsymbol{\phi}_*$ can be shown to be equal to $\boldsymbol{R}$ [defined in (5.5.19)], it follows from (5.5.33) that

$$\begin{aligned}\boldsymbol{R} = (\boldsymbol{\phi}_*\boldsymbol{T}\boldsymbol{\phi}_*)^+ &= \sum_{j=1}^{f_*}(\gamma_1 b_0\lambda_j + \gamma_2\lambda_j + 1)^{-1}\boldsymbol{v}_j\boldsymbol{v}_j' \\ &\quad + \sum_{j=f_*+1}^{h_*}(\gamma_2\lambda_j + 1)^{-1}\boldsymbol{v}_j\boldsymbol{v}_j' + \sum_{j=h_*+1}^{n-v}\boldsymbol{v}_j\boldsymbol{v}_j'. \qquad (5.5.34)\end{aligned}$$

From (5.5.19) and (5.5.34), the estimator (5.5.26) can now be written as

$$\begin{aligned}\tilde{\boldsymbol{\tau}} &= \hat{\boldsymbol{\tau}} + \boldsymbol{r}^{-\delta}\boldsymbol{\Delta}(\boldsymbol{T}\boldsymbol{R} - \hat{\boldsymbol{T}}\hat{\boldsymbol{R}})\boldsymbol{y} \\ &= \hat{\boldsymbol{\tau}} + \boldsymbol{r}^{-\delta}\boldsymbol{N}\boldsymbol{D}\left(\sum_{j=1}^{f_*}\xi_j^{-1}z_j\boldsymbol{v}_j + \sum_{j=f_*+1}^{h_*}\zeta_j^{-1}z_j\boldsymbol{v}_j\right), \qquad (5.5.35)\end{aligned}$$

where

$$\xi_j = (\gamma_1 b_0 + \gamma_2)\lambda_j + 1 \quad \text{for } j = 1, 2, ..., f_*,$$

$$\zeta_j = \gamma_2\lambda_j + 1 \quad \text{for } j = f_* + 1, f_* + 2, ..., h_*,$$

$$z_j = \frac{(\gamma_1 - \hat{\gamma}_1)b_0 + \gamma_2 - \hat{\gamma}_2}{(\hat{\gamma}_1 b_0 + \hat{\gamma}_2)\lambda_j + 1}\boldsymbol{v}_j'\boldsymbol{y} \quad \text{for } j = 1, 2, ..., f_* \qquad (5.5.36)$$

and

$$z_j = \frac{\gamma_2 - \hat{\gamma}_2}{\hat{\gamma}_2\lambda_j + 1}\boldsymbol{v}_j'\boldsymbol{y} \quad \text{for } j = f_* + 1, f_* + 2, ..., h_*, \qquad (5.5.37)$$

because, under (5.5.28) and (5.5.29), $\boldsymbol{G}'\boldsymbol{G}\boldsymbol{\phi}_*\boldsymbol{D}'\boldsymbol{D} = \boldsymbol{D}'\boldsymbol{D}\boldsymbol{\phi}_*\boldsymbol{G}'\boldsymbol{G}$, which implies that $\boldsymbol{\Delta}\boldsymbol{G}'\boldsymbol{G}\boldsymbol{v}_j = b_0\boldsymbol{\Delta}\boldsymbol{D}'\boldsymbol{D}\boldsymbol{v}_j$ for $j = 1, 2, ..., f_*$.

The relevant representation of $\widetilde{\boldsymbol{c}'\boldsymbol{\tau}} = \boldsymbol{c}'\tilde{\boldsymbol{\tau}}$ is then obvious. Note, however, that if $\boldsymbol{c}'\boldsymbol{\tau}$ is a contrast (i.e., $\boldsymbol{c}'\boldsymbol{1}_v = 0$) and the design satisfies the conditions of Remark 5.5.2, then $\boldsymbol{c}'\boldsymbol{r}^{-\delta}\boldsymbol{N}\boldsymbol{D}\boldsymbol{v}_j = b_0^{-1}\boldsymbol{c}'\boldsymbol{r}^{-\delta}\boldsymbol{\Delta}\boldsymbol{G}'\boldsymbol{G}\boldsymbol{v}_j = 0$ for $j = 1, 2, ..., f_*$, giving the reduced formula

$$\widetilde{\boldsymbol{c}'\boldsymbol{\tau}} = \widehat{\boldsymbol{c}'\boldsymbol{\tau}} + \boldsymbol{c}'\boldsymbol{r}^{-\delta}\boldsymbol{N}\boldsymbol{D}\sum_{j=f_*+1}^{h_*}\zeta_j^{-1}z_j\boldsymbol{v}_j.$$

Also, if in addition to the conditions of Remark 5.5.2, it is assumed that $n_1 = n_2 = \cdots = n_a = n_0$ (which means that both $\mathcal{D}^*$ and $\mathcal{D}$ are proper block designs, i.e., the experiment has the OBS property; see Lemma 5.4.1), then $\boldsymbol{\Delta G'Gv}_j = \mathbf{0}$ for $j = 1, 2, ..., f_*$ and (5.5.35) is automatically reduced to

$$\tilde{\boldsymbol{\tau}} = \hat{\boldsymbol{\tau}} + \boldsymbol{r}^{-\delta}\boldsymbol{ND} \sum_{j=f_*+1}^{h_*} \zeta_j^{-1} z_j \boldsymbol{v}_j,$$

a formula similar to (3.7.40).

Now, Lemma 3.7.2 can be extended as follows.

Lemma 5.5.3. *Let the model of the variables observed in an NB design satisfying the conditions* (5.5.28) *and* (5.5.29) *be as in* (5.2.7), *and suppose that the values of* γ_1 *and* γ_2 *in* (5.2.11) *and* (5.2.12) *are unknown, except that they satisfy the limits given in Theorem* 5.5.1 (*to secure that* $\boldsymbol{T}$ *is* p.d.), *the same being satisfied by their estimates* (*to secure that* $\hat{\boldsymbol{T}}$ *is* p.d.). *Furthermore, let the random variables* $\{z_j\}$ *defined in* (5.5.36) *and* (5.5.37) *satisfy the conditions*

$$\mathrm{E}(z_j) = 0, \quad \mathrm{E}(z_j z_{j'}) = 0 \quad \textit{and} \quad \mathrm{Var}(z_j) < \infty$$

for all j *and* $j' \neq j$ $(= 1, 2, ..., h_*)$ *and for all admissible values of* γ_1 *and* γ_2. *Then, the estimator* $\tilde{\boldsymbol{\tau}}$ *has the properties*

$$\mathrm{E}(\tilde{\boldsymbol{\tau}}) = \mathrm{E}(\hat{\boldsymbol{\tau}}) = \boldsymbol{\tau} \tag{5.5.38}$$

and

$$\begin{aligned}\mathrm{Cov}(\tilde{\boldsymbol{\tau}}) &= \mathrm{Cov}(\hat{\boldsymbol{\tau}}) \\ &\quad + \boldsymbol{r}^{-\delta}\boldsymbol{ND}\Bigg[\sum_{j=1}^{f_*} \xi_j^{-2}\mathrm{Var}(z_j)\boldsymbol{v}_j\boldsymbol{v}_j' \\ &\quad + \sum_{j=f_*+1}^{h_*} \zeta_j^{-2}\mathrm{Var}(z_j)\boldsymbol{v}_j\boldsymbol{v}_j'\Bigg]\boldsymbol{D}'\boldsymbol{N}'\boldsymbol{r}^{-\delta}. \end{aligned} \tag{5.5.39}$$

From Lemma 5.5.3, it follows immediately that for any function $\boldsymbol{c}'\boldsymbol{\tau}$,

$$\mathrm{E}(\widetilde{\boldsymbol{c}'\boldsymbol{\tau}}) = \boldsymbol{c}'\boldsymbol{\tau} \tag{5.5.40}$$

and

$$\begin{aligned}\mathrm{Var}(\widetilde{\boldsymbol{c}'\boldsymbol{\tau}}) &= \mathrm{Var}(\widehat{\boldsymbol{c}'\boldsymbol{\tau}}) + \sum_{j=1}^{f_*} \xi_j^{-2}(\boldsymbol{c}'\boldsymbol{r}^{-\delta}\boldsymbol{ND}\boldsymbol{v}_j)^2\mathrm{Var}(z_j) \\ &\quad + \sum_{j=f_*+1}^{h_*} \zeta_j^{-2}(\boldsymbol{c}'\boldsymbol{r}^{-\delta}\boldsymbol{ND}\boldsymbol{v}_j)^2\mathrm{Var}(z_j), \end{aligned} \tag{5.5.41}$$

the component $\sum_{j=1}^{f_*}\xi_j^{-2}(\boldsymbol{c}'\boldsymbol{r}^{-\delta}\boldsymbol{N}\boldsymbol{D}\boldsymbol{v}_j)^2\text{Var}(z_j)$ disappearing if $\boldsymbol{c}'\mathbf{1}_v = 0$ and the design satisfies the conditions of Remark 5.5.2.

To proceed further, it will be helpful first to note that, on account of (5.5.34), one can write

$$\boldsymbol{y}'\boldsymbol{R}\boldsymbol{G}'\boldsymbol{G}\boldsymbol{R}\boldsymbol{y} = \sum_{j=1}^{f_*}\xi_j^{-2}b_0\lambda_j(\boldsymbol{v}_j'\boldsymbol{y})^2, \tag{5.5.42}$$

$$\boldsymbol{y}'\boldsymbol{R}\boldsymbol{D}'\boldsymbol{D}\boldsymbol{R}\boldsymbol{y} = \sum_{j=1}^{f_*}\xi_j^{-2}\lambda_j(\boldsymbol{v}_j'\boldsymbol{y})^2 + \sum_{j=f_*+1}^{h_*}\zeta_j^{-2}\lambda_j(\boldsymbol{v}_j'\boldsymbol{y})^2 \tag{5.5.43}$$

and

$$\boldsymbol{y}'\boldsymbol{R}\boldsymbol{R}\boldsymbol{y} = \sum_{j=1}^{f_*}\xi_j^{-2}(\boldsymbol{v}_j'\boldsymbol{y})^2 + \sum_{j=f_*+1}^{h_*}\zeta_j^{-2}(\boldsymbol{v}_j'\boldsymbol{y})^2 + \sum_{j=h_*+1}^{n-v}(\boldsymbol{v}_j'\boldsymbol{y})^2. \tag{5.5.44}$$

Thus, the estimators $\hat{\gamma}_1$ and $\hat{\gamma}_2$, as obtainable from the equations (5.5.20), and the random variables $\{z_j\}$, defined in (5.5.36) and (5.5.37), can be expressed as functions of $\boldsymbol{y}$ solely through the variables

$$x_j = \boldsymbol{v}_j'\boldsymbol{y} \quad \text{for} \quad j = 1, 2, ..., n - v. \tag{5.5.45}$$

Next, Lemma 3.7.3 and Corollary 3.7.1 can be used to obtain the following corollary.

Corollary 5.5.1. *Let the random variables $\{x_j\}$ defined in* (5.5.45) *have mutually independent symmetric distributions around zero, in the sense that x_j and $-x_j$ are distributed identically and for each j $(= 1, 2, ..., n - v)$ independently. Further, let each of the statistics $\hat{\gamma}_1$ and $\hat{\gamma}_2$ be an even function of any x_j, in the sense that it is invariant under the change of x_j to $-x_j$ for any j. Then, the joint distribution of the random variables $\{z_j\}$, defined in* (5.5.36) *and* (5.5.37), *is symmetric around zero with regard to each z_j, in the sense that the distribution is invariant under the change of z_j to $-z_j$ for any j. Hence,* $\text{E}(z_j) = 0$ *for all j and* $\text{E}(z_j z_{j'}) = 0$ *for all $j \neq j'$ $(= 1, 2, ..., h_*)$, provided that these expectations exist.*

With these results the following main theorem (similar to Theorem 3.7.3) can be proved.

Theorem 5.5.2. *Let, for an NB design satisfying the conditions* (5.5.28) *and* (5.5.29), *the observed vector $\boldsymbol{y}$ have the model* (5.2.7) *with properties* (5.2.8) *and* (5.2.9), *and suppose that the ratios γ_1 and γ_2 appearing in* (5.2.11) *and* (5.2.12) *are unknown. Further, let the distribution of $\boldsymbol{y}$ be such that it induces the random variables $\{x_j\}$ defined in* (5.5.45) *to have mutually independent symmetric distributions around zero. Under these assumptions, if the statistics $\hat{\gamma}_1$ and $\hat{\gamma}_2$*

used to estimate γ_1 *and* γ_2*, respectively, are completely expressible in terms of even functions of* $\{x_j\}$*, i.e., depend on* $\boldsymbol{y}$ *solely through such functions, then, for all values of* γ_1 *and* γ_2 *and of their estimators satisfying the conditions* γ_1, $\hat{\gamma}_1 \geq 0$ *and* γ_2, $\hat{\gamma}_2 > -1/k_{\max}$, *the estimator* $\tilde{\boldsymbol{\tau}}$ *defined in* (5.5.26) *has the properties* (5.5.38) *and* (5.5.39), *and hence,* $\widetilde{\boldsymbol{c}'\boldsymbol{\tau}}$ *has the properties as in* (5.5.40) *and* (5.5.41), *for any* $\boldsymbol{c}$.

Proof. This theorem can be proved in a way similar to the proof of Theorem 3.7.3. □

It should also be noted that Theorem 5.5.2 is general in the sense that it applies, under its assumptions, to any estimators of γ_1 and γ_2 depending on the observed vector $\boldsymbol{y}$ through even functions of $\{x_j\}$ only.

Remark 5.5.3. In connection with the assumptions of Theorem 5.5.2, the following results are obtainable.

(a) The random variables $\{x_j\}$ defined in (5.5.45) have the properties

$$E(x_j) = 0 \tag{5.5.46}$$

and

$$E(x_j x_{j'}) = \begin{cases} \sigma_1^2 \xi_j & \text{for} \quad j = j' = 1, 2, \ldots, f_*, \\ \sigma_1^2 \zeta_j & \text{for} \quad j = j' = f_* + 1, f_* + 2, \ldots, h_*, \\ \sigma_1^2 & \text{for} \quad j = j' = h_* + 1, h_* + 2, \ldots, n - v, \\ 0 & \text{for} \quad j \neq j', \end{cases} \tag{5.5.47}$$

resulting from the properties (5.2.8) and (5.2.10) of $\boldsymbol{y}$ and the definitions of the vectors $\{\boldsymbol{v}_j\}$.

(b) If $\boldsymbol{y}$ has an n-variate normal distribution, then $\boldsymbol{x} = [x_1, x_2, \ldots, x_{n-v}]'$, with $x_j = \boldsymbol{v}_j'\boldsymbol{y}$, also has an $(n - v)$-variate normal distribution, which automatically implies, on account of (5.5.46) and (5.5.47), that its elements have mutually independent symmetric distributions around zero.

(c) Because the functions of the random vector $\boldsymbol{y}$ appearing in the equations (5.5.20) are, as shown in (5.5.42), (5.5.43) and (5.5.44), completely expressible in terms of the squares $x_j^2 = (\boldsymbol{v}_j'\boldsymbol{y})^2$, $j = 1, 2, \ldots, n - v$, the statistics $\hat{\gamma}_1$ and $\hat{\gamma}_2$ obtained from the solution of these equations (or their equivalence) satisfy the conditions of Theorem 5.5.2, provided that their values are not below the lower limits assumed for them.

It follows from Theorem 5.5.2 that if the unknown values of γ_1 and γ_2 appearing in (5.5.3) are replaced by their estimators $\hat{\gamma}_1$ and $\hat{\gamma}_2$ obtainable in accordance with the conditions of this theorem, then the unbiasedness of the estimators of $\boldsymbol{\tau}$ and $\boldsymbol{c}'\boldsymbol{\tau}$ established in Theorem 5.5.1 is not violated (see also Kłaczyński, Molińska and Moliński, 1994), but the variance of the estimator of $\boldsymbol{c}'\boldsymbol{\tau}$ is increased, as can be seen from (5.5.41). Although the exact formula of $\text{Var}(z_j) = E(z_j^2)$ is in general intractable, it can be approximated using the same approach as that applied in Section 3.7.3. However, for $j = f_* + 1, f_* + 2, ..., h_*$, the approximation of $\mathrm{E}(z_j^2)$ can be exactly copied from formulae (3.7.53)–(3.7.55), the matter is slightly more involved for $j = 1, 2, ..., f_*$. Let it be considered first.

Note that, because for $j = 1, 2, ..., f_*$ the variable z_j as a function of $\hat{\gamma}_1$ and $\hat{\gamma}_2$ is

$$z_j = \frac{(\gamma_1 - \hat{\gamma}_1)b_0 + \gamma_2 - \hat{\gamma}_2}{(\hat{\gamma}_1 b_0 + \hat{\gamma}_2)\lambda_j + 1} x_j,$$

its partial derivatives with respect to $\hat{\gamma}_1$ and $\hat{\gamma}_2$ are

$$\frac{\partial z_j}{\partial \hat{\gamma}_1} = -\frac{(\gamma_1 b_0 \lambda_j + \gamma_2 \lambda_j + 1)b_0}{(\hat{\gamma}_1 b_0 \lambda_j + \hat{\gamma}_2 \lambda_j + 1)^2} x_j \quad \text{and} \quad \frac{\partial z_j}{\partial \hat{\gamma}_2} = -\frac{\gamma_1 b_0 \lambda_j + \gamma_2 \lambda_j + 1}{(\hat{\gamma}_1 b_0 \lambda_j + \hat{\gamma}_2 \lambda_j + 1)^2} x_j,$$

respectively. From them, using Taylor's series expansion, one can write

$$\begin{aligned} z_j \equiv z_j(\hat{\gamma}_1, \hat{\gamma}_2) &= z_j(\gamma_1, \gamma_2) + (\hat{\gamma}_1 - \gamma_1) \left.\frac{\partial z_j}{\partial \hat{\gamma}_1}\right|_{\hat{\gamma}_1 = \gamma_1, \hat{\gamma}_2 = \gamma_2} \\ &\quad + (\hat{\gamma}_2 - \gamma_2) \left.\frac{\partial z_j}{\partial \hat{\gamma}_2}\right|_{\hat{\gamma}_1 = \gamma_1, \hat{\gamma}_2 = \gamma_2} + \cdots \\ &= 0 - \frac{(\hat{\gamma}_1 - \gamma_1)\vartheta_j}{\xi_j \lambda_j} x_j - \frac{\hat{\gamma}_2 - \gamma_2}{\xi_j} x_j + \cdots, \end{aligned}$$

to obtain the approximations

$$z_j \cong -\xi_j^{-1} b_0 (\hat{\gamma}_1 - \gamma_1) x_j - \xi_j^{-1} (\hat{\gamma}_2 - \gamma_2) x_j$$

and, hence,

$$z_j^2 \cong \xi_j^{-2} [b_0(\hat{\gamma}_1 - \gamma_1) + \hat{\gamma}_2 - \gamma_2]^2 x_j^2. \tag{5.5.48}$$

Applying now to the expectation of (5.5.48) Theorems 2b.3(i) and (iii) of Rao (1973), one obtains

$$\mathrm{E}(z_j^2) \cong \xi_j^{-2} \mathrm{E}\{[b_0(\hat{\gamma}_1 - \gamma_1) + \hat{\gamma}_2 - \gamma_2]^2 \mathrm{E}(x_j^2 | \hat{\gamma}_1, \hat{\gamma}_2)\}. \tag{5.5.49}$$

The conditional expectation $\mathrm{E}(x_j^2 | \hat{\gamma}_1, \hat{\gamma}_2)$ appearing in (5.5.49) is usually not readily available when $\hat{\gamma}_1$ and $\hat{\gamma}_2$ are obtained by solving the equations (5.5.20) involving the variables $\{x_j\}$, as can be seen from (5.5.42)–(5.5.44). However, if the numbers of x_j's used in calculating the statistics $\hat{\gamma}_1$ and $\hat{\gamma}_2$ are large, i.e., in the case of solving the equations (5.5.20) the number $n - v$ is large, then the statistical dependence between x_j^2 and $\hat{\gamma}_1$, $\hat{\gamma}_2$ can be ignored, and instead of $\mathrm{E}(x_j^2 | \hat{\gamma}_1, \hat{\gamma}_2)$, the unconditional expectation $\mathrm{E}(x_j^2)$ can be used. Accepting this simplification, the approximation formula (5.5.49) can be replaced by

$$\begin{aligned} \mathrm{E}(z_j^2) &\cong \xi_j^{-2} \mathrm{E}\{b_0(\hat{\gamma}_1 - \gamma_1) + \hat{\gamma}_2 - \gamma_2]^2 \mathrm{E}(x_j^2)\} \\ &= \xi_j^{-1} \sigma_1^2 \{b_0^2 \mathrm{E}[(\hat{\gamma}_1 - \gamma_1)^2] + 2b_0 \mathrm{E}[(\hat{\gamma}_1 - \gamma_1)(\hat{\gamma}_2 - \gamma_2)] \\ &\quad + \mathrm{E}[(\hat{\gamma}_2 - \gamma_2)^2)]\}, \end{aligned} \tag{5.5.50}$$

on account of (5.5.47).

As to the approximation of $\mathrm{E}(z_j^2)$ for $j = f_* + 1, f_* + 2, ..., h_*$, following exactly the procedure described in Section 3.7.3, one can write

$$z_j^2 \cong \left(\frac{\hat{\gamma}_2 - \gamma_2}{\gamma_2\lambda_j + 1}\right)^2 x_j^2 = \frac{(\hat{\gamma}_2 - \gamma_2)^2}{\zeta_j^2} x_j^2,$$

from which [again by Theorems 2b.3(i) and (iii) of Rao (1973)]

$$\mathrm{E}(z_j^2) \cong \zeta_j^{-2}\mathrm{E}[(\hat{\gamma}_2 - \gamma_2)^2 \mathrm{E}(x_j^2|\hat{\gamma}_2)]. \tag{5.5.51}$$

Applying the same reasoning as that for $j \le f_*$, i.e., using the unconditional expectation $\mathrm{E}(x_j^2)$ instead of $\mathrm{E}(x_j^2|\hat{\gamma}_2)$ under the assumption that $n - v$ is sufficiently large, one would be allowed to replace the approximation (5.5.51) by

$$\mathrm{E}(z_j^2) \cong \zeta_j^{-1}\sigma_1^2\mathrm{E}[(\hat{\gamma}_2 - \gamma_2)^2], \tag{5.5.52}$$

on account of (5.5.47).

Certainly, the expectations $\mathrm{E}[(\hat{\gamma}_1 - \gamma_1)^2]$ (the MSE of $\hat{\gamma}_1$), $\mathrm{E}[(\hat{\gamma}_2 - \gamma_2)^2)]$ (the MSE of $\hat{\gamma}_2$) and $\mathrm{E}[(\hat{\gamma}_1 - \gamma_1)(\hat{\gamma}_2 - \gamma_2)]$ appearing in (5.5.50) and (5.5.52) become $\mathrm{Var}(\hat{\gamma}_1)$, $\mathrm{Var}(\hat{\gamma}_2)$ and $\mathrm{Cov}(\hat{\gamma}_1, \hat{\gamma}_2)$, respectively, provided that the statistics $\hat{\gamma}_1$ and $\hat{\gamma}_2$ are some unbiased estimators of γ_1 and γ_2, respectively. Leaving these expectations as they are in the approximation formulae (5.5.50) and (5.5.52), it becomes possible to approximate the variance (5.5.41) as

$$\begin{aligned}\mathrm{Var}(\widetilde{\boldsymbol{c}'\boldsymbol{\tau}}) \cong{}& \mathrm{Var}(\widehat{\boldsymbol{c}'\boldsymbol{\tau}}) + \sigma_1^2 \sum_{j=1}^{f_*}\{b_0^2\mathrm{E}[(\hat{\gamma}_1 - \gamma_1)^2] \\ &+ 2b_0\mathrm{E}[(\hat{\gamma}_1 - \gamma_1)(\hat{\gamma}_2 - \gamma_2)] + \mathrm{E}[(\hat{\gamma}_2 - \gamma_2)^2]\}\xi_j^{-3}(\boldsymbol{c}'\boldsymbol{r}^{-\delta}\boldsymbol{N}\boldsymbol{D}\boldsymbol{v}_j)^2 \\ &+ \sigma_1^2\mathrm{E}[(\hat{\gamma}_2 - \gamma_2)^2] \sum_{j=f_*+1}^{h_*} \zeta_j^{-3}(\boldsymbol{c}'\boldsymbol{r}^{-\delta}\boldsymbol{N}\boldsymbol{D}\boldsymbol{v}_j)^2. \end{aligned} \tag{5.5.53}$$

However, if $\boldsymbol{c}'\boldsymbol{\tau}$ is a contrast (i.e., $\boldsymbol{c}'\boldsymbol{1}_v = 0$) and the design satisfies the conditions of Remark 5.5.2, then $\boldsymbol{c}'\boldsymbol{r}^{-\delta}\boldsymbol{N}\boldsymbol{D}\boldsymbol{v}_j = 0$ for $j = 1, 2, ..., f_*$ and, hence, (5.5.53) reduces to

$$\mathrm{Var}(\widetilde{\boldsymbol{c}'\boldsymbol{\tau}}) \cong \mathrm{Var}(\widehat{\boldsymbol{c}'\boldsymbol{\tau}}) + \sigma_1^2\mathrm{E}[(\hat{\gamma}_2 - \gamma_2)^2] \sum_{j=f_*+1}^{h_*} \zeta_j^{-3}(\boldsymbol{c}'\boldsymbol{r}^{-\delta}\boldsymbol{N}\boldsymbol{D}\boldsymbol{v}_j)^2. \tag{5.5.54}$$

Now, to make the approximation (5.5.53) or (5.5.54) applicable, the expectations appearing there need to be evaluated or approximated. If the distribution of $\boldsymbol{y}$ is assumed to be normal and the statistics $\hat{\gamma}_1$ and $\hat{\gamma}_2$ are obtained by solving the equations (5.5.20) [or, equivalently, (5.5.23)], i.e., by the MML (REML) method, then $\mathrm{E}[(\hat{\gamma}_1 - \gamma_1)^2]$, $\mathrm{E}[(\hat{\gamma}_2 - \gamma_2)^2)]$ and $\mathrm{E}[(\hat{\gamma}_1 - \gamma_1)(\hat{\gamma}_2 - \gamma_2)]$ can be replaced by the relevant asymptotic moments obtainable from the inverse of the appropriate information matrix. Noting that the information matrix associated with the MML (REML) estimation of γ_1, γ_2 and σ_1^2 is here

$$\frac{1}{2}\begin{bmatrix} \mathrm{tr}[(\boldsymbol{RG'G})^2] & \mathrm{tr}(\boldsymbol{RG'GRD'D}) & \sigma_1^{-2}\mathrm{tr}(\boldsymbol{RG'G}) \\ \mathrm{tr}(\boldsymbol{RD'DRG'G}) & \mathrm{tr}[(\boldsymbol{RD'D})^2] & \sigma_1^{-2}\mathrm{tr}(\boldsymbol{RD'D}) \\ \sigma_1^{-2}\mathrm{tr}(\boldsymbol{RG'G}) & \sigma_1^{-2}\mathrm{tr}(\boldsymbol{RD'D}) & \sigma_1^{-4}(n-v) \end{bmatrix}$$

(see Patterson and Thompson, 1971, p. 554), and taking its inverse, it can be found that

$$\mathrm{Var_{as}}(\hat{\gamma}_1) = \frac{2}{b_{00} - \dfrac{b_{02}(b_{02}b_{11} - b_{01}b_{12}) + b_{01}(b_{01}b_{22} - b_{02}b_{12})}{b_{11}b_{22} - b_{12}^2}}, \tag{5.5.55}$$

$$\mathrm{Var_{as}}(\hat{\gamma}_2) = \frac{2}{b_{11} - \dfrac{b_{12}(b_{00}b_{12} - b_{01}b_{02}) + b_{01}(b_{01}b_{22} - b_{02}b_{12})}{b_{00}b_{22} - b_{02}^2}} \tag{5.5.56}$$

and

$$\mathrm{Cov_{as}}(\hat{\gamma}_1, \hat{\gamma}_2) = \frac{2}{b_{01} - \dfrac{b_{00}(b_{11}b_{22} - b_{12}^2) + b_{02}(b_{01}b_{12} - b_{02}b_{11})}{b_{01}b_{22} - b_{02}b_{12}}}, \tag{5.5.57}$$

where, with $\vartheta_j = b_0\lambda_j$,

$$\begin{aligned}
b_{00} &= \mathrm{tr}[(\boldsymbol{RG'G})^2] = \textstyle\sum_{j=1}^{f_*} \xi_j^{-2}\vartheta_j^2, \\
b_{01} &= \mathrm{tr}(\boldsymbol{RG'GRD'D}) = \textstyle\sum_{j=1}^{f_*} \xi_j^{-2}\vartheta_j\lambda_j, \\
b_{02} &= \mathrm{tr}(\boldsymbol{RG'G}) = \textstyle\sum_{j=1}^{f_*} \xi_j^{-1}\vartheta_j, \\
b_{11} &= \mathrm{tr}[(\boldsymbol{RD'D})^2] = \textstyle\sum_{j=1}^{f_*} \xi_j^{-2}\lambda_j^2 + \sum_{j=f_*+1}^{h_*} \zeta_j^{-2}\lambda_j^2, \\
b_{12} &= \mathrm{tr}(\boldsymbol{RD'D}) = \textstyle\sum_{j=1}^{f_*} \xi_j^{-1}\lambda_j + \sum_{j=f_*+1}^{h_*} \zeta_j^{-1}\lambda_j, \\
b_{22} &= n - v,
\end{aligned}$$

on account of (5.5.34).

Thus, when under the normality assumption the ratios γ_1 and γ_2 are estimated by the MML (REML) method, the variance (5.5.53) can further be approximated as

$$\begin{aligned}
\mathrm{Var}(\widehat{\boldsymbol{c}'\boldsymbol{\tau}}) \cong\ & \mathrm{Var}(\widetilde{\boldsymbol{c}'\boldsymbol{\tau}}) + \sigma_1^2 \sum_{j=1}^{f_*} [\lambda_j^{-2}\vartheta_j^2 \mathrm{Var_{as}}(\hat{\gamma}_1) \\
& + 2\lambda_j^{-1}\vartheta_j \mathrm{Cov_{as}}(\hat{\gamma}_1, \hat{\gamma}_2) + \mathrm{Var_{as}}(\hat{\gamma}_2)]\xi_j^{-3}(\boldsymbol{c}'\boldsymbol{r}^{-\delta}\boldsymbol{ND}\boldsymbol{v}_j)^2 \\
& + \sigma_1^2 \mathrm{Var_{as}}(\hat{\gamma}_2) \sum_{j=f_*+1}^{h_*} \zeta_j^{-3}(\boldsymbol{c}'\boldsymbol{r}^{-\delta}\boldsymbol{ND}\boldsymbol{v}_j)^2.
\end{aligned} \tag{5.5.58}$$

In particular, if the design satisfies the conditions of Remark 5.5.2 and the equality $n_1 = n_2 = \cdots = n_a = n_0$ holds, which implies that also $k_{(1)} = k_{(2)} = \cdots = k_{(a)} = k \; (= n_0/b_0)$, then for any function $\boldsymbol{c}'\boldsymbol{\tau}$, the variance of $\widetilde{\boldsymbol{c}'\boldsymbol{\tau}}$ can be approximated as in (5.5.54) with $\mathrm{E}[(\hat{\gamma}_2 - \gamma_2)^2]$ replaced by

$$\mathrm{Var}_{\mathrm{as}}(\hat{\gamma}_2) = \frac{2}{\sum_{j=f_*+1}^{h_*}\left(\frac{\lambda_j}{\zeta_j}\right)^2 - \frac{1}{n-v-f_*}\left(\sum_{j=f_*+1}^{h_*}\frac{\lambda_j}{\zeta_j}\right)^2}, \tag{5.5.59}$$

i.e., as

$$\mathrm{Var}(\widetilde{\boldsymbol{c}'\boldsymbol{\tau}}) \cong \mathrm{Var}(\widehat{\boldsymbol{c}'\boldsymbol{\tau}}) + \sigma_1^2 \mathrm{Var}_{\mathrm{as}}(\hat{\gamma}_2) \sum_{j=f_*+1}^{h_*} \zeta_j^{-3}(\boldsymbol{c}'\boldsymbol{r}^{-\delta}\boldsymbol{N}\boldsymbol{D}\boldsymbol{v}_j)^2, \tag{5.5.60}$$

with $\mathrm{Var}_{\mathrm{as}}(\hat{\gamma}_2)$ given in (5.5.59). It may be noted that this particular case is the one most common among the resolvable block designs.

It should also be emphasized that the approximations considered here are equivalent to those suggested by Kackar and Harville (1984), when applied to the variance (5.5.41). As to the accuracy of those approximations, taking into account the asymptotic properties of the MML (REML) estimators of γ_1 and γ_2 or, equivalently, those of the iterated MINQUEs of these parameters (see, e.g., Brown, 1976; Rao and Kleffe, 1988, Chapter 10), it can be observed that, under the normality assumption, the estimators $\hat{\gamma}_1$ and $\hat{\gamma}_2$ obtained by solving the equations (5.5.20) [or (5.5.23)] are asymptotically unbiased and efficient (i.e., with the smallest possible limiting variance). Thus, it can be concluded that the approximation (5.5.58) [and, in particular, (5.5.60)] approaches the exact value of $\mathrm{Var}(\widetilde{\boldsymbol{c}'\boldsymbol{\tau}})$ as $n - v$ tends to infinity. However, because in practical applications the increase of n over v is possible within some limits only, the formula (5.5.58) [and (5.5.60)] will always remain merely an approximation. Not much is known about the closeness of this approximation. It remains to be investigated.

5.5.4 Testing hypotheses in a nested block design

As in the analysis of a block design with one stratum of blocks, also in the case of an NB design, i.e., with two block strata, a researcher may wish to test hypotheses concerning contrasts of treatment parameters. In any case, to apply an exact F test of a hypothesis of the type (3.7.60), i.e., concerning a set of ℓ linearly independent contrasts $\boldsymbol{U}'\boldsymbol{\tau}$, one would need to have their BLUEs, $\widehat{\boldsymbol{U}'\boldsymbol{\tau}}$, together with the covariance matrix, $\mathrm{Cov}(\widehat{\boldsymbol{U}'\boldsymbol{\tau}})$, of them. As it is known (see Sections 5.2.2 and 5.5.1), for the case of an NB design, they are available in general only when the true values of γ_1 and γ_2 are known. In practice, however, they are usually not known and have to be estimated. Then, instead of $\boldsymbol{U}'\hat{\boldsymbol{\tau}}$, one has to use $\boldsymbol{U}'\tilde{\boldsymbol{\tau}}$, where $\tilde{\boldsymbol{\tau}}$ is obtainable from (5.5.26). The covariance matrix of $\boldsymbol{U}'\tilde{\boldsymbol{\tau}}$ is then of the form $\mathrm{Cov}(\boldsymbol{U}'\tilde{\boldsymbol{\tau}}) = \boldsymbol{U}'\mathrm{Cov}(\tilde{\boldsymbol{\tau}})\boldsymbol{U}$, where $\mathrm{Cov}(\tilde{\boldsymbol{\tau}})$ is given in (5.5.39). Unfortunately, in this form, it is usually not readily available and cannot be used directly in a test of the hypothesis

$$H_0 : \boldsymbol{U}'\boldsymbol{\tau} = \boldsymbol{q}, \tag{5.5.61}$$

for the same reason as in Section 3.7.4. To derive an approximate test, one can proceed similarly as described in that section, taking however into account the difference between the covariance matrix given in (3.7.45) and that in (5.5.39). The resulting differences in the derivation of the required approximate test of the hypothesis (5.5.61) for the case of an NB design will now be described, assuming that the conditions of Lemma 5.5.2 are satisfied by the design.

First, note that, similarly to (3.7.64), the covariance matrix (5.5.39) can be approximated by replacing the usually intractable exact formula of

$$\mathrm{Var}(z_j) = \mathrm{E}(z_j^2)$$

by its approximation given in (5.5.50) for $j = 1, 2, ..., f_*$ and in (5.5.52) for $j = f_* + 1,\ f_* + 2, ..., h_*$. This approximation can be written as

$$\begin{aligned}\mathrm{Cov}(\tilde{\boldsymbol{\tau}}) \;&\cong\; \mathrm{Cov}(\hat{\boldsymbol{\tau}}) \\ &\quad + \sigma_1^2 \boldsymbol{r}^{-\delta}\boldsymbol{N}\boldsymbol{D}\left\{\sum_{j=1}^{f_*} \xi_j^{-3}\mathrm{E}\{[b_0(\hat{\gamma}_1 - \gamma_1) + \hat{\gamma}_2 - \gamma_2]^2\}\boldsymbol{v}_j\boldsymbol{v}_j'\right. \\ &\qquad \left. + \sum_{j=f_*+1}^{h_*} \zeta_j^{-3}\mathrm{E}[(\hat{\gamma}_2 - \gamma_2)^2]\boldsymbol{v}_j\boldsymbol{v}_j'\right\}\boldsymbol{D}'\boldsymbol{N}'\boldsymbol{r}^{-\delta} \\ &= \sigma_1^2\boldsymbol{C}_{\mathrm{ap}}^{-1} \quad \text{(say)},\end{aligned} \tag{5.5.62}$$

where

$$\begin{aligned}\mathrm{E}\{[b_0(\hat{\gamma}_1 - \gamma_1) + \hat{\gamma}_2 - \gamma_2]^2\} \;&=\; b_0^2\mathrm{E}[(\hat{\gamma}_1 - \gamma_1)^2] + 2b_0\mathrm{E}[(\hat{\gamma}_1 - \gamma_1)(\hat{\gamma}_2 - \gamma_2)] \\ &\quad + \mathrm{E}[(\hat{\gamma}_2 - \gamma_2)^2].\end{aligned}$$

In practice, the matrix $\boldsymbol{C}_{\mathrm{ap}}^{-1}$ becomes applicable after replacing the unknown values of γ_1 and γ_2 in the formulae for ξ_j and ζ_j by their estimates, i.e., in the form

$$\begin{aligned}\hat{\boldsymbol{C}}_{\mathrm{ap}}^{-1} \;&=\; \hat{\boldsymbol{C}}_c^{-1} + \boldsymbol{r}^{-\delta}\boldsymbol{N}\boldsymbol{D}\left\{\sum_{j=1}^{f_*}\hat{\xi}_j^{-3}\mathrm{E}\{[b_0(\hat{\gamma}_1 - \gamma_1) + \hat{\gamma}_2 - \gamma_2]^2\}\boldsymbol{v}_j\boldsymbol{v}_j'\right. \\ &\qquad \left. + \sum_{j=f_*+1}^{h_*}\hat{\zeta}_j^{-3}\mathrm{E}[(\hat{\gamma}_2 - \gamma_2)^2]\boldsymbol{v}_j\boldsymbol{v}_j'\right\}\boldsymbol{D}'\boldsymbol{N}'\boldsymbol{r}^{-\delta},\end{aligned} \tag{5.5.63}$$

where $\hat{\boldsymbol{C}}_c^{-1} = (\boldsymbol{\Delta}\hat{\boldsymbol{T}}^{-1}\boldsymbol{\Delta}')^{-1}$, as used in (5.5.26), and

$$\hat{\xi}_j = (\hat{\gamma}_1 b_0 + \hat{\gamma}_2)\lambda_j + 1 \quad \text{and} \quad \hat{\zeta}_j = \hat{\gamma}_2\lambda_j + 1,$$

the former for $j = 1, 2, ..., f_*$ and the latter for $j = f_* + 1, f_* + 2, ..., h_*$. With the matrix (5.5.63), one can enter the formula (3.7.65) for Y, the quadratic form to be then used as the numerator in the suggested approximate F type test defined in (3.7.75). Its use, however, requires proper approximation of the distribution of Y, obtainable following the approach adopted in Section 3.7.4, but with some modification following from the difference between (3.7.66) and (5.5.63).

Thus, the aim is to approximate the distribution of the quadratic form

$$\sigma_1^{-2}(\boldsymbol{U}'\tilde{\boldsymbol{\tau}} - \boldsymbol{q})'(\boldsymbol{U}'\hat{\boldsymbol{C}}_{\text{ap}}^{-1}\boldsymbol{U})^{-1}(\boldsymbol{U}'\tilde{\boldsymbol{\tau}} - \boldsymbol{q}) = Y \quad \text{(say)}, \tag{5.5.64}$$

where $\hat{\boldsymbol{C}}_{\text{ap}}^{-1}$ is as in (5.5.63), by the distribution of an appropriately scaled χ^2 variable having approximately the same expectation and variance as those of Y. For this purpose, consider first these moments conditional at given $\hat{\gamma}_1$ and $\hat{\gamma}_2$, assuming as usual that the distribution of $\boldsymbol{y}$ is multivariate normal. They receive, then, similar formulae as those in Section 3.7.4, which under H_0 reduce to

$$\text{E}(Y|\hat{\gamma}_1, \hat{\gamma}_2) = \sigma_1^{-2}\text{tr}[(\boldsymbol{U}'\hat{\boldsymbol{C}}_{\text{ap}}^{-1}\boldsymbol{U})^{-1}\boldsymbol{U}'\text{Cov}(\tilde{\boldsymbol{\tau}})\boldsymbol{U}] \tag{5.5.65}$$

and

$$\text{Var}(Y|\hat{\gamma}_1, \hat{\gamma}_2)$$

$$= 2\sigma_1^{-4}\text{tr}[(\boldsymbol{U}'\hat{\boldsymbol{C}}_{\text{ap}}^{-1}\boldsymbol{U})^{-1}\boldsymbol{U}'\text{Cov}(\tilde{\boldsymbol{\tau}})\boldsymbol{U}(\boldsymbol{U}'\hat{\boldsymbol{C}}_{\text{ap}}^{-1}\boldsymbol{U})^{-1}\boldsymbol{U}'\text{Cov}(\tilde{\boldsymbol{\tau}})\boldsymbol{U}], \tag{5.5.66}$$

analogously to (3.7.67) and (3.7.68), respectively. To obtain these moments unconditionally, the formulae (3.7.69) can be used accordingly. However, to make their application practically feasible, the matrices $\text{Cov}(\tilde{\boldsymbol{\tau}})$ and $(\boldsymbol{U}'\hat{\boldsymbol{C}}_{\text{ap}}^{-1}\boldsymbol{U})^{-1}$ are to be approximated. For the first matrix, the approximation (5.5.62) is available. To approximate the second matrix, one can proceed similarly as in Section 3.7.4, using the lower order terms of Taylor's series expansion of that matrix. Here, this procedure means to apply the approximation formula

$$\begin{aligned}
(\boldsymbol{U}'\hat{\boldsymbol{C}}_{\text{ap}}^{-1}\boldsymbol{U})^{-1} \cong{}& (\boldsymbol{U}'\boldsymbol{C}_{\text{ap}}^{-1}\boldsymbol{U})^{-1} + (\hat{\gamma}_1 - \gamma_1)\left.\frac{\partial(\boldsymbol{U}'\hat{\boldsymbol{C}}_{\text{ap}}^{-1}\boldsymbol{U})^{-1}}{\partial\hat{\gamma}_1}\right|_{\hat{\gamma}_1=\gamma_1, \hat{\gamma}_2=\gamma_2} \\
&+ (\hat{\gamma}_2 - \gamma_2)\left.\frac{\partial(\boldsymbol{U}'\hat{\boldsymbol{C}}_{\text{ap}}^{-1}\boldsymbol{U})^{-1}}{\partial\hat{\gamma}_2}\right|_{\hat{\gamma}_1=\gamma_1, \hat{\gamma}_2=\gamma_2} \\
={}& (\boldsymbol{U}'\boldsymbol{C}_{\text{ap}}^{-1}\boldsymbol{U})^{-1} \\
&- (\boldsymbol{U}'\boldsymbol{C}_{\text{ap}}^{-1}\boldsymbol{U})^{-1}\left\{(\hat{\gamma}_1 - \gamma_1)\left.\frac{\partial(\boldsymbol{U}'\hat{\boldsymbol{C}}_{\text{ap}}^{-1}\boldsymbol{U})}{\partial\hat{\gamma}_1}\right|_{\hat{\gamma}_1=\gamma_1, \hat{\gamma}_2=\gamma_2}\right. \\
&\left.+ (\hat{\gamma}_2 - \gamma_2)\left.\frac{\partial(\boldsymbol{U}'\hat{\boldsymbol{C}}_{\text{ap}}^{-1}\boldsymbol{U})}{\partial\hat{\gamma}_2}\right|_{\hat{\gamma}_1=\gamma_1, \hat{\gamma}_2=\gamma_2}\right\}(\boldsymbol{U}'\boldsymbol{C}_{\text{ap}}^{-1}\boldsymbol{U})^{-1},
\end{aligned}$$

where

$$\frac{\partial(U'\hat{C}_{\rm ap}^{-1}U)}{\partial\hat{\gamma}_1} = U'\frac{\partial\hat{C}_{\rm ap}^{-1}}{\partial\hat{\gamma}_1}U \quad \text{and} \quad \frac{\partial(U'\hat{C}_{\rm ap}^{-1}U)}{\partial\hat{\gamma}_2} = U'\frac{\partial\hat{C}_{\rm ap}^{-1}}{\partial\hat{\gamma}_2}U.$$

It will be sufficient to consider its application for the following two most common cases.

Case A. Suppose that in the considered NB design (satisfying the conditions of Lemma 5.5.2) both of the designs, $\mathcal{D}^*$ and $\mathcal{D}$, are either connected or disconnected of the same degree (see Definition 2.2.6 and Lemma 2.3.2), giving $f = \tilde{h}_1$. Then, the matrix (5.5.63) can be written, in terms of the vectors $\{u_j\}$ defined in (3.7.35), as

$$\begin{aligned}\hat{C}_{\rm ap}^{-1} &= r^{-\delta} \\ &\quad + r^{-\delta}N\Bigg\{\sum_{j=1}^{f_*}\frac{(\hat{\gamma}_1 b_0+\hat{\gamma}_2)\hat{\xi}_j^2+\lambda_j \mathrm{E}\{[(\hat{\gamma}_1-\gamma_1)b_0+\hat{\gamma}_2-\gamma_2]^2\}}{\hat{\xi}_j^3}u_ju_j' \\ &\qquad + \sum_{j=f_*+1}^{h_*}\frac{\hat{\gamma}_2\hat{\zeta}_j^2+\lambda_j\mathrm{E}[(\hat{\gamma}_2-\gamma_2)^2]}{\hat{\zeta}_j^3}u_ju_j' \\ &\qquad + (\hat{\gamma}_1 b_0+\hat{\gamma}_2)\sum_{j=h_*+1}^{b}u_ju_j'\Bigg\}N'r^{-\delta}\end{aligned}$$

and, hence, its partial derivatives as

$$\begin{aligned}\frac{\partial\hat{C}_{\rm ap}^{-1}}{\partial\hat{\gamma}_1} &= b_0 r^{-\delta}N\Bigg\{\sum_{j=1}^{f_*}\frac{\hat{\xi}_j^2-3\lambda_j^2\mathrm{E}\{[(\hat{\gamma}_1-\gamma_1)b_0+\hat{\gamma}_2-\gamma_2]^2\}}{\hat{\xi}_j^4}u_ju_j' \\ &\qquad + \sum_{j=h_*+1}^{b}u_ju_j'\Bigg\}N'r^{-\delta}\end{aligned} \tag{5.5.67}$$

and

$$\begin{aligned}\frac{\partial\hat{C}_{\rm ap}^{-1}}{\partial\hat{\gamma}_2} &= r^{-\delta}N\Bigg\{\sum_{j=1}^{f_*}\frac{\hat{\xi}_j^2-3\lambda_j^2\mathrm{E}\{[(\hat{\gamma}_1-\gamma_1)b_0+\hat{\gamma}_2-\gamma_2]^2\}}{\hat{\xi}_j^4}u_ju_j' \\ &\qquad + \sum_{j=f_*+1}^{h_*}\frac{\hat{\zeta}_j^2-3\lambda_j^2\mathrm{E}[(\hat{\gamma}_2-\gamma_2)^2]}{\hat{\zeta}_j^4}u_ju_j' \\ &\qquad + \sum_{j=h_*+1}^{b}u_ju_j'\Bigg\}N'r^{-\delta}.\end{aligned} \tag{5.5.68}$$

Case B. Suppose that in the considered NB design (satisfying the conditions of Lemma 5.5.2) $\mathcal{D}$ is a connected block design, whereas $\mathcal{D}^*$ is disconnected of any degree (connected if the degree is 0), giving $f_* = a - 1$. Then, the matrix (5.5.63) can be written, in terms of $\{\boldsymbol{u}_j\}$, as

$$\begin{aligned}\hat{\boldsymbol{C}}_{\text{ap}}^{-1} &= \boldsymbol{r}^{-\delta} \\ &+ \boldsymbol{r}^{-\delta}\boldsymbol{N}\Bigg\{\sum_{j=1}^{f_*}\frac{(\hat{\gamma}_1 b_0 + \hat{\gamma}_2)\hat{\xi}_j^2 + \lambda_j \mathrm{E}\{[(\hat{\gamma}_1 - \gamma_1)b_0 + \hat{\gamma}_2 - \gamma_2]^2\}}{\hat{\xi}_j^3}\boldsymbol{u}_j\boldsymbol{u}_j' \\ &\qquad + \sum_{j=f_*+1}^{h_*}\frac{\hat{\gamma}_2\hat{\zeta}_j^2 + \lambda_j \mathrm{E}[(\hat{\gamma}_2 - \gamma_2)^2]}{\hat{\zeta}_j^3}\boldsymbol{u}_j\boldsymbol{u}_j' \\ &\qquad + \hat{\gamma}_2\sum_{j=h_*+1}^{b-1}\boldsymbol{u}_j\boldsymbol{u}_j' + (\hat{\gamma}_1 b_0 + \hat{\gamma}_2)b^{-1}\mathbf{1}_b\mathbf{1}_b'\Bigg\}\boldsymbol{N}'\boldsymbol{r}^{-\delta}\end{aligned}$$

and, hence, its partial derivatives as

$$\begin{aligned}\frac{\partial\hat{\boldsymbol{C}}_{\text{ap}}^{-1}}{\partial\hat{\gamma}_1} &= b_0\boldsymbol{r}^{-\delta}\boldsymbol{N}\Bigg\{\sum_{j=1}^{f_*}\frac{\hat{\xi}_j^2 - 3\lambda_j^2\mathrm{E}\{[(\hat{\gamma}_1 - \gamma_1)b_0 + \hat{\gamma}_2 - \gamma_2]^2\}}{\hat{\xi}_j^4}\boldsymbol{u}_j\boldsymbol{u}_j' \\ &\qquad + b^{-1}\mathbf{1}_b\mathbf{1}_b'\Bigg\}\boldsymbol{N}'\boldsymbol{r}^{-\delta}\end{aligned} \tag{5.5.69}$$

and

$$\begin{aligned}\frac{\partial\hat{\boldsymbol{C}}_{\text{ap}}^{-1}}{\partial\hat{\gamma}_2} &= \boldsymbol{r}^{-\delta}\boldsymbol{N}\Bigg\{\sum_{j=1}^{f_*}\frac{\hat{\xi}_j^2 - 3\lambda_j^2\mathrm{E}\{[(\hat{\gamma}_1 - \gamma_1)b_0 + \hat{\gamma}_2 - \gamma_2]^2\}}{\hat{\xi}_j^4}\boldsymbol{u}_j\boldsymbol{u}_j' \\ &\qquad + \sum_{j=f_*+1}^{h_*}\frac{\hat{\zeta}_j^2 - 3\lambda_j^2\mathrm{E}[(\hat{\gamma}_2 - \gamma_2)^2]}{\hat{\zeta}_j^4}\boldsymbol{u}_j\boldsymbol{u}_j' \\ &\qquad + \sum_{j=h_*+1}^{b}\boldsymbol{u}_j\boldsymbol{u}_j'\Bigg\}\boldsymbol{N}'\boldsymbol{r}^{-\delta},\end{aligned}$$

exactly as in (5.5.68).

With these results, the moments (5.5.65) and (5.5.66) can obtain more suitable approximate forms

$$\mathrm{E}(Y|\hat{\gamma}_1, \hat{\gamma}_2) \cong \ell - (\hat{\gamma}_1 - \gamma_1)\mathrm{tr}(\boldsymbol{A}_1) - (\hat{\gamma}_2 - \gamma_2)\mathrm{tr}(\boldsymbol{A}_2), \tag{5.5.70}$$

$$\begin{aligned}\mathrm{Var}(Y|\hat{\gamma}_1, \hat{\gamma}_2) &\cong 2[\ell - 2(\hat{\gamma}_1 - \gamma_1)\mathrm{tr}(\boldsymbol{A}_1) - 2(\hat{\gamma}_2 - \gamma_2)\mathrm{tr}(\boldsymbol{A}_2) \\ &\quad + (\hat{\gamma}_1 - \gamma_1)^2\mathrm{tr}(\boldsymbol{A}_1\boldsymbol{A}_1) + 2(\hat{\gamma}_1 - \gamma_1)(\hat{\gamma}_2 - \gamma_2)\mathrm{tr}(\boldsymbol{A}_1\boldsymbol{A}_2) \\ &\quad + (\hat{\gamma}_2 - \gamma_2)^2\mathrm{tr}(\boldsymbol{A}_2\boldsymbol{A}_2)],\end{aligned} \tag{5.5.71}$$

where

$$A_1 = (U'C_{\text{ap}}^{-1}U)^{-1}U'\frac{\partial \hat{C}_{\text{ap}}^{-1}}{\partial \hat{\gamma}_1}\Bigg|_{\hat{\gamma}_1=\gamma_1,\hat{\gamma}_2=\gamma_2} U \tag{5.5.72}$$

and

$$A_2 = (U'C_{\text{ap}}^{-1}U)^{-1}U'\frac{\partial \hat{C}_{\text{ap}}^{-1}}{\partial \hat{\gamma}_2}\Bigg|_{\hat{\gamma}_1=\gamma_1,\hat{\gamma}_2=\gamma_2} U, \tag{5.5.73}$$

ℓ being the rank of the $v \times \ell$ matrix U. Depending on the designs $\mathcal{D}^*$ and $\mathcal{D}$ that compose the considered NB design, the partial derivatives appearing in (5.5.72) and (5.5.73) will have for Case A the forms given in (5.5.67) and (5.5.68) and for Case B the forms given in (5.5.69) and (5.5.68), respectively. If other than any of these two cases is to be considered, no difficulty should exist in finding the appropriate formulae for those derivatives.

Also, it may be noted that if, as a particular instance of Case B, the considered NB design satisfies the conditions of Remark 5.5.2, then for $j = 1, 2, ..., f_*$ the equalities

$$U'r^{-\delta}Nu_j = \lambda_j^{-1/2}U'r^{-\delta}NDv_j = \lambda_j^{-1/2}b_0^{-1}U'r^{-\delta}\mathcal{R}Gv_j = \mathbf{0} \tag{5.5.74}$$

hold. Evidently, (5.5.74) implies that $A_1 = \mathbf{O}$ and

$$\begin{aligned} A_2 \;=\; & (U'C_{\text{ap}}^{-1}U)^{-1}U'r^{-\delta}N\Bigg\{\sum_{j=f_*+1}^{h_*}\frac{\zeta_j^2 - 3\lambda_j^2\mathrm{E}[(\hat{\gamma}_2-\gamma_2)^2]}{\zeta_j^4}u_ju_j' \\ & + \sum_{j=h_*+1}^{b-1} u_ju_j'\Bigg\}N'r^{-\delta}U, \end{aligned} \tag{5.5.75}$$

with

$$U'C_{\text{ap}}^{-1}U = U'C_c^{-1}U + U'r^{-\delta}ND\sum_{j=f_*+1}^{h_*}\zeta_j^{-3}\mathrm{E}[(\hat{\gamma}_2-\gamma_2)^2]v_jv_j'D'N'r^{-\delta}U,$$

the term $\sum_{j=h_*+1}^{b-1} u_ju_j'$ in (5.5.75) disappearing when the design $\mathcal{D}^*$ is connected.

Now, extending the formulae (3.7.69) to the forms

$$\mathrm{E}(Y) = \mathrm{E}_{\hat{\gamma}_1,\hat{\gamma}_2}[\mathrm{E}(Y|\hat{\gamma}_1,\hat{\gamma}_2)]$$

and

$$\mathrm{Var}(Y) = \mathrm{E}_{\hat{\gamma}_1,\hat{\gamma}_2}[\mathrm{Var}(Y|\hat{\gamma}_1,\hat{\gamma}_2)] + \mathrm{Var}_{\hat{\gamma}_1,\hat{\gamma}_2}[\mathrm{E}(Y|\hat{\gamma}_1,\hat{\gamma}_2)],$$

and applying to them the approximations (5.5.70) and (5.5.71), with the substitutions $\mathrm{E}(\hat{\gamma}_1) = \gamma_1$, $\mathrm{E}(\hat{\gamma}_2) = \gamma_2$, $\mathrm{Var}(\hat{\gamma}_1) = \mathrm{E}[(\hat{\gamma}_1-\gamma_1)^2] = \mathrm{Var}_{\text{as}}(\hat{\gamma}_1)$, $\mathrm{Var}(\hat{\gamma}_2) =$

$\mathrm{E}[(\hat{\gamma}_2-\gamma_2)^2] = \mathrm{Var}_{\mathrm{as}}(\hat{\gamma}_2)$ and $\mathrm{Cov}(\hat{\gamma}_1,\hat{\gamma}_2) = \mathrm{E}[(\hat{\gamma}_1-\gamma_1)(\hat{\gamma}_2-\gamma_2)] = \mathrm{Cov}_{\mathrm{as}}(\hat{\gamma}_1,\hat{\gamma}_2)$, justified asymptotically, it is possible to approximate the required moments as in Section 3.7.4. The results so obtained are

$$\mathrm{E}(Y) \cong \ell = \mathrm{E}_{\mathrm{ap}}(Y) \quad \text{(say)}$$

and

$$\begin{aligned}\mathrm{Var}(Y) &\cong 2\ell + \mathrm{Var}_{\mathrm{as}}(\hat{\gamma}_1)c_{11} + 2\mathrm{Cov}_{\mathrm{as}}(\hat{\gamma}_1,\hat{\gamma}_2)c_{12} + \mathrm{Var}_{\mathrm{as}}(\hat{\gamma}_2)c_{22} \\ &= \mathrm{Var}_{\mathrm{ap}}(Y) \quad \text{(say)},\end{aligned}$$

where

$$c_{11} = 2\mathrm{tr}(\boldsymbol{A}_1\boldsymbol{A}_1) + [\mathrm{tr}(\boldsymbol{A}_1)]^2, \quad c_{22} = 2\mathrm{tr}(\boldsymbol{A}_2\boldsymbol{A}_2) + [\mathrm{tr}(\boldsymbol{A}_2)]^2$$

and

$$c_{12} = 2\mathrm{tr}(\boldsymbol{A}_1\boldsymbol{A}_2) + \mathrm{tr}(\boldsymbol{A}_1)\mathrm{tr}(\boldsymbol{A}_2),$$

and $\mathrm{Var}_{\mathrm{as}}(\hat{\gamma}_1)$, $\mathrm{Var}_{\mathrm{as}}(\hat{\gamma}_2)$ and $\mathrm{Cov}_{\mathrm{as}}(\hat{\gamma}_1,\hat{\gamma}_2)$ are given in (5.5.55), (5.5.56) and (5.5.57), respectively, whereas the matrices $\boldsymbol{A}_1$ and $\boldsymbol{A}_2$ are defined in (5.5.72) and (5.5.73), respectively. These results can be used to approximate the distribution of Y by that of $g\chi^2_d$, where g is a scale factor and χ^2_d stands for a random variable having the central χ^2 distribution with d d.f. Proceeding as in Section 3.7.4, one obtains the required extensions of (3.7.73) and (3.7.74) as

$$\begin{aligned} g &= \frac{\mathrm{Var}_{\mathrm{ap}}(Y)}{2\mathrm{E}_{\mathrm{ap}}(Y)} \\ &= \frac{2\ell + \mathrm{Var}_{\mathrm{as}}(\hat{\gamma}_1)c_{11} + 2\mathrm{Cov}_{\mathrm{as}}(\hat{\gamma}_1,\hat{\gamma}_2)c_{12} + \mathrm{Var}_{\mathrm{as}}(\hat{\gamma}_2)c_{22}}{2\ell} \end{aligned} \tag{5.5.76}$$

and

$$\begin{aligned} d &= \frac{2[\mathrm{E}_{\mathrm{ap}}(Y)]^2}{\mathrm{Var}_{\mathrm{ap}}(Y)} \\ &= \frac{2\ell^2}{2\ell + \mathrm{Var}_{\mathrm{as}}(\hat{\gamma}_1)c_{11} + 2\mathrm{Cov}_{\mathrm{as}}(\hat{\gamma}_1,\hat{\gamma}_2)c_{12} + \mathrm{Var}_{\mathrm{as}}(\hat{\gamma}_2)c_{22}}, \end{aligned} \tag{5.5.77}$$

respectively.

These results show that to approximate the distribution, under H_0, of the quadratic form Y given in (5.5.64) by that of $g\chi^2_d$, one can use for g and d the results (5.5.76) and (5.5.77), respectively, where $gd = \ell$. Then, the statistic Y/ℓ can be treated as having approximately the distribution of χ^2_d/d, if H_0 is true.

As to the denominator of the suggested approximate F type statistic, exactly the same quadratic form $\boldsymbol{y}'\boldsymbol{R}_{\mathrm{red}}\boldsymbol{y} = \boldsymbol{y}'\boldsymbol{\psi}_1\boldsymbol{y}$ as in (3.7.75) can be used, following the same argument as that considered in Section 3.7.4.

Thus, in the analysis of an NB design satisfying the conditions of Lemma 5.5.2, the hypothesis (5.5.61) can be tested approximately using the statistic

$$F = \frac{n-b-h}{\ell}\,\frac{(\boldsymbol{U}'\tilde{\boldsymbol{\tau}}-\boldsymbol{q})'(\boldsymbol{U}'\hat{\boldsymbol{C}}_{\mathrm{ap}}^{-1}\boldsymbol{U})^{-1}(\boldsymbol{U}'\tilde{\boldsymbol{\tau}}-\boldsymbol{q})}{\boldsymbol{y}'\boldsymbol{R}_{\mathrm{red}}\boldsymbol{y}}, \tag{5.5.78}$$

with the matrix $\hat{C}_{\rm ap}^{-1}$ now in the form (5.5.63), being an appropriate extension of that in (3.7.66). The distribution of this statistic under the normality of $\boldsymbol{y}$ can be, justifiably, approximated by the central F distribution with d and $n-b-h$ d.f., where d is now as given in (5.5.77), thus being a relevant extension of that in (3.7.74). However, if the considered NB design satisfies the conditions of Remark 5.5.2, then $\boldsymbol{U}'\hat{\boldsymbol{C}}_{\rm ap}^{-1}\boldsymbol{U}$ in (5.5.78) reduces to

$$\boldsymbol{U}'\hat{\boldsymbol{C}}_{\rm ap}^{-1}\boldsymbol{U} = \boldsymbol{U}'\hat{\boldsymbol{C}}_{c}^{-1}\boldsymbol{U} + \boldsymbol{U}'\boldsymbol{r}^{-\delta}\boldsymbol{N}\boldsymbol{D}\sum_{j=f_*+1}^{h_*}\hat{\zeta}_j^{-3}\mathrm{E}[(\hat{\gamma}_2-\gamma_2)^2]\boldsymbol{v}_j\boldsymbol{v}_j'\boldsymbol{D}'\boldsymbol{N}'\boldsymbol{r}^{-\delta}\boldsymbol{U},$$

and the formula (5.5.77) for d simplifies to

$$d = \frac{2\ell^2}{2\ell + \mathrm{Var}_{\rm as}(\hat{\gamma}_2)\{2\mathrm{tr}(\boldsymbol{A}_2\boldsymbol{A}_2) + [\mathrm{tr}(\boldsymbol{A}_2)]^2\}},$$

where $\boldsymbol{A}_2$ is of the form (5.5.75). If in addition the equality $n_1 = n_2 = \cdots = n_a = n_0$ holds (which also implies constant block sizes in $\mathcal{D}^*$), then $\mathrm{Var}_{\rm as}(\hat{\gamma}_2)$ obtains the simple form given in (5.5.59).

Finally, it should be mentioned that (as noticed in Section 3.7.4) if the contrasts $\boldsymbol{U}'\boldsymbol{\tau}$ in (5.5.61) obtain the BLUEs in the intra-block analysis with full efficiency, then the test based on the statistic (5.5.78) coincides with the exact F test in the intra-block analysis.

5.6 Concluding remarks on nested block designs

The general theory concerning the analysis of NB designs presented in this chapter reveals the possibilities of obtaining BLUEs for interesting treatment parametric functions either under the overall randomization model (Section 5.2) or its submodels related to the various strata of the nested classification of available experimental units (Section 5.3). Only in some special cases (Section 5.3.5) one can obtain the BLUEs under the overall model. Usually for an interesting function, the BLUE of it can be obtained under one or more submodels. For contrasts of treatment parameters, the submodels to be searched are the intra-block, inter-block-intra-superblock and inter-superblock submodels.

There may be cases when for contrasts of interest the BLUEs exist under all three submodels, or only under one or two of them. Their existence depends on the relations of the contrasts to the relevant C-matrices and, in the case of the second or third submodel, on some additional conditions.

For estimation under the second model (Section 5.3.2), the additional condition relates the considered contrast to the departures of the incidence matrices $\boldsymbol{N}_h$ from relevant orthogonal structures, modified by possible block-size inequalities within the component designs $\mathcal{D}_h$, $h = 1, 2, ..., a$ [Corollary 5.3.1(b)]. One extreme case is when all $\mathcal{D}_h$ are orthogonal ($\boldsymbol{N}_h = n_h^{-1}\boldsymbol{r}_h\boldsymbol{k}_h'$). Then, no contrasts can be estimated under the inter-block-intra-superblock submodel [Corollary

5.3.1(a)]. The other extreme case is when for any $\mathcal{D}_h$ for which $\boldsymbol{N}_h \neq n_h^{-1}\boldsymbol{r}_h\boldsymbol{k}_h'$, the block sizes are equal. Then, any contrast generated by

$$\tilde{\boldsymbol{C}}_2 = \tilde{\boldsymbol{N}}_0\boldsymbol{k}^{-\delta}\tilde{\boldsymbol{N}}_0'$$

obtains the BLUE under this submodel (Remark 5.3.1). Although the first extreme case would be rather uncommon, the second may happen often in practice, as, e.g., in Example 5.3.1. Another common design case is that in which the columns of each $\boldsymbol{N}_h$ are linearly independent. In such an NB design, the existence of BLUEs does not depend on the block sizes [Remark 5.3.2(b)]. To this class of NB designs, in particular, the so-called α-designs belong (see, e.g., Patterson, Williams and Hunter, 1978; John, 1987, Section 4.8).

As to the third submodel (Section 5.3.3), the additional condition relates the considered contrast to the difference between the departure of the incidence matrix $\boldsymbol{N}$ from the relevant orthogonal structure and such departures of the $\boldsymbol{N}_h$ incidence matrices, this being modified by corresponding possible block-size inequalities, and to the departure of the incidence matrix $\mathfrak{R}$ from the relevant orthogonal structure modified by possible inequalities among superblock sizes [Corollary 5.3.3(b)]. Again, two extreme cases can be visualized: the first, when the design $\mathcal{D}$ is orthogonal, the second, when all block sizes are equal and all superblock sizes are equal. In the first of these cases, no contrast can be estimated under the inter-superblock submodel [Corollary 5.3.3(a)]; in the second case, any contrast generated by

$$\tilde{\boldsymbol{C}}_3 = \mathfrak{R}_0\boldsymbol{n}^{-\delta}\mathfrak{R}_0'$$

will obtain the BLUE under this submodel, provided that $\mathfrak{R}_0 \neq \mathbf{O}$. Here, in fact, the first extreme case can be considered as common for all resolvable block designs, as illustrated by Example 5.3.4. On the other hand, some factorial experiments may be designed in such a way that $\mathfrak{R} \neq n^{-1}\boldsymbol{r}\boldsymbol{n}'$, as, e.g., in Example 5.3.2. If in such a case the columns of the incidence matrix $\mathfrak{R}$ are linearly independent (as in that example), then for any contrast generated by $\boldsymbol{C}_3$, the BLUE exists under this submodel, independently of the block and superblock sizes [Remark 5.3.4(b)]. It exists always for any such contrast when the block and superblock sizes are constant [Corollary 5.3.3(d)], as in Example 5.3.3.

The total-area submodel (Section 5.3.4) provides BLUEs for the general parametric mean, or its multiplicities, only. This function cannot be estimated under any other submodel. It obtains the BLUE under the overall model if and only if the block sizes are constant within any connected subdesign of $\mathcal{D}^*$, and the sizes of the superblocks are constant within any of the connected subdesigns of $\mathcal{D}$ [Corollary 5.3.5(d)]. This situation is illustrated by Examples 5.3.3 and 5.3.4.

If the condition of constant block sizes and constant superblock sizes holds, then, as shown in Section 5.4, the NB design induces the OBS property of the experiment (see Lemma 5.4.1). If, in addition, the matrix $\boldsymbol{N}\boldsymbol{N}'\boldsymbol{r}^{-\delta}\mathfrak{R}\mathfrak{R}'$ is symmetric, then the NB design is GB (Lemma 5.4.2). These OBS and GB properties are usually desirable, because they allow the estimation of contrasts within different strata to be considered in a uniform way (Theorem 5.4.1).

Notably, if a contrast is estimated under more than one submodel, i.e., within two or three strata, then it is desirable to combine the obtained information on the contrast from the relevant strata. This procedure is considered in Section 5.5. To make use of that section, one has to realize that the theory of estimating the vector of treatment parameters $\boldsymbol{\tau}$ and a parametric function $\mathbf{c}'\boldsymbol{\tau}$, presented in Sections 5.5.1 and 5.5.2, is applicable to any NB design, regardless of the constructions of the underlying designs $\mathcal{D}^*$ and $\mathcal{D}$. The approximation formula (5.5.53) for the variance of the resulting estimator of $\mathbf{c}'\boldsymbol{\tau}$ can, however, be applied under certain conditions only. These conditions are given in Lemma 5.5.2. They are also to be satisfied when proceeding to tests of hypotheses, as considered in Section 5.5.4. Furthermore, to apply the simplified approximation formula (5.5.54), and the relevant simplified test procedure, it is required that the NB design satisfies the additional conditions given in Remark 5.5.2, i.e., that all its component designs, $\mathcal{D}_h$, $h = 1, 2, ..., a$, are proper with the same number of blocks, and its superblock design, $\mathcal{D}$, is connected and orthogonal. The requirements of Lemma 5.5.2 are met by many, though not all, NB designs commonly used in practice, and those of Remark 5.5.2 are satisfied by most of the NB designs constructed according to the concept of resolvability. Note, in particular, that among the traditional NB designs, the well-known lattice designs belong to this class, as well as all other proper resolvable incomplete block designs (see, e.g., John, 1987, Sections 3.4 and 4.7–4.10). In fact, the wide class of α-resolvable block designs originated by Bose (1942) and generalized by Shrikhande and Raghavarao (1964), later used and extended further by various authors, in particular, by Kageyama (1972, 1973, 1976a), Patterson and Williams (1976), Williams (1976), Williams, Patterson and John (1976, 1977), and Ceranka, Kageyama and Mejza (1986), is a subclass of the large class of NB designs satisfying the conditions of Remark 5.5.2, with the exception of those resolvable block designs that are composed of component designs, $\mathcal{D}_h$, which are nonproper (of unequal block sizes) or differ among themselves with regard to the number of blocks, or both. Thus, the most general class of resolvable block designs considered by Kageyama (1976a), that which allows for unequal block sizes and unequal block numbers in the component designs, exceeds the above class of NB designs satisfying the conditions of Remark 5.5.2. Therefore, an approximation formula more general than that given in (5.5.53), or in (5.5.62), is still needed. Interesting results in this direction have been obtained by Kenward and Roger (1997). Another approach suitable for a special case of nested multidimensional block designs has been considered by Srivastava and Beaver (1986).

Finally, special attention should be given to NB designs that are GB in the sense of Definition 3.6.1, extended in Section 5.4. From Sections 5.4 and 5.5, it has become evident that NB designs with the GB property are particularly attractive from the point of view of analytical simplicity, this applying especially to proper and resolvable block designs. Further studies worth undertaking in connection with GB NB designs should include relevant extensions of the results given in Sections 3.8.2–3.8.5.

Appendix
A. Subspaces and Projections

Although it is assumed that the reader is familiar with the terminology and results of linear algebra used in the basic literature on experimental design, it will be useful to recall some notions and results concerning the vector space theory, particularly those related to projections on subspaces (as mentioned in Section 1.5).

A.1 Linear vector subspaces

It may be useful to recall first the basic terminology and results concerning real vector spaces and their linear subspaces.

Definition A.1.1. The set of all n-component vectors of real numbers is called the n-dimensional real vector space and is denoted by $\mathcal{R}^n$.

If not otherwise stated, the $n \times 1$ vector $[x_1, x_2, ..., x_n]'$ will simply be denoted by $\boldsymbol{x}$, and $\boldsymbol{x} \in \mathcal{R}^n$ will be written to indicate that $\boldsymbol{x}$ "is in" $\mathcal{R}^n$.

Definition A.1.2. A set of n-component vectors is a linear subspace of $\mathcal{R}^n$ if for any two vectors in the set, their sum is in the set, and if for any vector in the set, its product by a scalar is in the set.

For more on these concepts see, e.g., Rao (1973, Section 1a) or Schott (1997, Chapter 2).

In the geometrical interpretation of a real vector space, the vector $\boldsymbol{x} = [x_1, x_2, ..., x_n]'$ is considered as the direct line segment from the origin O to the point P (say) with coordinates $x_1, x_2, ..., x_n$ in an n-dimensional Euclidean space. In this space, the product $\boldsymbol{x}'\boldsymbol{y} = \sum_{i=1}^{n} x_i y_i$, called the inner (or scalar) product of two vectors $\boldsymbol{x} = [x_1, x_2, ..., x_n]'$ and $\boldsymbol{y} = [y_1, y_2, ..., y_n]'$, plays an important role in defining some geometrical concepts.

Definition A.1.3. The norm (or length) of a vector $\boldsymbol{x}$, written $||\boldsymbol{x}||$, is defined to be $||\boldsymbol{x}|| = (\boldsymbol{x}'\boldsymbol{x})^{1/2}$.

Definition A.1.4. The distance between two vectors $\boldsymbol{x}$ and $\boldsymbol{y}$ (also called Euclidean distance) is defined to be $||\boldsymbol{x} - \boldsymbol{y}|| = [(\boldsymbol{x} - \boldsymbol{y})'(\boldsymbol{x} - \boldsymbol{y})]^{1/2}$.

Definition A.1.5. The angle α between two nonzero vectors $\boldsymbol{x}$ and $\boldsymbol{y}$ is defined by $\cos\alpha = \boldsymbol{x}'\boldsymbol{y}/(||\boldsymbol{x}||||\boldsymbol{y}||)$, which lies in the closed interval $[-1, 1]$.

Definition A.1.6. For any two vectors $\boldsymbol{x}$ and $\boldsymbol{y}$ such that $\boldsymbol{y} \neq \boldsymbol{0}$, the vector $||\boldsymbol{y}||^{-2}(\boldsymbol{x}'\boldsymbol{y})\boldsymbol{y}$ is called the projection of $\boldsymbol{x}$ on $\boldsymbol{y}$.

Now, further useful concepts can be introduced.

Definition A.1.7. Two vectors $\boldsymbol{x}$ and $\boldsymbol{y}$ are said to be orthogonal if $\boldsymbol{x}'\boldsymbol{y} = 0$ (i.e., their inner product is zero).

It is evident from Definitions A.1.6 and A.1.7 that two nonzero vectors are orthogonal if and only if the projection of either one on the other is $\boldsymbol{0}$. If this happens, one of the vectors is perpendicular to the other (see Definition A.1.5). Hence, the algebraic concept of orthogonality has its geometrical interpretation in the sense of perpendicularity. The notation $\boldsymbol{x} \perp \boldsymbol{y}$ is used to indicate that $\boldsymbol{x}$ and $\boldsymbol{y}$ are orthogonal (perpendicular).

Definition A.1.8. A vector $\boldsymbol{x}$ is said to be normalized if $\boldsymbol{x}'\boldsymbol{x} = 1$.

Definition A.1.9. Two vectors are said to be orthonormal if they are normalized and orthogonal.

It can also be said that a set of vectors is orthogonal (orthonormal) if all vectors in the set are pairwise orthogonal (orthonormal).

The concept of orthogonality between two vectors can be extended to this type of relation between a vector and a linear subspace of $\mathcal{R}^n$. Of particular interest is a linear subspace corresponding to a given matrix.

Definition A.1.10. Let $\boldsymbol{A} = [\boldsymbol{a}_1 : \boldsymbol{a}_2 : \cdots : \boldsymbol{a}_m]$ be an $n \times m$ matrix such that its m columns are in $\mathcal{R}^n$. The linear subspace spanned by these columns, i.e., the set of all vectors that are linear combinations of $\boldsymbol{a}_1, \boldsymbol{a}_2, ..., \boldsymbol{a}_m$, is called the column space (range) of $\boldsymbol{A}$; it is written $\mathcal{C}(\boldsymbol{A})$. The rank of $\boldsymbol{A}$ is then called the dimension of $\mathcal{C}(\boldsymbol{A})$, which can be written $\text{rank}(\boldsymbol{A}) = \dim[\mathcal{C}(\boldsymbol{A})]$.

Note that one can write $\mathcal{C}(\boldsymbol{A}) \subset \mathcal{R}^n$ if $\boldsymbol{A}$ has n rows.

Definition A.1.11. Let $\mathcal{C}(\boldsymbol{A}) \subset \mathcal{R}^n$ and $\boldsymbol{x} \in \mathcal{R}^n$. Then, $\boldsymbol{x}$ is said to be orthogonal to $\mathcal{C}(\boldsymbol{A})$, which is denoted by $\boldsymbol{x} \perp \mathcal{C}(\boldsymbol{A})$, if $\boldsymbol{x}$ is orthogonal to every vector in $\mathcal{C}(\boldsymbol{A})$.

Relevant results are given in the following five lemmas, which can be proved as in Scheffé (1959, Appendix I).

Lemma A.1.1. *If $\mathcal{C}(\boldsymbol{A}) \subset \mathcal{R}^n$, then a vector $\boldsymbol{x} \in \mathcal{R}^n$ is orthogonal to $\mathcal{C}(\boldsymbol{A})$ if and only if $\boldsymbol{x}$ is orthogonal to each column of $\boldsymbol{A}$.*

Lemma A.1.2. *Let $\mathcal{C}(\boldsymbol{A}) \subset \mathcal{R}^n$ and $\boldsymbol{x} \in \mathcal{R}^n$. There exist vectors $\boldsymbol{y}$ and $\boldsymbol{z}$ such that $\boldsymbol{x} = \boldsymbol{y} + \boldsymbol{z}$, where $\boldsymbol{y} \in \mathcal{C}(\boldsymbol{A})$ and $\boldsymbol{z} \perp \mathcal{C}(\boldsymbol{A})$. This decomposition is unique in the sense that if there exist $\boldsymbol{y}^* \in \mathcal{C}(\boldsymbol{A})$ and $\boldsymbol{z}^* \perp \mathcal{C}(\boldsymbol{A})$ such that $\boldsymbol{x} = \boldsymbol{y}^* + \boldsymbol{z}^*$, then necessarily, $\boldsymbol{y} = \boldsymbol{y}^*$ and $\boldsymbol{z} = \boldsymbol{z}^*$.*

This lemma allows the following useful concept to be introduced.

Definition A.1.12. Given a vector $\boldsymbol{x} \in \mathcal{R}^n$, then, the vector $\boldsymbol{y} \in \mathcal{C}(\boldsymbol{A}) \subset \mathcal{R}^n$ defined in Lemma A.1.2, which is such that $(\boldsymbol{x} - \boldsymbol{y}) \perp \mathcal{C}(\boldsymbol{A})$, is called the orthogonal projection of $\boldsymbol{x}$ on $\mathcal{C}(\boldsymbol{A})$.

Note that if $\boldsymbol{A}$ is reduced to a single vector $\boldsymbol{a}$, then Definition A.1.12 reduces to Definition A.1.6.

Lemma A.1.3. *Given a fixed $\mathcal{C}(\boldsymbol{A}) \subset \mathcal{R}^n$, a fixed vector $\boldsymbol{x} \in \mathcal{R}^n$ and a variable vector $\boldsymbol{y} \in \mathcal{C}(\boldsymbol{A})$, then, $\|\boldsymbol{x} - \boldsymbol{y}\|$ has a minimum value, which is attained if and only if $\boldsymbol{y}$ is the orthogonal projection of $\boldsymbol{x}$ on $\mathcal{C}(\boldsymbol{A})$.*

Lemma A.1.4. *If $\mathcal{C}(\boldsymbol{A}) \subset \mathcal{R}^n$, $\boldsymbol{x} \in \mathcal{R}^n$ and $\boldsymbol{x}$ is orthogonal to every $\boldsymbol{y}$ orthogonal to $\mathcal{C}(\boldsymbol{A})$, then $\boldsymbol{x} \in \mathcal{C}(\boldsymbol{A})$.*

This lemma implies the use of the following concept.

Definition A.1.13. Let $\mathcal{C}(\boldsymbol{A}) \subset \mathcal{R}^n$ and $\mathcal{C}(\boldsymbol{B}) \subset \mathcal{R}^n$. If $\mathcal{C}(\boldsymbol{A}) \subset \mathcal{C}(\boldsymbol{B})$, i.e., $\boldsymbol{x} \in \mathcal{C}(\boldsymbol{A})$ implies $\boldsymbol{x} \in \mathcal{C}(\boldsymbol{B})$, then the totality of vectors in $\mathcal{C}(\boldsymbol{B})$ that are orthogonal to $\mathcal{C}(\boldsymbol{A})$ is called the orthogonal complement (orthocomplement) of $\mathcal{C}(\boldsymbol{A})$ in $\mathcal{C}(\boldsymbol{B})$.

To this definition, a relevant result is the following lemma.

Lemma A.1.5. *The orthogonal complement of $\mathcal{C}(\boldsymbol{A})$ in $\mathcal{C}(\boldsymbol{B})$ is a linear subspace of $\mathcal{R}^n$ [which means that it can be written as $\mathcal{C}(\boldsymbol{C}) \subset \mathcal{R}^n$ for some matrix $\boldsymbol{C}$]. If $\text{rank}(\boldsymbol{A}) = s$ and $\text{rank}(\boldsymbol{B}) = r$, then the dimension of this subspace is $r - s$ [i.e., $\text{rank}(\boldsymbol{C}) = r - s$].*

It can also be shown that if $\mathcal{C}(\boldsymbol{C})$ denotes the orthogonal complement of $\mathcal{C}(\boldsymbol{A})$ in $\mathcal{C}(\boldsymbol{B})$, then the orthogonal complement of $\mathcal{C}(\boldsymbol{C})$ in $\mathcal{C}(\boldsymbol{B})$ is $\mathcal{C}(\boldsymbol{A})$. Lemma A.1.4 may be considered as following from this remark when $r = n$.

Definition A.1.13 applies in particular to the case in which $\mathcal{C}(\boldsymbol{A})$ is any linear subspace of $\mathcal{R}^n$ and $\mathcal{C}(\boldsymbol{B})$ is taken to be $\mathcal{R}^n$. The orthogonal complement of $\mathcal{C}(\boldsymbol{A})$ in $\mathcal{R}^n$ is then written as $\mathcal{C}^{\perp}(\boldsymbol{A})$.

Another set of vectors corresponding to an $n \times m$ matrix $\boldsymbol{A}$ is the totality of vectors $\boldsymbol{x}$ for which $\boldsymbol{A}\boldsymbol{x} = \boldsymbol{0}$. This equality is equivalent to the relation $\boldsymbol{x} \perp \mathcal{C}(\boldsymbol{A}')$, as follows from Definition A.1.10 and Lemma A.1.1. But, according to Definition A.1.13 and Lemma A.1.5, the totality of such vectors $\boldsymbol{x}$, the orthogonal complement of $\mathcal{C}(\boldsymbol{A}')$, is a linear subspace of $\mathcal{R}^n$.

Definition A.1.14. Let an $n \times m$ matrix $\boldsymbol{A}$ be of rank r. The linear subspace of all vectors orthogonal to $\mathcal{C}(\boldsymbol{A}')$, i.e., the orthogonal complement of $\mathcal{C}(\boldsymbol{A}')$, is called the null space (kernel) of $\boldsymbol{A}$; it is written $\mathcal{N}(\boldsymbol{A})$. The dimension of $\mathcal{N}(\boldsymbol{A})$ is called the nullity of $\boldsymbol{A}$, and it is equal to $m - r$.

Thus, to every $n \times m$ matrix $\boldsymbol{A}$ there correspond two fundamental linear subspaces of $\mathcal{R}^n$, $\mathcal{C}(\boldsymbol{A})$ and $\mathcal{N}(\boldsymbol{A})$. Similarly, to $\boldsymbol{A}'$ there correspond $\mathcal{C}(\boldsymbol{A}')$ and $\mathcal{N}(\boldsymbol{A}')$. Evidently, $\mathcal{N}(\boldsymbol{A}) = \mathcal{C}^{\perp}(\boldsymbol{A}')$ and $\mathcal{N}(\boldsymbol{A}') = \mathcal{C}^{\perp}(\boldsymbol{A})$. An implication for a symmetric matrix is obvious. Also, it follows that $\dim[\mathcal{C}(\boldsymbol{A})] + \dim[\mathcal{N}(\boldsymbol{A})] = m$.

Now, turning to the type of problems announced in Lemma A.1.2, the following concepts become useful.

Definition A.1.15. Let $\mathcal{C}(\boldsymbol{A}) \subset \mathcal{R}^n$ and $\mathcal{C}(\boldsymbol{B}) \subset \mathcal{R}^n$. The set of all vectors in $\mathcal{R}^n$, which can be expressed as $\boldsymbol{x} = \boldsymbol{y} + \boldsymbol{z}$ with $\boldsymbol{y} \in \mathcal{C}(\boldsymbol{A})$ and $\boldsymbol{z} \in \mathcal{C}(\boldsymbol{B})$, is called the sum of $\mathcal{C}(\boldsymbol{A})$ and $\mathcal{C}(\boldsymbol{B})$; it is denoted by the symbol $\mathcal{C}(\boldsymbol{A}) + \mathcal{C}(\boldsymbol{B})$. It is called the direct sum of $\mathcal{C}(\boldsymbol{A})$ and $\mathcal{C}(\boldsymbol{B})$ if the decomposition $\boldsymbol{x} = \boldsymbol{y} + \boldsymbol{z}$ is unique for every vector in the set; the direct sum is denoted by $\mathcal{C}(\boldsymbol{A}) \oplus \mathcal{C}(\boldsymbol{B})$.

Definition A.1.16. Let $\mathcal{C}(\boldsymbol{A}) \subset \mathcal{R}^n$ and $\mathcal{C}(\boldsymbol{B}) \subset \mathcal{R}^n$. The set of all vectors in $\mathcal{R}^n$ belonging to both of the linear subspaces is called the intersection of $\mathcal{C}(\boldsymbol{A})$ and $\mathcal{C}(\boldsymbol{B})$ and is denoted by $\mathcal{C}(\boldsymbol{A}) \cap \mathcal{C}(\boldsymbol{B})$.

Lemma A.1.6. *If $\mathcal{C}(\boldsymbol{A}) \subset \mathcal{R}^n$ and $\mathcal{C}(\boldsymbol{B}) \subset \mathcal{R}^n$, then $\mathcal{C}(\boldsymbol{A}) + \mathcal{C}(\boldsymbol{B})$, $\mathcal{C}(\boldsymbol{A}) \oplus \mathcal{C}(\boldsymbol{B})$ and $\mathcal{C}(\boldsymbol{A}) \cap \mathcal{C}(\boldsymbol{B})$ are linear subspaces of $\mathcal{R}^n$. Then, the dimensionality relation*

$$\dim[\mathcal{C}(\boldsymbol{A}) + \mathcal{C}(\boldsymbol{B})] + \dim[\mathcal{C}(\boldsymbol{A}) \cap \mathcal{C}(\boldsymbol{B})] = \operatorname{rank}(\boldsymbol{A}) + \operatorname{rank}(\boldsymbol{B})$$

holds. It reduces to

$$\dim[\mathcal{C}(\boldsymbol{A}) \oplus \mathcal{C}(\boldsymbol{B})] = \operatorname{rank}(\boldsymbol{A}) + \operatorname{rank}(\boldsymbol{B}), \quad \textit{if} \quad \mathcal{C}(\boldsymbol{A}) \cap \mathcal{C}(\boldsymbol{B}) = \{\boldsymbol{0}\}.$$

[In the latter case, the linear subspaces $\mathcal{C}(\boldsymbol{A})$ and $\mathcal{C}(\boldsymbol{B})$ are said to be virtually disjoint.]

The proof of this lemma is left to the reader. (See also Rao, 1973, Section 1a, or Schott, 1997, Section 2.9.)

Note that now the result of Lemma A.1.2 can be written in the form $\mathcal{R}^n = \mathcal{C}(\boldsymbol{A}) \oplus \mathcal{C}^{\perp}(\boldsymbol{A})$, with $\dim[\mathcal{C}(\boldsymbol{A})] + \dim[\mathcal{C}^{\perp}(\boldsymbol{A})] = n$, for any $\mathcal{C}(\boldsymbol{A}) \subset \mathcal{R}^n$.

Also note that the orthogonal complement of $\mathcal{C}(\boldsymbol{A})$ in $\mathcal{C}(\boldsymbol{B})$ (Definition A.1.13) can be written $\mathcal{C}^{\perp}(\boldsymbol{A}) \cap \mathcal{C}(\boldsymbol{B})$.

Use can also be made of the following two results, which can be proved as in Seber (1980, Section 1.4).

Lemma A.1.7. *If $\mathcal{C}(\boldsymbol{A}) \subset \mathcal{R}^n$ and $\mathcal{C}(\boldsymbol{B}) \subset \mathcal{R}^n$, then $[\mathcal{C}(\boldsymbol{A}) \cap \mathcal{C}(\boldsymbol{B})]^{\perp} = \mathcal{C}^{\perp}(\boldsymbol{A}) + \mathcal{C}^{\perp}(\boldsymbol{B})$.*

Lemma A.1.8. *If $\mathcal{C}(\boldsymbol{A})$, $\mathcal{C}(\boldsymbol{B})$ and $\mathcal{C}(\boldsymbol{C})$ are all in $\mathcal{R}^n$, and $\mathcal{C}(\boldsymbol{B}) \subset \mathcal{C}(\boldsymbol{A})$, then*

$$\mathcal{C}(\boldsymbol{A}) \cap [\mathcal{C}(\boldsymbol{B}) + \mathcal{C}(\boldsymbol{C})] = \mathcal{C}(\boldsymbol{B}) + \mathcal{C}(\boldsymbol{A}) \cap \mathcal{C}(\boldsymbol{C}).$$

To complete this section, it should be mentioned that results given here can be generalized when using a more general inner product, of the form $\boldsymbol{y}'\boldsymbol{M}\boldsymbol{x}$, where $\boldsymbol{M}$ is a positive definite (p.d.) matrix. Then, in particular, the ordinary norm (Definition A.1.3) can be generalized to an $\boldsymbol{M}$-norm, defined for a vector $\boldsymbol{x}$ as

$$||\boldsymbol{x}||_M = (\boldsymbol{x}'\boldsymbol{M}\boldsymbol{x})^{1/2}$$

(see, e.g., Rao and Mitra, 1971, p. 45). This concept leads to a generalization of the ordinary Euclidean distance between two vectors (Definition A.1.4) to an $\boldsymbol{M}$-distance, defined for a pair of vectors $\boldsymbol{x}$ and $\boldsymbol{y}$ as

$$||\boldsymbol{x}-\boldsymbol{y}||_M = [(\boldsymbol{x}-\boldsymbol{y})'\boldsymbol{M}(\boldsymbol{x}-\boldsymbol{y})]^{1/2}.$$

Then, also, the orthogonality concept (Definition A.1.7) is generalized to the concept of $\boldsymbol{M}$-orthogonality, expressed by saying that vectors $\boldsymbol{x}$ and $\boldsymbol{y}$ are $\boldsymbol{M}$-orthogonal when they satisfy the condition $\boldsymbol{x}'\boldsymbol{M}\boldsymbol{y} = 0$. Generalizing in the same way the concept of normalization (Definition A.1.8), a set of vectors $\boldsymbol{x}_1, \boldsymbol{x}_2, ..., \boldsymbol{x}_m$ can be called $\boldsymbol{M}$-orthonormal if the vectors satisfy the condition $\boldsymbol{x}_i'\boldsymbol{M}\boldsymbol{x}_{i'} = \delta_{ii'}$ (the Kronecker delta) for $i, i' = 1, 2, ..., m$.

A.2 Projection operators

In this section, some notions and results concerning the theory of a class of matrices called projection operators (projectors) will be recalled.

Definition A.2.1. Let $\mathcal{C}(\boldsymbol{A})$ and $\mathcal{C}(\boldsymbol{B})$ be two virtually disjoint linear subspaces such that $\mathcal{R}^n = \mathcal{C}(\boldsymbol{A}) \oplus \mathcal{C}(\boldsymbol{B})$, so that any vector $\boldsymbol{x} \in \mathcal{R}^n$ has the unique decomposition $\boldsymbol{x} = \boldsymbol{y} + \boldsymbol{z}$, with $\boldsymbol{y} \in \mathcal{C}(\boldsymbol{A})$ and $\boldsymbol{z} \in \mathcal{C}(\boldsymbol{B})$ (see Definition A.1.15). Then, the mapping $\boldsymbol{P} : \boldsymbol{x} \to \boldsymbol{y}$ (i.e., the transformation $\boldsymbol{P}\boldsymbol{x} = \boldsymbol{y}$) is called the projection of $\boldsymbol{x}$ on $\mathcal{C}(\boldsymbol{A})$ along $\mathcal{C}(\boldsymbol{B})$.

Note that this definition is more general than Definition A.1.12. As will be seen, the two definitions coincide when $\mathcal{C}(\boldsymbol{B}) = \mathcal{C}^{\perp}(\boldsymbol{A})$.

Basic results concerning the projection operator $\boldsymbol{P}$ are given in the following theorem.

Theorem A.2.1 [1c.4(i)-(iv) in Rao, 1973]. *With the notation used in Definition* A.2.1, *the following hold:*

(i) $\boldsymbol{P}$ *is a linear transformation and is unique (i.e.,* $\boldsymbol{P}$ *can be represented by a matrix that is denoted by the same symbol* $\boldsymbol{P}$, *and is necessarily unique).*

(ii) $\boldsymbol{P}$ *is an idempotent matrix (i.e.,* $\boldsymbol{P}^2 = \boldsymbol{P}$*).*

(iii) $\boldsymbol{I}_n - \boldsymbol{P}$ *is a projector on* $\mathcal{C}(\boldsymbol{B})$ *along* $\mathcal{C}(\boldsymbol{A})$.

(iv) *Let* $\mathcal{R}^n = \mathcal{C}(\boldsymbol{A}_1) \oplus \mathcal{C}(\boldsymbol{A}_2) \oplus \cdots \oplus \mathcal{C}(\boldsymbol{A}_k)$. *Then, there exist such matrices* $\boldsymbol{P}_1, \boldsymbol{P}_2, ..., \boldsymbol{P}_k$ *that* (a) $\boldsymbol{P}_i^2 = \boldsymbol{P}_i$, (b) $\boldsymbol{P}_i\boldsymbol{P}_{i'} = \boldsymbol{O}$ *for* $i \neq i'$, *and* (c) $\boldsymbol{I}_n = \boldsymbol{P}_1 + \boldsymbol{P}_2 + \cdots + \boldsymbol{P}_k$.

Definition A.2.2. Let $\mathcal{C}(\boldsymbol{A}) \subset \mathcal{R}^n$, and let $\mathcal{C}^{\perp}(\boldsymbol{A})$ be the orthogonal complement of $\mathcal{C}(\boldsymbol{A})$ in $\mathcal{R}^n$ (see Definition A.1.13 and the discussion thereafter, as well as the remark following Lemma A.1.6). Then, the operator $\boldsymbol{P}$, which projects onto $\mathcal{C}(\boldsymbol{A})$ along $\mathcal{C}^{\perp}(\boldsymbol{A})$ is called an orthogonal projector.

Theorem A.2.2 [1c.4(v) in Rao, 1973]. *Let the inner product in $\mathcal{R}^n$ be $\boldsymbol{y}'\boldsymbol{M}\boldsymbol{x}$, where $\boldsymbol{M}$ is a p.d. matrix. Then, $\boldsymbol{P}$ is an orthogonal projector (one could say $\boldsymbol{M}$-orthogonal projector) if and only if*

$$\boldsymbol{P}^2 = \boldsymbol{P} \quad \textit{and} \quad \boldsymbol{MP} \ \textit{is symmetric.}$$

Remark A.2.1. It can easily be checked that if $\boldsymbol{P}^2 = \boldsymbol{P}$ and $\boldsymbol{MP}$ is symmetric, then $\boldsymbol{P}$ projects onto $\mathcal{C}(\boldsymbol{P})$ along $\mathcal{C}(\boldsymbol{I}_n - \boldsymbol{P})$, which means that $\mathcal{C}(\boldsymbol{A})$ and $\mathcal{C}^{\perp}(\boldsymbol{A})$ can then be identified with $\mathcal{C}(\boldsymbol{P})$ and $\mathcal{C}(\boldsymbol{I}_n - \boldsymbol{P})$, respectively.

Note that if $\boldsymbol{M} = \boldsymbol{I}_n$, then $\boldsymbol{P}$ is an orthogonal projector if and only if $\boldsymbol{P}$ is idempotent and symmetric.

Now, it will be interesting to have the explicit expression of an orthogonal projector.

Theorem A.2.3 [1c.4(vi) in Rao, 1973]. *Let $\mathcal{C}(\boldsymbol{A}) \subset \mathcal{R}^n$, and let the inner product in $\mathcal{R}^n$ be as in Theorem A.2.2. Then, the orthogonal ($\boldsymbol{M}$-orthogonal) projector on $\mathcal{C}(\boldsymbol{A})$ is*

$$\boldsymbol{P} = \boldsymbol{A}(\boldsymbol{A}'\boldsymbol{M}\boldsymbol{A})^{-}\boldsymbol{A}'\boldsymbol{M} = \boldsymbol{P}_{A(M)} \quad (\textit{say}),$$

which is unique for any choice of the generalized inverse (g-inverse) involved. (See also Rao and Kleffe, 1988, Section 1.3.)

Thus, if $\boldsymbol{M} = \boldsymbol{I}_n$, the orthogonal projector on $\mathcal{C}(\boldsymbol{A})$ has the explicit expression in the form

$$\boldsymbol{P} = \boldsymbol{A}(\boldsymbol{A}'\boldsymbol{A})^{-}\boldsymbol{A}' = \boldsymbol{P}_A \quad \text{(say)},$$

again invariant for any choice of $(\boldsymbol{A}'\boldsymbol{A})^{-}$, a g-inverse of $\boldsymbol{A}'\boldsymbol{A}$.

It will be useful also to recall some other known results concerning orthogonal projectors. Although the following results are given here for projectors of the type $\boldsymbol{P}_A$, they can easily be generalized to the type $\boldsymbol{P}_{A(M)}$ case.

Corollary A.2.1. *If $\boldsymbol{P}_A$ is the orthogonal projector on $\mathcal{C}(\boldsymbol{A}) \subset \mathcal{R}^n$, then $\boldsymbol{I}_n - \boldsymbol{P}_A$ is the orthogonal projector on $\mathcal{C}^{\perp}(\boldsymbol{A})$.*

This result follows from Theorem A.2.1 and Definition A.2.2. See also Remark A.2.1.

Theorem A.2.4. *If $\boldsymbol{P}_A$ is the orthogonal projector on $\mathcal{C}(\boldsymbol{A})$ and $\boldsymbol{P}_B$ is the orthogonal projector on $\mathcal{C}(\boldsymbol{B})$, then $\boldsymbol{P} = \boldsymbol{P}_B - \boldsymbol{P}_A$ is an orthogonal projector if and only if $\boldsymbol{P}_B\boldsymbol{P}_A = \boldsymbol{P}_A\boldsymbol{P}_B = \boldsymbol{P}_A$, in which case, $\boldsymbol{P}$ is the orthogonal projector on $\mathcal{C}^{\perp}(\boldsymbol{A}) \cap \mathcal{C}(\boldsymbol{B})$, i.e., on the orthogonal complement of $\mathcal{C}(\boldsymbol{A})$ in $\mathcal{C}(\boldsymbol{B})$.*

Proof. This result follows from Theorem 5.1.3 in Rao and Mitra (1971), on account of Lemma A.1.7 here. □

Remark A.2.2. If $\boldsymbol{P}_A$ and $\boldsymbol{P}_B$ are as in Theorem A.2.4, then $\boldsymbol{P}_B\boldsymbol{P}_A = \boldsymbol{P}_A$ $(= \boldsymbol{P}'_A = \boldsymbol{P}_A\boldsymbol{P}_B)$ if and only if $\mathcal{C}(\boldsymbol{A}) \subset \mathcal{C}(\boldsymbol{B})$, as can easily be seen on account of Theorem A.2.3 and Lemma 2.2.6 in Rao and Mitra (1971, p.22).

Finally, to facilitate the interpretation of the projector considered in Theorem A.2.4, it may be useful to represent the intersection $\mathcal{C}^{\perp}(\boldsymbol{A}) \cap \mathcal{C}(\boldsymbol{B})$, i.e., the orthogonal complement of $\mathcal{C}(\boldsymbol{A})$ in $\mathcal{C}(\boldsymbol{B})$, as the column space of a properly chosen matrix (see also Lemma A.1.5). For this reason, the following result is helpful.

Theorem A.2.5. *Let $\mathcal{C}(\boldsymbol{B}) \subset \mathcal{R}^n$ and $\mathcal{C}(\boldsymbol{A}) \subset \mathcal{C}(\boldsymbol{B})$. If $\boldsymbol{P}_A$ is the orthogonal projector on $\mathcal{C}(\boldsymbol{A})$, then $\mathcal{C}^{\perp}(\boldsymbol{A}) \cap \mathcal{C}(\boldsymbol{B}) = \mathcal{C}[(\boldsymbol{I}_n - \boldsymbol{P}_A)\boldsymbol{B}]$.*

Proof. First note that if $\boldsymbol{x} \in \mathcal{C}^{\perp}(\boldsymbol{A}) \cap \mathcal{C}(\boldsymbol{B})$, then (on account of Corollary A.2.1) $\boldsymbol{x} = (\boldsymbol{I}_n - \boldsymbol{P}_A)\boldsymbol{x} = (\boldsymbol{I}_n - \boldsymbol{P}_A)\boldsymbol{B}\boldsymbol{u}$ for some $\boldsymbol{u}$, i.e., $\boldsymbol{x} \in \mathcal{C}[(\boldsymbol{I}_n - \boldsymbol{P}_A)\boldsymbol{B}]$. Conversely, if $\boldsymbol{x} \in \mathcal{C}[(\boldsymbol{I}_n - \boldsymbol{P}_A)\boldsymbol{B}]$, then, for some $\boldsymbol{u}$, $\boldsymbol{x} = (\boldsymbol{I}_n - \boldsymbol{P}_A)\boldsymbol{B}\boldsymbol{u}$ and, hence, $\boldsymbol{P}_B\boldsymbol{x} = \boldsymbol{x}$, because (by Remark A.2.2) $\boldsymbol{P}_B(\boldsymbol{I}_n - \boldsymbol{P}_A)\boldsymbol{B} = (\boldsymbol{I}_n - \boldsymbol{P}_A)\boldsymbol{P}_B\boldsymbol{B} = (\boldsymbol{I}_n - \boldsymbol{P}_A)\boldsymbol{B}$, and $(\boldsymbol{I}_n - \boldsymbol{P}_A)\boldsymbol{x} = \boldsymbol{x}$, as $\boldsymbol{I}_n - \boldsymbol{P}_A$ is idempotent. Thus, $\boldsymbol{x} \in \mathcal{C}^{\perp}(\boldsymbol{A}) \cap \mathcal{C}(\boldsymbol{B})$ (see Remark A.2.1). □

Further appendices, related to mathematical tools used in constructions of designs, will appear (as mentioned in Section 1.5) at the end of Volume II.

References

AGRAWAL, H. L. and J. PRASAD (1983). On construction of balanced incomplete block designs with nested rows and columns. *Sankhyā Ser. B* **45**, 345-350.

ANSCOMBE, F. J. (1948). Contribution to the discussion on D. G. Champernowne's "Sampling theory applied to autoregressive sequences." *J. Roy. Statist. Soc. Ser. B* **10**, 239.

ATIQULLAH, M. (1961). On a property of balanced designs. *Biometrika* **48**, 215-218.

BAILEY, R. A. (1981). A unified approach to design of experiments. *J. Roy. Statist. Soc. Ser. A* **144**, 214-223.

BAILEY, R. A. (1991). Strata for randomized experiments. *J. Roy. Statist. Soc. Ser. B* **53**, 27-78.

BAILEY, R. A. (1994). General balance: Artificial theory or practical relevance? In: T. Caliński and R. Kala (eds.), *Proc. Int. Conf. Linear Statist. Inference LINSTAT'93*. Kluwer Academic Publishers, Dordrecht, 171-184.

BAILEY, R. A., D. C. GOLDREI and D. F. HOLT (1984). Block deigns with block size two. *J. Statist. Plann. Inference* **10**, 257-263.

BAILEY, R. A. and C. A. ROWLEY (1987). Valid randomization. *Proc. Roy. Soc. London Ser. A* **410**, 105-124.

BAILEY, R. A. and C. A. ROWLEY (1990). General balance and treatment permutations. *Linear Algebra Appl.* **127**, 183-225.

BAILEY, R. A. and T. P. SPEED (1986). Rectangular lattice designs: Efficiency factors and analysis. *Ann. Statist.* **14**, 874-895.

BAKSALARY, J. K., A. DOBEK and R. KALA (1980). A necessary condition for balance of a block design. *Biom. J.* **22**, 47-50.

BAKSALARY, J. K. and P. D. PURI (1988). Criteria for the validity of Fisher's condition for balanced block designs. *J. Statist. Plann. Inference* **18**, 119-123.

BAKSALARY, J. K. and Z. TABIS (1985). Existence and construction of connected block designs with given vectors of treatment replications and block sizes. *J. Statist. Plann. Inference* **12**, 285-293.

BANERJEE, S. and S. KAGEYAMA (1990). Existence of α-resolvable nested incomplete block designs. *Utilitas Math.* **38**, 237-243.

BANERJEE, S. and S. KAGEYAMA (1993). Methods of constructing nested partially balanced incomplete block designs. *Utilitas Math.* **43**, 3-6.

BHATTACHARYA, C. G. (1978). Yates type estimators of a common mean. *Ann. Inst. Statist. Math.* **30**, 407-414.

BOSE, R. C. (1942). A note on the resolvability of balanced incomplete block designs. *Sankhyā* **6**, 105-110.

BOSE, R. C. (1950). Least squares aspects of analysis of variance. *Mimeo Series 9*, Institute of Statistics, University of North Carolina, Chapel Hill.

BOSE, R. C., W. H. CLATWORTHY and S. S. SHRIKHANDE (1954). Tables of Partially Balanced Designs with Two Associate Classes. *North Carolina Agric. Exp. Station Tech. Bull.* **107**.

BOSE, R. C. and S. S. SHRIKHANDE (1960). On the composition of balanced incomplete block designs. *Canad. J. Math.* **12**, 177-188.

BOX, G. E. P. (1954). Some theorems on quadratic forms applied in the study of analysis of variance problems, I. Effect of inequality of variance in the one-way classification. *Ann. Math. Statist.* **25**, 290-302.

BROWN, K. G. (1976). Asymptotic behavior of MINQUE-type estimators of variance components. *Ann. Statist.* **4**, 746-754.

BROWN, L. D. and A. COHEN (1974). Point and confidence estimation of a common mean and recovery of inter-block information. *Ann. Statist.* **2**, 963-976.

BRZESKWINIEWICZ, H. (1994). Experiment with split-plot generated by PBIB designs. *Biom. J.* **36**, 557-570.

CALIŃSKI, T. (1971). On some desirable patterns in block designs (with discussion). *Biometrics* **27**, 275-292.

CALIŃSKI, T. (1972). On non-orthogonal experiments. *Biom. J.* **14**, 73-84.

CALIŃSKI, T. (1977). On the notion of balance in block designs. In: J. R. Barra, F. Brodeau, G. Romier and B. van Cutsem (eds.), *Recent Developments in Statistics.* North-Holland, Amsterdam, 365-374.

CALIŃSKI, T. (1993a). Balance, efficiency and orthogonality concepts in block designs. *J. Statist. Plann. Inference* **36**, 283-300.

CALIŃSKI, T. (1993b). The basic contrasts of a block experimental design with special reference to the notion of general balance. *Listy Biometryczne—Biometr. Lett.* **30**, 13-38.

CALIŃSKI, T. (1994). On the randomization theory of experiments in nested block designs. *Listy Biometryczne—Biometr. Lett.* **31**, 45-77.

CALIŃSKI, T. (1996a). The basic contrasts of a block design with special reference to the recovery of inter-block information. In: A. Pázman and V. Witkovský (eds.), *Tatra Mountains Mathematical Publications, Vol. 7: PROBASTAT'94, Smolenice.* Mathematical Institute, Bratislava, 23-37.

CALIŃSKI, T. (1996b). On the existence of BLUEs under a randomization model for the randomized block design (with discussion). *Listy Biometryczne—Biometr. Lett.* **33**, 1-23.

CALIŃSKI, T. (1997). Recovery of inter-block information when the experiment is in a nested block design. *Listy Biometryczne—Biometr. Lett.* **34**, 9-26.

CALIŃSKI, T. and B. CERANKA (1974). Supplemented block designs. *Biom. J.* **16**, 299-305.

CALIŃSKI, T., B. CERANKA and S. MEJZA (1980). On the notion of efficiency of a block design. In: W. Klonecki, A. Kozek and J. Rosiński (eds.), *Mathematical Statistics and Probability Theory. Lecture Notes in Statistics* **2**. Springer-Verlag, New York, 47-62.

CALIŃSKI, T., S. GNOT and A. MICHALSKI (1998). On admissibility of the intra-block and inter-block variance component estimators. *Listy Biometryczne—Biometr. Lett.* **35**, 11-26.

CALIŃSKI, T. and S. KAGEYAMA (1988). A randomization theory of intrablock and interblock estimation. *Technical Report No. 230*, Statistical Research Group. Hiroshima University, Hiroshima, Japan.

CALIŃSKI, T. and S. KAGEYAMA (1991). On the randomization theory of intra-block and inter-block analysis. *Listy Biometryczne—Biometr. Lett.* **28**, 97-122.

CALIŃSKI, T. and S. KAGEYAMA (1996a). The randomization model for experiments in block designs and the recovery of inter-block information. *J. Statist. Plann. Inference* **52**, 359-374.

CALIŃSKI, T. and S. KAGEYAMA (1996b). Block designs: Their combinatorial and statistical properties. In: S. Ghosh and C. R. Rao (eds.), *Handbook of Statistics, Vol. 13: Design and Analysis of Experiments.* Elsevier Science, Amsterdam, 809-873.

CERANKA, B. (1973). Experimental incomplete block designs: Theory and application. In: W. Oktaba et al. (eds.), *Third Methodological Colloquium on Agro-Biometry.* Polish Academy of Sciences and the Polish Biometric Society, Warszawa, 143-212 (in Polish).

CERANKA, B. (1975). Balanced block designs with different block sizes. In: W. Oktaba et al. (eds.), *Fifth Methodological Colloquium on Agro-Biometry.* Polish Academy of Sciences and the Polish Biometric Society, Warszawa, 95-107 (in Polish).

CERANKA, B. (1976). Balanced block designs with unequal block sizes. *Biom. J.* **18**, 499-504.

CERANKA, B. (1983). Planning of experiments in C-designs. *Scientific Dissertations* **136**, *Annals of Poznań Agricultural University*, Poland.

CERANKA, B. (1984). Construction of partially efficiency balanced block designs. *Calcutta Statist. Assoc. Bull.* **33**, 165-172.

CERANKA, B. and H. CHUDZIK (1984). On construction of some augmented block designs. *Biom. J.* **26**, 849-857.

CERANKA, B., S. KAGEYAMA and S. MEJZA (1986). A new class of C-designs. *Sankhyā Ser. B* **48**, 199-206.

CERANKA, B. and M. KOZŁOWSKA (1983). On C-property in block designs. *Biom. J.* **25**, 681-687.

CERANKA, B. and M. KOZŁOWSKA (1984). Some methods of constructing C-designs. *J. Statist. Plann. Inference* **9**, 253-258.

CERANKA, B. and S. MEJZA (1979). On the efficiency factor for a contrast of treatment parameters. *Biom. J.* **21**, 99-102.

CERANKA, B. and S. MEJZA (1980). A new proposal for classification of block designs. *Studia Sci. Math. Hungar.* **15**, 79-82.

CERANKA, B., S. MEJZA and P. WIŚNIEWSKI (1979). Analysis of block designs with recovery of inter-block information. *Roczniki Akademii Rolniczej w Poznaniu, Algorytmy Biometryczne i Statystyczne* **8**, 3-27 (in Polish).

CHAKRABARTI, M. C. (1962). *Mathematics of Design and Analysis of Experiments.* Asia Publishing House, Bombay.

CLATWORTHY, W. H. (1973). *Tables of Two-Associate-Class Partially Balanced Designs. NBS Applied Mathematics Series* **63**. U.S. Department of Commerce, National Bureau of Standards.

COCHRAN, W. G. and G. M. COX (1957). *Experimental Designs*, 2nd ed. Wiley, New York.

COHEN, A. and H. B. SACKROWITZ (1989). Exact tests that recover interblock information in balanced incomplete block designs. *J. Amer. Statist. Assoc.* **84**, 556-559.

COLBOURN, C. J. and M. J. COLBOURN (1983). Nested triple systems. *Ars Combin.* **16**, 27-34.

CORBEIL, R. R. and S. R. SEARLE (1976). Restricted maximum likelihood estimation of variance components in the mixed model. *Technometrics* **18**, 31-38.

CORSTEN, L. C. A. (1976). Canonical correlation in incomplete blocks. In: S. Ikeda, T. Hayakawa, H. Hudimoto, M. Okamoto, M. Siotani and S. Yamamoto (eds.), *Essays in Probability and Statistics.* Shinko Tsusho, Tokyo, Japan, 125-154.

COX, D. R. (1958). *Planning of Experiments.* Wiley, New York.

CUNNINGHAM, E. P. and C. R. HENDERSON (1968). An iterative procedure for estimating fixed effects and variance components in mixed model situation. *Biometrics* **24**, 13-25.

DARROCH, J. N. and S. D. SILVEY (1963). On testing more than one hypothesis. *Ann. Math. Statist.* **34**, 555-567.

DAS, M. N. and D. K. GHOSH (1985). Balancing incomplete block designs. *Sankhyā Ser. B* **47**, 67-77.

DEMPSTER, A. P. (1969). *Elements of Continuous Multivariate Analysis.* Addison-Wesley, Reading, Massachusetts.

DEY, A., U. S. DAS and A. K. BANERJEE (1986). Constructions of nested balanced incomplete block designs. *Calcutta Statist. Assoc. Bull.* **35**, 161-167.

EL-SHAARAWI, A., R. L. PRENTICE and K. R. SHAH (1975). Marginal procedures for mixed models with reference to block designs. *Sankhyā Ser. B* **37**, 91-99.

FEDERER, W. T. (1955). *Experimental Design: Theory and Application.* Macmillan, New York.

FEINGOLD, M. (1985). A test statistic for combined intra- and inter-block estimates. *J. Statist. Plann. Inference* **12**, 103-114.

FEINGOLD, M. (1988). A more powerful test for incomplete block designs. *Commun. Statist. Part A—Theor. Meth.* **17**, 3107-3119.

FINNEY, D. J. (1960). *An Introduction to the Theory of Experimental Design.* The University of Chicago Press, Chicago.

FISHER, R. A. (1925). *Statistical Methods for Research Workers.* Oliver Boyd, Edinburgh.

FISHER, R. A. (1926). The arrangement of field experiments. *J. Ministry Agriculture* **33**, 503-513.

FISHER, R. A. (1935). *The Design of Experiments.* Oliver Boyd, Edinburgh.

FISHER, R. A. (1940). An examination of the different possible solutions of a problem in incomplete blocks. *Ann. Eugen.* **10**, 52-75.

FISHER, R. A. and F. YATES (1963). *Statistical Tables for Biological, Agricultural and Medical Research*, 6th ed. Hafner, New York (1st ed. 1938).

GENSTAT 5 COMMITTEE (1996). *Genstat 5 Release 3.2 Command Language Manual.* Numerical Algorithms Group, Oxford.

GHOSH, S. and C. R. RAO (1996). *Handbook of Statistics, Vol. 13: Design and Analysis of Experiments.* Elsevier Science, Amsterdam.

GIESBRECHT, F. G. (1986). Analysis of data from incomplete block designs. *Biometrics* **42**, 437-448.

GNOT, S. (1976). The mean efficiency of block designs. *Math. Operat.-forschung Statist.* **7**, 75-84.

GRAF-JACCOTTET, M. (1977). Comparative classification of block designs. In: J. R. Barra, F. Brodeau, G. Romier and B. van Cutsem (eds.), *Recent Developments in Statistics.* North-Holland, Amsterdam, 471-474.

GRAYBILL, F. A. and R. B. DEAL (1959). Combining unbiased estimators. *Biometrics* **15**, 543-550.

GRAYBILL, F. A. and V. SESHADRI (1960). On the unbiasedness of Yates' method of estimation using inter-block information. *Ann. Math. Statist.* **29**, 786-787.

GRAYBILL, F. A. and D. L. WEEKS (1959). Combining inter-block and intra-block information in balanced incomplete blocks. *Ann. Math. Statist.* **30**, 799-805.

GUPTA, S. C. (1984). An algorithm for constructing nests of 2 factors designs with orthogonal factorial structure. *J. Statist. Comput. Simul.* **20**, 59-79.

GUPTA, S. and S. KAGEYAMA (1994). Optimal complete diallel crosses. *Biometrika* **81**, 420-424.

HARVILLE, D. A. (1977). Maximum likelihood approaches to variance component estimation and to related problems. *J. Amer. Statist. Assoc.* **72**, 320-338.

HEDAYAT, A. and W. T. FEDERER (1974). Pairwise and variance balanced incomplete block designs. *Ann. Inst. Statist. Math.* **26**, 331-338.

HERZBERG, A. M. and H. P. WYNN (1982). A general approach to Yates' inter-block and intra-block analysis. *Utilitas Math.* **21B**, 189-203.

HINKELMANN, K. and O. KEMPTHORNE (1994). *Design and Analysis of Experiments, Vol. I: Introduction to Experimental Design.* Wiley, New York.

HOMEL, R. J. and J. ROBINSON (1975). Nested partially balanced incomplete block designs. *Sankhyā Ser. B* **37**, 201-210.

HOUTMAN, A. M. and T. P. SPEED (1983). Balance in designed experiments with orthogonal block structure. *Ann. Statist.* **11**, 1069-1085.

JAMES, A. T. and G. N. WILKINSON (1971). Factorization of the residual operator and canonical decomposition of nonorthogonal factors in the analysis of variance. *Biometrika* **58**, 279-294.

JARRETT, R. G. (1977). Bounds for the efficiency factor of block designs. *Biometrika* **64**, 67-72.

JIMBO, M. and S. KURIKI (1983). Constructions of nested designs. *Ars Combin.* **16**, 275-285.

JOHN, J. A. (1987). *Cyclic Designs.* Chapman and Hall, London.

JOHN, J. A. and E. R. WILLIAMS (1995). *Cyclic and Computer Generated Designs*, 2nd ed. Chapman and Hall, London.

JOHN, P. W. M. (1980). *Incomplete Block Designs.* Marcel Dekker, New York.

JONES, R. M. (1959). On a property of incomplete blocks. *J. Roy. Statist. Soc. Ser. B* **21**, 172-179.

KACKAR, R. N. and D. A. HARVILLE (1981). Unbiasedness of two-stage estimation and prediction procedures for mixed linear models. *Commun. Statist. Part A—Theor. Meth.* **10**, 1249-1261.

KACKAR, R. N. and D. A. HARVILLE (1984). Approximations for standard errors of estimators of fixed and random effects in mixed linear models. *J. Amer. Statist. Assoc.* **79**, 853-862.

KAGEYAMA, S. (1972). A survey of resolvable solutions of balanced incomplete block designs. *Internat. Statist. Rev.* **40**, 269-273.

KAGEYAMA, S. (1973). On μ-resolvable and affine μ-resolvable balanced incomplete block designs. *Ann. Statist.* **1**, 195-203.

KAGEYAMA, S. (1974). Reduction of associate classes for block designs and related combinatorial arrangements. *Hiroshima Math. J.* **4**, 527-618.

KAGEYAMA, S. (1976a). Resolvability of block designs. *Ann. Statist.* **4**, 655-661. Addendum: *Bull. Inst. Math. Statist.* **7**(5)(1978), 312.

KAGEYAMA, S. (1976b). Constructions of balanced block designs. *Utilitas Math.* **9**, 209-229.

KAGEYAMA, S. and Y. MIAO (1997). Nested designs with block size five and subblock size two. *J. Statist. Plann. Inference* **64**, 125-139.

KAGEYAMA, S. and Y. MIAO (1998a). A construction for resolvable designs and its generalizations. *Graph and Combin.* **14**, 11-24.

KAGEYAMA, S. and Y. MIAO (1998b). Nested designs of subblock size four. *J. Statist. Plann. Inference* **73**, 1-5.

KAGEYAMA, S. and P. D. PURI (1985a). Properties of partially efficiency-balanced designs. *Bull. Inform. Cyber.* (formerly *Bull. Math. Statist.*) **21**, 19-28.

KAGEYAMA, S. and P. D. PURI (1985b). A new special class of PEB designs. *Commun. Statist. Part A—Theor. Meth.* **14**, 1731-1744.

KAISER, L. D. (1989). Nonexistence of BLUE's under a randomization model for the randomized block design. *J. Statist. Plann. Inference* **22**, 63-69.

KALA, R. (1981). Projectors and linear estimation in general linear models. *Commn. Statist. Part A—Theor. Meth.* **10**, 849-873.

KALA, R. (1991). Elements of the randomization theory. III. Randomization in block experiments. *Listy Biometryczne—Biometr. Lett.* **28**, 3-23 (in Polish).

KASSANIS, B. and A. KLECZKOWSKI (1965). Inactivation of a strain of tobacco necrosis virus and of the RNA isolated from it, by ultraviolet radiation of different wavelengths. *Photochem. Photobiol.* **4**, 209-214.

KEMPTHORNE, O. (1952). *The Design and Analysis of Experiments.* Wiley, New York.

KEMPTHORNE, O. (1955). The randomization theory of experimental inference. *J. Amer. Statist. Assoc.* **50**, 946-967.

KEMPTHORNE, O. (1956). The efficiency factor of an incomplete block design. *Ann. Math. Statist.* **27**, 846-849.

KEMPTHORNE, O. (1975). Inference from experiments and randomization. In: J. N. Srivastava (ed.), *A Survey of Statistical Design and Linear Models.* North-Holland, Amsterdam, 303-331.

KEMPTHORNE, O. (1977). Why randomize? *J. Statist. Plann. Inference* **1**, 1-25.

KENWARD, M. G. and J. H. ROGER (1997). Small sample inference for fixed effects from restricted maximum likelihood. *Biometrics* **53**, 983-997.

KHATRI, C. G. and K. R. SHAH (1974). Estimation of location parameters from two linear models under normality. *Commun. Statist.* **3**, 647-663.

KHATRI, C. G. and K. R. SHAH (1975). Exact variance of combined inter- and intra-block estimates in incomplete block designs. *J. Amer. Statist. Assoc.* **70**, 402-406.

KIEFER, J. (1958). On the nonrandomized optimality and randomized nonoptimality of symmetrical designs. *Ann. Math. Statist.* **29**, 675-699.

KŁACZYŃSKI, K., A. MOLIŃSKA and K. MOLIŃSKI (1994). Unbiasedness of the estimator of the function of expected value in the mixed linear model. *Biom. J.* **36**, 185-191.

KLECZKOWSKI, A. (1960). Interpreting relationships between the concentration of plant viruses and numbers of local lesions. *J. Gen. Microbiol.* **4**, 53-69.

KSHIRSAGAR, A. M. (1958). A note on incomplete block designs. *Ann. Math. Statist.* **29**, 907-910.

LONGYEAR, J. Q. (1981). A survey of nested designs. *J. Statist. Plann. Inference* **5**, 181-187.

MARGOLIN, B. H. (1982). Blocks, randomized complete. In: S. Kotz and N. L. Johnson (eds.), *Encyclopedia of Statistical Sciences, Vol. 1.* Wiley, New York, 288-292.

MARTIN, F. B. and G. ZYSKIND (1966). On combinability of information from uncorrelated linear models by simple weighting. *Ann. Math. Statist.* **37**, 1338-1347.

MEJZA, I. (1996). Control treatments in incomplete split-plot designs. In: A. Pázman and V. Witkovský (eds.), *Tatra Mountains Mathematical Publications, Vol. 7: PROBASTAT'94, Smolenice.* Mathematical Institute, Bratislava, 69-77.

MEJZA, S. (1978). Use of inter-block information to obtain uniformly better estimators of treatment contrasts. *Math. Operat.-forschung Statist. Ser. Statist.* **9**, 335-341.

MEJZA, S. (1985). On testing hypothese in a mixed linear model for incomplete block designs. *Scand. J. Statist.* **12**, 241-247.

MEJZA, S. (1987). Experiments in incomplete split-plot designs. In: T. Pukkila and S. Puntanen (eds.), *Proc. Second Int. Tampere Conf. Statist.*. Dept. Math. Sci., University of Tampere, 575-584.

MEJZA, S. (1992). On some aspects of general balance in designed experiments. *Statistica* **52**, 263-278.

MEJZA, S. and S. KAGEYAMA (1995). On the optimality of certain nested block designs under a mixed effects model. In: C. P. Kitsos and W. G. Müller (eds.), *Moda 4—Advances in Model-Oriented Data Analysis.* Physica-Verlag, Heidelberg, 157-164.

MEJZA, S. and S. KAGEYAMA (1998). Some statistical properties of nested block designs. In: A. C. Atkinson, L. Pronzato and H. P. Wynn (eds.), *Moda 5—Advances in Model-Oriented Data Analysis and Experimental Design.* Physica-Verlag, Heidelberg, 231-238.

MONTGOMERY, D. C. (1984). *Design and Analysis of Experiments*, 2nd ed. Wiley, New York.

MORGAN, J. P. (1996). Nested designs. In: S. Ghosh and C. R. Rao (eds.), *Handbook of Statistics, Vol. 13: Design and Analysis of Experiments.* Elsevier Science, Amsterdam, 939-976.

NAIR, K. R. (1944). The recovery of inter-block information in incomplete block designs. *Sankhyā* **6**, 383-390.

NAIR, K. R. and C. R. RAO (1942). A note on partially balanced incomplete block designs. *Sci. and Culture* **7**, 568-569.

NAIR, K. R. and C. R. RAO (1948). Confounding in asymmetrical factorial experiments. *J. Roy. Statist. Soc. Ser. B* **10**, 109-131.

NELDER, J. A. (1954). The interpretation of negative components of variance. *Biometrika* **41**, 544-548.

NELDER, J. A. (1965a). The analysis of randomized experiments with orthogonal block structure. I. Block structure and the null analysis of variance. *Proc. Roy. Soc. Lond. Ser. A* **283**, 147-162.

NELDER, J. A. (1965b). The analysis of randomized experiments with orthogonal block structure. II. Treatment structure and the general analysis of variance. *Proc. Roy. Soc. Lond. Ser. A* **283**, 163-178.

NELDER, J. A. (1968). The combination of information in generally balanced designs. *J. Roy. Statist. Soc. Ser. B* **30**, 303-311.

NELDER, J. A. (1977). A reformulation of linear models. *J. Roy. Statist. Soc. Ser. A* **140**, 48-76.

NEYMAN, J. (1923). Próba uzasadnienia zastosowań rachunku prawdopodobieństwa do doświadczeń polowych (Sur les applications de la théorie des probabilités aux expérience agricoles: Essay de principes). *Roczniki Nauk Rolniczych* **10**, 1-51. [A partial English translation by D. M. DABROWSKA and T. P. SPEED (1990). *Statist. Sci.* **5**, 465-472.]

NEYMAN, J. (1935), with cooperation of K. IWASZKIEWICZ and S. KOLODZIEJCZYK. Statistical problems in agricultural experimentation (with discussion). *J. Roy. Statist. Soc. Suppl.* **2**, 107-180.

NIGAM, A. K. and P. D. PURI (1982). On partially efficiency balanced designs—II. *Commun. Statist. Part A—Theor. Meth.* **11**, 2817-2830.

NIGAM, A. K., P. D. PURI and V. K. GUPTA (1988). *Characterizations and Analysis of Block Designs.* Wiley Eastern, New Delhi.

OGAWA, J. (1961). The effect of randomization on the analysis of randomized block design. *Ann. Inst. Statist. Math.* **13**, 105-117.

OGAWA, J. (1963). On the null-distribuation of the F-statistic in a randomized balanced incomplete block design under the Neyman model. *Ann. Math. Statist.* **34**, 1558-1568.

OGAWA, J. (1974). *Statistical Theory of the Analysis of Experimental Designs.* Marcel Dekker, New York.

OLSEN, A., J. SEELY and D. BIRKES (1976). Invariant quadratic unbiased estimation for two variance components. *Ann. Statist.* **4**, 878-890.

PATTERSON, H. D. and E. A. HUNTER (1983). The efficiency of incomplete block designs in national list and recommended list cereal variety trials. *J. Agric. Sci.* **101**, 427-433.

PATTERSON, H. D. and R. THOMPSON (1971). Recovery of inter-block information when block sizes are unequal. *Biometrika* **58**, 545-554.

PATTERSON, H. D. and R. THOMPSON (1975). Maximum likelihood estimation of components of variance. In: L. C. A. Corsten and T. Postelnicu (eds.), *Proc. 8th Int. Biometric Conf.* Editura Academiei, Bucuresti, 197-207.

PATTERSON, H. D. and E. R. WILLIAMS (1976). A new class of resolvable incomplete block designs. *Biometrika* **63**, 83-92.

PATTERSON, H. D., E. R. WILLIAMS and E. A. HUNTER (1978). Block designs for variety trials. *J. Agric. Sci.* **90**, 395-400.

PEARCE, S. C. (1960). Supplemented balance. *Biometrika* **47**, 263-271.

PEARCE, S. C. (1963). The use and classification of non-orthogonal designs (with discussion). *J. Roy. Statist. Soc. Ser. A* **126**, 353-377.

PEARCE, S. C. (1964). Experimenting with blocks of natural size. *Biometrics* **20**, 699-706.

PEARCE, S. C. (1970). The efficiency of block designs in general. *Biometrika* **57**, 339-346.

PEARCE, S. C. (1971). Precision in block experiments. *Biometrika* **58**, 161-167.

PEARCE, S. C. (1976). Concurrences and quasi-replication: An alternative approach to precision in designed experiments. *Biom. J.* **18**, 105-116.

PEARCE, S. C. (1983). *The Agricultural Field Experiment: A Statistical Examination of Theory and Practice.* Wiley, Chichester.

PEARCE, S. C., T. CALIŃSKI and T. F. DE C. MARSHALL (1974). The basic contrasts of an experimental design with special reference to the analysis of data. *Biometrika* **61**, 449-460.

PREECE, D. A. (1967). Nested balanced incomplete block designs. *Biometrika* **54**, 479-486.

PREECE, D. A. (1977). Orthogonality and designs: A terminological muddle. *Utilitas Math.* **12**, 201-223.

PREECE, D. A. (1982). Balance and designs: Another terminological tangle. *Utilitas Math.* **21C**, 85-186.

PURI, P. D. and A. K. NIGAM (1975a). On patterns of efficiency balanced designs. *J. Roy. Statist. Soc. Ser. B* **37**, 457-458.

PURI, P. D. and A. K. NIGAM (1975b). A note on efficiency balanced designs. *Sankhyā Ser. B* **37**, 457-460.

PURI, P. D. and A. K. NIGAM (1977a). Partially efficiency balanced designs. *Commun. Statist. Part A—Theor. Meth.* **6**, 753-771.

PURI, P. D. and A. K. NIGAM (1977b). Balanced block designs. *Commun. Statist. Part A—Theor. Meth.* **6**, 1171-1179.

RAGHAVARAO, D. (1971). *Constructions and Combinatorial Problems in Design of Experiments.* Wiley, New York. Reprinted (1988) by Dover with some addendum.

RAO, C. R. (1947). General methods of analysis for incomplete block designs. *J. Amer. Statist. Assoc.* **42**, 541-561.

RAO, C. R. (1956). On the recovery of inter-block information in varietal trials. *Sankhyā* **17**, 105-114.

RAO, C. R. (1959). Expected values of mean squares in the analysis of incomplete block experiments and some comments based on them. *Sankhyā* **21**, 327-336.

RAO, C. R. (1970). Estimation of heteroscedastic variances in linear models. *J. Amer. Statist. Assoc.* **65**, 161-172.

RAO, C. R. (1971a). Estimation of variance and covariance components—MINQUE theory. *J. Multivariate Anal.* **1**, 257-275.

RAO, C. R. (1971b). Minimum variance quadratic unbiased estimation of variance components. *J. Multivariate Anal.* **1**, 445-456.

RAO, C. R. (1972). Estimation of variance and covariance components in linear models. *J. Amer. Statist. Assoc.* **67**, 112-115.

RAO, C. R. (1973). *Linear Statistical Inference and Its Applications*, 2nd ed. Wiley, New York.

RAO, C. R. (1974). Projectors, generalized inverses and the BLUEs. *J. Roy. Statist. Soc. Ser. B* **36**, 442-448.

RAO, C. R. (1979). MINQUE theory and its relation to ML and MML estimation of variance components. *Sankhyā Ser. B* **41**, 138-153.

RAO, C. R. and J. KLEFFE (1988). *Estimation of Variance Components and Applications.* North-Holland, Amsterdam.

RAO, C. R. and S. K. MITRA (1971). *Generalized Inverse of Matrices and Its Applications.* Wiley, New York.

RAO, V. R. (1958). A note on balanced designs. *Ann. Math. Statist.* **29**, 290-294.

RASCH, D. and G. HERRENDÖRFER (1986). *Experimental Design: Sample Size Determination and Block Designs.* D. Reidel, Dordrecht.

ROY, J. and K. R. SHAH (1962). Recovery of interblock information. *Sankhyā Ser. A* **24**, 269-280.

RUBIN, D. B. (1990). Comment: Neyman (1923) and causal inference in experiments and observational studies. *Statist. Sci.* **5**, 472-480.

SAHA, G. M. (1976). On Caliński's patterns in block designs. *Sankhyā Ser. B* **38**, 383-392.

SCHEFFÉ, H. (1959). *The Analysis of Variance.* Wiley, New York.

SCHOTT, J. R. (1997). *Matrix Analysis for Statistics.* Wiley, New York.

SEBER, G. A. F. (1980). *The Linear Hypothesis: A General Theory.* Charles Griffin, London.

SESHADRI, V. (1963). Constructing uniformly better estimators. *J. Amer. Statist. Assoc.* **58**, 172-175.

SHAH, B. V. (1960). Balanced factorial experiments. *Ann. Math. Statist.* **31**, 502-514.

SHAH, K. R. (1964a). Use of inter-block information to obtain uniformly better estimates. *Ann. Math. Statist.* **35**, 1064-1078.

SHAH, K. R. (1964b). On a local property of combined inter- and intra-block estimators. *Sankhyā Ser. A* **26**, 87-90.

SHAH, K. R. (1971). Use of truncated estimator of variance ratio in recovery of inter-block information. *Ann. Math. Statist.* **42**, 816-819.

SHAH, K. R. (1975). Analysis of block designs: A review article. *Gujarat Statist. Rev.* **2**, 1-11.

SHAH, K. R. (1992). Recovery of interblock information: An update. *J. Statist. Plann. Inference* **30**, 163-172.

SHAH, K. R. and B. K. SINHA (1989). *Theory of Optimal Designs. Lecture Notes in Statistics* **54**. Springer-Verlag, New York.

SHRIKHANDE, S. S. and D. RAGHAVARAO (1964). Affine α-resolvable incomplete block designs. In: C. R. RAO (ed.), *Contributions to Statistics*. Pergamon Press, Statistical Publishing Society, Calcutta, 471-480.

SPEED, T. P. (1990). Introductory remarks on Neyman (1923). *Statist. Sci.* **5**, 463-464.

SPEED, T. P. and R. A. BAILEY (1987). Factorial dispersion models. *Int. Statist. Rev.* **55**, 261-277.

SRIVASTAVA, J. N. and R. J. BEAVER (1986). On the superiority of the nested multidimensional block designs, relative to the classical incomplete block designs. *J. Statist. Plann. Inference* **13**, 133-150.

STEIN, C. (1966). An approach to the recovery of inter-block information in balanced incomplete block designs. In: F. N. David (ed.), *Research Papers in Statistics*. Wiley, New York, 351-366.

STREET, D. J. (1996). Block and other designs used in agriculture. In: S. Ghosh and C. R. Rao (eds.), *Handbook of Statistics, Vol. 13: Design and Analysis of Experiments*. Elsevier Science, Amsterdam, 759-808.

THORNETT, M. L. (1982). The role of randomization in model-based inference. *Austral. J. Statist.* **24**, 137-145.

TOCHER, K. D. (1952). The design and analysis of block experiments (with discussion). *J. Roy. Statist. Soc. Ser. B* **14**, 45-100.

VARTAK, M. N. (1963). Disconnected balanced designs. *J. Indian Statist. Assoc.* **1**, 104-107.

VERDOOREN, L. R. (1980). On estimation of variance components. *Statist. Neerlandica* **34**, 83-106.

WHITE, R. F. (1975). Randomization in the analysis of variance. *Biometrics* **31**, 555-571.

WILLIAMS, E. R. (1975). Efficiency-balanced designs. *Biometrika* **62**, 686-688.

WILLIAMS, E. R. (1976). Resolvable paired-comparison designs. *J. Roy. Statist. Soc. Ser. B* **38**, 171-174.

WILLIAMS, E. R., H. D. PATTERSON and J. A. JOHN (1976). Resolvable designs with two replications. *J. Roy. Statist. Soc. Ser. B* **38**, 296-301.

WILLIAMS, E. R., H. D. PATTERSON and J. A. JOHN (1977). Efficient two-replicate resolvable designs. *Biometrics* **33**, 713-717.

YATES, F. (1935). Complex experiments (with discussion). *J. Roy. Statist. Soc. Suppl.* **2**, 181-247.

YATES, F. (1936a). Incomplete randomized blocks. *Ann. Eugen.* **7**, 121-140.

YATES, F. (1936b). A new method of arranging variety trials involving a large number of varieties. *J. Agric. Sci.* **26**, 424-455.

YATES, F. (1939). The recovery of inter-block information in variety trials arranged in three-dimensional lattices. *Ann. Eugen.* **9**, 136-156.

YATES, F. (1940a). Lattice squares. *J. Agric. Sci.* **30**, 672-687.

YATES, F. (1940b). The recovery of inter-block information in balanced incomplete block designs. *Ann. Eugen.* **10**, 317-325.

YATES, F. (1965). A fresh look at the basic principles of the design and analysis of experiments. In: L. M. LeCam and J. Neyman (eds.), *Proc. 5th Berkeley Symp. Math. Statist. and Probability, Vol. 4*. University of California Press, Berkeley, 777-790.

ZYSKIND, G. (1967). On canonical forms, non-negative covariance matrices and best and simple least squares linear estimators in linear models. *Ann. Math. Statist.* **38**, 1092-1109.

ZYSKIND, G. (1975). Error structures, projections and conditional inverses in linear model theory. In: J. N. Srivastava (ed.), *A Survey of Statistical Design and Linear Models*. North-Holland, Amsterdam, 647-663.

Author Index

Subject Index

Lecture Notes in Statistics

For information about Volumes 1 to 79, please contact Springer-Verlag

Vol. 80: M. Fligner, J. Verducci (Eds.), Probability Models and Statistical Analyses for Ranking Data. xxii, 306 pages, 1992.

Vol. 81: P. Spirtes, C. Glymour, R. Scheines, Causation, Prediction, and Search. xxiii, 526 pages, 1993.

Vol. 82: A. Korostelev and A. Tsybakov, Minimax Theory of Image Reconstruction. xii, 268 pages, 1993.

Vol. 83: C. Gatsonis, J. Hodges, R. Kass, N. Singpurwalla (Editors), Case Studies in Bayesian Statistics. xii, 437 pages, 1993.

Vol. 84: S. Yamada, Pivotal Measures in Statistical Experiments and Sufficiency. vii, 129 pages, 1994.

Vol. 85: P. Doukhan, Mixing: Properties and Examples. xi, 142 pages, 1994.

Vol. 86: W. Vach, Logistic Regression with Missing Values in the Covariates. xi, 139 pages, 1994.

Vol. 87: J. Müller, Lectures on Random Voronoi Tessellations.vii, 134 pages, 1994.

Vol. 88: J. E. Kolassa, Series Approximation Methods in Statistics. Second Edition, ix, 183 pages, 1997.

Vol. 89: P. Cheeseman, R.W. Oldford (Editors), Selecting Models From Data: AI and Statistics IV. xii, 487 pages, 1994.

Vol. 90: A. Csenki, Dependability for Systems with a Partitioned State Space: Markov and Semi-Markov Theory and Computational Implementation. x, 241 pages, 1994.

Vol. 91: J.D. Malley, Statistical Applications of Jordan Algebras. viii, 101 pages, 1994.

Vol. 92: M. Eerola, Probabilistic Causality in Longitudinal Studies. vii, 133 pages, 1994.

Vol. 93: Bernard Van Cutsem (Editor), Classification and Dissimilarity Analysis. xiv, 238 pages, 1994.

Vol. 94: Jane F. Gentleman and G.A. Whitmore (Editors), Case Studies in Data Analysis. viii, 262 pages, 1994.

Vol. 95: Shelemyahu Zacks, Stochastic Visibility in Random Fields. x, 175 pages, 1994.

Vol. 96: Ibrahim Rahimov, Random Sums and Branching Stochastic Processes. viii, 195 pages, 1995.

Vol. 97: R. Szekli, Stochastic Ordering and Dependence in Applied Probability. viii, 194 pages, 1995.

Vol. 98: Philippe Barbe and Patrice Bertail, The Weighted Bootstrap. viii, 230 pages, 1995.

Vol. 99: C.C. Heyde (Editor), Branching Processes: Proceedings of the First World Congress. viii, 185 pages, 1995.

Vol. 100: Wlodzimierz Bryc, The Normal Distribution: Characterizations with Applications. viii, 139 pages, 1995.

Vol. 101: H.H. Andersen, M.Højbjerre, D. Sørensen, P.S.Eriksen, Linear and Graphical Models: for the Multivariate Complex Normal Distribution. x, 184 pages, 1995.

Vol. 102: A.M. Mathai, Serge B. Provost, Takesi Hayakawa, Bilinear Forms and Zonal Polynomials. x, 378 pages, 1995.

Vol. 103: Anestis Antoniadis and Georges Oppenheim (Editors), Wavelets and Statistics. vi, 411 pages, 1995.

Vol. 104: Gilg U.H. Seeber, Brian J. Francis, Reinhold Hatzinger, Gabriele Steckel-Berger (Editors), Statistical Modelling: 10th International Workshop, Innsbruck, July 10-14th 1995. x, 327 pages, 1995.

Vol. 105: Constantine Gatsonis, James S. Hodges, Robert E. Kass, Nozer D. Singpurwalla(Editors), Case Studies in Bayesian Statistics, Volume II. x, 354 pages, 1995.

Vol. 106: Harald Niederreiter, Peter Jau-Shyong Shiue (Editors), Monte Carlo and Quasi-Monte Carlo Methods in Scientific Computing. xiv, 372 pages, 1995.

Vol. 107: Masafumi Akahira, Kei Takeuchi, Non-Regular Statistical Estimation. vii, 183 pages, 1995.

Vol. 108: Wesley L. Schaible (Editor), Indirect Estimators in U.S. Federal Programs. viii, 195 pages, 1995.

Vol. 109: Helmut Rieder (Editor), Robust Statistics, Data Analysis, and Computer Intensive Methods. xiv, 427 pages, 1996.

Vol. 110: D. Bosq, Nonparametric Statistics for Stochastic Processes. xii, 169 pages, 1996.

Vol. 111: Leon Willenborg, Ton de Waal, Statistical Disclosure Control in Practice. xiv, 152 pages, 1996.

Vol. 112: Doug Fischer, Hans-J. Lenz (Editors), Learning from Data. xii, 450 pages, 1996.

Vol. 113: Rainer Schwabe, Optimum Designs for Multi-Factor Models. viii, 124 pages, 1996.

Vol. 114: C.C. Heyde, Yu. V. Prohorov, R. Pyke, and S. T. Rachev (Editors), Athens Conference on Applied Probability and Time Series Analysis Volume I: Applied Probability In Honor of J.M. Gani. viii, 424 pages, 1996.

Vol. 115: P.M. Robinson, M. Rosenblatt (Editors), Athens Conference on Applied Probability and Time Series Analysis Volume II: Time Series Analysis In Memory of E.J. Hannan. viii, 448 pages, 1996.

Vol. 116: Genshiro Kitagawa and Will Gersch, Smoothness Priors Analysis of Time Series. x, 261 pages, 1996.

Vol. 117: Paul Glasserman, Karl Sigman, David D. Yao (Editors), Stochastic Networks. xii, 298, 1996.

Vol. 118: Radford M. Neal, Bayesian Learning for Neural Networks. xv, 183, 1996.

Vol. 119: Masanao Aoki, Arthur M. Havenner, Applications of Computer Aided Time Series Modeling. ix, 329 pages, 1997.

Vol. 120: Maia Berkane, Latent Variable Modeling and Applications to Causality. vi, 288 pages, 1997.

Vol. 121: Constantine Gatsonis, James S. Hodges, Robert E. Kass, Robert McCulloch, Peter Rossi, Nozer D. Singpurwalla (Editors), Case Studies in Bayesian Statistics, Volume III. xvi, 487 pages, 1997.

Vol. 122: Timothy G. Gregoire, David R. Brillinger, Peter J. Diggle, Estelle Russek-Cohen, William G. Warren, Russell D. Wolfinger (Editors), Modeling Longitudinal and Spatially Correlated Data. x, 402 pages, 1997.

Vol. 123: D. Y. Lin and T. R. Fleming (Editors), Proceedings of the First Seattle Symposium in Biostatistics: Survival Analysis. xiii, 308 pages, 1997.

Vol. 124: Christine H. Müller, Robust Planning and Analysis of Experiments. x, 234 pages, 1997.

Vol. 125: Valerii V. Fedorov and Peter Hackl, Model-oriented Design of Experiments. viii, 117 pages, 1997.

Vol. 126: Geert Verbeke and Geert Molenberghs, Linear Mixed Models in Practice: A SAS-Oriented Approach. xiii, 306 pages, 1997.

Vol. 127: Harald Niederreiter, Peter Hellekalek, Gerhard Larcher, and Peter Zinterhof (Editors), Monte Carlo and Quasi-Monte Carlo Methods 1996, xii, 448 pages, 1997.

Vol. 128: L. Accardi and C.C. Heyde (Editors), Probability Towards 2000, x, 356 pages, 1998.

Vol. 129: Wolfgang Härdle, Gerard Kerkyacharian, Dominique Picard, and Alexander Tsybakov, Wavelets, Approximation, and Statistical Applications, xvi, 265 pages, 1998.

Vol. 130: Bo-Cheng Wei, Exponential Family Nonlinear Models, ix, 240 pages, 1998.

Vol. 131: Joel L. Horowitz, Semiparametric Methods in Econometrics, ix, 204 pages, 1998.

Vol. 132: Douglas Nychka, Walter W. Piegorsch, and Lawrence H. Cox (Editors), Case Studies in Environmental Statistics, viii, 200 pages, 1998.

Vol. 133: Dipak Dey, Peter Müller, and Debajyoti Sinha (Editors), Practical Nonparametric and Semiparametric Bayesian Statistics, xv, 408 pages, 1998.

Vol. 134: Yu. A. Kutoyants, Statistical Inference For Spatial Poisson Processes, vii, 284 pages, 1998.

Vol. 135: Christian P. Robert, Discretization and MCMC Convergence Assessment, x, 192 pages, 1998.

Vol. 136: Gregory C. Reinsel, Raja P. Velu, Multivariate Reduced-Rank Regression, xiii, 272 pages, 1998.

Vol. 137: V. Seshadri, The Inverse Gaussian Distribution: Statistical Theory and Applications, xi, 360 pages, 1998.

Vol. 138: Peter Hellekalek, Gerhard Larcher (Editors), Random and Quasi-Random Point Sets, xi, 352 pages, 1998.

Vol. 139: Roger B. Nelsen, An Introduction to Copulas, xi, 232 pages, 1999.

Vol. 140: Constantine Gatsonis, Robert E. Kass, Bradley Carlin, Alicia Carriquiry, Andrew Gelman, Isabella Verdinelli, Mike West (Editors), Case Studies in Bayesian Statistics, Volume IV, xvi, 456 pages, 1999.

Vol. 141: Peter Müller, Brani Vidakovic (Editors), Bayesian Inference in Wavelet Based Models, xi, 394 pages, 1999.

Vol. 142: György Terdik, Bilinear Stochastic Models and Related Problems of Nonlinear Time Series Analysis: A Frequency Domain Approach, xi, 258 pages, 1999.

Vol. 143: Russell Barton, Graphical Methods for the Design of Experiments, x, 208 pages, 1999.

Vol. 144: L. Mark Berliner, Douglas Nychka, and Timothy Hoar (Editors), Case Studies in Statistics and the Atmospheric Sciences, x, 208 pages, 1999.

Vol. 145: James H. Matis and Thomas R. Kiffe, Stochastic Population Models, viii, 220 pages, 2000.

Vol. 146: Wim Schoutens, Stochastic Processes and Orthogonal Polynomials, xiv, 163 pages, 2000.

Vol. 147: Jürgen Franke, Wolfgang Härdle, and Gerhard Stahl, Measuring Risk in Complex Stochastic Systems, xvi, 272 pages, 2000.

Vol. 148: S.E. Ahmed and Nancy Reid, Empirical Bayes and Likelihood Inference, x, 200 pages, 2000.

Vol. 149: D. Bosq, Linear Processes in Function Spaces: Theory and Applications, xv, 296 pages, 2000.

Vol. 150: Tadeusz Caliński and Sanpei Kageyama, Block Designs: A Randomization Approach, Volume I: Analysis, ix, 313 pages, 2000.

Vol. 151: Håkan Andersson and Tom Britton, Stochastic Epidemic Models and Their Statistical Analysis: ix, 152 pages, 2000.

Vol. 152: David Ríos Insua and Fabrizio Ruggeri, Robust Bayesian Analysis: xiii, 435 pages, 2000.

Vol. 153: Parimal Mukhopadhyay, Topics in Survey Sampling, x, 303 pages, 2000.